Bacterial Growth and Lysis

Metabolism and Structure of the
Bacterial Sacculus

FEDERATION OF EUROPEAN MICROBIOLOGICAL SOCIETIES SYMPOSIUM SERIES

Recent FEMS Symposium volumes published by Plenum Press

1990 • DENITRIFICATION IN SOIL AND SEDIMENT
Edited by Niels Peter Revsbech and Jan Sørensen
(FEMS Symposium No. 56)

1991 • *CANDIDA* AND CANDIDAMYCOSIS
Edited by Emel Tümbay, Heinz P. R. Seeliger, and Özdem Anğ
(FEMS Symposium No. 50)

1991 • MICROBIAL SURFACE COMPONENTS AND TOXINS IN RELATION TO PATHOGENESIS
Edited by Eliora Z. Ron and Shlomo Rottem
(FEMS Symposium No. 51)

1991 • GENETICS AND PRODUCT FORMATION IN *STREPTOMYCES*
Edited by Simon Baumberg, Hans Krügel, and Dieter Noack
(FEMS Symposium No. 55)

1991 • THE BIOLOGY OF *ACINETOBACTER:* Taxonomy, Clinical Importance, Molecular Biology, Physiology, Industrial Relevance
Edited by K. J. Towner, E. Bergogne-Bérézin, and C. A. Fewson
(FEMS Symposium No. 57)

1991 • MOLECULAR PATHOGENESIS OF GASTROINTESTINAL INFECTIONS
Edited by T. Wadström, P. H. Mäkelä, A.-M. Svennerholm, and H. Wolf-Watz
(FEMS Symposium No. 58)

1992 • MOLECULAR RECOGNITION IN HOST–PARASITE INTERACTIONS
Edited by Timo K. Korhonen, Tapani Hovi, and P. Helena Mäkelä
(FEMS Symposium No. 61)

1992 • THE RELEASE OF GENETICALLY MODIFIED MICROORGANISMS—REGEM 2
Edited by Duncan E. S. Stewart-Tull and Max Sussman
(FEMS Symposium No. 63)

1993 • BACTERIAL GROWTH AND LYSIS: Metabolism and Structure of the Bacterial Sacculus
Edited by M. A. de Pedro, J.-V. Höltje, and W. Löffelhardt
(FEMS Symposium No. 65)

A Continuation Order Plan is available for this series. A continuation order will bring delivery of each new volume immediately upon publication. Volumes are billed only upon actual shipment. For further information please contact the publisher.

Bacterial Growth and Lysis

Metabolism and Structure of the Bacterial Sacculus

Edited by

M. A. de Pedro

Centro de Biología Molecular
CSIC-UAM
Madrid, Spain

J.-V. Höltje

Max-Planck-Institut für Entwicklungsbiologie
Tübingen, Germany

and

W. Löffelhardt

Institut für Allgemeine Biochemie
Universität Wien
Wien, Austria

PLENUM PRESS • NEW YORK AND LONDON

Library of Congress Cataloging in Publication Data

Bacterial growth and lysis: metabolism and structure of the bacterial sacculus / edited by M. A. de Pedro, J.-V. Höltje, W. Löffelhardt.
p. cm.—(Federation of European Microbiological Societies symposium series; no. 65)
"Proceedings of a symposium held under the auspices of the Federation of European Microbiological Societies, held April 5–10, 1992, in Mallorca, Spain"—T.p. verso.
Includes bibliographical references and index.
ISBN 0-306-44401-1
1. Peptidoglycans—Metabolism—Congresses. 2. Bacterial growth—Congresses. 3. Microbial metabolism—Congresses. I. De Pedro, M. A. II. Höltje, J.-V. (Joachim-Volker), date. III. Löffelhardt, W. IV. Federation of European Microbiological Societies. V. Series: FEMS symposium; no. 65.
QR92.P39B33 1993 92-44630
589.9′08751—dc20 CIP

Proceedings of a symposium held under the auspices of the
Federation of European Microbiological Societies,
April 5–10, 1992, in Mallorca, Spain

ISBN 0-306-44401-1

A Division of Plenum Publishing Corporation
233 Spring Street, New York, N.Y. 10013

Printed in the United States of America

PREFACE

This volume is based on a FEMS Symposium entitled "Bacterial Growth and Lysis: Metabolism and Structure of the Bacterial Sacculus" held at the Monastery of Lluc (Mallorca, Spain) on 5-10 April, 1992. The goals of the symposium were to assess the present state of knowledge on the structure and physiology of the bacterial murein sacculus, and to develop new hypotheses and strategies to promote further development of the field. Consequently, the contributions compiled in this volume include broadly different approaches, from the introduction of new analytical methods to the presentation of provocative models for cell wall growth and division. Structural, biochemical, and genetic aspects are widely covered with special emphasis on the enzymology and regulation of murein hydrolases (autolysins). Comprehensive reviews on bacterial S-layers and yeast cell walls are included to stimulate conceptual cross-feeding with these closely related topics. We believe that this book will provide the reader with a useful and up-to-date review of the topic. We would feel deeply rewarded by any positive influence this book may have on the future progress of the field, whereby all the scientific credit for it should be given to the authors of the excellent contributions presented.

We would like to express our gratitude to the Federation of European Microbiology Societies (FEMS), the Austrian Ministry for Science and Research, the Mixed Committee for Scientific-Technical Cooperation Austria-Spain, the "Comisión Interministerial de Ciencia y Tecnología" (Spain), the "Consejo Superior de Investigaciones Científicas" (Spain), the Spanish Ministry for Foreign Affairs, and "Laboratorios Beecham S.A.-SmithKline Beecham Pharmaceuticals" of Spain for sponsoring the symposium. Thanks are also due to Professors R.G. Board (Bath, U.K.), C.A. Fewson (Glasgow, U.K.), and C. Nombela (Madrid, Spain) , to Dr. A. Torrellas (Madrid, Spain), and to Mrs. G. García Cossio (Madrid, Spain) for their help and friendly advice in the organization of the meeting. Finally we thank all our colleagues, especially Mrs. J. de la Rosa, for their enthusiastic help in the practical aspects of the realization of the symposium and in the preparation of this book.

M.A. de Pedro
J.-V. Höltje
W. Löffelhardt

Madrid, Tübingen and Wien

CONTENTS

2. BIOSYNTHESIS AND MATURATION

3. CELL WALL HYDROLASES - REGULATION AND CONTROL

4. PENICILLIN INTERACTIVE PROTEINS

5. MORPHOGENESIS AND SEPTUM FORMATION

6. CELL GROWTH

THE SACCULUS AFTER FOUR DECADES - SEEN FROM SOME DISTANCE *

Uli Schwarz

Biochemistry Department
Max-Planck-Institut für Entwicklungsbiologie
Spemannstrasse 35
D-7400 Tübingen, FRG

INTRODUCTION

Anthony van Leeuwenhoek, in a letter to the Royal Society, described the little animals he saw in his microscope. In his note he specifically mentioned that these little creatures were bonded by some sort of structure and expressed his hope to resolve the question of what held them together (from Salton, 1960). It was about 200 years later, in 1887, that the first publication appeared on the biochemical properties of a structure which the author Livio Vincenzi, believed to be the cell wall of the "fission yeast" *Bacillus subtilis* (Vincenzi, L., 1887). It took almost another 70 years until isolated bacterial envelopes and finally the chemically pure shape maintaining structure of the bacterial cell wall, the sacculus, were available for structural and chemical characterization (Review: Salton, M.R.J., 1960)

THE CLASSICAL MODEL OF THE SACCULUS

In the following a sketch of the evolvement of our knowledge on the structure of the sacculus of the rod-shaped *E. coli* cell will be given and our

*This contribution is dedicated to E.H. Beachey, E.W. Goodell and Y. Hirota who are no longer with us. They had given a lot to science and to myself.

Bacterial Growth and Lysis, Edited by M.A. de Pedro *et al.*, Plenum Press, New York, 1993

changes of view of its biosynthesis will be summarized in a model. Such models must explain how new building blocks can be inserted into the existing sacculus whereby not only its mechanical stability is maintained but also its general form, its diameter. It is suggested that the copying of the preexisting sacculus, its replication, could be an important feature of the morphogenetic control of the sacculus when it grows.

The original analysis of the polymer from which the sacculus is woven, either murein or peptidoglycan, depending on taste (Weidel and Pelzer, 1964), indicated a rather simple structure. Paper chromatography of the lysozyme degradation products of *E. coli* murein indicated four major building blocks. Their chemical characterization led to a model of murein structure in which the fabric of murein consists of polysaccharide chains of undefined length, interlinked through one type of peptide bridge (Weidel and Pelzer, 1964). The model did not include statements on the arrangement of the glycan chains, regular or not, parallel or perpendicular to the long axis of the cell. This was our textbook knowledge until the mid-eighties. However, over the years we became more and more convinced that the assumption of a very simply structured monolayered murein was not at all in agreement with the complexity of the set of murein hydrolases, a list of at least nine such enzymes had accumulated, found to be involved in murein metabolism (Höltje and Schwarz, 1985).

Studies on murein metabolism and on murein structure made the function and possible role of some of the enzymes understandable, but still leaving many questions on murein structure open. Studies on the pool of low molecular weight murein intermediates revealed that in a growing and dividing cell murein biosynthesis is paralleled by continuous murein hydrolysis. The liberated muropeptides are reused for murein biosynthesis. This turnover of murein is extensive, about half of the murein of a cell undergoes turnover within one generation (Goodell, 1985)

The recycling of murein, as will be discussed later, is a logical consequence of the mode of insertion of new murein subunits in the growing sacculus. It explains the existence of some of the murein hydrolases present in abundance. In addition, it had become clear that the activity of a lytic transglycosylase found in *E. coli* could explain why the glycan chains in the sacculus are devoid of a reducing end. Instead the chain ends have a 1,6-anhydro muramic acid residue, the reaction product of a lytic transglycosylase (Höltje et al., 1975).

THE REVISED MODEL OF THE SACCULUS

Thus, we had indeed begun to understand the role of some of the enzymes dealing with murein metabolism. However, we still felt quite uncomfortable about the discrepancy between the complexity of the set of murein synthetases and murein hydrolases known and the assumed simplicity of murein structure in *E. coli*. A careful reexamination of murein chemistry using the modern tool of high pressure liquid chromatography has finally solved this paradox. B. Glauner designed a procedure to separate the murein subunits obtained from pure sacculi after complete enzymatic degradation on a reverse phase column, yielding a pattern of astounding complexity. Instead of the few types of murein subunits known so far we found a collection of about 80 different compounds. They all were isolated, purified and their chemical structure was determined; all turned out to be native components of the sacculus (Glauner et al, 1988)

Some of the novel murein constituents are of special relevance for our conceptions about murein structure and metabolism. One of them is a trimer of the disaccharide subunits interlinked by peptide side chains. In situ the disaccharide units must belong to three different glycan chains. In a strictly monolayered murein - the accepted model at this time - such subunits could exist only at chain ends if one accepts the available physico-chemical data on murein structure based on X-ray diffraction studies, on infra-red spectroscopy and ^{15}N-nuclear magnetic resonance relaxation studies (Labischinski et al., 1985) However, eighty per cent of the trimers were found within glycan chains. This is possible only if one of the three glycan chains interlinked by this trimeric element lies outside the plane of the two others. Logically, this rules out the notion of *E. coli* murein being exclusively monolayered. Also pulse-chase experiments strongly support the assumption of a multi-storey murein at least at sites of murein synthesis. Crossbridges in murein have different life spans. The peptide links formed between newly incorporated glycan strands are relatively stable whereas crossbridges between new and preexisting glycan strands are cleaved and reformed during the processing of murein. New and preexisting murein have a different fate (Schwarz and Glauner, 1988). All data available so far indicate that new murein is added to the inner side of the preexisting murein. Through an inside-to-outside growth mechanism the newly added murein becomes part of the continuous stress-bearing layer. This involves cleavage and reformation of bonds both in preexisting and between new and preexisting murein. According to an elegant growth model of *E. coli* murein by J.-V. Höltje (this volume) the integration of new units into the shape-maintaining stress bearing layer of murein involves the twisting of a preexisting murein, when new subunits are integrated into the stress bearing layer of the sacculus. Due to this slight twist, part of the preexisting murein is moved outside the plane of this layer to the periphery of the sacculus where it is subject to the turnover process described previously.

MORPHOGENETIC CONTROL OF MUREIN SYNTHESIS

One of the most interesting but quite neglected questions related to murein biosynthesis is its morphogenetic control. How can new subunits be integrated into the existing sacculus without distortion of its shape and diameter? Actually this question is completely open to speculation, however, a very logical possibility can be experimentally tested. Is it possible that copying - the replication of preexisting murein - is an essential parameter in the morphogenetic control of murein synthesis?

The structural data on murein, and studies on the arrangement of the glycan chains in the sacculus, using electromicroscopy, can be summarized in a model in which the glycan chains surround the sacculus in a belt-like fashion with a long range orientation perpendicular to its long axis (Verwer et al, 1978). Assuming an average chain length of 30 disaccharide units it would require about 100 chains, in tandem, to span the sacculus. In consequence, the diameter of the sacculus, of the cell, would be defined by the number of glycan chains and by their length. If those two parameters would be kept constant during replication, the proper diameter of the cell would be maintained during the enlargement of the sacculus in a growing cell.

It is quite likely that murein synthesis depends on a matrix of preexisting murein and it should be possible to test, experimentally, whether the matrix indeed is copied such that both number and length of chains in the matrix are mirrored by the product of synthesis. In *E. coli* the matrix, the murein to be copied, can be manipulated experimentally by starving, under osmotic protection, a mutant defective in the biosynthesis of 2,6-diaminopimelic acid. Under those conditions, the murein is degraded by autolytic enzymes, the rod-shaped bacteria are converted into round spheroplasts, the potential for murein biosynthesis, however, remaining unaffected. If the missing amino acid is added again to such spheroplasts, murein synthesis is resumed immediately, a new sacculus is formed. It reflects the spherical shape of the matrix at its time of synthesis (Schwarz and Leutgeb, 1971).

These spherical, mechanically stable cells, in fact, can regain rod-shape, but only after many generation times in a not at all understood process of trial and error. In the experiment, the shape of the matrix is copied, a spherical matrix yields a spherical copy. What about the reproduction of the chemical composition of the matrix modified by starvation? At the time when the experiment was originally done there were no methods available to measure number and length of glycan chains in murein. Also the original experiment was, in fact, not intended to answer the question we are dealing with now. The study has to be repeated to test the assumption that murein is replicated, that murein biosynthesis involves a more or less exact copying of a preexisting matrix. Pilot experiments indicate indeed that in spheroplasts with a degraded matrix significantly shorter glycan chains are synthesized compared with rod shaped cells with an intact murein (Obermann, W., unpublished). Quite likely, the copying process is imperfect and proof-reading enzymes as is common in nucleic acid metabolism may guarantee the exact length of the glycan chains. We know already that during synthesis glycan chains are produced which are a bit longer than finally found in murein. This could, in fact, be achieved by lytic transglycoslylases as found in *E. coli*. They act as exoenzymes (Beachey et al, 1981) might use preexisting glycan chains as a measuring stick, clipping of hanging chain ends.

It is very clear that the control of morphogenesis of E. coli requires much more than just a copying of number and length of glycan chains. The replication of glycan chains, for example, would allow the ordered elongation of the *E. coli* rod. But the site of division has to be defined and new cell ends, hemispherical polar caps, have to be formed. Their morphogenesis could be achieved by a slight modification of the copying process as described (see J.-V. Höltje, this volume).

One parameter whose control, at first sight, might be difficult to explain, the topology and number of the peptide bridges crosslinking the glycan chains in murein, might not require specific control. Due to the three-dimensional structure or murein only a limited number of peptide side chains are available for crosslinking and what potentially can be crosslinked will be crosslinked. This statement agrees with the extent of crosslinkage as determined in sacculi (Glauner et al, 1988). For other obvious questions we do not have any answer, not even guesses. For example, according to which principles is the shape of the cell optimized? Would the genetic apparatus of the cell allow the production of principally different shapes of the cell? What defines the optimal length and diameter of the cell?

HERITAGE OF STRUCTURAL INFORMATION

What in fact is postulated is the heritage of structural information stored in the sacculus for the daughter cells. This is the cells contribution to the enormous task of maintaining the shape and integrity of the sacculus although every generation and a new sacculus has to be built. When the sacculus has to be generated de novo, without significant structure information, as happens during the reversion of spheroplasts (Landman, 1968) it takes many generations instead of just the one in a normally dividing cell. The possibility of a transfer of structural information of a cell besides from that of its chromosomes to daughter cells has been discussed for quite some time (Reviews: Sonneborn, 1963; Frankel, 1974; Nijhout, 1990). However it has never gained very broad attention especially at the present time when the overwhelming achievements of genetics have led to the unproven assumption that the genome directly encodes also higher levels of cellular organisation including those responsible for the maintenance of shape (Nijhout, 1990). A heritage of structural information can be shown to occur in protozoa (Review: Sonneborn, 1970) thus in this respect bacteria would not be at all unique. However they offer, opposed to other systems, the opportunity to study the mechanisms underlying the replication of structural information in molecular terms.

REFERENCES

Beachey, E.H., Keck, W., de Pedro, M.A. and Schwarz, U. (1981) Exoenzymatic activity of transglycosylase from *Escherichia coli*. Eur. J. Biochem. 116, 355 - 358.

Frankel, J. (1974) Positional information in unicellular organisms. J. Theor. Biol. 47, 439 - 481.

Glauner, B., Höltje, J.-V. and Schwarz, U. (1988) The composition of the murein of *Escherichia coli*. J. Biol. Chem. 263, 10088 - 10095.

Glauner, B. and Höltje, J.-V. (1990) Growth pattern of the murein sacculus of *Escherichia coli*. J. Biol. Chem. 265, 18988 - 18996.

Goodell, E.W. (1985) Recycling of murein by *Escherichia coli*. J. Bacteriol. 163, 305 - 310.

Höltje, J.-V., Mirelman, D., Sharon, N. and Schwarz, U. (1975) Novel type of murein transglycosylase in *Escherichia coli*.. J. Bacteriol. 124, 1067 - 1076.

Höltje, J.-V. and Schwarz, U. (1985) In "Molecular Cytology of *Escherichia coli*" (Nanninga, N. Ed.) p. 77 - 119, Academic Press Inc., London.

Labischinski, M., Barnickel, G., Naumann, D. and Keller, P. (1985) Conformational and topical aspects of the three-dimensional architecture of bacterial peptidoglycan. Ann.Microbiol. (Paris), 136A, 45 - 50.

Landman, O.E. (1968) in "Microbial protoplasts, spheroplasts and L-forms" (Guze, L.B., Ed.), pp. 319 - 332, Williams and Wilkins Co., Baltimore, Maryland.

Nijhout, H.F. (1990) Metaphors and the role of genes in development. Bioessays 12, 441 - 446.

Salton, M.R.J. (1960) Microbial cell walls. John Wiley and Sons, Inc., New York - London.

Schwarz, U. and Leutgeb, W. (1971) Morphogenetic aspects of murein structure and biosynthesis. J. Bacteriol. 106, 588 - 595.

Schwarz, U. and Glauner, B. (1988) Murein structure data and their relevance for the understanding of murein metabolism in *Escherichia coli*. In "Antibiotic inhibition of bacterial cell surface assembly and

function" (Actor, P., Daneo-Moore, L., Higgins, M.L., Salton, M.R.J. and Shockman, G.D. Eds.), pp. 306 - 311, Society for Microbiology, Washington.
Sonneborn, T.M. (1963) Does preformed cell structure play an essential role in cell heredity? In "The nature of biological diversity" (Allen, J.M., Ed.), pp. 165 - 221. Mc Graw - Hill Inc., New York.
Sonneborn, T.M. (1970). Gene action in development. Proc. Roy. Soc. Lond., B. 176, 347 - 366.
Verwer, R.W.H., Nanninga, N., Keck, W. and Schwarz, U. (1978) Arrangement of glycan chains in the sacculus of *Escherichia coli*. J. Bacteriol. 136, 723 - 729.
Vincenzi, L. (1887) Über die chemischen Bestandtheile der Spaltpilze. Z. für physiologische Chemie, 11, 181 - 183.
Weidel, W. and Pelzer, H. (1964) Bagshaped macromolecules - a new outlook on bacterial cell walls. Adv. Enzymol. 26, 193 - 232.

BIOPHYSICAL AND BIOCHEMICAL STUDIES ON THE FINE STRUCTURE OF THE SACCULI FROM *ESCHERICHIA COLI* AND *STAPHYLOCOCCUS AUREUS*

Harald Labischinski,[1] Michael Hochberg,[2] Thomas Sidow,[1] Heinrich Maidhof[1], Uta Henze[1], Brigitte Berger-Bächi[3], and Jörg Wecke[1]

[1]Robert Koch-Institute, D-1000 Berlin 65, Germany
[2]SUNY Institute of Technology, Utica, NY 13504-3050, USA
[3]Institute of Medical Microbiology, CH-8028 Zürich, Switzerland

INTRODUCTION

The general principles of the chemical structure of the bacterial sacculus and the basic conformational features of its building bricks, - oligosaccharide chains bearing peptide side chains for crosslinking - are well known for a long time (for review see e.g. Schleifer and Kandler, 1972; Labischinski et al., 1985). However, these data gave only rough ideas how the real giant macromolecule murein (synonym: peptidoglycan) is organized to fulfill its functions like protecting the bacterial cell against its own osmotic pressure, maintaining shape, yet allowing growth and division of the cell. For these purposes the sacculus must confer rigidity, flexibility and structural dynamics at the same time. Obviously, the topology of the peptides to be crosslinked plays an important role for constructing a covalently linked sacculus. The enzymes engaged in this job are anchored in the cytoplasmic membrane and should crosslink peptides only, if these come in close spatial contact. For this reason, the helical nature of the murein glycan chains, which forces consecutive peptides along a sugar strand to point approximately to the left, downward, right and upward, largely limits the theoretical constructions of stress bearing, covalently linked networks (topotypes, see Labischinski et al., 1985; Labischinski and Johannsen, 1986; Koch, 1991). To differentiate between the still large number of possible arrangements, experimental data are urgently needed, but difficult to obtain for large, low-ordered macromolecular assemblies like the murein sacculus. This report will focus on two of the few experimental approaches available - small angle neutron solution scattering and HPLC-techniques to study enzymatically produced murein degradation products -, which recently contributed to our knowledge on the architecture of the gram-negative sacculus and on the interrelationships between peptidoglycan crosslinking fine structure and ß-lactam resistance in staphylococci, respectively.

The gram-negative sacculus: mono- or more-than-single-layered?

It is known for 25 years that the gram-negative sacculus, as compared to its gram-positive counterpart, is extremely thin. If it represents a monolayered architecture, or just a modification of the gram-positive scheme, comprising few, but more than one, layers is still a matter of debate. Fig. 1 very schematically depicts some simplified sketches of models under consideration, in the order of increasing amount of murein per cell postulated. These models range from single-layered up to multi- (e.g. triple-) layered throughout the surface of the sacculus (models 1 and 5 in Fig. 1), but also include topological variants as e.g. multilayered polar caps and septa (model 4) or the existence of patches of multilayered material as proposed by Höltje and Glauner (1990) or the "make-before-break" strategy (model 3) discussed in detail by Koch (1990).

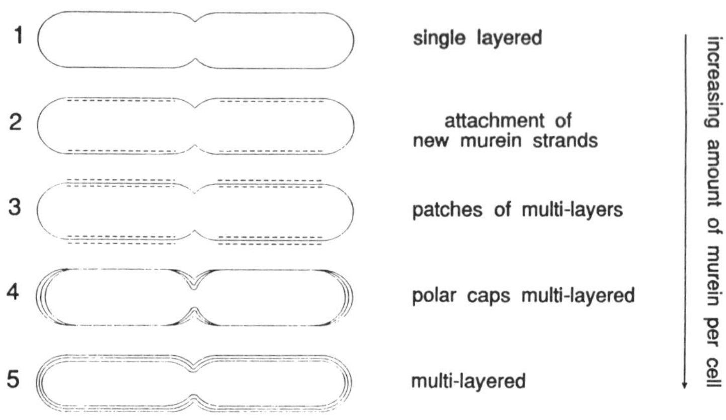

Figure 1. Simplified sketches of models for the thickness profiles of the *Escherichia coli* sacculus

We have recently published the results of a neutron scattering study (Labischinski et al., 1991) which directly showed that the sacculi of exponentially growing *Escherichia coli* W7 contain a basic single layer covered on about 20 - 25 % of its surface by two additional layers, which may or may not be arranged centrosymetrically to the main stress bearing layer. Together with the data of Wientjes et al. (1991) suggesting that there is not enough murein present to cover even two complete layers of a gram-negative sacculus, models 1, 5, and, at least in its strict version, model 2 (see Fig. 1) can already be excluded. However, since the neutron scattering method averages over the complete surface of the sacculus, it does not allow to differentiate between models 3, 4 or possible combinations thereof. We, therefore, decided to perform analogous scattering experiments using cells which appeared to be enriched in either polar cap material (using minicells as a modelsystem) or in cylindrical wall (using either *fts*Z-mutants or nalidixic acid treated bacteria), as well as bacteria with selectively blocked PBP2, which are well known to grow as large spheres and eventually lyse. Since the necessary neutron scattering facilities were not available until early 1992, we collected some corresponding X-ray data on minicell-preparations and sacculi isolated from cells treated with 1 µg/ml mecillinam (PBP2-specific) for 2 hours after which the cells appeared to be rounded, but not completely spherical. Fig. 2 shows the scattering density profiles of these new preparations, which simply correspond to the distribution of mass across the thickness of the sacculus, averaged over the entire sacculus surface. While the density profiles obtained earlier for sacculi from exponentially growing cells revealed a maximum (not average!) thickness of 7.5 nm (corresponding to the stress bearing layer of 2.5 nm plus two additonal layers on 20 - 25 % of its surface of again 2.5 nm thickness, each, see Labischinski et al., 1991) the maximum thickness of the *E. coli* minicell sacculi was much larger (up to 20 nm c.f. Fig. 2).

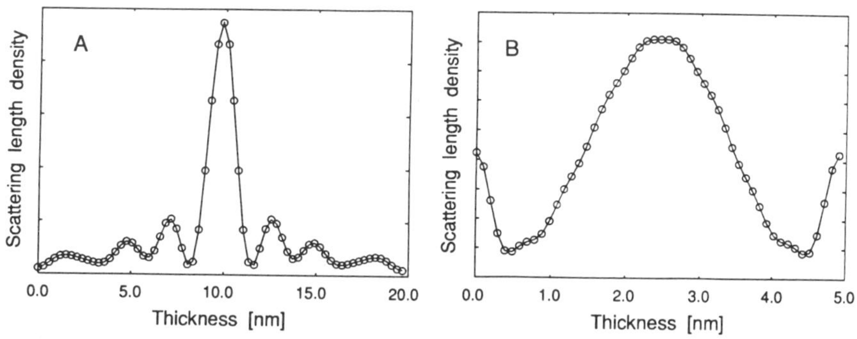

Figure 2. Average density cross-section profiles (arbitrary units) across the thickness of *E. coli* sacculi. A: The average cross-section of minicells revealed a single layer (peak in the middle of the figure)covered by small amounts of additonal material (bumps to the right and left of the main layer). Besides showing additional wall remnants to a total maximum thickness of approx. 20 nm, the profile is very similar to that observed earlier for sacculi from exponentially growing cells (Labischinski et al., 1991) B: The thickness density profile of sacculi from mecillinam treated cells: essentially a monolayer structure.

This increase in maximum thickness was, however, due to a very low percentage of wall material (less than 10 %, c.f. Fig. 2A) while the bulk of the material shows a similar organization as in sacculi from exponentially growing cells: the minicells did also contain one single layer covered partly by additional layers. This fits observations by Nanninga (personal communication) that electron micrographs of unstained *E. coli* sacculi, recorded using a mass-dependent contrast, did not reveal differences in mass per area between polar cap and cylindrical wall regions. One can only speculate, if the low amount of additional material at the minicell sacculi surface is also found in real polar caps of normal cells; in any case it might well represent septal "safety" material which could act as an additional "security measure" when the thin wall material at the leading edge is split during cell division. This is reminiscent to the situation found in staphylococci: cell separation involves splitting of the cross wall whereby considerable amounts of wall material comprising the entire, concentrically arranged "splitting system" are turned over and lost into the culture medium (Giesbrecht et al., 1976).

The mass distribution across the thickness of the sacculi isolated from mecillinam treated cells is shown in Fig. 2b. It must be mentioned that the profile shown is the result of a first approximation run and needs some further optimization. Nevertheless, one can already recognize that the mecillinam treated sacculi are very close to the single layer model with very few remnants of the additional layers found in exponentially growing cells. Considering that PBP2 seems to be the PBP performing the bulk of cylindrical wall synthesis in *E. coli* (Wientjes and Nanninga, 1991), the very low amount of more-than-single layered material in the PBP2 blocked cells might be due to the activity of PBP3 engaged in septal wall synthesis.

From all these findings the following tentative view of the *E. coli* sacculus can be extracted:

- The experimental data are consistent with one single, essentially stress-bearing layer covering the complete surface of the bacterium.

- The *E. coli* sacculus nevertheless represents essentially a more-than-single layered architecture, where relatively small regions (approx. 20 - 25 %) of both, polar cap and cylindrical wall, contain triple layered areas (patch-like multilayer architecture).

- In the polar cap and/or septal wall regions, at least under certain conditions small patches with even more layers might be (transiently?) present.

- An architecture close to a "pure" single-layer structure could only be detected at conditions (PBP2-inhibition) which resulted in loss of correct wall morphology and, eventually, cell death.

- A "classical" multilayer arrangement corresponding to the usual scheme found in gram-positive organisms is obviously not present in *E. coli*. The "patch-like" multilayer architecture might yet serve similar functions and phenomena as the gram-postive wall, adapted to the much lower osmotic pressure of the gram-negatives: it could allow for turnover processes, septum-like structures at the leading edge of the constricting cell wall and partial stress bearing by more than one layer as a general safety device, e.g. against bulging of the wall, functions not available by just attaching new murein as single strands at the wall growth areas.

If one considers the cell separation process of staphylococci, where during cross wall formation the utmost external parts of the cross wall are apparently only synthesized for the sake that they have to be solubilized during cell separation, one has to differentiate in staphylococci between constitutive, permanent parts of the cell wall and transitory parts of the cell wall, which are only attached to the permanent parts of the cross wall. However, presently there are no indications, if also in *E. coli* such a dual wall system is existent and to what extent it might contribute to the experimentally observed partial multilayer appearance of the *E. coli* sacculus.

CORRELATIONS BETWEEN STAPHYLOCOCCAL MUREIN FINE STRUCTURE AND ß-LACTAM RESISTANCE

In contrast to *E. coli*, the murein of *Staphylococcus aureus* reveals an exceptional high degree of crosslinking (values between 70 and 90 %) which has previously been shown to be possible only because of the existence of a relatively long and very flexible pentaglycine interpeptide bridge (Labischinski et al., 1985). Due to this additional structural feature, peptides can be crosslinked which are otherwise separated too far to be recognized by the membrane bound transpeptidases. Compared to the situation in *E. coli*, much less details are known about the four penicillin binding proteins (PBP) of *S. aureus*; nevertheless it is clear that all four are candidates for ß-lactam-sensitive transpeptidation reactions (Reynolds, 1988; Beise et al., 1988; Snowden and Perkins, 1991) by crosslinking the free amino group of the interpeptide bridge glycine to the carboxyl group of the D-alanine at position 4 of another muropeptide side chain by a peptide bond, thereby releasing the D-alanine in position 5 of the same peptide chain. One might, therefore, anticipate that the staphylococcal PBP should be optimized to handle the unique glycine pentapeptide bridge. The same should be true for those wall lytic enzymes and structures involved in the morphogenesis and division of *Staphylococcus aureus*, as far as their substrate and/or binding sites involve the crosslinked muropeptides.

If all PBP, or certain subsets of the PBP including PBP1, (Reynolds, 1988; Beise et al., 1988) are blocked, the staphylococci die and, depending on a number of other factors, may or may not lyse subsequently (for details see Giesbrecht et al., 1992 a, b). Staphylococci are known as being notorious in developing resistance against antibiotics (for review see Lyon and Skurray, 1987). Of particular clinical importance worldwide is the spread of the so-called methicillin-resistant *Staphylococcus aureus* (MRSA) strains, which, as a rule, are multiresistant against almost all clinically important antibiotics with the, yet, lucky exception of some glycopeptide-antibiotics. The basis for their resistance against virtually all ß-lactam antibiotics is the occurence of an additional low affinity PBP, called PBP 2a or 2′, for which there is no allele equivalent in susceptible strains, suggesting a non-staphylococcal origin of the protein (for review, see Hackbarth and Chambers, 1989). If this turns to be true, the question arises, how this "non-staphylococcal PBP" can produce a kind of staphylococcal cell wall sufficient for survival under conditions where the original PBP 1-4 are already completely inhibited.

Architecture of the Murein Sacculus Produced by PBP 2a

Using a set of isogenic strains (kindly provided by Brigitte Berger-Bächi, University Zürich) which differed only by the presence of the mec-determinant coding for PBP 2a, we first analyzed the amount of cell wall produced by the usual set of PBP 1-4 as compared to that constructed by PBP 2a by measuring the incorporation of the cell wall specific marker ^{14}C-N-acetyl-glucosamine (Fig. 3).

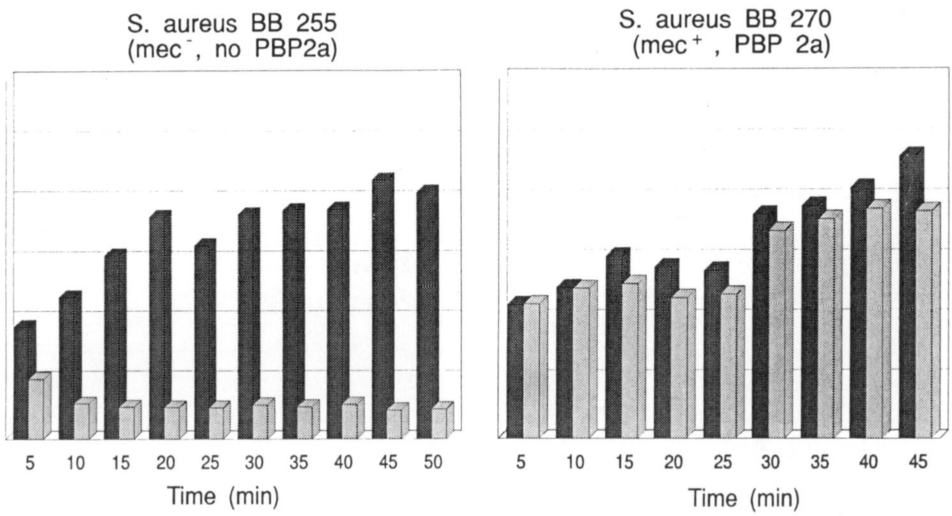

Figure 3. Incorporation of the cell wall specific label ^{14}C-N-acetylglucosamine in *S. aureus* due to PBP2a. Dark rey bars: control values; light grey bars: label incorporated after addition of 1 µg/ml penicillin G at zero time.

Inspection of the results given in Fig. 3 shows that even under conditions, when the normal PBPs were inhibited to a degree that stopped cell wall synthesis almost completely within the first minutes after addition of the ß-lactam, the isogenic strain expressing PBP 2a still produced nearly the control amount of cell wall. However, examination of this strain by electron microscopy already revealed that severe changes had taken place under these conditions (Fig. 4): cell separation was retarded, cross walls appeared irregularly thickened and partially misplaced without the appearance of the typical splitting system (c.f. Giesbrecht et al., 1976, 1992b), and the surface of the peripheral cell wall showed large amounts of fuzzy wall material. It was interesting to note that all these variations, especially the inhibition of cell separation, could not be observed in untreated cells of this strain.

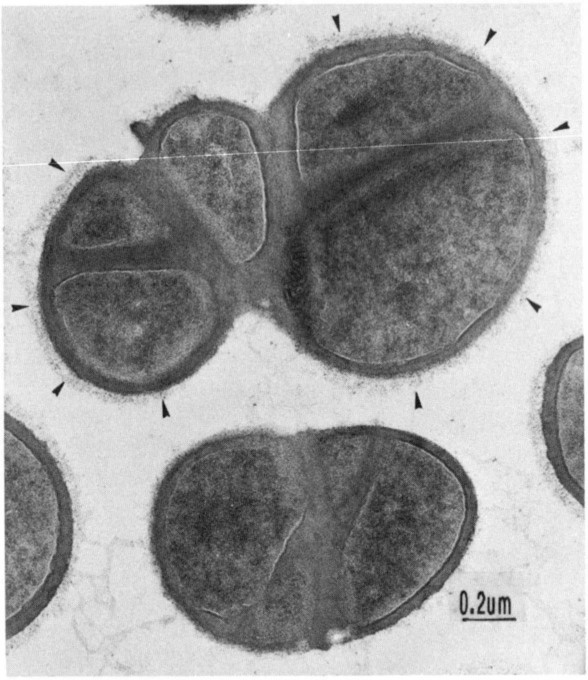

Figure 4. Thin section of methicillin resistant strain *S. aureus* BB270 grown in the presence of 1 μg/ml penicillin for 4 h: irregular division, fuzzy like material covers the surface of the cell (arrows).

Concomitantly, a reduction of wall turnover was observed (data not shown). In line with these observations of a drastically altered cell wall architecture is the finding of deJonge, Tomasz and coworkers (1991), who first demonstrated that the wall chemistry changed after at least partial inhibition of the normal set of PBP in a PBP2a producing strain, by using a reversed phase HPLC-technique to separate muramidase-digested peptidoglycan, which indicated a very low crosslinked murein quality. We repeated their experiments using strain BB270 at conditions, where PBP1 to 4 should be completely saturated (5 μg/ml penicillin G), but using a gel-permeation type HPLC-column to separate the muropeptides after complete muramidase digestion according to size (c.f. Snowden and Perkins, 1991), in order to be able to determine the degree of crosslinking (Fig. 5). As expected, the peptidoglycan synthezised with the help of the usual PBP 1-4 showed the typical broad staphylococcal pattern, in which larger oligomers (comprising up to more than 30 subunits) with a maximum corresponding to the hexamer were observed (degree of crosslinking approx. 75 %). However the murein synthesized after inhibition of PBP 1-4 showed a very peculiar behaviour: the most prominent peaks were the mono- and dimers in about the same quantity, very similar to what one observes for gram-negative bacteria; there were in addition some tri-, tetra- and pentamers, but no contribution of the typical gram-positive oligomers (crosslinking approx. 49 %). Thus, although a sufficient amount of wall material was made in the PBP 2a producing strain even at large penicillin-concentrations, its quality more closely resembled a gram-negative wall, albeit allowing survival of the bacterium.

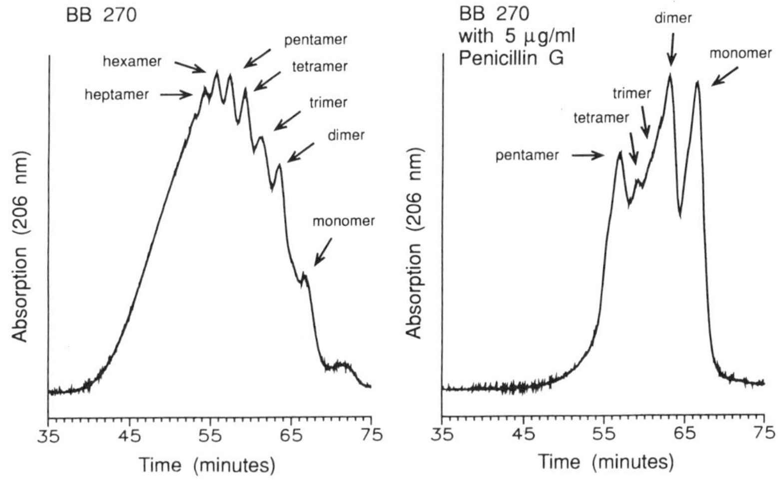

Figure 5. Comparison between the gel-permeation HPLC-profiles of muramidase digested cell walls isolated from strain BB270, producing PBP2a grown without (left) or with (right) addition of 5 μg/ml penicillin G, a concentration effectively blocking the normal set of staphylococcal PBP1-4, but not PBP2a. The control cells (left side of the fig.) revealed a typical staphylococcal profile in which the muropeptide-oligomers dominate. The profile produced from cells grown under almost sole action of PBP2a is more reminiscent to the situation observed in gram-negative bacteria (monomers and dimers as most prominent peaks).

The Influence of *fem*A and *fem*B on Staphylococcal Wall Architecture and Resistance

Although the existence of the additional PBP2a is the pre-requisite for resistance, it is known that several other chromosomal factors are necessary for the expression of resistance, which are not related to the mec-determinant and do not influence the expression of the mecA gene coding for PBP2a. This was first shown by Berger-Bächi (1983; for review see Berger-Bächi et al., 1990) who described the existence of several chromosomal factors called *fem*A, *fem*B, *fem*C and *fem*D, which after inactivation by transposon mutagenesis led to reduced expression of methicillin resistance (*fem* = factor for the expression of methicillin resistance). These mutants are quite analogous to the so called *aux* mutants described by Tomasz (1990). Since the reduction of resistance expression is most prominent in the *fem*A and *fem*B mutants, we will restrict ourselves to the description of these factors. Only very recently the first hints to their physiological function were obtained. Amongst the first observations was the finding that autolysis and turnover of these mutants were reduced (deJonge et al., 1991; Maidhof et al., 1991). This could, formally, be caused by either changes in the autolytic enzymes themselves or in the cell wall substrate, or both. However, our studies (Maidhof et al., 1991) already suggested that alterations in the cell wall substrate were the responsible factors: not only was the degradation of *fem*A-mutants slowed down if externally added lysostaphin was used; more important, we could demonstrate that the glycine content was drastically reduced in the interpeptide bridge of *fem*A-mutant cell walls (40 - 60 % reduction, depending on the strain used). The same study also revealed that *fem*A is involved in proper construction of the pentaglycine bridge not only in MRSA-strains but also in ß-lactam-sensitive strains. *fem*A-inactivation in these strains resulted in hypersensitivity against many antibiotics, including the ß-lactams and glycopeptide-antibiotics (Maidhof et al., 1991; Witte and Labischinski, unpublished observations).

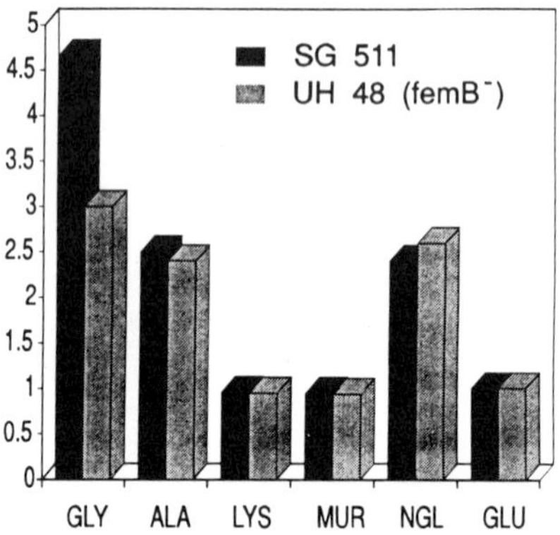

Figure 6. Cell wall amino compound composition of *S. aureus* strain SG511 and its respective *fem*B-inactivated mutant. The molar masses of each amino compound were normalized to glutamic acid (GLU). N-acetylmuramic acid (MUR) and N-acetyl-glucosamine (NGL) are also shown. The only significant difference was found in the glycine content (GLY), which was reduced by almost 40 % in the *fem*B-mutant, similar to the situation observed earlier in *fem*A mutants (Maidhof et al., 1991).

Interestingly, *fem*A and *fem*B are transcribed on a polycistronic mRNA. We, therefore, also analyzed the amino acid composition of the cell walls of a recently constructed *fem*B mutant (Henze et al., 1992). As shown in Fig. 6, a reduction of 40 % in glycine content was also observed for this mutant. The morphology of the *fem*A and *fem*B mutants is evident from the electron micrographs in Fig. 7. These cells are impressive because of their highly aberrant placement of cross walls and heavy retardation of cell separation, but do survive under these conditions. Inspection of their crosslinking profile still shows a typical "gram-positive" pattern (Fig. 8), although there was a clear shift to lower oligomers as compared to the parent strain (corresponding to a significantly lowered crosslinking value), in line with the theoretical prediction that the long pentaglycine chain is necessary to reach the high crosslinking index typical for *S. aureus*.

Thus, taking together these data with those described in the preceeding section on PBP2a the following model about the relationship between staphylococcal interpeptide

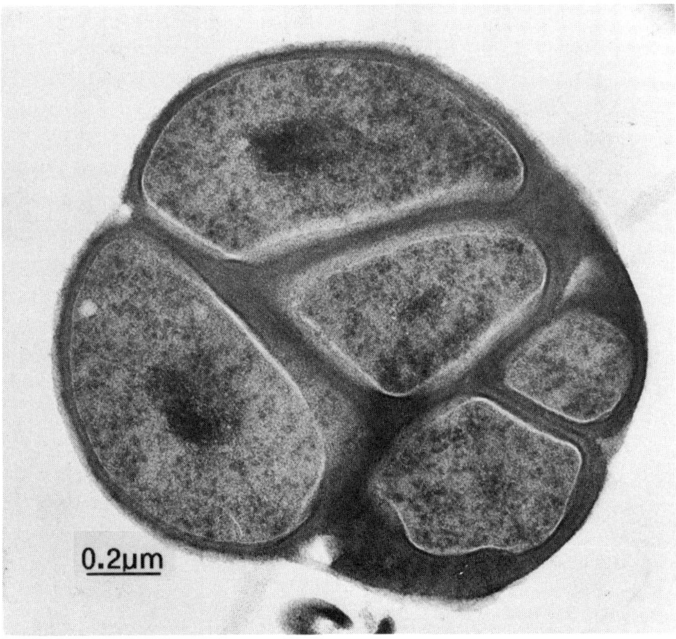

Figure 7. Thin section of a *femB* mutant of *S. aureus*; pseudomulticellular staphylococcus caused by inhibition of cell separation.

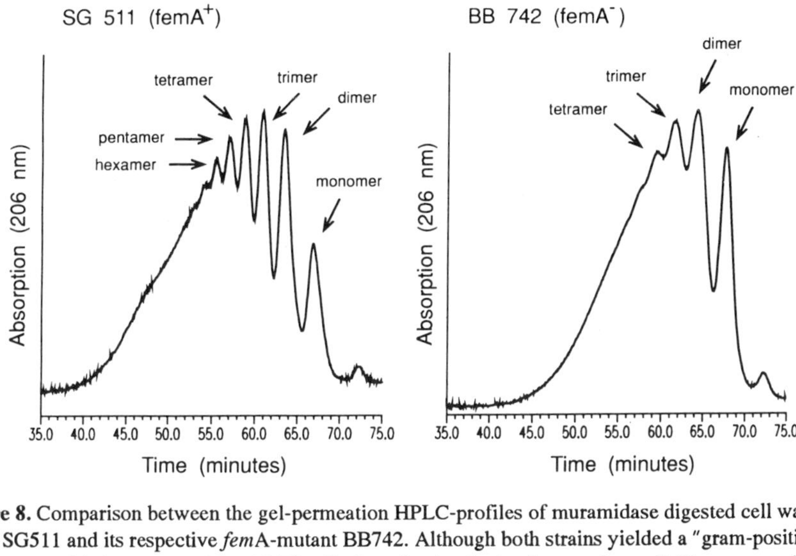

Figure 8. Comparison between the gel-permeation HPLC-profiles of muramidase digested cell walls from strain SG511 and its respective *femA*-mutant BB742. Although both strains yielded a "gram-positive" type size pattern of their muropeptides (cf. Fig. 5), the reduction in the degree of crosslinking as revealed by the shift to longer retention times (equivalent to lower mass of the peptides) in the *femA*-mutant is obvious.

bridge architecture, crosslinking capability of PBP1 - 4 and PBP2a, and ß-lactam resistance can be proposed (Fig. 9):

- The so called *fem* or *aux*-factors assumed to be directly connected to the expression of methicillin resistance must, at least in the case of *fem*A and *fem*B, be considered as factors involved in pentaglycine interpeptide bridge formation.

- Their insertional inactivation by transposon mutagenesis leads to a length reduction of the unique staphylococcal interpeptide bridge, which in turn results in reduced murein crosslinking, by a gradual shift towards smaller oligomers up to the monomer fraction, reduced susceptibility to lytic and autolytic enzymes which attack and/or recognize this part of the murein network, and morphogenetic errors in crosswall formation.

- When the normal set of staphylococcal PBP1 - 4 is blocked by ß-lactam antibiotics, the non-staphylococcal PBP2a is able to produce a cell wall of sufficient quality for survival. However, a hypocrosslinked "gram-negative type" murein is synthesized, with essentially only low oligomeric peptide chains present. This cell wall material is not a good substrate for at least some of the staphylococcal murein hydrolases; morphogenetic disturbances including retarded cell separation is one of the consequences.

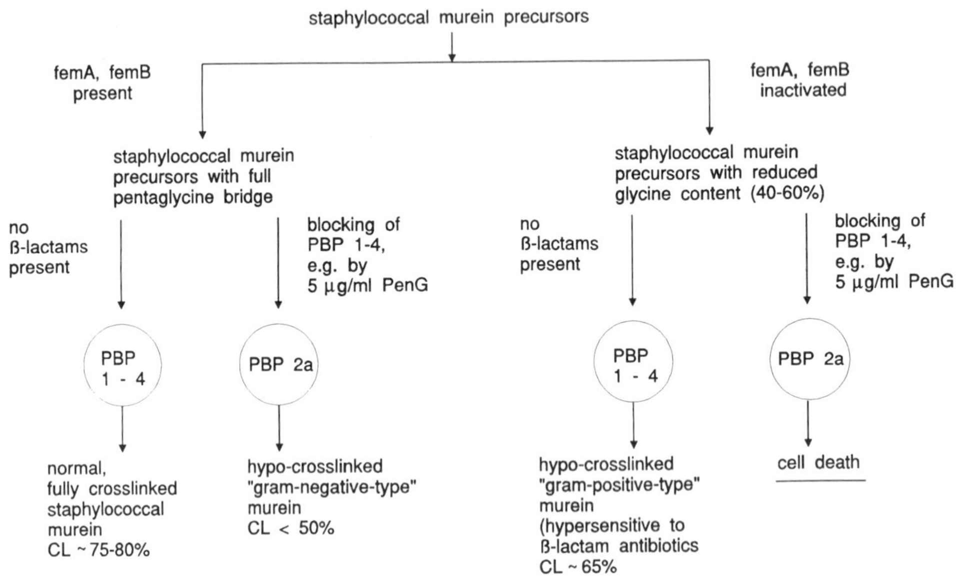

Figure 9. Tentative model for the interplay of *fem*A, *fem*B, PBP1-4, and PBP2a in constructing a cell wall of correct, "staphylococcal" type. The consequences of inactivation of the various factors are indicated at the bottom of the figure. For details see text.

- If the pentaglycine bridge formation is disturbed by *fem*A or *fem*B inactivation, and, in addition, PBP1 - 4 will be inhibited, even PBP2a is no more capable of synthesizing a functional wall. Eventually, cell death is the consequence.

These findings do not only reveal the vital role of the interpeptide fine architecture for the staphylococcal cell wall. Based on the finding that *fem*A or *fem*B inactivation causes loss of resistance expression in MRSA-strains and induces hypersensitivity to various antibiotics in sensitive staphylococci, potential inhibitors of *fem*A and/or *fem*B might also become important staphylococci- specific drugs for therapeutic applications.

SUMMARY

Two extremes in bacterial cell wall architecture were investigated using direct and indirect structural methods like neutron scattering, electron microscopy, chemical wall analysis and HPLC of muropeptides: the very thin, low-crosslinked *E. coli* sacculus with respect to its thickness profile across its surface, and the thick, high-crosslinked *S. aureus* murein with respect to the role of its unique pentaglycine interpeptide bridge for crosslinking, wall architecture, and resistance against ß-lactam antibiotics.

For *E. coli* a more-than-single layered arrangement is observed, which nevertheless comprises only one essentially stress bearing layer on about 75 - 80 % of its surface. All data are compartible with a patch-like triple layer architecture for both , cylindrical and polar cap regions of the wall. In additon some hints to the existence of transitory material, possibly synthesized just for the sake to be released during cell division in a process reminescent to that observed in staphylococci (Giesbrecht et al., 1976), are obtained. A pure single-layer architecture, as approximately detected in mecillinam-treated *E. coli*, seems not to be compatible with correct wall morphogenesis and cell division.

For *S. aureus* the unique pentaglycine interpeptide bridge is shown to be essential for murein crosslinking, autolysine efficacy and normal sensitivity towards the action of ß-lactam antibiotics. Mutants in *fem*A and *fem*B genes, involved in correct formation of the interpeptide bridge, produce walls which cannot be correctly handled by the normal set of PBP1-4 and by autolytic enzymes involved in turnover and cell separation; these mutants, which produce a less crosslinked cell wall, also reveal a drastically increased sensitivity towards ß-lactam antibiotics. In MRSA strains, the characteristic additional "non-staphylococcal" PBP2a is not adapted to the pentaglycine-chain bearing murein substrate;consequently, when the normal set of PBP1-4 is inactivated, a hypocrosslinked "gram-negative" type wall is produced, which, however, allows survival of the cell. Additional inactivation of *fem*A or *fem*B in these strains leads to cell death. *fem*A and *fem*B might, therefore, serve as specific anti-staphylococcal targets in chemotherapy, especially for the multiresistant MRSA-strains.

ACKNOWLEDGEMENTS

The experimental help of Margaretha Gross and Ingrid Warme with preparation of cell walls, of Bernhard Reinicke with amino acid analysis, Angelika Brauer with preparation of the figures, and Ellen Albrecht with typing of the manuscript is gratefully acknowledged. Parts of this work were supported by the *Deutsche Forschungsgemeinschaft* (LA 476/3-1) and by the *National Science Foundation* (grant No. 91-0390).

REFERENCES

Beise, F., Labischinski, H. and Giesbrecht, P. (1988) Selective inhibition of penicillin-binding proteins and its effects on growth and architecture of *Staphylococcus aureus*. FEMS Microbiol. Lett. 55, 195-202.

Berger-Bächi, B., (1983) Insertional inactivation of staphylococcal methicillin resistance by Tn*551*. J. Bacteriol. 154, 479-487.

Berger-Bächi, B., Strässle, A., Barberiso-Maino, L., Tesch, W., Ryffel, C. and Kayser, F.H. (1990) *Fem*A, a host-mediated trans-acting factor essential for methicillin resistance in *Staphylococcus aureus*, in "Molecular Biology of the Staphylococci" (Novick, R.P., Ed.), pp. 509-520. VCH Publishers, Inc.

deJonge, B.L.M., deLencastre, H. and Tomasz, A. (1991) Suppression of autolysis and cell wall turnover in heterogeneous Tn*551* mutants of a methicillin-resistant *Staphylococcus aureus* strain. J. Bacteriol. 173, 1105-1110.

Giesbrecht, P., Franz, M., Krüger, D., Labischinski, H. and Wecke J. (1992a) Bacteriolysis of staphylococci is only a side-effect of penicillin-induced death, in "50 years of penicillin application - history and trends" (Kleinkauf, H. and von Döhren, H., Eds.), in press. Walter de Gruyter, Berlin.

Giesbrecht, P., Kersten, T., Blümel, P. and Wecke, J. (1992b) Penicillin-induced bacteriolysis of staphylococci as a post-mortem consequence of murosome-mediated killing via wall perforations and attempts to imitate these processes without applying antibiotics. In this volume.

Giesbrecht, P., Wecke, J. and Reinicke, B. (1976) On the morphogenesis of the cell wall of staphylococci. Int. Rev. Cytol. 44, 225-318.

Hackbarth, C.J. and Chambers, H.F. (1989) Methicillin-resistant staphylococci: genetics and mechanisms of resistance. Antimicrob. Agents Chemother. 33, 991-994.

Henze, U.

Höltje, J.-V. and Glauner, B. (1990) Structure and metabolism of the murein sacculus. Res. Microbiol. 141, 75-89.

Koch, A.L. (1991) Effective growth by the simplest means: the bacterial way. Through evolution, bacteria prove the adage "stress builds character". ASM News 57, 633-637.

Koch, A.L. (1990) Additonal aguments for the key role of "smart" autolysins in the enlargement of the wall of gram-negative bacteria. Res. Microbiol. 141, 529-541.

Labischinski, H., Barnickel, G., Naumann, D. and Keller, P. (1985) Conformational and topological asepcts of the three dimensional architecture of bacterial peptidoglycan. Ann. Inst. Pasteur Microbiol. 136A, 45-50.

Labischinski, H., Goodell, E.W., Goodell, A. and Hochberg, M.L. (1991) Direct proof of a "more-than-single-layered" peptidoglycan architecture of *Escherichia coli* W7: a neutron small-angle scattering study. J. Bacteriol. 173, 751-756.

Labischinski, H. and Johannsen, L. (1986) On the relationships between conformational and biological properties of murein, in "Biological Properties of Peptidoglycan" (Seidl, H.P. and Schleifer, K.-H., Ed.), pp. 37-42. DeGruyter u. Co., Berlin - New York.

Lyon, B. and Skurray, R. (1987) Antimicrobial resistance of *Staphylococcus aureus*: genetic basis. Microbiol. Rev. 51, 88-134.

Maidhof, H., Reinicke, B., Blümel, P., Berger-Bächi, B. and Labischinski, H. (1991) *fem*A, which encodes a factor essential for expression of methicillin resistance, affects glycine content of peptidoglycan in methicillin-resistant and methicillin-susceptible *Staphylococcus aureus* strains. J. Bacteriol. 173, 3507-3513.

Reynolds, P.E. (1988) The essential nature of staphylococcal penicillin-binding proteins, in "Antibiotic Inhibition of Bacterial Cell Surface Assembly and Function" (Acotr, P., Daneo-Moore, L., Higgins, M., Salton, M.R.J. and Shockman, G.D., Ed.), pp. 343-351. American Society for Microbiology, Washington, DC.

Schleifer, K.H. and Kandler, O. (1972) Peptidoglycan types of bacterial cell walls and their taxonomic implications. Bacteriol. Rev. 36, 407-477.

Snowden, M.A. and Perkins, H.R. (1991) Cross-linking and O-acetylation of peptidoglycan in *Staphylococcus aureus* (strains H and MR-1) grown in the presence of sub-growth-inhibitory concentrations of ß-lactam antibiotics. J. Gen. Microbiol. 137, 1661-1666.

Tomasz, A. (1990) Auxiliary genes assisting in the expression of methicillin resistance in *Staphylococcus aureus*, in "Molecular Biology of the Staphylococci" (Novick, R.P., Ed.), pp. 565-583. VCH Publishers, Inc.

Wientjes, F.B. and Nanninga, N. (1991) On the role of the high molecular weight penicillin-binding proteins in the cell cycle of *Escherichia coli*. Res. Microbiol. 142, 333-344.

Wientjes, F.B., Woldringh, C.L. and Nanninga, N. (1991) Amount of peptidoglycan in cell walls of gram-negative bacteria. J. Bacteriol. 173, 7684-7691.

NEW MASS SPECTROMETRIC METHODS FOR PEPTIDOGLYCAN ANALYSIS

Günter Allmaier* and Erich R. Schmid

Institute for Analytical Chemistry, University of Vienna, Währingerstr. 38, A-1090 Vienna, Austria

INTRODUCTION

During the last two decades, mass spectrometry has encountered a series of dramatic advances in desorption/ionization techniques and experimental strategies which have improved its capability to characterize the structure of underivatized peptidoglycan monomers and oligomers. These techniques, collectively referred to as soft ionization methods, provide molecular weight and especially in combination with tandem mass spectrometer structural information for a wide range of polar and thermally labile compounds. Complementary information to the well established techniques such as two-dimensional nuclear magnetic resonance, gas-phase Edman degradation and amino sugar analysis can be obtained. One of these important improvements was the introduction of fast atom bombardment combined with a four-sector tandem mass spectrometer for the analysis of non-reduced peptidoglycan monomers by Martin et al. (1987). Recently, several other new desorption/ionization techniques have been applied to the characterization of large, thermally labile biomolecules. Cf-252 plasma desorption (PD), matrix assisted laser desorption (MALD) time-of-flight mass spectrometry and electrospray (ESI) mass spectrometry are important new techniques. All offer relative ease of sample preparation, high sensitivity, accurate molecular weight determination and the possibility to analyze molecules with molecular masses to 50 kDa and beyond. This paper offers a comparison of the performance of the last three techniques when applied to the direct analysis of reduced peptidoglycan monomers and oligomers and the extent of information which may be acquired.

EXPERIMENTAL

Materials

All samples were obtained from the murein of *E.coli strain* MC6-RP3 and JE-5684 or the cyanelles of *Cyanophora paradoxa*. The peptidoglycan monomer and oligomer mixtures were prepared following an adapted version of Glauner et al. (1988). The final purification step based on high performance liquid chromatography (HPLC) with a salt gradient was changed

Bacterial Growth and Lysis, Edited by M.A. de Pedro *et al.*, Plenum Press, New York, 1993

to a quite different gradient consisting of water/acetonitrile/trifluoroacetic acid which is more suitable for the described mass spectrometric techniques (Pittenauer et al., 1992a). All chemicals were analytical grade products purchased from Merck, Fluka, Pierce or Biorad.

Plasma Desorption Mass Spectrometry

All plasma desorption mass spectra were recorded on a BioIon 20 time-of-flight mass spectrometer (ABI, Foster City, CA) with a flight tube length of 150 mm. The positive ion spectra were accumulated for 1.5×10^6 - 2×10^7 fission events at 17 kv or 19 kV accelerating voltage. In the negative mode spectra were acquired for 3×10^6 fission events at -15 kV. Mass calibration was based on the H^+, Na^+ or pump oil ions for positive and on H^- and CN^- ions for negative ion spectra. The lyophilized peptidoglycan samples were prepared as described by Pittenauer et al. (1992a) and Allmaier et al. (1992).

Matrix Assisted Laser Desorption Mass Spectrometry

Dihydroxy benzoic acid was chosen as matrix due to the strong absorption at the selected laser wavelength. The sample was prepared by mixing 0.7 μl of freshly prepared matrix solution (10 μg/1 μl in 30% acetonitrile, 0.1% trifluoroacetic acid) with 0.7 μl of sample solution (solvent: 30% acetonitrile, 0.1% trifluoroacetic acid) directly on the aluminium target. Before insertion into the mass spectrometer the sample was dried at ambient temperature. The matrix assisted laser desorption mass spectra were generated using a BioIon 10 time-of-flight mass spectrometer (ABI, Foster City, CA) modified to UV laser desorption as described by Roepstorff et al. (1991). The applied laser was a VSL-337 ND nitrogen laser (Laser Science Inc., Boston, MA) operating at a wavelength of 337 nm. Ions were accelerated to a potential of 20 kV before entering a field-free drift region of 1 m length. The mass spectrum reported here was obtained from the signal averaging of 60 laser shots.

Electrospray Ionization and Tandem Mass Spectrometry

The electrospray mass spectra were obtained on a VG Bio Q mass spectrometer (VG Biotech, Altrincham, UK), an instrument which consists of a dedicated electrospray ion source followed by a triple quadrupole mass analyzing system. The samples were dissolved in 50 μl 50% aequeous acetonitrile with 1% formic acid or 1% ammonia added for positive or negative ion mode. Peptidoglycan monomer solutions (10 μl) were introduced via a Rheodyne loop into the ion source at a flow rate of 2-4 μl min^{-1}. The applied electrospray potential was in the range of 50 - 80 V. The extracting cone voltage (V_C) was adjusted to between 45 and 100 V. The V_C is essentially similar to the voltage difference between the nozzle and skimmer described by Smith et al. (1990). Spectra were recorded with a scan rate of 10 - 12 seconds acquired by operating the data system as a multi-channel analyzer and all useful scans summed to obtain the final spectrum. Calibration of the mass spectrometer was performed from a separate introduction using polyethylene glycole 300/600/1000. Results are quoted as a mean average molecular mass.

The tandem mass spectrometric experiments were performed in the following way. The protonated molecular ions were generated by electrospray ionization and mass selected by MS 1 (Quadrupole #1). The rf-only quadrupole (Quadrupole #2) was used for collision-activated dissociation with Argon as collision gas at a collision energy of 2eV and at a collision cell pressure of 4 x 10^{-3} mbar. The so-formed ions were then mass analyzed by scanning MS 2 (Quadrupole #3) representing the so-called daughter ion spectrum. Ten scans were averaged for representative mass spectra.

RESULTS AND DISCUSSION

Figure 1 shows the positive and negative ion PD mass spectra of the peptidoglycan monomer Tetra. The HPLC purified monomer could be easily analyzed in the negative mode without any special sample preparation. The monomers as well as oligomers yield mainly deprotonated molecular ions with low abundance fragment ions. As seen in Figure 1 (b) the mass accuracy is quite sufficient (Δm = -0.01 u). The positive ion mass spectra of monomers are dominated by protonated molecular ions after applying our special sample preparation technique (Allmaier et al., 1992) to remove the excess of alkali ions which are tightly bound to the sugar moiety of the peptidoglycans, otherwise generating mainly cationized molecular ions. In case of unknown compounds this would complicate the molecular weight determination. In the upper mass range only few fragment ions could be detected (Figure 1(a), $^m/z$ 925.1, 742.0). The mass difference between calculated and found $[MH]^+$ is relatively small (Figure 1(a), Δm = + 0.05 u). Sensitivity is in the negative ion mode superior in practical all cases. The largest so far detected oligomer was the trimer Tetra-Tetra-Tetra at $^m/z$ 2790.7 (Δm = -0.10 u). Peptidoglycan oligomers do not exhibit any useful fragment ions. From the present studies it can be concluded that mass accuracy is sufficient, sensitivity in the negative ion mode is in the range of 200 - 1000 picomoles and mass range is theoretically unlimited. PD mass spectrometry has the clear disadvantage of critical sample purity and preparation, which could be partly circumvented by the sandwich technique. The structural information content of PD mass spectra is limited and no on-line combination with high performance separation techniques is possible.

The positive ion MALD mass spectrum of Tetra is shown in Figure 2. The spectrum is dominated by the $[MNa]^+$ ion and exhibits additionally low abundant $[MH]^+$ and $[MK]^+$ ions. Calculated and observed $^m/z$ value of $[MNa]^+$ is in good agreement. No fragment ions could be detected. The molecular weight assignment for unknown peptidoglycans is complicated by the fact that only low abundant protonated molecular ions were observed. This problem can not be solved by the technique applied in PD mass spectrometry. If there is no need to get $[MH]^+$ ions, the sample purity is not critical. A disadvantage in the moment is the non-existing structural information as well as no on-line combination possibility with HPLC or capillary zone electrophoresis (CZE). The extrem sensitivity (0.1 - 10 picomoles) of MALD seems to be a particularly advantage for peptidoglycan oligomer detection.

The ESI mass spectrum (Figure 3) shows the single protonated ion signal for the peptidoglycan Tetra. The negative ion ESI mass spectrum exhibits only a deprotonated molecular ion. No doubly or triply charged ion signals were obtained by ESI mass spectrometry, despite this is characteristic for the technique. Low intensity fragment ions were detected in both modes. The mass accuracy was found to be not as good as for PD or MALD in our experiments. The sensitivity goes from 10 - 1000 picomoles depending strongly on the structure of the

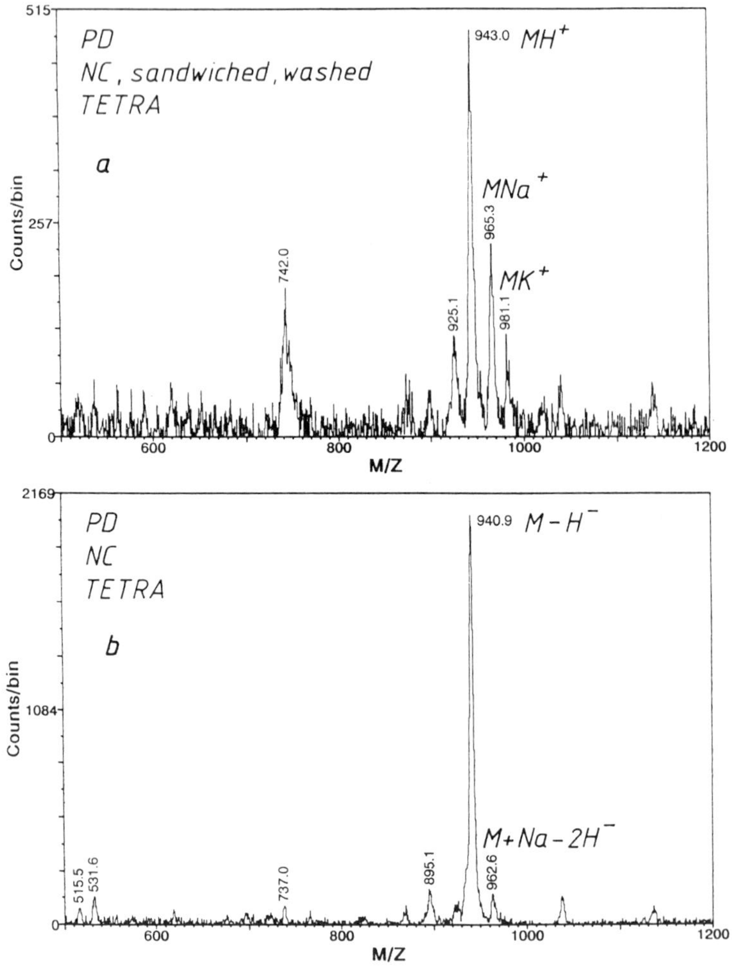

Figure 1. Positive ion (a) PD mass spectrum of the nitrocellulose-deposited, afterwards nitrocellulose-sandwiched and spin washed peptidoglycan monomer Tetra, $[MH]^+ = 942.95$ (calc.). Negative ion (b) PD mass spectrum of the nitrocellulose-deposited unwashed peptidoglycan momomer Tetra, $[M-H]^- = 940.93$ (calc.). The monomer was isolated from *E.coli* strain MC6-RP3 and both spectra were background subtracted.

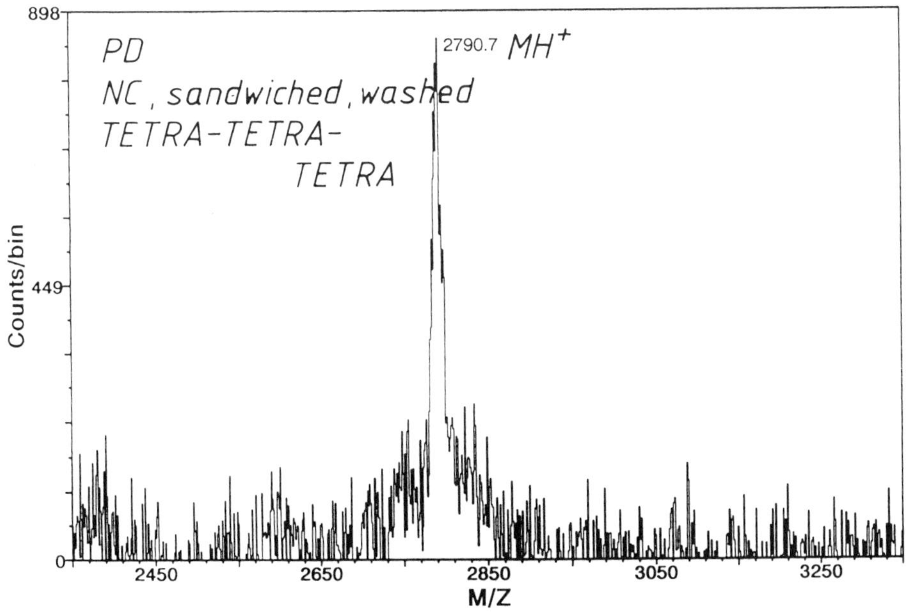

Figure 2. Positive ion PD mass spectrum of the nitrocellulose-deposited, afterwards nitrocellulose-sandwiched and spin washed peptidoglycan trimer Tetra-Tetra-Tetra, $[MH]^+$ = 2790.80 (calc.), isolated from cyanelles of *C.paradoxa*. The spectrum was background subtracted.

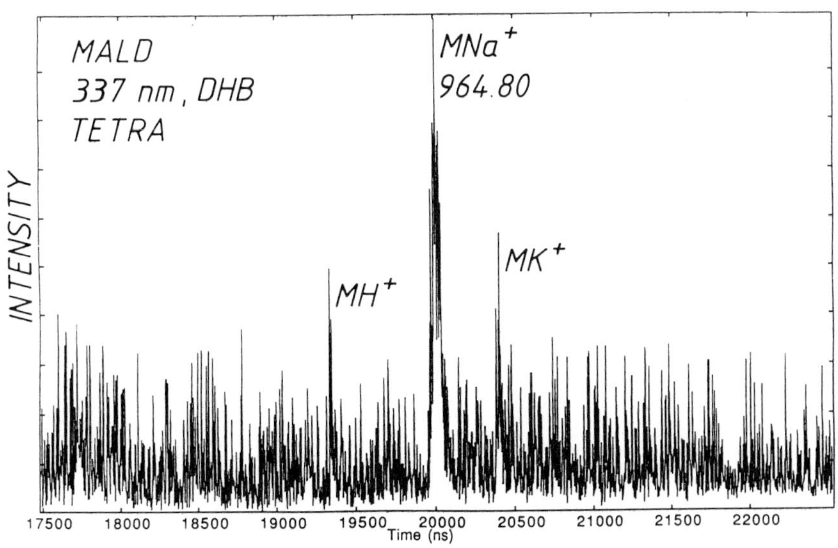

Figure 3. Positive ion MALD mass spectrum of the peptidoglycan momomer Tetra, $[MNa]^+$ = 964.93 (calc.), isolated from *E.coli* strain MC6-RP3. Dihydroxy benzoic acid was used as matrix.

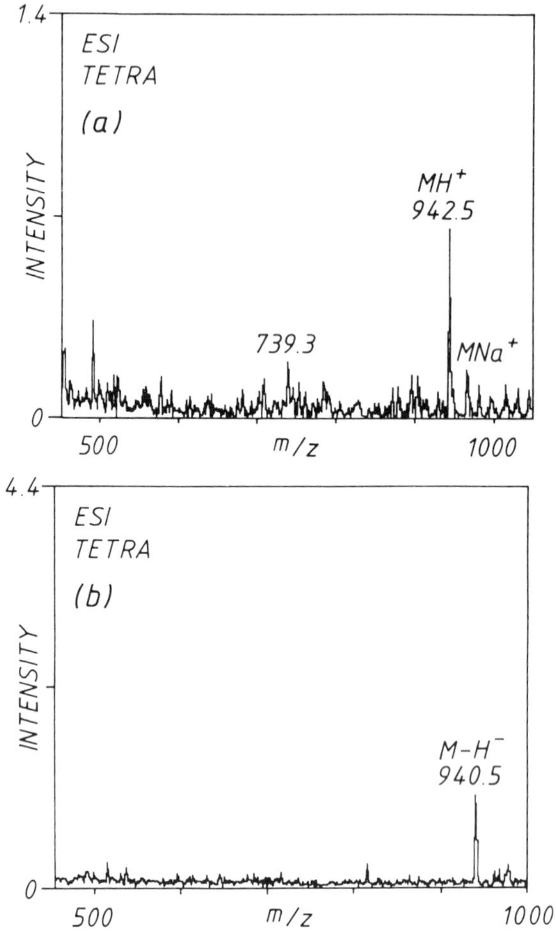

Figure 4. Positive ion (a) and negative ion (b) ESI mass spectrum of the peptidoglycan monomer Tetra, $[MH]^+$ = 942.95 (calc.) and $[M-H]^-$ = 940.93 (calc.) isolated from *E.coli* strain MC6-RP3.

peptidoglycan. From our experience it is clear that ESI has some clear advantages, particularly it has no problem with salt contaminations due to the fact of liquid phase sample introduction into the desorption/ionization region. A further advantage is the possibility to combine ESI with tandem mass spectrometric experiments as demonstrated in Figure 5. The protonated molecular ion of Tetra (generated by ESI) was mass selected by the first mass analyzer (MS1) and transferred into a collision chamber. In that region the intact molecular ion underwent a collision-activated dissociation induced by collision with Argon atoms forming a large number of structural characteristic fragment ions. These ions were then mass analyzed by a second analyzer (MS2). The so obtained fragment ions were used for strucutre elucidation as described in detail by Pittenauer et al. (1992b). However, for very complex peptidoglycan mixtures high performance separation (e.g., HPLC or CZE) can be combined easily on-line with ESI-MS or ESI-MS/MS. The latter combination may be considered the most sophisticated analytical strategy to characterize peptidoglycan digests.

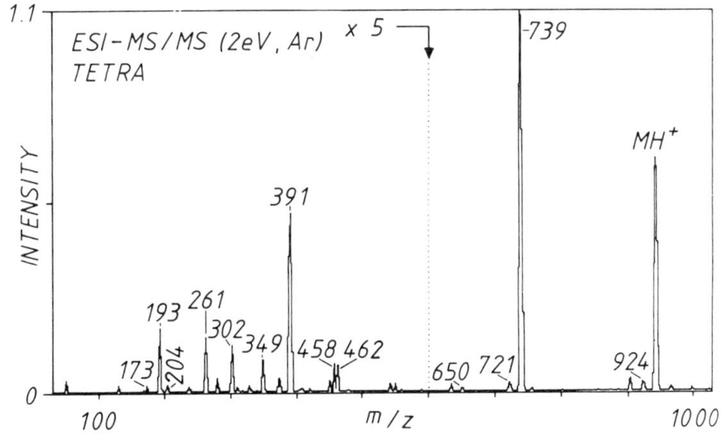

Figure 5. Positive ion ESI-MS/MS spectrum of the $[MH]^+$ molecular ion of the peptidoglycan Tetra.

ACKNOWLEDGEMENTS

Financial support from the "Jubiläumsfond der Österreichischen Nationalbank" (grant No. 4044 to G.A.) and the EMBO short term fellowship (ASTF 6719 to G.A.) is gratefully acknowledged. We thank W.Löffelhardt, M.C. Rodriguez and M.A. de Pedro for the peptidoglycan samples. The authors are indebted to A.Ashcroft for her assistance to obtain the ESI tandem mass spectrum and especially to P.Roepstorff for his assistance with and use of the MALD time-of-flight instrument.

REFERENCES

Allmaier, G., Rodriguez, M.C. and Pittenauer, E. (1992) Optimization of sample deposition for plasma desorption mass spectrometry. Rapid Commun. Mass Spectrom. 6, 284-288.

Glauner, B., Höltje, J.V., and Schwarz, U. (1988) The composition of the murein of *Eschericha coli.* J.Biol.Chem. 263, 10088-10095.

Martin, S.A., Rosenthal, R.S. and Biemann, K. (1987) Fast atom bombardment mass spectrometry and tandem mass spectrometry of biologically active peptidoglycan monomers from *Neisseria gonorrhoea.* J.Biolog.Chem. 252, 7514-7522.

Pittenauer, E., Rodriguez, M.C., de Pedro, M.A., Allmaier, G., and Schmid, E.R. (1992 a) HPLC and Cf-252 plasma desorption mass spectrometry of muropeptides isolated from *E.coli.*, see this volume.

Pittenauer, E., Allmaier, G. and Schmid, E.R. (1992 b) Structure elucidation of peptidoglycan monomers by fast atom bombardment - and electrospray ionization - tandem mass spectrometry. see this volume.

Roepstoff, P., Klarskov, K., Andersen, J., Mann, M., Vorm, O., Etienne, G. and Parello, J. (1991) Mass spectrometry of proteins: studies of parvalbumins by plasma desorption, laser desorption and electrospray mass spectrometry. Int.J.Mass Spectrom. Ion Proc. 111, 151-172.

Smith, R.D., Loo, J.A. Barinaga, C.J., Edmonds, C.G. and Udseth, H.R. (1990) Collision activation and collision-activated dissociation of large multiply charged polypeptides and proteins produced by electrospray ionization. J.Am.Soc.Mass Spectrom. 1, 53-65.

HPLC AND ^{252}Cf PLASMA DESORPTION-MASS SPECTROMETRY OF MUROPEPTIDES ISOLATED FROM E. COLI

Ernst Pittenauer,[1*] Marta Caparros Rodriguez,[2] Miguel A. dePedro,[2] Günter Allmaier,[1] and Erich R. Schmid[1]

[1]Institute for Analytical Chemistry
University of Vienna
Vienna, Austria
[2]Centro de Biologia Molecular
Universidad Autonoma de Madrid
Madrid, Spain

INTRODUCTION

Since the introduction of mass spectrometry (MS) as an analytical tool several soft ionization techniques, suitable for the analysis of very polar and non-volatile compounds like peptides, proteins, glycopeptides, oligosaccharides and oligonucleotides have been developed. The most important mass spectrometric techniques for these classes of compounds proved to be ^{252}Cf plasma desorption-(PD-MS), fast atom bombardment-(FAB-MS), matrix assisted laser desorption-(MALD-MS) and electrospray ionization-mass spectrometry (ESI-MS).

The analysis of muramidase-digested and reduced peptidoglycans by reversed-phase high-performance liquid chromatography (Glauner and Schwarz, 1983) in combination with wet chemical and enzymatic degradation techniques (Glauner *et al.*, 1988) is now a well established strategy for muropeptide characterization. All these chromatographic separations are based on a non-volatile alkali salt buffer-containing mobile phases, which are incompatible with most mass spectrometric techniques mentioned above.

So far primarily only non-reduced peptidoglycan monomers have been characterized by FAB-MS (Martin *et al.*, 1984; Martin *et al.*, 1986; Martin *et al.*, 1987), few reduced monomers by PD-MS (Allmaier *et al.*, 1992) and N-acetylmuramoyl-L-alanine amidase-digested pneumococcal peptidoglycan by PD-MS (Garcia-Bustos *et al.*, 1987; Garcia-Bustos *et al.*, 1988). In the recent work an off-line combination of liquid chromatography and plasma desorption-mass spectrometry as screening technique for peptidoglycan monomers and oligomers is described.

EXPERIMENTAL

HPLC. HPLC separation of Chalaropsis B-muramidase-digested and sodium borohydride-reduced peptidoglycans from *Escherichia coli* (strains MC6-RP3 and JE5684; additionally, MC6-RP3 was grown in the presence of S-methyl-D-cysteine, D-methionine, D-valine, D-phenylalanine, D-tryptophane and lanthionine) was performed on a HP 1090 high performance liquid chromatograph (Hewlett Packard, Palo Alto, CA, USA) at a detection wavelength of 210 nm. The peptidoglycan preparations were separated with a C_{18} (4.6 x 250 mm ID; 3 μm particle size; 120 Å pore size) reversed phase Hypersil column (Shandon Southern, Runcorn, UK). The elution gradient was linear from 100 % water/0.05 % trifluoroacetic acid to 20 % CH_3CN/0.035 % trifluoroacetic acid from 10 to 100 min at a flow rate of 0.5 ml/min. The column temperature was maintained at 30°C.

Sample preparation for PD-MS. The usefulness of mass spectrometric data is very dependent on the sample preparation technique selected. Usually such glycopeptides contain significant amounts of alkali salts, which complicate molecular weight determination by mass spectrometry, even after chromatographic purification. For sample preparation the well established nitrocellulose adsorption technique (Jonsson *et al.*, 1986), where an aluminized polyester backing is coated with nitroclluulose, was chosen. Lyophilized HPLC-fractions were dissolved in 5 μl of water/0.1 % trifluoroacetic acid and directly deposited on nitrocellulose supports. After positive- and negative-ion measurement the recently reported nitrocellulose sandwich-technique, where a thin film of nitrocellulose is sprayed over the nitrocellulose-deposited peptidoglycan sample before rinsing the target with ultrapure water/0.1 % trifluoroacetic acid, was used (Allmaier *et al.*, 1992). This procedure guarantees retention of the weakly adsorbing glycopeptides on the nitrocellulose support, even of the the very hydrophilic peptidoglycan monomers, while excess alkali is selectively removed by the rinsing procedure and therefore unequivocal molecular weight determination by PD-MS can be obtained.

PD-MS. A BioIon 20 plasma desorption time-of-flight mass spectrometer (Applied Biosystems, Foster City, CA, USA) with a 10 μCi ^{252}Cf ion source was operated under the following conditions: accelerating voltage, +17kV and -15kV, and spectrum accumulation time, 1.5x10^6 for positive- and 3x10^6 for negative-ion spectra. Mass calibration was based on H^+ and Na^+ for positive- and on H^- and CN^- for negative ion spectra.

RESULTS AND DISCUSSION

Until recently chromatographic separation of peptidoglycan preparations was usually performed using a non-volatile saltbuffer-based mobile phase for getting optimum chromatographic resolution (Glauner and Schwarz, 1983; Glauner *et al.*, 1988). The chromatogramm shown in Figure 1 clearly demonstrates the high resolving power of a water-acetonitrile gradient performed with a Hypersil 3 μm-column, where the two isomeric Tetra-Tri (linkage type: DAP-DAP and Ala-DAP) as well as Tetra-Tetra-Tetra and Tetra-Tri-Lys-Arg, which elute very close to one another with the same type of HPLC-column when using a saltbuffer-methanol-gradient, are well separated. Only for the late eluting anhydro-dimers, which are indistinguishable by plasma desorption mass spectrometry based on molecular weight determination, the corresponding isomers show insufficient separation under these HPLC-conditions.

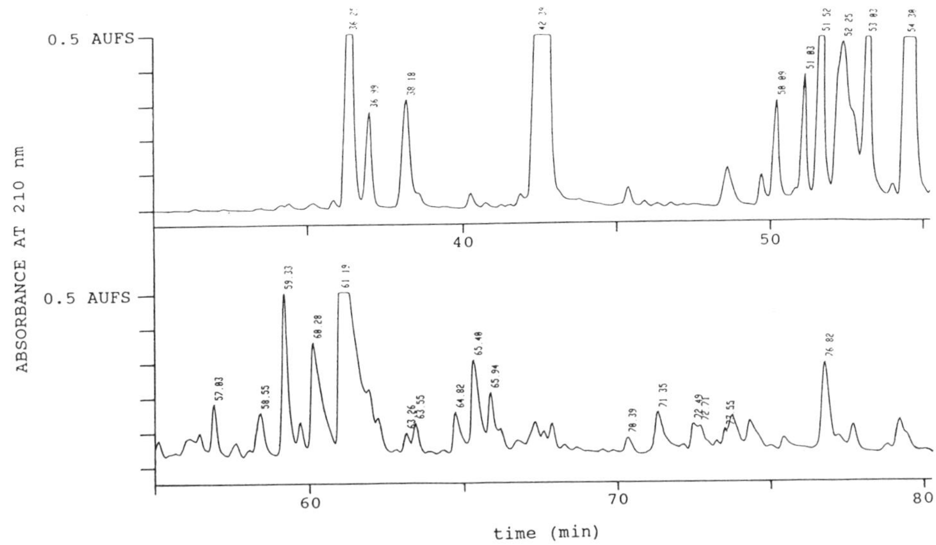

Figure 1. Section part of the reversed phase-high performance liquid chromatogramm of Chalaropsis B-muramidase-digested and sodium borohydride-reduced peptidoglycan of *E. coli*-strain MC6-RP3 using a water-acetonitrile gradient and a Hypersil 3 μm-column (only the part of the chromatogramm containing muropeptide peaks is shown).

The retention times and the average molecular weights of the peptidoglycan monomers, dimers and trimers separated and fractionated from a single injection by the above described HPLC-method are given in Table 1.

Plasma desorption mass spectrometry of monomeric as well as dimeric muropeptides mainly yield either singly protonated and/or cationized (typically by sodium and potassium) molecular ions. Besides, for peptidoglycan monomers some low abundant fragment ions could be observed in the positive- (e. g. $[MH-H_2O]^+$, $[MH-COOH]^+$, $[MH-GlcNAc]^+$, $[MH-GlcNAc-GlcNAc]^+$) as well as in the negative-ion mass spectra (e. g. $[M-H-H_2O]^-$, $[M-H-COOH]^-$, $[M-H-GlcNAc]^-$, $[M-H-GlcNAc-GlcNAc]^-$) in case of higher sample amounts used. Oligomeric peptidoglycans do not exhibit significant fragment ions. A typical example of effective removal of sodium and potassium cations, for which the nitrocellulose-sandwich technique, applied in our laboratory (Allmaier et al., 1992), was used, is given in Figure 2.

Table 1. Retention time and corresponding (de)protonated molecular ions of muropeptides from E. coli MC6-RP3 as determined by positive- and negative-ion plasma desorption mass spectrometry [found values with an asterisk are due to sodium cation attachment of the molecular ion].

Rt (min)		averaged molecular weight (calc.)	PDMS $[M+H]^+$ (found)	PDMS $[M-H]^-$ (found)
36.25	Tri	870.90	872.30	869.26
36.99	Di	698.68	699.66	697.47
38.18	Tri-Gly	927.91	928.00	926.35
42.39	Tetra	941.94	942.85	940.45
50.09	Tri-Tri	1723.79	1744.41*	-
51.52	Tetra-Tri	1794.83	1795.22	1793.63
52.25	Tri-Lys-Arg	1155.27	1156.69	1154.00
53.03	Tetra-Tri	1794.83	1816.81*	-
54.30	Tetra-Tetra	1864.17	1867.25	1864.63
59.33	Tetra-Tetra-Tetra	2789.79	2790.25	2787.83
60.28	Tri-Tri-Lys-Arg	2008.16	2009.07	2007.50
61.19	Tetra-Tri-Lys-Arg	2079.24	2080.28	2078.08
63.26	Tetra-Tri (Anh)	1776.81	1777.21	-
65.40	Tetra-Tetra-Tri-Lys-Arg	3001.42	3002.72	-
65.94	Tetra-Tetra (Anh)	1846.15	1868.08*	-
71.35	Tri-Lys-Arg (Anh)	1137.25	1137.07	-
72.49	Tetra-Tri-Lys-Arg (Anh)	2061.22	2062.09	-

Whereas for the non-rinsed sample the protonated molecular ion is only a peak of minor intensity, alkali-adduct ions are dominating the positive-ion mass spectrum shown in Figure 2 (a) prior to this procedure. Without using this sandwiching technique for this very hydrophilic muropeptide no such abundant protonated molecular ion would be detectable after the rinsing procedure as shown in Figure 2 (b).

Another example of peptidoglycan monomer isolated from the same *E. coli*-strain is the pronase E-degradation product of the murein-lipoprotein conjugate, Tri-Lys-Arg. This very basic compound strongly adsorbes to the nitrocellulose support and exhibits especially after final rinsing significant fragment ions corresponding to the stepwise loss of the two sugar moieties and afterwards to complex fragmentation of the peptide-moiety.

A different strain of *E. coli*, JE 5684, which turned out to be a carboxypeptidase-deficient mutant, produces a significant amount of disaccharide-pentapeptide, which represents the biosynthetic precursor molecule for Tetra and Tri. The murein of this strain therefore exhibits a slightly different HPLC-pattern with some additional peaks due to pentapeptide containing monomers and oligomers. For the oligomers a typical example is Tetra-Penta (Figure 4), a dimer which is of minor abundance in the wild type strain.

Completely new monomeric and oligomeric muropeptides were obtained by growing *E. coli* MC6-RP3 in the presence either of a non-physiological diamino acid (lanthionine), which substi-

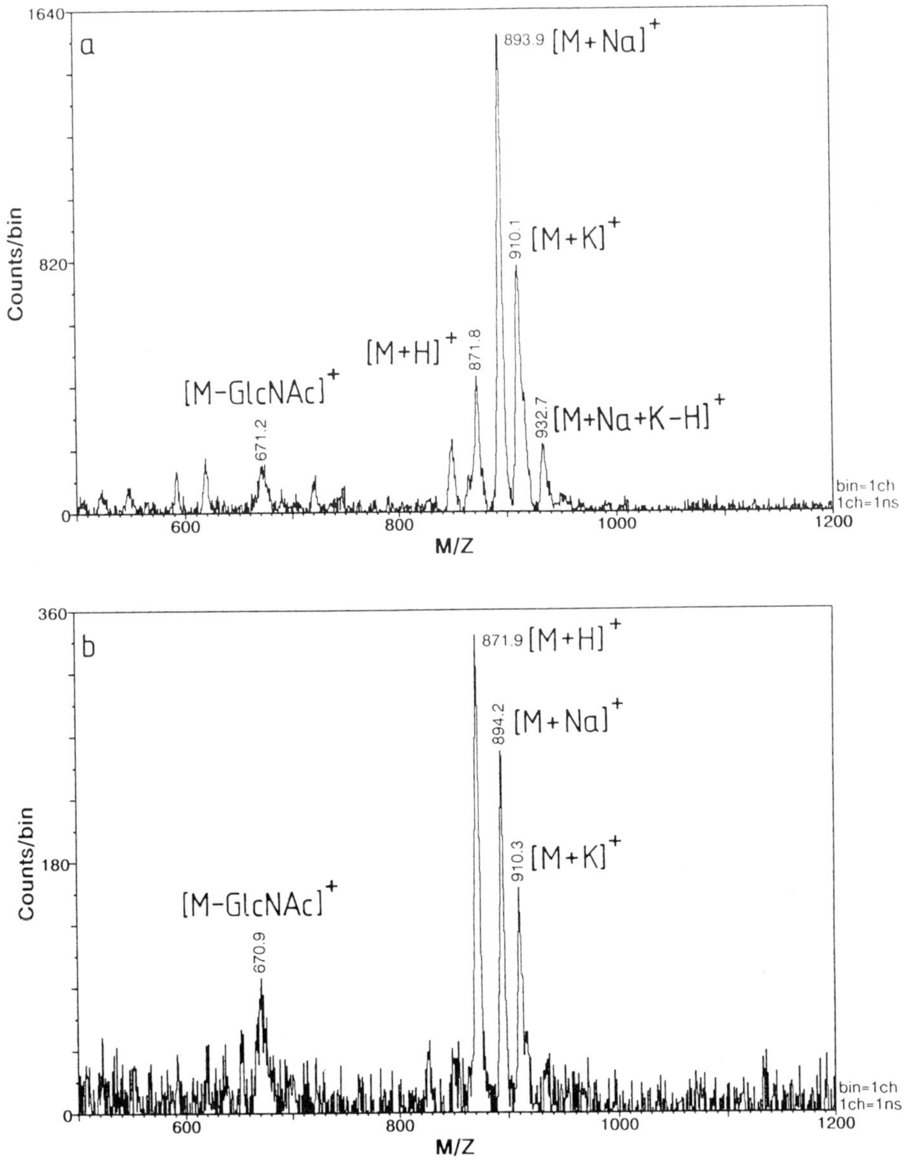

Figure 2. Positive-ion PD-mass spectrum of the nitrocellulose-deposited (a) and afterwards nitrocellulose-sandwiched and rinsed (b) peptidoglycan monomer Tri, $[M+H]^+ = 871.91$ (calc.), isolated from *E. coli* MC6-RP3.

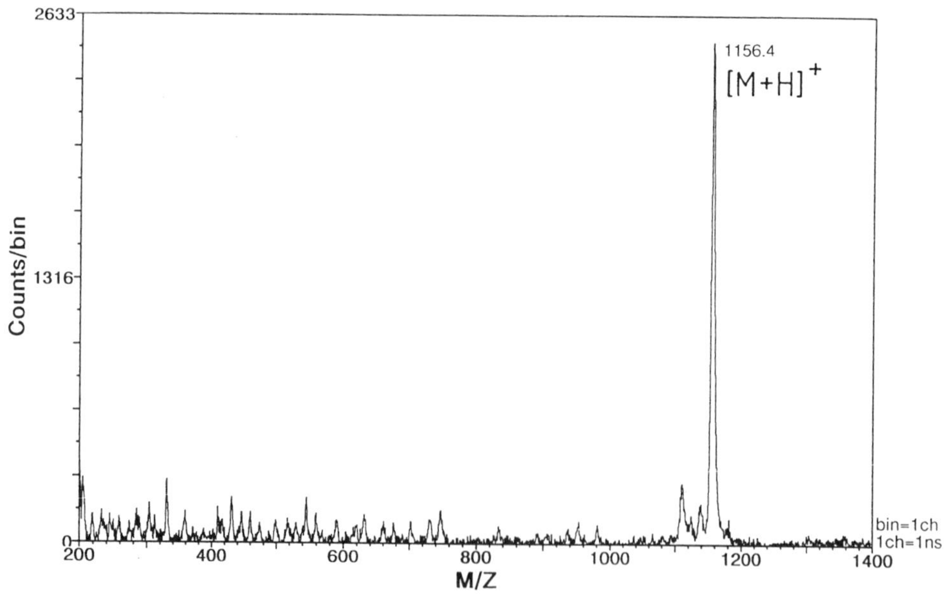

Figure 3. Positive-ion PD-mass spectrum of the nitrocellulose-sandwiched and rinsed peptidoglycan monomer Tri-Lys-Arg, $[M+H]^+$ = 1156.27 (calc.), isolated from *E. coli* MC6-RP3.

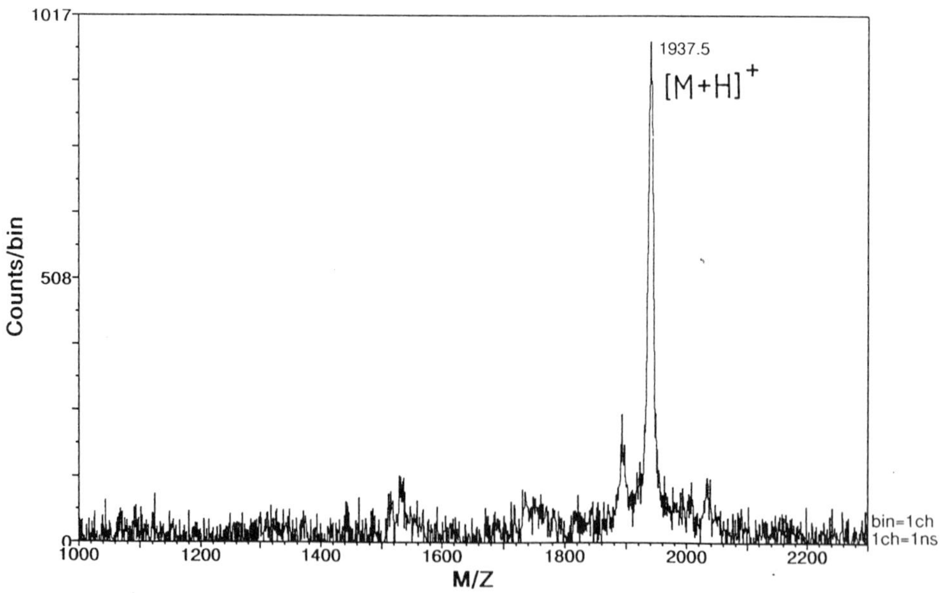

Figure 4. Positive-ion PD-mass spectrum of the nitrocellulose-sandwiched and rinsed peptidoglycan dimer Tetra-Penta, $[M+H]^+$ = 1937.95 (calc.), isolated from *E. coli* JE 5684.

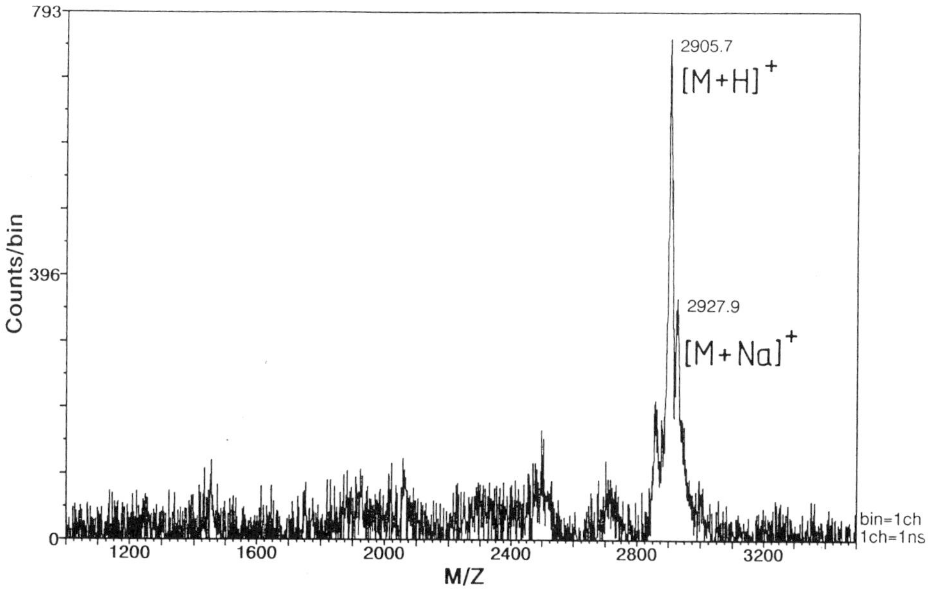

Figure 5. Positive-ion PD-mass spectrum of a novel peptidoglycan trimer (nitrocellulose-sandwiched and rinsed), isolated from *E. coli* MC6-RP3 grown in the presence of D-tryptophane: Tetra-Tetra-Tri-Trp, $[M+H]^+$ = 2805.98 (calc.).

tutes for diaminopimelic acid, or of non-physiological D-amino acids (S-methyl-D-cysteine, D-methionine, D-valine, D-phenyl- alanine and D-tryptophane), which substitute for the C-terminal D-alanine of Tetra. An example of a trimer, where one D-alanine is exchanged by D-tryptophane, is given in Figure 5.

Although the incorporation of these unusual D-amino acids into the *E. coli*-murein is not complete, from the corresponding HPLC-chromatogramms modified monomers, dimers and trimers could be detected by plasma desorption-mass spectrometry. The data discussed in this paper clearly demonstrate that reversed phase HPLC in off-line combination with PD-MS is an excellent analytical tool for the rapid and accurate screening of known and novel muropeptides.

ACKNOWLEDGEMENT

This work was supported by a grant from the Jubiläumsfonds der österreichischen Nationalbank (grant no. 4044).

REFERENCES

Allmaier, G., Rodriguez, M. C. and Pittenauer, E. (1992) Optimization of sample deposition for plasma desorption mass spectrometry of peptidoglycan monomers. Rapid Commun. Mass Spectrom. 6, 284-288.

Garcia-Bustos, J. F., Chait, B. T. and Tomasz, A. (1987) Strucure of the peptide network of pneumococcal peptidoglycan. J. Biol. Chem. 262, 15400-15405.

Garcia-Bustos J. F., Chait B. T. and Tomasz A. (1988) Altered peptidoglycan structure in pneumococcal transformant resistent to penicillin. J. Bacteriol. 170, 2143-2147.

Glauner, B. and Schwarz, U. (1983) The analysis of murein composition with high-pressure-liquid-chromatography, in "The Target of Penicillin" (Hakenbeck, R., Höltje, J.-V. and Labischinski, H. Ed.), pp. 29-34. Walter de Gruyter, Berlin.

Glauner, B., Höltje, J-V. and Schwarz, U. (1988) The composition of the murein of *Escherichia coli*. J. Biol. Chem. 263, 10088-10095.

Jonsson, G. P. Hedin, A. B., Hakansson, P. L., Sundqvist, B. U. R., Säve, G. S., Nielsen P. F., Roepstorff, P., Johansson, K.-E., Kamenski, I. and Lindberg, M. S. L. (1986) Plasma desorption mass spectrometry of peptides and proteins adsorbed on nitrocellulose. Anal. Chem. 58, 1084-1087.

Martin, S. A., Karnovsky, M. L., Krueger, J. M. Pappenheimer, J. R. and Biemann, K. (1984) Peptidoglycans as promotors of slow-wave sleep. I. Structure of the sleep-promoting factors isolated from human urine. J. Biol. Chem. 259, 12652-12658.

Martin, S. A., Biemann, K. and Rosenthal, R. S. (1986) Biological properties of peptidoglycans by fast atom bombardment mass spectrometry, in "Biological Properties of Peptidoglycans" (Seidl, P. H. and Schleifer, K. H., Ed.), pp. 49-54. Water de Gruyter, Berlin-New York.

Martin, S. A., Rosenthal, R. S. and Biemann, K. (1987) Fast atom bombardment and tandem mass spectrometry of biologically active peptidoglycan monomers from *Neisseria gonorrhoeae*. J. Biol. Chem. 262, 7514-7522.

STRUCTURE ELUCIDATION OF PEPTIDOGLYCAN MONOMERS BY FAST ATOM BOMBARDMENT- AND ELECTROSPRAY IONIZATION-TANDEM MASS SPECTROMETRY

Ernst Pittenauer,* Günter Allmaier and Erich R. Schmid

Institute for Analytical Chemistry
University of Vienna, Währinger Straße 38
A-1090 Vienna, Austria

INTRODUCTION

Fast atom bombardment-mass spectrometry (FAB-MS) in combination with tandem mass spectrometry (MS/MS) is now a well established technique for the primary structure elucidation of complex biomolecules like peptides, oligosaccharides, lipids and glycolipids.

In contrast, data on analysis of glycopeptides, like peptidoglycan monomers (Martin *et al.*, 1987; Martin, 1988; Tomasic *et al.*, 1988; Johannsen *et al.*, 1989) or O-glycosylated peptides (Medzihradsky *et al.*, 1990; Settineri *et al.*, 1990), by using multisector tandem mass spectrometer operated under high energy collision conditions are very scarce. Although it was demonstrated in several examples that especially hybrid instruments could also be useful for MS/MS of precursor ions above m/z 1000 only peptides or oligosaccharides have been analyzed (Gaskell and Reilly, 1988; Mathews *et al.*, 1989; Poulter and Taylor, 1989; Thorne *et al.*, 1989; Alexander *et al.*, 1990; Bean *et al.*, 1991). No such example of primary structure elucidation of a glycopeptide is known to us when either hybrid- or triple-quadrupole-mass spectrometers were used, which are typically operated under low energy collision conditions.

In the present paper we describe our first results on selected reduced and subsequently highly purified peptidoglycan monomers isolated from *Escherichia coli* and the cyanelles of *Cyanophora paradoxa*. Further, the site of modification of two muropeptides and one lactylpeptide is also determined by tandem mass spectrometry.

EXPERIMENTAL

HPLC. All reduced peptidoglycan monomers investigated

(selected samples from *Escherichia coli* and *Cyanophora paradoxa*) were finally purified by using a HP 1090 liquid chromatograph (Hewlett Packard, Palo Alto, CA, USA) under similar conditions as described (Pittenauer *et al.*, 1992). The stationary phase used was a C_{18} (3.9 x 300 mm ID; 10 μm particle size; 120 Å pore size) reversed phase μBondapak column (Waters, Milford, MA, USA).

FABMS and FAB-MS/MS. All measurements were performed on a Finnigan MAT H-SQ 30 hybrid tandem mass spectrometer (Finnigan MAT, Bremen, Germany) of BEqQ-configuration (B = magnetic sector, E = electrostatic sector, q = rf-only quadrupole collision cell, Q = quadrupole mass filter) coupled to a SS 300 data sytem. All compounds were ionized by FAB. Xenon atoms from a Model 11NF saddle field fast-atom gun (Ion Tech, Teddington, UK), were used as primary beam at 8-10 keV. The secondary ions were accelerated to an energy of 3 keV. For mass calibration of MS I Ultramark 2500F dissolved in glycerol, for MS II cesium iodide dissolved in glycerol was used. All samples were dissolved in 3 μl of water/0.1 % trifluoroacetic acid and added to 3 μl of the liquid matrix. For all experiments glycerol was used as matrix. For acquiring FAB-mass spectra the magnet was scanned exponential from m/z 30-1100 with a scan speed of 10 s/decade. For MS/MS, the parent ion, $[M+H]^+$, was selected with a resolution of approximately 1000 (10 % valley definition) using BE and subjected to collision induced dissociation (CID) with argon as collision gas in the quadrupole collision cell (q). The collision energy (laboratory frame-of-reference) used in all experiments was E_{LAB} = 45 eV, the collision cell pressure was approximately $6x10^{-6}$ mbar corresponding to near single collision conditions, and the resolution of Q was near unit resolution. The precursor ion beam was attenuated to 50 % of its original value by the introduction of collision gas in q. Daughter-ion mass spectra were acquired by scanning Q linear over the mass range m/z 30-1100 at a scanrate of 30 s/100 u. All data shown represent single scans.

ESI-MS and ESI-MS/MS. All measurements were performed on a VG Bio-Q triple quadrupole instrument (VG Biotech, Altrincham, UK) of Q_1qQ_2-configuration (Q_1 = quadrupole mass filter, q = rf-only quadrupole, Q_2 = either operated as rf-only quadrupole or as quadrupole mass filter). The peptidoglycan monomer TETRA was dissolved in 50 μl of a 50 % aqueous solution of acetonitril containing 1 % formic acid. 10 μl of this solution was introduced into the mass spectrometer by loop-injection at a flow rate of 2 μl/min. For CID-experiments argon was used as collision gas at a target gas pressure of $4x10^{-3}$ mbar corresponding to multiple collision conditions. For acquiring CID-mass spectra Q_2 was scanned in 12 s from m/z 30 to 1030. The MS/MS-spectrum shown (Allmaier and Schmid, 1992) represents an average of ten scans.

RESULTS AND DISCUSSION

All muropeptides listed in Table 1 exhibit intense protonated molecular ions in their FAB-mass spectra. Besides, some minor adduct ions with sodium and potassium are observed above the quasimolecular ion peaks.

Table 1. Primary structures of the peptidoglycan monomers characterized by FAB-MS/MS (* isolated from E. coli and C. paradoxa; X indicates a yet unknown modification covalently linked to the α-COOH of iGlu for the C. paradoxa muropeptides; abbreviations used are: GlcNAc, N-acetylglucosamin; MurNAc, N-acetylmuramic acid).

TRI*	GlcNAc-MurNAc-Ala-iGlu-DAP
TETRA*	GlcNAc-MurNAc-Ala-iGlu-DAP-Ala
modified TRI	GlcNAc-MurNAc-Ala-iGlu(X)-DAP
modified TETRA	GlcNAc-MurNAc-Ala-iGlu(X)-DAP-Ala

Due to the liquid matrix glycerol used for these measurements, several matrix cluster ions, which avoid unambigous identification of fragment ions and therefore complete primary sequence-determination of these glycopeptides, are observed in the mass region below m/z 500 (Figure 1).

Figure 2 (a) exhibits the CID-mass spectrum obtained with the hybrid instrument of the monoisotopic $[M+H]^+$ ion of unmodified TETRA isolated from C. paradoxa (m/z 942). This spectrum shows very prominent loss of the N-acetylglucosamin moiety (m/z 739 and m/z 721), corresponding to fragmentation around the O-glycosidic bond. In the low mass region m/z 204 also indicates the presence of a GlcNAc moiety; corresponding fragments thereof are observed at m/z 138, 126, 84 and 60 (Laine et al., 1988). Although the loss of the reduced GlcNAc-moiety is a process of minor abundance, it is still clearly detectable (m/z 532). In contrast, the fragmentation pattern of the lactylpeptide-part is much more complex

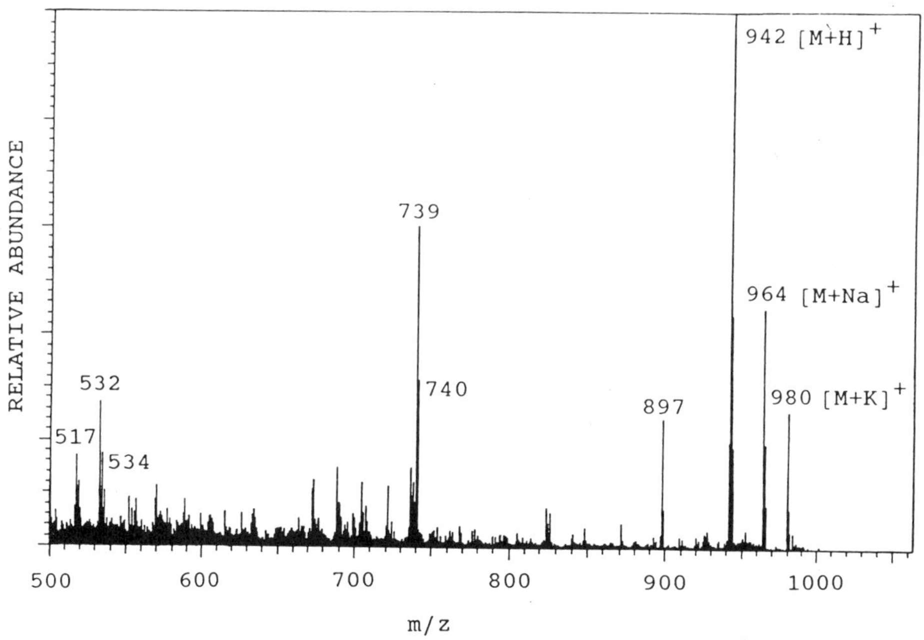

Figure 1. FAB-mass spectrum of reduced TETRA, $[M+H]^+ = 942.41$ (calc.), isolated from E. coli.

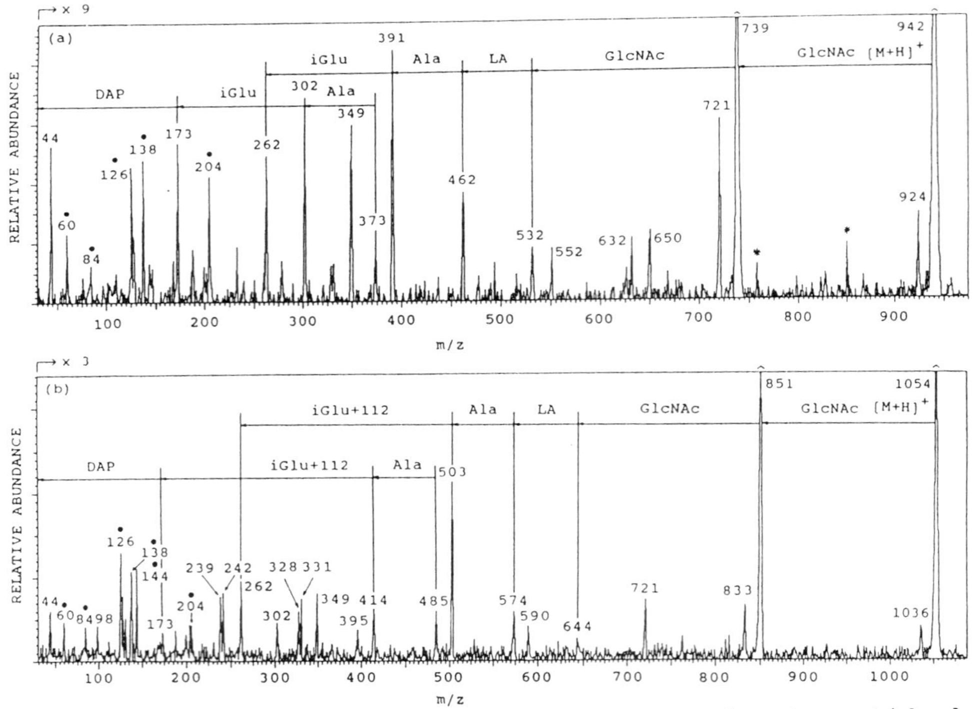

Figure 2. FAB-MS/MS mass spectra of (a) the reduced peptidoglycan monomer TETRA, $[M+H]^+$ = 942.5 (found) and of (b) the reduced, yet unknown modified, peptidoglycan monomer TETRA, $[M+H]^+$ = 1054.5 (found), isolated from the cyanelles of *C. paradoxa*. Ions labelled with an asterisk are present due to glycerol loss from a background peak with the same nominal mass as the parent ion. Ions labeled with "•" are due to the GlcNAc-moiety). Abbreviations used are: DAP (diaminopimelic acid), LA (lactic acid), iGlu (glutamic acid linked with the γ-carboxylic acid function to DAP).

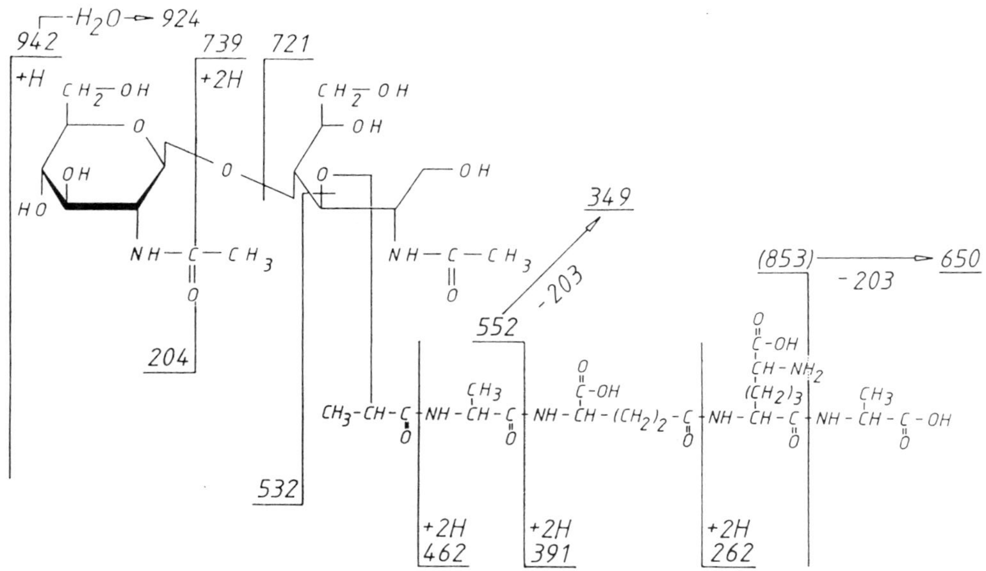

Figure 3. Structure and observed sequence ions for the reduced peptidoglycan monomer TETRA, $[M+H]^+$ = 942, detected in the FAB- and ESI-MS/MS-spectrum. The numerical values refer to the mass of the fragment produced by peptide backbone cleavage. "+H" or "+2H" indicates hydrogens transferred to the the ion species observed in the mass spectrum.

than that of the sugar part. Bond-cleavage takes mainly place at the peptide bond, typically observed under low energy collision conditions, with charge retention either at the amino-terminal end or at the C-terminal end of the lactylpeptide by transfer of two protons. The mass difference between m/z 391 - m/z 262 and m/z 302 - m/z 173 (Δm = 129) indicates that the glutamic acid moiety is not amidated (in case of amidation Δm = 128). The ion m/z 173 identifies the diaminopimelic acid-moiety also as not amidated. No fragmentation processes are observed involving C-C-bond cleavage, like side-chain cleavages of the amino acids or two-bond fissions of the sugar-units, which are both only observed under high energy collision conditions (Martin *et al.*, 1987; Martin, 1988) where E_{LAB} > 1keV.

In comparison, the electrospray ionization-MS/MS-spectrum of TETRA exhibits quite similar daughter ion pattern, with some exceptions (Allmaier and Schmid, 1992). The ions m/z 902, 458 and 193, which are not observed in the FAB-MS/MS spectrum, are maybe due to to multiple collision conditions.

Another example of a peptidoglycan monomer, the modified TETRA, isolated from *C. paradoxa* cyanelle peptidoglycan, which shows a significantly higher retention time in reversed phase-HPLC compared to unmodified TETRA, exhibits a protonated molecular ion at m/z 1054.5 in the FAB-mass spectrum (data not shown). From this mass shift of +112 u an yet unknown modification located either at the sugar- or at the lactylpeptide-moiety, indicating a modifying molecule of molecular weight of 130 u, was detected. Collision induced dissociation of the protonated molecular ion [see Figure 2 (b)] clearly indicates the modification to be located on the glutamic acid moiety as deduced from the mass difference of m/z 503 - m/z 262 and m/z 485 - m/z 173. In the same way the site of modification could also be determined for a modified lactyltripeptide isolated from the same microorganism (Pfanzagl *et al.*, 1992).

Although it is not possible to differentiate between glutamic and isoglutamic acid moieties by tandem mass spectrometry under low energy collision conditions, this technique is a very powerful tool for the primary structure characterization of peptidoglycan monomers. A clear localization of modification sites of peptidoglycan monomers is possible by hybrid tandem mass spectrometry, even when the precursor ion is above m/z 1000. The structural elucidation of the modifying moiety will be subject of further investigations.

ACKNOWLEDGEMENTS

The authors are grateful to W. Stanek and J. Dolezal for technical assistance. The H-SQ 30-tandem mass spectrometer was made available by the Austrian "Fonds zur Förderung der wissenschaftlichen Forschung". Further, we thank A. E. Ashcroft for her help obtaining the ES-tandem mass spectrum, M. Caparros and B. Pfanzagl for kindly providing the samples.

REFERENCES

Alexander, A. J., Thibault, P., Boyd, R. K., Curtis, J. M. and Rinehart, K. L. (1990) Collision induced dissociation of peptide ions. Part 3. Comparison of results obtained using sector-quadrupole hybrids with those from tandem double focusing instruments. Int. J. Mass Spectrom. Ion Processes 98, 107-134.

Bean, M. F., Carr, S. A., Thorne, G. C. Reilly, M. H., Gaskell, S. J. (1991) Tandem mass spectrometry of peptides using hybrid and four-sector instruments: a comparative study. Anal. Chem. 63, 1473-1481.

Allmaier, G. and Schmid, E. R. (1992) New mass spectrometric techniques for peptidoglycan analysis, see this volume.

Gaskell, S. J. and Reilly, M. H. (1988) Hybrid tandem mass spectrometry of peptides above mass 1000. Rapid Commun. Mass Spectrom. 2, 188-191.

Johannsen, L., Rosenthal, R. S., Martin, S. A., Cady, A. B., Obal, O., Guinand, M. and Krueger J. M. (1989) Somnogenic activity of O-acetylated and dimeric muramyl-peptides. Inf. Immun. 57, 2726-2732.

Laine, R. A., Pamidimukkala, K. M., French, A. D., Hall, R. W., Saeed, A. A., Rakesh, K. J. and Khushi L. M. (1988) Linkage position in oligosaccharides by fast atom bombardment ionization , collision-activated dissociation, tandem mass spectrometry and molecular modeling. L-Fucosyl$_p$-(α->X)-D-N-acetyl-D-glucosaminyl$_p$-(ß->3)- D-galactosyl$_p$-(ß1-O-methyl) where X= 3, 4, or 6. J. Am. Chem. Soc. 110, 6931-6939.

Martin, S. A., Rosenthal, R. S. and Biemann, K. (1987) Fast atom bombardment and tandem mass spectrometry of biologically active peptidoglycan monomers from *Neisseria gonorrhoeae*. J. Biol. Chem. 262, 7514-7522.

Martin, S. A. (1988) Mass spectrometry of peptidoglycans, in "Antibiotic Inhibition of Bacterial Cell Surface Assembly and Function" (Actor, P., Daneo-Moore, L., Higgins, M. L., Salton, M. R. J. and Shockman, G. D., Ed.), pp. 129-145. American Society for Microbiology, Washington, DC-Boston.

Mathews, W. R., Runge, T. A., Haroldsen, P. E. and Gaskell, S. J. (1989) Characterization of impurities in a synthetic renin substrate peptide by fast-atom bombardment mass spectrometry and hybrid tandem mass spectrometry. Rapid Commun. Mass Spectrom. 3, 314-319.

Medzihradsky, K. F., Gillece-Castro, B. L., Settineri, C. A., Townsend, R. R., Masiarz, F. R. and Burlingame, A. L. (1990) Structure determination of O-linked glycopeptides by tandem mass spectrometry. Biomed. Environ. Mass Spectrom. 19, 777-781.

Pfanzagl, B., Pittenauer, E., Allmaier, G., Martinez, J., Plaimauer, B., Berenguer, J., Schmid, E., de Pedro, M. A. and Löffelhardt, W. (1992) Investigations on structure and biosynthesis of cyanelle murein from *Cyanophora paradoxa*, see this volume.

Pittenauer, E., Rodriguez, M. C., de Pedro, M. A., Allmaier, G. and Schmid, E. R. (1992) HPLC and plasma desorption mass spectrometry of muropeptides isolated from *Escherichia coli*, see this volume.

Poulter, L. and Taylor, L. C. E. (1989) A comparison of low and high energy collisionally activated decomposition MS-MS for peptide sequencing. Int. J. Mass Spectrom. Ion Processes 91, 183-197.

Settineri, C. A., Medzihradszky, K. F., Masiarz, F. R., Burlingame, A. L., Chu, C. and Geoge-Nascimento, C. (1990) Characterization of O-glycosylation sites in recombinant B-chain of platelet-derived growth factor expressed in yeast using liquid secondary ion mass spectrometry and Edman sequence analysis. Biomed. Environ. Mass Spectrom. 19, 665-676.

Tomasic, J., Sesartic, L., Martin, S. A., Valinger, Z. and Ladesic, B. (1988) Comparative susceptibility of a peptidoglycan monomer from *Brevibacterium divaricatum* and its anhydromuramyl analogue to hydrolysis with N-acetylmuramyl-L-alanine amidase. Isolation and characterization of anhydromuramyl-peptidoglycan monomer. J. Chromatogr. 440, 405-414.

Thorne, G. C., Ballard, K. D. and Gaskell, S. J. (1990) Metastable decomposition of peptide $[M+H]^+$ ions via rearrangement involving loss of the C-terminal amino acid residue. J. Am. Soc. Mass Spectrom. 1, 249-257.

INVESTIGATIONS ON STRUCTURE AND BIOSYNTHESIS OF CYANELLE MUREIN FROM *Cyanophora paradoxa*

Beatrix Pfanzagl,[1] Ernst Pittenauer,[2] Günter Allmaier,[2] Jorge Martinez,[3] Jose Berenguer,[3] Erich R. Schmid,[2] Miguel A. de Pedro[3] and Wolfgang Löffelhardt[1]

[1]Institut für Allgemeine Biochemie der Universität Wien und Ludwig Boltzmann-Forschungsstelle für Biochemie, A-1030 Vienna, Austria, [2]Institut für Analytische Chemie der Universität Wien, A-1090 Vienna, Austria and [3]Centro de Biologia Molecular, Universidad Autonoma de Madrid, E-28049 Madrid, Spain

INTRODUCTION

The endosymbiotic origin of plastids and mitochondria from prokaryotic invaders into originally heterotrophic and anaerobic protoeukaryotic host cells is now generally accepted (Margulis, 1981). However, examples for intermediary or side line forms in this long-lasting and presumably still ongoing process that e. g. combine the function and genetic complexity of an organelle with the distinct prokaryotic structural feature of the peptidoglycan wall could be found to date only for the evolutionary line leading to plastids. The cyanelles from the photoautotrophic protist *Cyanophora paradoxa* have originally been considered as endosymbiotic cyanobacteria since they resemble them in morphology, thylakoid structure (Giddings *et al.*, 1983), the presence of carboxysomes (Mangeney and Gibbs, 1987) and of a lysozyme-sensitive murein sacculus between their inner and outer envelope membranes (Schenk, 1970).

With the onset of molecular biological investigations in *C. paradoxa* (Herdman and Stanier, 1977; Mucke *et al.*, 1980) it became clear that the cyanelle genome is smaller by a factor of about 25 than genomes from unicellular free-living cyanobacteria. Thus the cyanelles in fact represent the plastids of *C. paradoxa* and have kept their denomination - the diminutive reflecting the reduction in genome size - to account for their wall which is unique among cell organelles. Preliminary biochemical investigations on cyanelle murein revealed amino sugars and amino acids typical for Gram-negative bacteria (Aitken and Stanier, 1979). The sensitivity of *C. paradoxa* to ß-lactam antibiotics (Berenguer *et al.*, 1987; Kies, 1988) was explained by the presence in the cyanelle envelope of seven penicillin-binding proteins (PBPs) in the size range from 110 to 35kDa (Berenguer *et al.*, 1987). Recently the biosynthesis of UDP-muramylpentapeptide could be shown to occur exclusively in the cyanelle stroma, whereas indirect evidence was obtained for a periplasmic localization of enzymes that modify and degrade peptidoglycan like *DD*- and

Bacterial Growth and Lysis, Edited by M.A. de Pedro *et al.*, Plenum Press, New York, 1993

LD-carboxypeptidase and endopeptidase (Plaimauer *et al.*, 1991).

Meanwhile about 70 % of the 127kb cyanelle genome have been sequenced (for a review see Bohnert and Löffelhardt, 1992). In addition to the genes for the photosynthetic, transcription and translation apparatus known to reside on plastid DNAs, cyanelles encode more than 30 polypeptides that are the products of nuclear genes in higher plants. One of them is an analog to SecY from *E. coli* which it can replace in thermosensitive mutants (Löffelhardt *et al.*, 1992). However the coding sites for more than 20 enzymes of *C. paradoxa* expected to be involved in cyanelle peptidoglycan biosynthesis are still unknown. In this paper we present data on the fine structure of the cyanelle murein sacculus.

MATERIALS AND METHODS

Preparation of Cyanelle Peptidoglycan. Both strains of *Cyanophora paradoxa* were grown as described (Bohnert *et al.*, 1982), except that the cells were harvested in the exponential growth phase. Cyanelle peptidoglycan was purified from whole cells following a protocol developed for *E. coli* (Höltje *et al.*, 1975). α-Amylase from *Bacillus subtilis* and pronase E from *Streptomyces griseus* were purchased from Boehringer Mannheim. The peptidoglycan thus obtained was still contaminated with other SDS insoluble material. The peptidoglycan was digested over night with 20mg/ml *Chalaropsis* muramidase (prepared as described by Hash and Rothlauf, 1967) in 50mM potassium phosphate buffer at pH4.9 to produce muropeptides.

Size Exclusion Chromatography of Muropeptides. Muropeptides were fractionated according to size by application to a *Biogel*-P6 column (700 x 16mm) from Biorad and eluted with 100mM LiCl in 20mM potassium phosphate buffer pH6. Fractions absorbing at 214nm were analyzed by HPLC.

Reduction and ß-Elimination of Muropeptides. For reduction muropeptides were incubated with 8mg/ml $NaBH_4$ in 25mM borate buffer pH8 for 20min. ß-Elimination was achieved by incubation of unreduced muropeptides at pH11 for 8 hours at 37^oC (Tipper, 1968). Both reactions were stopped by adjusting the pH to 4.5 with *o*-phosphoric acid.

HPLC of Muropeptides. For the separation of muropeptides the HPLC-system developed by B. Glauner (Glauner, 1988; Glauner *et al.*, 1988) was used (Hypersil ODS column, 250 x 4mm, 5μm). Reduced muropeptides were eluted with a linear gradient of methanol in potassium phosphate buffer at pH5.1 at a flow rate of 0.5ml/min. Unreduced muropeptides were eluted at pH3.6 at a flow rate of 1ml/min.

Desalting of Muropeptides for PDMS. The muropeptides collected from the methanol gradient in phosphate buffer were desalted for PDMS using a RP18 column and eluting with a gradient of acetonitrile in water containing trifluoroacetic acid (see Pittenauer *et al.*, this volume).

Plasma Desorption Mass Spectrometry (PD-MS) and Fast Atom Bombardment/Tandem Mass Spectrometry (FAB/MS-MS). Measurements were performed as described elsewhere (Pittenauer *et al.*, this volume; Allmaier and Schmid, this volume).

Amino Acid Analysis and Dinitrophenylation. A procedure was followed as described by Caparros *et al.*, 1992).

Determination of Glycan Chain Length. The average length of glycan chains was determined by the method of Schindler *et al.*, 1976. Peptidoglycan containing 60 nmoles of diaminopimelic acid or less was incubated in 25mM MOPS buffer pH7.4 and 20mM $MnCl_2$ with 0.74μCi (^{14}C)-UDP-*D*-galactose (New England Nuclear, 268mCi/mmole) and 0.25 U galactosyltransferase (Sigma) in a total volume of 110μl. After incubation for two hours at 37°C the peptidoglycan was washed several times until no radioactivity could be detected in the supernatant. The peptidoglycan was then digested with *Chalaropsis* muramidase as described and the radioactivity of the muropeptide solution was determined using Ecolume (ICN Biomedicals) as scintillation cocktail. The muropeptide solution obtained from a control incubated without galactosyltransferase did not contain any radioactivity above the background level. The diaminopimelic acid content of the muropeptide solution was determined after hydrolysis in 6N HCl at 110°C over night. After evaporation of the liquid aliquots of the hydrolysis products were resuspended in 50% acetic acid in water. 0.5ml ninhydrin reagent (250mg ninhydrin, 6ml acetic acid, 4ml 0.6M H_3PO_4) was added and the sample was heated in a boiling water bath for 5 minutes. Then the absorbance at 436nm was determined and compared to standards of 1 to 20μg diaminopimelic acid.

RESULTS

According to electron microscopy studies the outer membrane surrounding the cyanelles is only loosely attached to the peptidoglycan and is easily ruptured (Giddings *et al.*, 1983). Therefore it was not surprising that no lipoprotein covalently linked to cyanelle peptidoglycan could be detected. There is no difference in the muropeptide patterns obtained by HPLC from peptidoglycan treated with pronase and untreated peptidoglycan. Likewise, using the Bradford assay (Bradford, 1976), no protein could be detected in muropeptide preparations from peptidoglycan not treated with pronase.

The average length of the glycan chains of cyanelle peptidoglycan was determined to be 73 by addition of (^{14}C)-D-galactose to the C-4 of N-acetylglucosamine moieties at the chain ends using galactosyltransferase. Taking into account the low purity of the cyanelle peptidoglycan preparations this result should be considered an upper limit of the average chain length, due to the possibility that the enzyme reaction was impaired by contaminating material.

Comparison of the muropeptide patterns of cyanelles and *Escherichia coli* (Glauner, 1988) after HPLC shows that there are comparatively few cyanelle muropeptides eluting at low retention times (Figure 1). Many abundant cyanelle muropeptides elute later than the major components of *E. coli* peptidoglycan. By coelution of cyanelle and *E. coli* muropeptides at several pH-values five cyanelle muropeptides could be identified as Tri, Tetra, Tetra-Tri, Tetra-Tetra and Tetra-Tetra-Tri since their retention behavior was identical to that of the standard muropeptides. The relative amounts observed for Tri and Tetra, as well as Tetra-Tri and Tetra-Tetra, point to a higher activity of cyanelle carboxypeptidase II as compared to *E. coli*.

To explain the high retention times of the unidentified cyanelle muropeptides, several possibilities were considered. One possibility was high cross-linkage of cyanelle peptidoglycan resulting in a higher proportion of trimers and oligomers. Separation of muropeptides according to molecular weight with a Biogel-P6 column and HPLC analysis of the fractions thus obtained showed that this was not the case. Many of the muropeptides with high retention times turned out to be dimers, while trimers were much less abundant. The most likely explanation for the long retention of these muropeptides thus seemed to be that they were carrying some lipophilic substitution. O-Acetylation at the sugar moiety could be excluded, since incubation of cyanelle peptidoglycan at pH11 at 37°C over night to remove O-acetyl groups did not change its muropeptide pattern.

To further elucidate the structure of the unidentified muropeptides, individual peaks were submitted to PD-MS and amino acid analysis combined with dinitrophenylation.

They were collected from the salt gradient at pH5.1 and, if necessary, further purified from minor contaminants with another salt gradient at a different pH. For PD-MS they had to be desalted using a gradient of acetonitrile in water. PD-MS and dinitrophenylation confirmed the previous assignment of the muropeptides to molecular weight groups (Table 1). Through PD-MS the accurate molecular weights of the muropeptides were obtained, but no structural relevant fragmentation could be detected. Comparison of the molecular weights of the unknown muropeptides with the data from amino acid analysis clearly showed them to be derived from the already identified muropeptides (Tri, Tetra etc.) by a modification causing an increment in molecular weight of approximately 112 (Figure 1, table 1). No additional common amino acid could be detected.

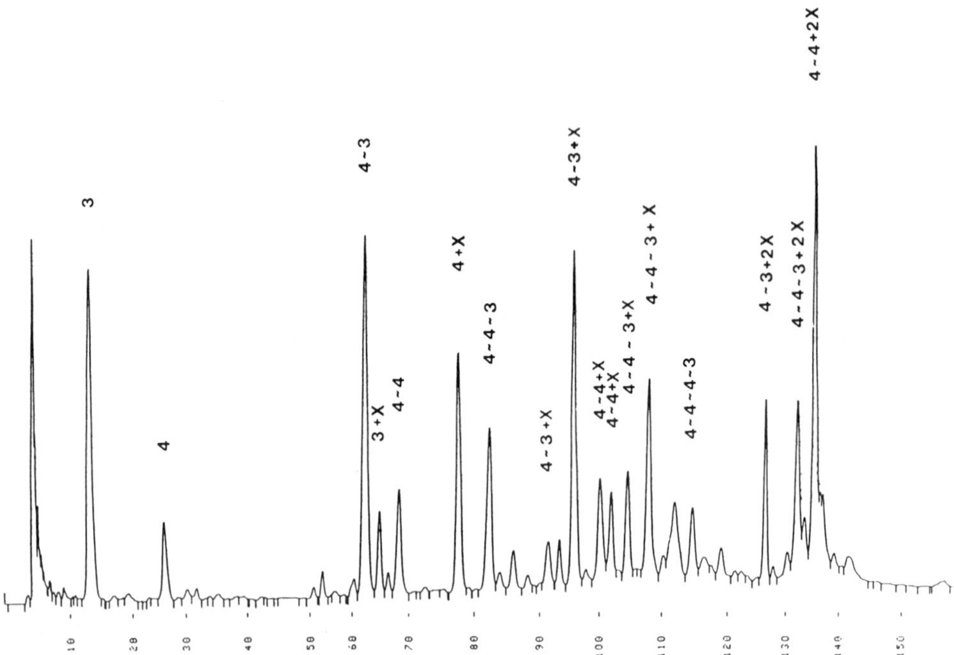

Figure 1. HPLC of cyanelle muropeptides at pH 5.1. The inferred structures of the individual muropeptides are indicated: 3 - disaccharide-tripeptide, 4 - disaccharide-tetrapeptide, X - substituent leading to an increase in molecular weight of 112.

To determine whether the substituent was located at the sugar or at the peptide moiety, the two monomeric muropeptides carrying the modification were isolated in their unreduced form and submitted to ß-elimination. The resulting lactyl peptides were purified with the gradient of acetonitrile in water. Their molecular weights obtained by PD-MS showed that the substituent was located at the peptide moiety. By FAB/MS-MS of the lactyl peptides it was confirmed that the increment in molecular weight was precisely 112. The fragmentation pattern showed that the substituent was located at the glutamic acid residue (Figure 2). This is in accordance with the observed low effect of variation of the pH of the elution buffer on the retention times of the modified muropeptides (Table 1), the pH-dependance of the retention times of muropeptides below pH7 being caused by the extent of dissociation of the carboxyl groups.

Table 1. Collection of data on isolated and analyzed cyanelle muropeptides. The calculated average protonated molecular ions, $[M+H]^+$, for unsubstituted muropeptides paralleled in *E. coli* are given in parentheses.

RT[1] (min)	m.w. group	$[M+H]^+$ (PDMS)	a.a.a. A/G	a.a.a. A/D	pH-sens. of RT (%)[3]	proposed structure
14.4	M	872.4 (871.8)	-		20-30	Tri (3)
26.7	M	943.2 (942.9)	-		20-30	Tetra (4)
62.0	D	1795.4 (1795.7)	-		20-30	4-3
64.4	M	984.1	0.98	1.38	<10	3+X
67.4	D	1867.1 (1866.7)	-		20-30	4-4
76.6	M	1054.7	1.96	2.37	<10	4+X
81.2	T	2719.5 (2719.5)	1.68	1.64	20-30	4-4-3
90.6	D	1906,6	1.54	1.73	10-20	4-3+X
95.3	D	1906.3	1.48	1.56	10-20	4-3+X
97.3	D	1978.7	1.87	2.37	10-20	4-4+X
99.4	D	1978.1	1.96	2.31	10-20	4-4+X
101.7	T	2830.1	1.58	1.98	20-30	4-4-3+X
104.9	T	2829.3 2870.1[2]	1.58	2.24	20-30	4-4-3+X (?)
122.7	D	2019.4 1977.5[2]	1.33	1.80	10	4-3+2X (?)
128.5	T	2942.8	1.63	2.55	10-20	4-4-3+2X
131.0	D	2088	2.01	2.12	10	4-4+2X

RT: retention time; m.w.: molecular weight; a.a.a.: amino acid analysis, A/G: ratio alanine/glutamic acid; A/D: ratio alanine/diaminopimelic acid; pH-sens.: pH-sensitivity; M: monomer; D: dimer; T: trimer.

[1]At pH 5.1

[2]Second prominent $[M+H]^+$ ion in the mass spectrum, presumably originating from a minor component of the muropeptide peak isolated by HPLC.

[3]The sensitivity of the retention times towards variation of the pH of the elution buffer was determined by comparing the retention times at pH 5.1 and pH 5.5. The retention time differences were calculated in % of the retention time at pH 5.1. They were grouped into three sensitivity classes.

The nature of the substituent is still unknown. High resolution FAB mass spectrometry of modified muropeptides will be applied to reveal its elemental composition and structure. From the peak areas of the HPLC pattern of cyanelle muropeptides it can be estimated that about half of the monomer subunits of cyanelle peptidoglycan are modified in exponentially growing cultures. Cells entering the stationary phase have a lower amount of modified muropeptides, especially of those with the inferred structures Tetra+X and Tetra-Tetra+2X (Figure 3). This might be due to ongoing activities of carboxypeptidase II and of an enzyme cleaving off the substituent while peptidoglycan synthesis is slowed down. The existence of an enzyme able to split off the substituent from intact peptidoglycan is indicated by the fact that this reduction in the amount of modified muropeptides also takes place in cells treated with 50mg/ml ampicillin (prevented from dying by keeping them in the dark), a concentration which has been shown to kill growing *Cyanophora* by inhibiting growth and division of the cyanelles (Berenguer *et al.*, 1987).

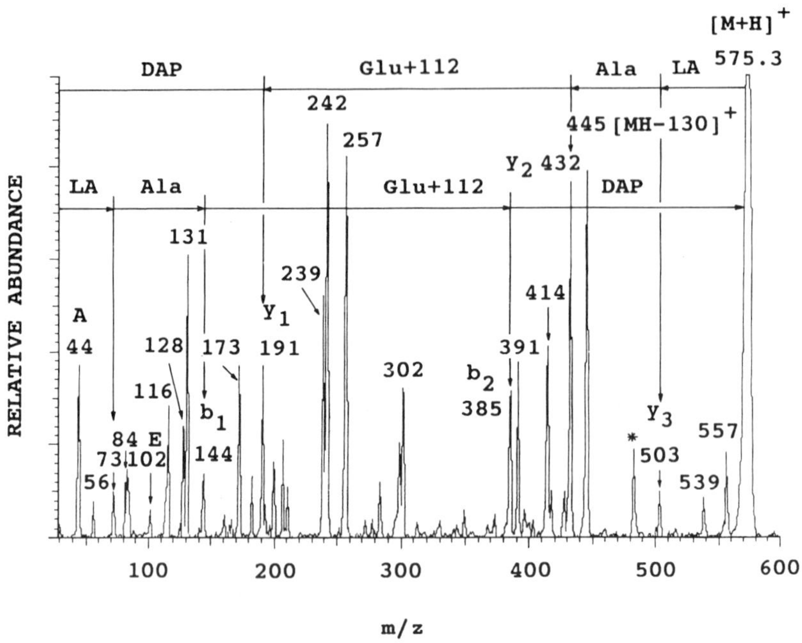

Figure 2. Daughter ion mass spectrum of the $[M+H]^+$ ion (m/z 575.3) of a modified lactyl-(alanyl-isoglutamyl-diaminopimelic acid) tripeptide indicating a substitution at the isoglutamyl moiety. The nomenclature of sequence-specific fragment ions is according to Biemann, 1990. The asterisk marks a matrix-related fragment ion.

DISCUSSION

Peptidoglycan and the enzymes involved in its biosynthesis are found among eukaryotes only in cyanelle-containing organisms. These comprise around 12 different species and have been grouped together under the denomination glaucophytes. Since *C. paradoxa* is the only member of this group that can be grown easily, the investigations concentrated on this organism and in addition on the strain LB555-UTEX. The second known isolate, strain 1555, differs in size and restriction pattern of cyanelle DNA (Breiteneder *et al.*, 1988) and in the organization of the nuclear rDNA units (Aryee and Löffelhardt, unpublished). However, number and molecular weights of PBPs correspond to those from strain LB555-UTEX (B. Pfanzagl, unpublished).

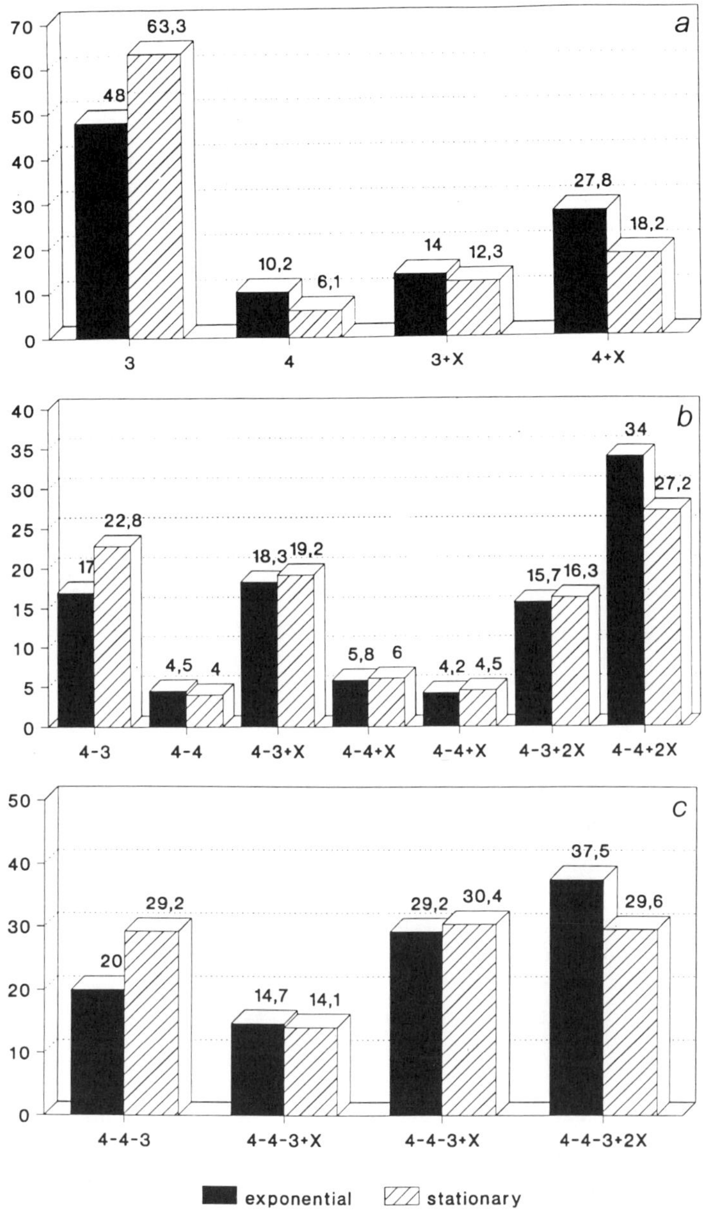

Figure 3. Amounts of the individual muropeptides in percentages of the whole amount of monomers (a), dimers (b) or trimers (c) in exponentially growing cultures and in cultures entering the stationary phase.

It would be interesting to compare the fine structure of the murein from cyanelles to that from their closer relatives or ancestors, the cyanobacteria. By that way the differences between organelle and cell wall might become apparent: obviously a function in osmotic stabilization is not required in the cytoplasm of *C. paradoxa* whereas the peptidoglycan seems to be essential in the division process for the free-living organism as well as for the semiautonomous organelle. At present it is known that cyanobacterial peptidoglycan corresponds to *E. coli* in overall amino acid and amino sugar composition but resembles murein from Gram-positives in the higher number of layers and in the presence of covalently-bound high molecular carbohydrates. (Woitzik *et al.*, 1988). Thus we relied upon the thoroughly characterized murein from *E. coli* for comparison.

Parallels between *E. coli* and cyanelle murein comprise the amino acid composition and the presence of five identical major muropeptides (Table 1). Differences are the lack of lipoprotein and the resulting arg-lys-muropeptides in the cyanelle envelope. Furthermore no indication for anhydro-compounds has been obtained thus far for cyanelles. The most conspicuous structural feature of cyanelle murein is the substitution at the *D*-glutamyl moiety. Possible candidates are an alcohol forming an ester bond with the α-carboxyl group or a (poly)amino compound which was missed during amino acid analysis. In either case a molecular weight of 130u is calculated. Indeed a corresponding protonated fragment ion (m/z 131) was observed in the FAB/MS-MS spectrum (Figure 2).

In the case of dimers with one and trimers with one or two substituted glutamic acid residues endopeptidase treatment would be required to determine which monomer units are modified. The distribution of the modification among the various muropeptides obviously is not statistical. Considering the high amounts of tetra+X and tetra-tetra+2X it seems to be concentrated on tetra-subunits, but on some of them in a lower percentage than on others. For example, about 50% of the tetra-tri dimers are unsubstituted, while most of the tetra-tetra dimers carry two substituents. This discrepancy makes it unlikely that the substituent is bound to glutamic acid residues in an additional step after the insertion of new peptidoglycan chains into the sacculus and crosslinking. A more probable explanation seems to be that already the pentapeptide side chains are at least partially modified, maybe at the precursor level, and that the substituents are later split off from some of the muropeptides. Tetra-tri would then accumulate as the result of the combined action of carboxypeptidase II and an enzyme removing the substituent.

ACKNOWLEDGEMENTS

This work was supported by a grant from the Fonds zur Förderung der wissenschaftlichen Forschung (P7832-CHE, W. L.). We are grateful for travel money obtained through the "Scientific-Technical Cooperation Austria-Spain".

REFERENCES

Aitken, A. and Stanier, R. Y. (1979) Characterization of peptidoglycan from the cyanelles of *Cyanophora paradoxa*. J. Gen. Microbiol. 212, 218-229.

Berenguer, J., Rojo, F., de Pedro, M. A., Pfanzagl, B. and Löffelhardt, W. (1987) Penicillin-binding proteins in the cyanelles of *Cyanophora paradoxa*, a eukaryotic photoautotroph sensitive to ß-lactam antibiotics. FEBS Lett. 224, 401-405.

Biemann, K. (1990) Nomenclature for peptide fragment ions (positive ions). Methods Enzymol. 193, 886-887.

Bohnert, H. J. and Löffelhardt, W. (1992) Molecular genetics of cyanelles from *Cyanophora paradoxa*, in "Algae and Symbioses"(Reisser, W., Ed.), pp. 379-397. Biopress, Bristol.

Bohnert, H. J., Crouse, E. J., Pouyet, J., Mucke, H. and Löffelhardt, W. (1982) The

subcellular location of DNA components from *Cyanophora paradoxa*, a flagellate containing endosymbiotic cyanelles. Eur. J. Biochem. 126, 381-388.
Bradford, M.M. (1976) A rapid and sensitive method for the quantification of microgram quantities of protein utilizing the principle of protein dye binding. Anal. Biochem. 72, 248-254.
Breiteneder, H., Seiser, C., Löffelhardt, W., Michalowski, C. and Bohnert, H. J. (1988) Physical map and protein gene map of cyanelle DNA from the second known isolate of *Cyanophora paradoxa* (Kies-strain). Curr. Genet. 13, 199-206.
Caparros, M., Pisabarro, A. G. and de Pedro, M. A. (1992) Effect of *D*-amino acids on structure and synthesis of peptidoglycan in *E. coli.* J. Bacteriol., in press.
Giddings, jr., T. H., Wasmann, C. and Staehelin, L. A. (1983) Structure of the thylakoids and envelope membranes of the cyanelles of *Cyanophora paradoxa*. Plant Physiol. 71, 409-419.
Glauner, B. (1988) Separation and quantification of muropeptides with HPLC. Anal. Biochem. 172, 451-464.
Glauner, B., Höltje, J.-V. and Schwarz, U. (1988) The composition of murein of *E. coli*. J. Biol. Chem. 263, 10088-10095.
Hash, J. H. and Rothlauf, M. V. (1967) The N,O-diacetyl-muramidase of *Chalaropsis* species; purification and cristallization. J. Biol. Chem. 242, 5586-5590.
Herdman, M. and Stanier, R. Y. (1977) The cyanelle: chloroplast or endosymbiotic prokaryote. FEMS Microbiol. Lett. 1, 7-12.
Höltje, J.V., Mirelman, D., Sharon, N. and Schwarz, U. (1975) Novel type of murein transglycosylase in *Escherichia coli*. J. Bacteriol. 124, 1067-1076.
Kies, L. (1988) The effect of penicillin on the morphology and ultrastructure of *Cyanophora* , *Gloeochaete* and *Glaucocystis* (Glaucocystophyceae) and their cyanelles. Endocyt. Cell Res. 5, 361-372.
Löffelhardt, W., Michalowski, C. B., Kraus, M., Neumann-Spallart. C., Jakowitsch, J., Flachmann, R. and Bohnert, H. J. (1992) Towards the complete structure of the *Cyanophora paradoxa* cyanelle genome, in "Endocytobiology V "(Ishikawa, H., Ed.), Tübingen University Press,in press.
Mangeney, E. and Gibbs, S. P. (1987) Immunocytochemical localization of ribulose-1,5-bisphosphate carboxylase/oxygenase in the cyanelles of *Cyanophora paradoxa* and *Glaucocystis nostochinearum*. Eur. J. Cell Biol. 43, 65-70.
Margulis, L. (1981) "Symbiosis and Cell Evolution", W. H. Freeman and Co., San Francisco.
Mucke, H., Löffelhardt, W. and Bohnert, H. J. (1980) Partial characterization of the genome of the "endosymbiotic" cyanelles from *Cyanophora paradoxa*. FEBS Lett. 111, 347-352.
Plaimauer, B., Pfanzagl, B., Berenguer, J., de Pedro, M. A. and Löffelhardt, W. (1991) Subcellular distribution of enzymes involved in the biosynthesis of cyanelle murein in the protist *Cyanophora paradoxa*. FEBS Lett. 284, 169-172.
Schenk, H. E. A. (1970) Nachweis einer lysozymempfindlichen Stützmembran der Endocyanellen von *Cyanophora paradoxa* (Korschikoff). Z. Naturforsch. 25b, 656.
Schindler, M., Mirelmann, D. and Schwarz, U. (1976) Quantitative determination of N-acetylglucosamine residues at the non-reducing ends of peptidoglycan chains by enzymic attachment of ^{14}C-*D*-Galactose. Eur. J. Biochem. 71, 131-134.
Tipper, D. J. (1968) Alkali-catalyzed elimination of *D*-lactic acid from muramic acid and its derivatives and the determination of muramic acid. Biochemistry 7, 1441-1449.
Woitzik, D., Weckesser, J. and Jürgens, U. J. (1988) Isolation and characterization of cell wall components of the unicellular cyanobacterium *Synechococcus* sp. PCC 6307. J. Gen. Microbiol. 134, 619-627.

NEW METHODS IN ELECTRON MICROSCOPY HELP ELUCIDATE THE STRUCTURE OF THE MUREIN SACCULUS AND THE DISTRIBUTION OF PENICILLIN-BINDING PROTEINS

Terry J. Beveridge

Department of Microbiology, College of Biological Science
University of Guelph, Guelph, Ontario Canada N1G 2W1

INTRODUCTION

Since bacteria are so small (usually about 1.5-2.5 μm^3), microbiologists have relied on microscopy to differentiate single bacteria from their neighbours, to determine cellular shape and form, to distinguish gram-negative cultures from gram-positive ones, and to detect the position of envelope adornments (e.g., flagella and capsules). These characteristics could usually be determined by light microscopy and it was not until the early 1950s that more detail was discovered by electron microscopy (Chapman and Hillier, 1953). The absence of a nucleus was confirmed, and the unequivocal detection of cell wall layers, and their contribution to cell division, outside a bilayered plasma membrane was established (Murray et al, 1965). These were halcyon days for microbial structure since electron microscopy was capable of discerning most of the structural bits and pieces which composed intact bacterial cells (Holt and Beveridge, 1982). Ribosomes and chromosomal DNA fibres could be discerned within the cytoplasm, and gram-positive and gram-negative bacteria could be distinguished by their cell envelope profiles. The former possessed an amorphous, electron dense wall above the plasma membrane, whereas the later possessed a more complex profile consisting of a thin peptidoglycan layer sandwiched between a plasma membrane and an outer membrane (Beveridge, 1981). Indeed, these early observations of envelope profiles by thin section firmly established the existence of a periplasmic space (i.e., a region encompassing or surrounding the cellular plasm) and, later, of Bayer's so-called "adhesion zones" (Bayer, 1974) in gram-negative bacteria.

During the last decade, it has become apparent that many of the conventional methods used for preserving and viewing bacterial structure are not as accurate as was first believed (Beveridge, 1988b; Stewart and Vigers, 1986). This was especially true for chemically fixed bacteria which are embedded in plastic and thin-sectioned for the physical loss of organics (lipids, nucleic acids, proteins, etc.), the denaturation of macromolecular complexes, and the redistribution of structure within cells were considerable (Graham and Beveridge, 1990*a*; Weibull *et al.*, 1983). For this reason, new emphasis has been placed on the use of rapid freezing or "cryo" techniques to

Bacterial Growth and Lysis, Edited by M.A. de Pedro
et al., Plenum Press, New York, 1993

better preserve biological structure (Dubochet *et al.*, 1987; Stewart and Vigers, 1986; Taylor and Glaeser, 1976). Certainly, the use of freeze-fracture and -etching has been used for many years, but the uncontrollable nature of the fracture line, the large grain size of the shadowing substance, and the great expense of freeze-etching equipment has made this technique of limited use. Newer methods such as cryo-sectioning (Dubochet *et al.*, 1983) and the thin frozen film technique (Dubochet *et al.*, 1987; Stewart and Vigers, 1986) offer many preservative advantages, but also require expensive cryoultramicrotomes or cryo-EM stages which only dedicated microscopy facilities possess. Furthermore, since these samples must remain frozen during EM viewing, they are not stained and are imaged by the differential contrast imparted to them by the mass difference between the cellular organics and the surrounding ice.

FREEZE-SUBSTITUTION

There is no doubt that cryo-presentation of bacteria is the better method of maintaining cell envelopes in their native form. They are frozen so rapidly with cooling rates greater than thousands of degrees per second (Robards and Sleytr, 1985), that molecular movement ceases within the structural components, molecular conformations are retained, and the possibility of enzymatic digestion of peptidoglycan, protein or lipid is negligible. At the same time, the envelope is encased in vitreous ice and therefore is not subjected to deformation by growing ice crystals (Robards and Sleytr, 1985). Yet, because very low temperatures and small sample sizes (to ensure rapid heat transfer during freezing) are necessary for cryo-preservation, there is great difficulty in specimen manipulation and eventual viewing by electron microcopy. Freeze-substitution overcomes many of these difficulties and the images produced by the technique are revolutionizing our perception of procaryotic structure (Amako *et al.*, 1983, 1984 and 1988; Amako and Takade, 1985; Beveridge and Graham, 1991; Graham *et al.*, 1991; Hobot *et al.*, 1984 and 1985).

Freeze-substitution is a cryo-technique which bridges the gap between conventional plastic embeddings and frozen material. Samples are rapidly frozen for good structural preservation and are chemically fixed by low temperature substitution to increase the molecular bonding within the cells. Once this has occurred, the sample can be embedded in plastic and thin-sectioned as in conventional processing. In addition, new methacrylates (Lowicryls) can be used for low temperature embedding with extremely good retention of chemical and conformational antigenicity (Armbruster *et al.*, 1982; Carlemalm *et al.*, 1982 and 1985). To obtain a clearer understanding of freeze-substitution and how it affects bacteria, it is necessary to explain the processing steps.

The films of bacteria are spread onto a surface and either plunged into an ultracold fluid cryogen (e.g., propane or ethane) or slammed onto a smooth metal surface (e.g., gold or copper). It is essential that the cryogen or the metal surface have extremely good heat transfer since they must cool the specimen so rapidly that crystalline ice cannot form. They must also be efficient "heat sinks" to ensure all of the latent heat of crystallization is removed from the bacteria so that localized melting and recrystallization of vitreous ice does not occur to cause structural damage (Robards and Sleytr, 1985; Stewart and Vigers, 1986). To ensure extremely low temperatures, the cryogen (or metal surface) is suspended (or cooled with) liquid nitrogen or liquid helium.

Our usual method (Graham and Beveridge, 1990*a*) of freezing is to suspend bacteria in a small volume of molten 2% (wt/vol) Noble agar which is spread as a 20-30 µm thin film on a cellulose-ester membrane (Gelman Sciences). A wedge-shaped portion of the membrane is immediately plunged into propane held at -196°C using custom designed equipment (Beveridge *et al.*, 1985). Once frozen, the membrane is placed cell-side down on frozen (-196°C) substitution medium consisting of 2% (wt/vol) osmium tetroxide and 2% uranyl acetate in anhydrous acetone in the presence of sodium aluminosilicate (pore diameter, 0.4 nm; Sigma Chemical Co.) as a molecular sieve (Graham and Beveridge, 1990b). The container (a sealed glass scintillation vial) is equilibrated to -80°C at which the substitution medium melts and chemical substitution proceeds for 72h. At this temperature, the ice surrounding the cells sublimes and is captured by the molecular filter, and the bacteria are chemically fixed and stained by the osmium tetroxide and uranyl acetate (Graham and Beveridge, 1990b). At no time has the frozen sample been subjected to a temperature warmer than -80°C so that native molecular conformations and molecule positions are effectively crosslinked in place. Also, all free water has been removed from the cells and replaced with acetone (from the substitution medium); the bacteria are ready for plastic embedding. Lowicryl can be used at -80°C and cured using ultraviolet light (Carlemalm *et al.*, 1985), or the samples can be brought to room temperature and more conventional plastics (e.g., Epon 812 or LR White) used. Thin-sectioning is performed on a conventional ultramicrotome.

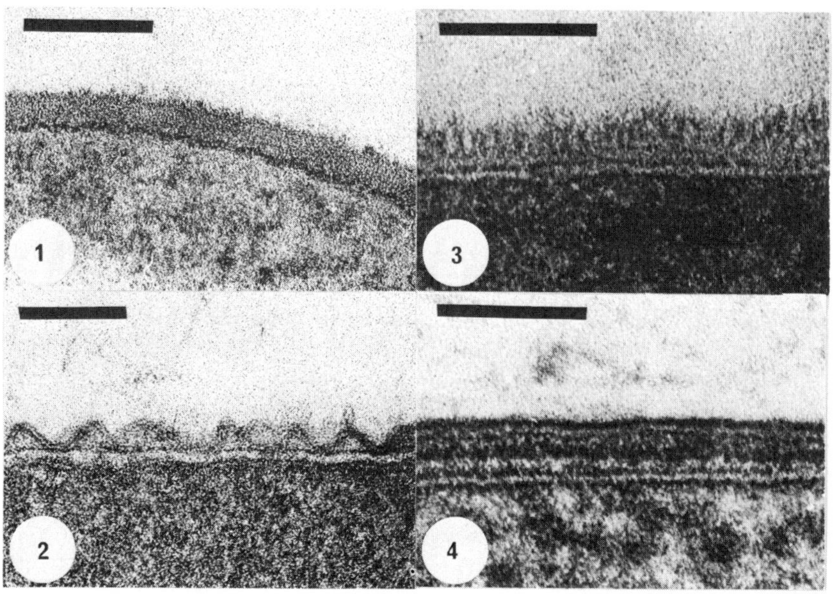

Figure 1. Thin-section of a conventionally processed *Bacillus subtilis* 168 cell envelope showing the amorphous cell wall typical of gram-positive bacteria.

Figure 2. Thin-section of a conventionally processed *Aquaspirillum serpens* VHL cell envelope showing the wavey outer membrane, the thin peptidoglycan (murein) layer, and the plasma membrane typical of gram-negative bacteria.

Figure 3. Thin-section of a freeze-substituted *B. subtilis* 168 envelope for direct comparison with Figure 1. Note the "picket-fence" arrangement of fibrils on the wall.

Figure 4. Thin-section of a freeze-substituted *Escherichia coli* D-12 (AB264) envelope showing the "periplasmic gel" and the intense staining of the outer face of the outer membrane. In all figures the bar = 100nm.

CELL ENVELOPE PROFILES

Thin-sections of conventionally processed bacteria have been instrumental in forming a general view of bacterial envelopes. For gram-positive bacteria, such as *Bacillus subtilis* 168, an amorphous cell wall ca. 25 nm thick lies on top of the plasma membrane (Fig. 1). The amorphous wall matrix consists, primarily, of peptidoglycan and teichoic acid (Beveridge and Murray, 1980). It is difficult to differentiate the membrane from the inner face of the wall because a periplasmic space is not evident; the wall is tightly opposed to the outer membrane face and shows-up as a thick electron dense line (Fig. 1).

For gram-negative bacteria, such as *Aquaspirillum serpens* VHL, a thin (ca. 2.5 nm) peptidoglycan layer is seen suspended above the plasma membrane in a periplasmic space which is 15-40 nm wide (Fig. 2). The variability in periplasmic width is because of the wavy character of the outer membrane (Fig. 2); it is difficult to believe that this is the natural state of the structure of gram-negative envelopes. Indeed, the impression is that the conventional processing for electron microscopy has somehow altered or extracted some of the "stuff" which makes up these enveloping structures.

Using radiolabelling of nucleic acids, peptidoglycan and lipopolysaccharide to monitor the extractive effects of conventional fixation-embedding versus freeze-substitution regimens on two isogenic strains of *E. coli* K-12, Graham and Beveridge (1990*a*) were able to show that 33-48% of the total labelled cellular constituents could be extracted during conventional procedures. Radiolabelled RNA and peptidoglycan in *B. subtilis* strains 168 and W23 showed similar trends. For these gram-positive and gram-negative bacteria, the dehydration step during processing was particularly damaging.

Freeze-substituted cells of the same bacteria retained much more of the radiolabels as long as the cells were frozen rapidly enough to ensure vitreous ice conditions (Graham and Beveridge, 1990*a*). The amount of extraction depends on the types of chemical fixatives used in the substitution medium; osmium tetroxide plus uranyl acetate are best for routine freeze-substitution (Graham and Beveridge, 1990b). Using this freeze-substitution method, *B. subtilis* walls have a "picket fence" appearance indicative of infrastructure beyond the amorphous matrix seen in conventional sections (cf. Fig. 1 and 3). The "pickets" extend halfway into the wall and are 15-20 nm in length making the entire wall thickness about 30-40 nm (Fig. 3). A periplasmic space is still not seen.

Gram-negative envelope profiles by freeze-substitution are even more interesting. The periplasmic space between the outer and plasma membranes is electron dense and is filled with the so-called "periplasmic gel" (Hobot *et al.*, 1984) which is 25-28 nm wide. In some select profiles of *E. coli* a darker region ca. 10 nm thick can be seen centrally in the gel; this would correspond to the region in which the peptidoglycan is expected to reside but it is ~4x the thickness seen in conventional embeddings (cf. Fig. 2 and 4). Not all gram-negative bacteria have thick, dense periplasmic gels like *E. coli* (Graham *et al.*, 1991). Some, such as *Pseudomonas aeruginosa* ATCC 9027 and *Vibrio cholerae* CA401, have intermediate density and others, such as *Caulobacter cresentus* CB15A and *Leptothrix discophora*, resemble those seen in conventional embeddings (Fig. 5).

Figure 4 is of the K-12 AB264 strain whose major source of lipopolysaccharide (LPS) electronegativity resides in the free phosphates of the lipid and core region and one KDO residue in the core (Ferris and Beveridge, 1984 and 1986b) since it does not have extended carbohydrate side-chains (O-antigen). LPS resides on the external face of the outer membrane (Funahara and Nikaido, 1980; Mühlradt and Golecki, 1985)

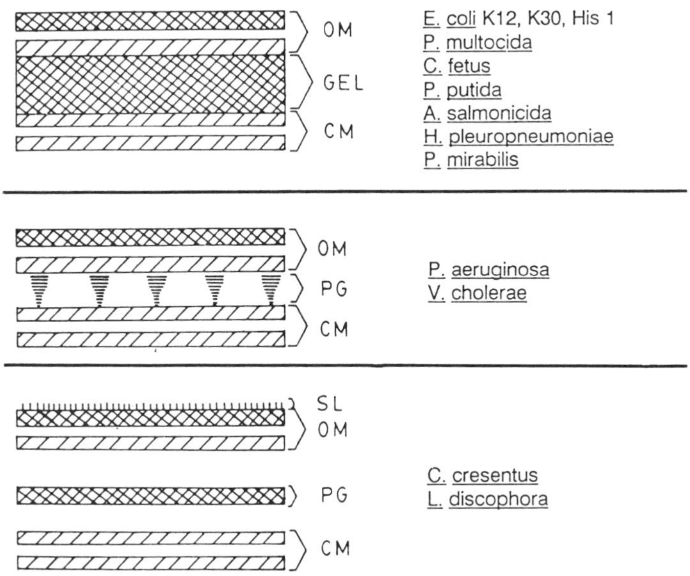

Figure 5. Schematic representation of envelope profiles of the gram-negative examined in Graham *et al.*'s 1991 study. OM, Outer membrane; GEL periplasmic gel; CM, cytoplasmic membrane; PG, peptidoglycan; SL, S layer, sheath, and/or capsule. (From Graham *et al.* 1991 with permission of the authors and the Amer. Soc. Microbiol.)

and has a high capacity to bind metal ions (Ferris and Beveridge, 1986a); it is this face which is most electron dense (Fig. 4) suggesting that freeze-substitution has retained outer membrane lipid asymmetry. Images of outer membranes containing long-chain, electronegative, O-antigenic LPS such as those on *Pseudomonas aeruginosa* PA01 show the chains as thin fibrils extending from the bilayer (Beveridge and Graham, 1991). Likewise, capsular polysaccharides are also well preserved and can extend hundreds of nm from the cell surface (Graham and Beveridge, 1991).

DOES FREEZE-SUBSTITUTION SHOW CELL WALL TURNOVER IN *BACILLUS SUBTILIS*?

Thin-sections of freeze-substituted, intact, exponentially-growing *B. subtilis* 168 reveals that the entire outer periphery of the cell wall is of the "picket-fence" structure (Fig. 6). It is tempting to speculate that freeze-substitution has captured the actual wall topography as autolysins clip bonds and remove polymers from the cell surface. In this respect, these surface polymers, consisting of both peptidoglycan plus secondary macromolecules, would be more loosely knit than the underlying fabric and would manifest themselves as a loose array of fibrils, i.e., the "picket-fence". Those polymers already excised from the surface would not be seen since they are soluble and would be removed during processing.

The juncture between the plasma membrane and the inner wall face consists of a heavily stained band ~10 nm thick (Fig. 3). In actively growing cells, new peptidoglycan must be continually being transported from membrane to inner wall face where it is incorporated into the pre-existing wall fabric. Because this peptidoglycan (which is in transit from membrane to wall) has not yet used its total bonding capacity to cement itself to wall peptidoglycan, it will have more free carboxylates to interact with staining heavy metals and will stain darker. It is possible that this is why a darker band is seen here. This band should be most apparent at the septa and along the cylindrical portions of the rod.

The wall is not of a constant thickness around the periphery of the cell (Fig. 6). The thickness of the wall at the cylindrical and polar regions were 30-35 nm, whereas those of the junctions between cylinder and pole or cylinder to septum were thicker (38-40; for all measurements $p \leq 0.001$). Interestingly, when these cells were suspended in non-saturating, limiting concentrations of polycationized ferritin to probe for strongly electronegative sites (Sonnenfeld *et al.*, 1985) these thicker wall junctions labelled whereas the others did not (Graham, personal communication). It is possible that differences in cross-linking and polymer density at these junctures accounts for their increased charge density.

Taken together, these freeze-substitution observations of the *B. subtilis* wall structure fit well with the modern perception (Koch and Doyle, 1985) of *Bacillus* wall polymer turnover which accommodates cell extension by growth and rapid polymer input at septa for cell division by fission.

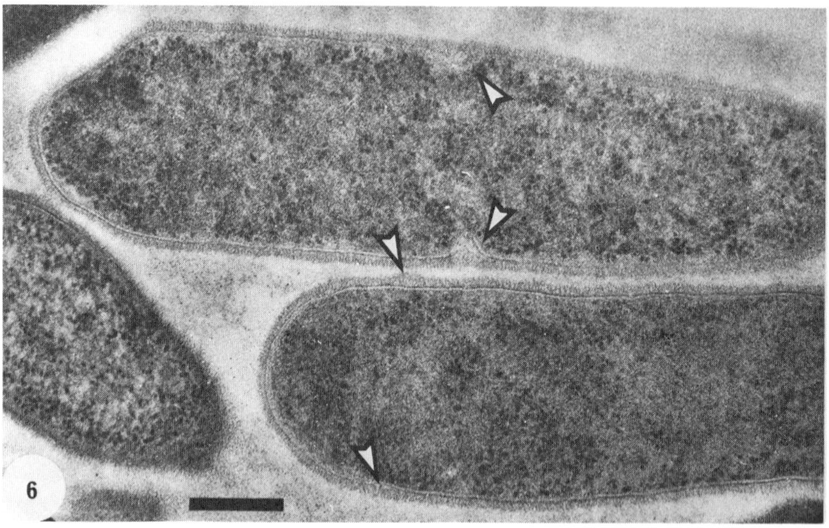

Figure 6. Thin section of a freeze-substituted *B. subtilis* 168. Note the thicker wall regions at the junctures of the side wall with polar caps and septum (arrows). Bar = 500 nm.

USE OF HIGH RESOLUTION ELECTRON DENSE PROBES TO STUDY THE STRUCTURE OF THE BACTERIAL SACCULUS

Gram Stain

Almost a decade ago, we developed a heavy metal reagent (potassium trichloro-[η^2-ethylene]-platinum (II) or TPt) capable of chemically complexing with crystal violet via metathetical anion exchange, forming a more neutral charge transfer complex with the pi bonds of the dye (Davies *et al.*, 1983). The platinum compound effectively replaces Gram's iodine during a Gram stain and makes possible the electron microscopy of the reaction deposit within cells. In this way, it was shown that the gram-positive and gram-negative attributes of eubacteria were due to the fundamental properties of the cell walls (Beveridge and Davies, 1983) and this also applies to gram-variable cells (Beveridge, 1990). Archaeobacteria have very different wall structures, often with unusual chemical constituents, than their eubacterial counterparts (König, 1988) and can stain much differently. For example, *Methanospirillum hungatei* stains gram-positive only at the filament termini because the staining reagents are excluded by the overlying sheath in all other regions of these chains of cells (Beveridge *et al.*, 1991). *Methanobacterium formicum*'s wall consists of glycan strands of pseudomurein (alternating β(1-3)-linked N-acetyl-D-glucosamine and N-acetyl-L-talosaminuronic acid residues) which are interwoven together in such a way as to resemble those of *B. subtilis* in thin section, although the wall is thinner (Fig. 7). Yet, even through *M. formicum* stains gram-positive, the reaction is not as intense as in *B. subtilis*. This is because there are fewer stain deposits in *M. formicum* and they are not spread throughout cytoplasm (as in *B. subtilis*), but reside against the inner face of the cell wall (Fig. 8). It is possible that pseudomurein reacts quite differently to the staining reagents than does peptidoglycan and that this accounts for the different staining response. Clearly, the reagents enter through the wall into the cell but the reaction deposits are segregated close to the membrane - wall interface.

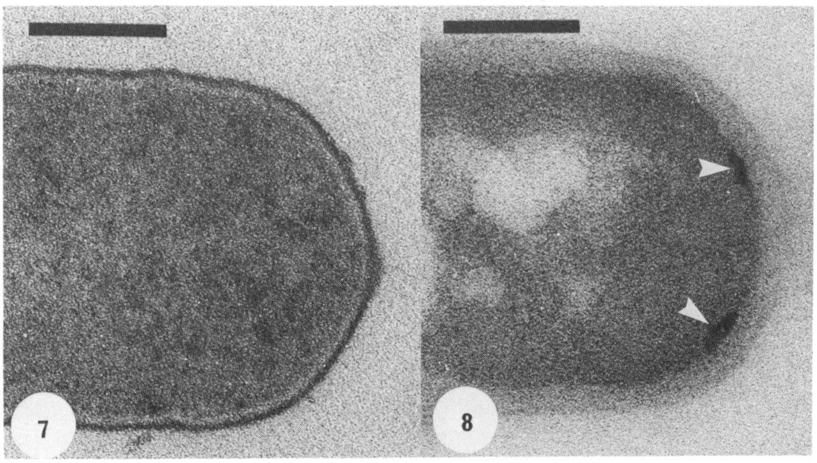

Figure 7. Thin-section of the polar region of *Methanobacterium formicum* showing the amorphous gram-positive wall of this archaeobacterium. **Figure 8.** Thin-section of the polar region of *M. formicum* after Gram staining with TPt. Note that the crystal violet-TPt staining deposit is not spread throughout the cell, but is located directly under the cell wall (arrows). In both figures the bar = 200 nm.

Mercury-Penicillin V

Penicillin binding proteins (PBPs) are necessary for eubacteria to grow and divide since they are responsible for the incorporation of newly synthesized peptidoglycan into the murein sacculus. *E. coli* possesses at least 10 PBPs (PBP1A, 1B, 1C, 2 to 8) which can be differentiated by radioactive penicillin on SDS-PAGE gels and, although exact attributes for some are difficult to ascertain, they seem all involved in maintenance and growth of the sacculus. Overall, PBP1A and 1B (transglycosylase/transpeptidase) are involved with polymerization of the growing peptidoglycan strands into the murein, PBP4, 5 and 6 (DD-carboxypeptidases) with cutting and shaping the murein at select growth regions, PBP3 (transglycylase/transpeptidase) with septum formation, and PBP2 (transpeptidase) with cell elongation (Park, 1987). Function(s) for PBPs 1C, 7 and 8 are still equivocal.

Great strides have been made using PBP-defective mutants and cloned overproducers to determine functional attributes for each PBP in a range of gram-positive and gram-negative bacteria. Yet, the exact positioning of PBPs within cells is not so clear. Breakage and fractionation of bacteria have led to the overall conclusion that PBPs are closely associated with the plasma membrane. For gram-positive cells whose walls are not separated by wide periplasmic spaces (Fig. 1 and 3), the involvement of PBPs in the initial stages of peptidoglycan metabolism is relatively easy to understand since the membrane and wall face are in close apposition. For gram-negative cells with relatively wide expanses of periplasmic space (Fig. 2) or gel (Fig. 4), it is more difficult to understand.

It seems reasonable to believe there must be relatively close juncture between the membrane-bound PBPs and the murein layer and this is part of the attraction of the existence of a periplasmic gel (Fig. 4). Yet, it is inconceivable that the entire gel can consist of intact murein; instead, the gel must consist of a composite of periplasmic enzymes, substances in transit to both the outside and inside of the cell, wall precursors (including those destined for both outer membrane and murein sacculus), and, of course, the murein layer itself. In gram-negatives, those PBPs responsible for growing peptidoglycan strand polymerization could be embedded in the plasma membrane without loss of function, but how do those responsible for clipping and reshaping peptidoglycan reach their catalytic sites if the murein layer is not closely adjacent to the membrane? It all depends how thick the murein is, on how many layers of peptidoglycan make up the murein, how each layer is conformed, and how close the entire meshwork sits above the membrane with its constituent PBPs. These details are not easily solved, but the consensus of opinion is that even if there is more than one peptidoglycan layer (Labischinski *et al.*, 1991), there can not be more than three complete layers (Beveridge, 1981) and, no matter how open their conformation, they could barely fill a periplasmic gel ~25 nm thick (Graham and Beveridge, 1990*a* and *b*; Hobot *et al.*, 1984). In *E. coli*, the outer membrane is cemented to the murein layer by covalent linkage of the outer membrane's lipoprotein and this puts severe limits on the possible distance between the outer membrane and the murein. The lipoprotein is a small α-coiled polypeptide containing three fatty acid residues that must reside in the hydrophobic domain of the outer membrane (DiRienzo *et al.*, 1978); even if most of the polypeptide exists on or below the inner face of this membrane, it could not extend more than 7.0 nm into the periplasmic gel. If the murein layer is 2.5 nm thick (Beveridge, 1981), or even 5 nm, we have a separation between murein and plasma membrane of 13.0-15.5 nm. It is very possible, then, that some PBPs must be capable of entering into the periplasmic gel to assist in processing the peptidoglycan. These may be a select few and they would be difficult to detect.

To distinguish the location of PBPs within intact bacteria with a high degree of resolution an electron microscopical technique is required. The easiest route is to use antibody-conjugated colloidal gold probes for PBPs on thin sections (Bayer *et al.*, 1990) but there are several disadvantages to this technique. Chemical fixation reduces antigenicity and only those PBPs exposed at the thin-section surface can be labelled. Small gold particles are still relatively large (~5nm) for exact localization and once coated with antibody become almost twice as large. Indirect antibody/protein A-colloidal gold techniques are even less resolving (Beveridge *et al.*, 1992).

For this reason we developed a new EM probe for PBP localization, mercury-penicillin V (Hg-PenV). This penicillin has the same antibiotic activity to penicillin V and similar PBP affinity (Paul *et al.*, 1992*a*, *b*). Since penicillins bind to PBPs through a covalent linkage, the Hg-PenV-PBP complex will be very stable and will not be disrupted by processing for electron microscopy. The electron scattering ability of Hg-PenV is imparted by the single Hg atom which is attached to the side group distant from the β-lactam ring (Paul *et al.*, 1992*a*). Each Hg by itself is beyond the normal detection limit of thin-section TEM because too much organic mass resides within the thickness of each sectioned bacterium, but clusters of Hg-PenV-PBPs should be seen.

It is believed that a single *E. coli* cell possesses ~2600 PBP molecules and about 70% of these are PBP5 (Park, 1987). Previous work has shown a typical *E. coli* K-12 (AB264) cell to be 4.83 μm long and 0.81 μm wide (Beveridge, 1988a). Using the formula $4\pi r^2 + 2\pi r(1+r)$, the surface area of the cell is ~14 x 10^6 nm^2. A median, longitudinal thin-section of the bacterium is ~60 nm thick or ~7.5% of the cell's thickness (Beveridge, 1992, 1988a); it contains 68 x 10^4 nm^2 or ~4.8% of the cell's plasma membrane. Accordingly, if all 2600 PBP molecules are equally distributed along the entire membrane in the cell, this thin-section would contain only ~125 PBP molecules and ~87 of them would be PBP5; for 125 equally distributed PBP molecules, we could expect 1 PBP per 5470 nm^2. This wide distribution would be difficult to detect using Hg-PenV unless the PPBs were arranged as clusters distributed over membrane and even then, only a small fraction of these would be exposed at the surface of a thin-section for immunogold labelling (Bayer *et al.*, 1990).

When *E. coli* is treated with Hg-PenV, electron dense deposits can be seen (see Fig.5 and 6, Paul *et al.*, this volume). These must be clusters of Hg-PenV-PBPs and they reside on the plasma membrane and in the periplasm. They are not only seen in over-producing PBP strains but also in wild-type strains (Paul *et al.*, 1992*a,b*). The very fact that Hg-PenV electron density can be distinguished in thin-sections argues that many of the PBPs reside together as clustered units and are not distributed as single entities over the cell periphery. Our experience with other metal stains suggests at least 10 Hg atoms would be required for detection in these thin-sections and the size of the deposits suggests many more could be present. Although not as plentiful as membrane-associated PBP aggregates, it is interesting to note that those PBPs which were detected in the periplasm were also as clusters (Fig. 6, Paul *et al.*, this volume).

CONCEPT OF THE PERIPLASM, PERIPLASMIC GEL AND PERIPLASMIC SPACE

Of all cell envelope constituents, the most confusing are those that reside between the physical wall structure and the plasma membrane. Gram-negative envelopes in this context are the easiest to understand since a tangible periplasmic space has always been seen (Murray *et al.*, 1965; Silva and Sousa, 1973). The periplasmic space is that region in the gram-negative envelope between the inner face of the outer membrane and the outer face of the plasma membrane (Fig. 2). The murein layer lies within this space. Gram-positive envelopes do not show a periplasmic space (Fig. 1) unless they are under high osmotic pressure to plasmolyse the cells and separate the wall and membrane.

The periplasm refers to the physical substance which lies within the periplasmic space. It is not so much a structural unit but is more a functional entity. It was first ascribed to gram-negative envelopes and since gram-positives have no periplasmic spaces, it was believed that they could not possess periplasms either. This is inconsistent with the functional attributes of the periplasm.

Periplasms contain a wide spectrum of enzymes ranging from degradative (phosphatases, proteases, endonucleases, etc.) to detoxifying (β-lactamases, aminoglycoside phosphorylases, etc.) varieties and binding proteins (for amino acids, sugars, vitamins and ions). They also contain substances which are in transit to the outside (wastes and secreted compounds) or inside (food stuffs, etc.) of cells and those macromolecules destined to be incorporated into cell walls. Now, we also have evidence that select PBPs can be found here too (Paul *et al.*, 1992*a* and *b*).

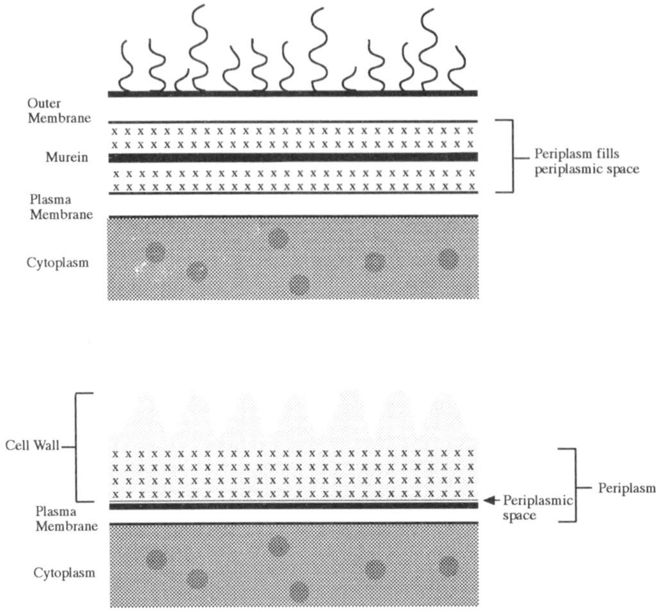

Figure 9. Schematic representation of the location of the periplasm in gram-positive (bottom) and gram-negative bacteria (top) after freeze-substitution.

Certainly, gram-positive envelopes have these functional attributes and have active periplasms since they possess the functional units attributed to them. Since they do not have physically "open" periplasmic spaces and since the periplasmic activity resides above the plasma membrane, gram-positive periplasms must reside in their cell walls (Fig. 9; Graham *et al.*, 1991).

The periplasmic gel of gram-negative envelopes is the "substance" within the periplasmic space which has been better preserved and retained by the freeze-substitution technique (Beveridge and Graham, 1991); it is the gram-negative periplasm. Although gram-positive envelopes are also better preserved by freeze-substitution, since the periplasm resides within the wall matrix it can still not be seen because there is already too much substance in place which scatters electrons effectively. Although some reports suggest that the periplasmic space occupies 20-40%

of the total cell volume (Stock *et al.*, 1977), freeze-substitution and the good retention of the periplasm as a gel suggest these estimates are too high (Graham *et al.*, 1991).

Acknowledgements

The work reported from my own laboratory was supported by an operating grant from the Medical Research Council of Canada (MRC). The PBP research was made possible by a MRC University-Industry Grant from which partial funding was supplied by Eli Lilly Canada, Inc. The electron microscopy was performed in the NSERC Guelph Regional STEM Facility of the Department of Microbiology, CBS, University of Guelph which obtains partial infrastructure funding through the Natural Sciences and Engineering Research Council of Canada. I am indebted to T. Paul of my laboratory for converting my crude diagrams of periplasms into the nice diagrams seen in Figure 9.

REFERENCES

Amako, K., and Takade, A. (1985) The fine structure of *Bacillus subtilis* revealed by the rapid-freezing and subsitution-fixation method. J. Electron Microsc. 34, 13-17.

Amako, K., Murata, K. and Umeda, A. (1983) Structure of the envelope of *Escherichia coli* observed by the rapid-freezing and substitution fixation method. Microbiol. Immunol. 27, 95-99.

Amako, K., Okada, K. and Miake, S. (1984) Evidence for the presence of a capsule in *Vibrio vulnificus*. J. Gen. Microbiol. 130, 2741-2743.

Amako, K., Meno, Y. and Takade, A. (1988) Fine structures of the capsules of *Klebsiella pneumoniae* and *Escherichia coli* K1. J. Bacteriol. 170, 4960-4962.

Armbruster, B.L., Carlemalm, E., Chiovetti, R., Garavito, R.M., Hobot, J.A., Kellenberger, E. and Villiger, W. (1982) Specimen preparation for electron microscopy using low temperature embedding resins. J. Microsc. 126, 77-85.

Bayer, M.E. (1968) Areas of adhesion between wall and membrane of *Escherichia coli*. J. Gen. Microbiol. 53, 395-404.

Bayer, M.H., Keck, W. and Bayer, M.E. (1990) Localization of penicillin-binding protein 1b in *Escherichia coli* immunoelectron microscopy and immunotransfer studies J. Bacteriol. 172, 125-135.

Beveridge, T.J. (1981) Ultrastructure, chemistry, and function of the bacterial wall. Int. Rev. Cytol. 72, 229-317.

Beveridge, T.J. (1988a) The bacterial surface: general considerations towards design and function. Can. J.Microbiol. 34, 363-372.

Beveridge, T.J. (1988b) Wall ultrastructure: how little we know, in "Antibiotic Inhibition of Bacterial cell Surface Assembly and Function" (Actor, P., Daneo-Moore, L., Higgins, M.L., Salton, M.R.J., and Shockman, G.D., eds.), p.3-20. Amer. Soc. Microbiol., Washington, D.C.

Beveridge, T.J. (1990) Mechanism of gram variability in select bacteria. J. Bacteriol. 172, 1609-1620.

Beveridge, T.J. and Davies, J.A. (1983) Cellular responses of *Bacillus subtilis* and *Escherichia coli* to the Gram stain. J. Bacteriol. 156, 846-858.

Beveridge, T.J. and Graham, L.L. (1991) Surface layers of bacteria. Microbiol. Rev. 55, 684-705.

Beveridge, T.J. and Murray, R.G.E. (1980) Sites of metal deposition in the cell wall of *Bacillus subtilis*. J. Bacteriol. 141, 876-887.

Beveridge, T.J., Harris, R. and Humphrey, R. (1985) Evaluation of conventional and low temperature techniques for the preservation of bacterial structure. Proc. Micros. Soc. Can. 12, 22-23.

Beveridge, T.J., Sprott, G.D. and Whippey, P. (1991) Ultrastructure, inferred porosity, and Gram-staining character of *Methanospirillum hungatei* filament termini describe a unique cell permeability for this archaeobacterium. J. Bacteriol. 173, 130-140.

Beveridge, T.J., Popkin, T.J. and Cole, R.M. (1992) Electron microscopy, in "Methods for General and Molecular Bacteriology" (Gerhardt, P., ed.), in press, Amer. Soc. Microbiol., Washington, D.C.

Carlemalm, E., Garavito, R.M. and Villiger, W. (1982) Resin development for electron microscopy and an analysis of embedding at low temperature. J. Microsc. 126, 123-143.

Carlemalm, E., Villiger, W., Hobot, J.A., Acetarin, J.D.and Kellenberger, E. (1985) Low temperature embedding with Lowicryl resins: two new formulations and some applications. J. Microsc. 140, 55-63.

Chapman, G.B. and Hillier, J. (1953) Electron microscopy of ultra-thin sections of bacteria. I. Cellular division in *Bacillus cereus*. J. Bacteriol. 66, 362-373.

Davies, J.A., Anderson, G.K., Beveridge, T.J. and Clark, H.C. (1983) Chemical mechanism of the Gram stain and the synthesis of a new electron-opaque marker for electron microscopy which replaces the iodine mordant of the stain. J. Bacteriol. 156, 837-845.

DiRienzo, J.M., Nakamura, K. and Inouye, M. (1978) The outer membrane of gram-negative bacteria:biosynthesis, assembly and functions. Annu. Rev. Biochem. 47, 481-532.

Dubochet, J., McDowall, A.W., Menge, B., Schmid, E.N. and Lickfeld, K.G. (1983) Electron microscopy of frozen-hydrated bacteria. J. Bacteriol. 155, 381-390.

Dubochet, J. Adrian, M., Chang, J.-J., Lepault, J. and McDowall, A.W. (1987) Cryoelectron microscopy of vitrified specimens, in "Cryotechniques in Biological Electron Microscopy" (Steinbrecht, R.A., and Zierold, K., eds.) p.114-131. Springer-Verlag KG, Berlin.

Ferris, F.G. and Beveridge, T.J. (1984) Binding of a paramagnetic metal cation to *Escherichia coli* K-12 outer membrane vesicles. FEMS Microbiol. Lett. 24, 43-46.

Ferris, F.G. and Beveridge, T.J. (1986*a*) Physicochemical roles of soluble metal cations in the outer membrane of *Escherichia coli*. Can. J. Microbiol. 32, 594-601.

Ferris, F.G. and Beveridge, T.J. (1986*b*) Site specificity of metallic ion binding in *Escherichia coli* K-12 lipopolysaccharide. Can. J. Microbiol. 32, 52-55.

Funahara, Y. and Nikaido, H. (1980) Asymmetric location of lipopolysaccharides on the outer membrane of *Salmonella typhimurium*. J. Bacteriol. 141, 1463-1465.

Graham, L.L. and Beveridge, T.J. (1990*a*) Evaluation of freeze-substitution and conventional embedding protocols for routine electron microscopic processing of eubacteria. J. Bacteriol. 172, 2141-2149.

Graham, L.L. and Beveridge, T.J. (1990*b*) Effect of chemical fixatives on accurate preservation of *Escherichia coli* and *Bacillus subtilis* structure in cells prepared by freeze-substitution. J. Bacteriol. 172, 2150-2159.

Graham, L.L., Beveridge, T.J. and Nanninga, N. (1991) Periplasmic space and the concept of the periplasm. Trends Biochem. Sci. 16, 328-329.

Graham, L.L., Harris,R., Villiger, W. and Beveridge, T.J. (1991) Freeze-substitution of gram-negative eubacteria: general cell morphology and envelope profiles. J. Bacteriol. 173, 1623-1633.

Hobot, J.A., Carlemalm, E., Villiger, W. and Kellenberger, E. (1984) Periplasmic gel:

a new concept resulting from the reinvestigation of bacterial cell envelope ultrastructure by new methods. J. Bacteriol. 160, 143-152.

Hobot, J.A., Villiger, W., Escaig, J., Maeder, M., Ryter, A. and Kellenberger, E. (1985) Shape and fine structure of nucleoids observed on sections of ultrarapidly frozen and cryosubstituted bacteria. J. Bacteriol. 162, 960-971.

Holt, S.C. and Beveridge, T.J. 1982. Electron microscopy: its development and application to microbiology. Can. J. Microbiol. 28, 1-53.

Koch, A.L. and Doyle, R.J. (1985) Inside-to-outside growth and the turnover of the gram-positive rod. J. Theor. Biol. 117, 137-157.

König, H. (1988) Archaeobacterial cell envelopes. Can. J. Microbiol. 34, 395-406.

Labischinski, H., Goodell, E.W., Goodell, A. and Hochberg, M.L. (1991) Direct proof of a "more-than-single-layered" peptidoglycan architecture of *Escherichia coli* W7. A neutron small-angle scattering study. J. Bacteriol. 173, 751-756.

Mühlradt, P.F. and Golecki, J.R. (1975) Asymmetrical distribution and artificial reorientation of lipopolysaccharide in the outer membrane bilayer of *Salmonella typhimurium*. Eur. J. Biochem. 51, 343-352.

Murray, R.G.E., Steed, P. and Elson, H.E. (1965) The location of the mucopeptide in sections of the cell wall of *Escherichia coli* and other gram-negative bacteria. Can. J. Microbiol. 11, 547-560.

Park, J.T. (1987) The murein sacculus, in "*Escherichia coli* and *Salmonella typhimurium*: Cellular and Molecular Biology" (Neidhardt, F.C., Ingraham, J.L., Low, K.B., Magasanik, B., Schaechter, M. and Umbarger, H.E., Eds.), pp. 23-30 Amer. Soc. Microbiol., Washington, D.C.

Paul, T.R., Halligan, N.G., Blaszczak, L., Parr, T.R., Jr. and Beveridge, T.J. (1992a) A new mercury-penicillin V derivative as a probe for ultrastructural location of penicillin-binding proteins in *Escherichia coli*. J. Bacteriol. (in press).

Paul, T.R., Beveridge, T.J., Halligan, N.G., Blaszczak, L.C. and Parr, T.R., Jr. (1992b) The use of a mercury penicillin derivative to localize penicillin-binding proteins in *Escherichia coli*, in "Bacterial Growth and Lysis: Metabolism and Structure of the Bacterial Sacculus" (de Pedro, M.A., Höltje, J.-V and Löffelhardt, W., eds). p. - . Plenum Pub. Corp., New York.

Robards, A.W. and Sleytr, U.B. (1985) Low temperature methods in biological electron microscopy. Elsevier Biomedical Press, Amsterdam.

Silva, M.T. and Sousa, J.C.F. (1973) Ultrastructure of the cell wall and cytoplasmic membrane of gram-negative bacteria with different fixation techniques. J. Bacteriol. 113, 953-962.

Sonnenfeld, E.M., Beveridge, T.J. and Doyle, R.J. (1985) Discontinuity of charge on cell wall poles of *Bacillus subtilis*. Can. J. Microbiol. 31, 875-877.

Stewart, M. and Vigers, G. (1986) Electron microscopy of frozen hydrated biological material. Nature (London) 319, 631-636.

Stock, J.B., Rauch, B. and Roseman, S. (1977) Periplasmic space in *Salmonella typhimurium* and *Escherichia coli*. J. Biol. Chem. 252, 7850-7861.

Taylor, K.A. and Glaeser, R.M. (1976) Electron microscopy of frozen hydrated biological specimens. J. Ultrastruct. Res. 55, 448-456.

Weibull, C., Christiansson, A. and Carlemalm E. (1983) Extraction of membrane lipids during fixation, dehydration and embedding of *Acholeplasma laidlawii* cells for electron microscopy. J. Microsc. 129, 201-207.

SYNTHESIS OF α AND β ANOMERS OF UDP-*N*-ACETYLMURAMIC ACID

Didier Blanot, Geneviève Auger, Dominique Liger, and Jean van Heijenoort

Enveloppes Bactériennes et Peptides
CNRS URA 1131
Université de Paris-Sud
91405 Orsay, France

INTRODUCTION

UDP-*N*-acetylmuramyl (UDP-MurNAc) derivatives are substrates for several cytoplasmics steps of the synthesis of bacterial peptidoglycan (Park, 1952). Their availability is a prerequisite for developing the detailed study of the synthetases catalyzing these reactions. Since they are not commercial compounds, they have to be prepared from bacterial cells in which they accumulate under specific conditions. However, such procedures are long and tedious, and yields are often low. An alternative approach is to chemically synthesize UDP-*N*-acetylmuramic acid on a large scale, and to use it as starting material for the *in vitro* enzymatic preparation of the other UDP-MurNAc precursors. In this communication, we wish to report the total synthesis of UDP-MurNAc.

RESULTS

The starting material for our synthesis was commercially available benzyl *N*-acetyl-4,6-*O*-benzylidene-muramic acid (Sigma compound B 5757). Esterification by diazomethane in methanol (Flowers and Jeanloz, 1963 ; Osawa and Jeanloz, 1965 *a*) gave the corresponding methyl ester, which was crystallized from methanol [yield : 82% ; m.p. 214-216° ; $[\alpha]_D$ +119° (*c* 0.95, chloroform)].

The benzylidene group was removed by boiling for 30 min in 60% acetic acid (Flowers and Jeanloz, 1963), and benzyl MurNAc methyl ester was purified by silica gel column chromatography in chloroform-methanol 15:1 (v/v) [yield : 60% ; m.p. 121-123°; $[\alpha]_D$ +153° (*c* 0.91, chloroform)].

Bacterial Growth and Lysis, Edited by M.A. de Pedro *et al.*, Plenum Press, New York, 1993

MurNAc methyl ester was prepared by removal of the benzyl group by catalytic hydrogenolysis. The reaction was carried out in glacial acid in the presence of 5% palladium on charcoal for 7h. The product was obtained as an oil (yield : 100%), as already described by Carroll (1963) and Osawa and Jeanloz (1965 *b*).

MurNAc methyl ester was then peracetylated by acetic anhydride in pyridine for 4.5h. Tri-*O*-acetyl-MurNAc methyl ester was obtained as an amorphous compound (yield : 89%). TLC in several solvent systems showed two spots, the lower one being much more intense than the upper one ; they probably correspond to the 9:1 mixture of α and β anomers described by Carroll (1963).

A phosphate group was introduced in position 1 of the peracetylated compound by treatment with 8 equivalents of molten phosphoric acid at 55° for 3h (MacDonald, 1962 ; Heymann *et al.*, 1968); saponification of the ester functions by lithium hydroxide gave the lithium salt of MurNAc-1-phosphate (yield : 22%). Reverse-phase HPLC (Fig.1A) revealed the presence of two peaks in a 46:54 ratio. The less polar peak co-eluted with MurNAc-1-phosphate prepared from natural UDP-MurNAc treated with snake venom pyrophosphatase (D. Liger and D. Blanot, unpublished results) ; it was therefore identified as the α anomer, the more polar peak being the β one. Integration of the two peaks and calculation relative to natural MurNAc-1-phosphate as a standard showed that the synthetic compound contained 71% of MurNAc-1-phosphate. It was used without further purification for the next reaction.

MurNAc-1-phosphate (42 μmol) was transformed into its tri-*n*-octylammonium salt and coupled in pyridine with uridine 5'-phosphoromorpholidate (42 μmol ; Sigma compound U 6000) according to Roseman *et al.* (1961) and Heymann *et al.* (1968). The reaction was carried out for 5 days at room temperature ; on the third day, one more equivalent of UMP-morpholidate was added. Analytical HPLC of the crude product (Fig.1B) revealed the presence of 5 main peaks. The corresponding compounds were purified by reverse-phase HPLC in ammonium formate buffers on μBondapak C_{18} and on Nucleosil $10C_{18}$ (Flouret *et al.*, 1981). Their identity was checked by comparison with authentic samples, by uridine, phosphate and MurNAc analyses, and by NMR spectroscopy. By order of elution, they were UDP (1), UMP (2), β-UDP-MurNAc (3), α-UDP-MurNAc (4), an unidentified compound (5) and UMP-morpholidate (retained on the analytical column). 9.5 μmol of α anomer (co-eluting with natural UDP-MurNAc) and 6.1 μmol of β anomer were obtained (yield : 37%). In ^{1}H-NMR (400 MHz, D_2O), the chemical shift of the anomeric proton of the MurNAc moiety was 5.64 ppm and 5.06 ppm for the α and β compounds, respectively.

UDP-MurNAc anomers were assayed as substrates for the incorporation of L-[^{14}C]Ala catalyzed by the L-alanine-adding enzyme in the conditions described by Liger *et al.* (1991). A partially purified preparation, originating from an over-producing strain of *Escherichia coli* (C. Parquet, D. Liger, D.

Blanot and J. van Heijenoort, unpublished results) was used. Both anomers were substrates of the reaction. K_m values for the α and β anomers were 0.13 mM and 2 mM, respectively. The former value was comparable with that determined with natural UDP-MurNAc (0.17 mM).

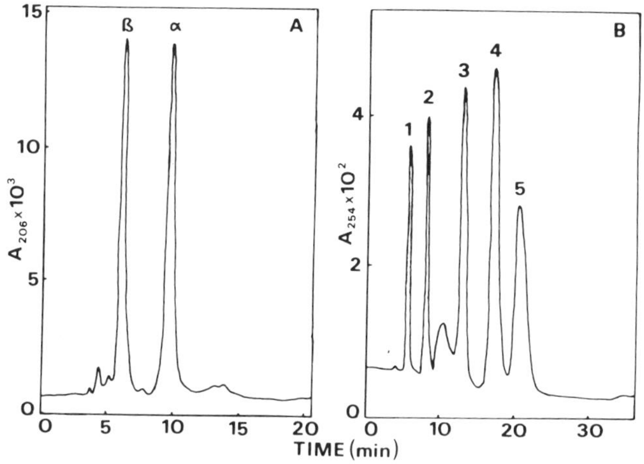

Figure 1. Analytical HPLC of MurNAc-1-phosphate (A) and crude UDP-MurNAc (B). Column : Nucleosil $5C_{18}$ 150 x 4.6 mm. Mobile phase : 50 mM ammonium phosphate pH 3.5. Flow rate : 0.5 ml/min. See text for the identification of the peaks.

DISCUSSION

A chemical synthesis of UDP-MurNAc was reported more than 20 years ago by Heymann *et al.* (1968). However, the description of the synthetic steps leading to MurNAc-1-phosphate was incomplete ; furthermore, the authors could not separate the two anomers of UDP-MurNAc. In this communication, we describe a facile synthesis in which the starting material (protected muramic acid) and the main reagent (uridine 5′-phosphoromorpholidate) are commercially available. Purification by HPLC offers a convenient way of separating α and β anomers. Synthetic UDP-MurNAc, which can be prepared on a large scale by our procedure, can serve as starting material for the *in vitro* enzymatic preparation of the other UDP-MurNAc precursors. These reactions can be carried out with a crude bacterial extract (Mengin-Lecreulx *et al.*, 1982 ; Liger *et al.*, 1991) or, better, with the corresponding synthetases partially purified from over-producing strains (Duncan *et al.*, 1990 ; Michaud *et al.*, 1990 ; Pratviel-Sosa *et al.*, 1991). One of the advantages of this approach is the easy preparation of highly radioactive precursors for studying the effect of analogs of the amino acid or dipeptide substrate of these synthetases.

ACKNOWLEDGEMENTS

This work was supported by grants from the Centre National de la Recherche Scientifique (URA 1131) and the Institut National de la Santé et de la Recherche Médicale (900314).

REFERENCES

Carroll, P.M. (1963) Acetyl derivatives of muramic acid. Nature 197, 694-695.

Duncan, K., van Heijenoort, J. and Walsh, C.T. (1990) Purification and characterization of the D-alanyl-D-alanine-adding enzyme from *Escherichia coli*. Biochemistry 29, 2379-2386.

Flouret, B., Mengin-Lecreulx, D. and van Heijenoort, J. (1981) Reverse-phase high-pressure liquid chromatography of uridine diphosphate *N*-acetylmuramyl peptide precursors of bacterial cell wall peptidoglycan. Anal. Biochem. 114, 59-63.

Flowers, H.M. and Jeanloz, R.W. (1963) The synthesis of 2-acetamido-3-*O*-(D-1-carboxyethyl)-2-deoxy-α-D-glucose (*N*-acetylmuramic acid) and of benzyl glycoside derivatives of 2-amino-3-*O*-(D-1-carboxyethyl)-2-deoxy-D-glucose (muramic acid). J. Org. Chem. 28, 2983-2986.

Heymann, H., Turdiu, R., Lee, B.K. and Barkulis, S.S. (1968) A synthesis of uridinediphospho-*N*-acetylmuramic acid and its use as an acceptor of L-[^{14}C]alanine. Biochemistry 7, 1393-1399.

Liger, D., Blanot, D. and van Heijenoort, J. (1991) Effect of various alanine analogues on the L-alanine-adding enzyme from *Escherichia coli*. FEMS Microbiol. Lett. 80, 111-116.

MacDonald, D.L. (1962) A new route to glycosyl phosphates. J. Org. Chem. 27, 1107-1109.

Mengin-Lecreulx, D., Flouret, B. and van Heijenoort, J. (1982) Cytoplasmic steps of peptidoglycan synthesis in *Escherichia coli*. J. Bacteriol. 151, 1109-1117.

Michaud, C., Mengin-Lecreulx, D., van Heijenoort, J. and Blanot, D. (1990) Over-production, purification and properties of the uridine-diphosphate-*N*-acetylmuramoyl-L-alanyl-D-glutamate : *meso*-2,6-diaminopimelate ligase from *Escherichia coli*. Eur. J. Biochem. 194, 853-861.

Osawa, T. and Jeanloz, R.W. (1965 *a*) An improved, stereoselective synthesis of 2-amino-3-*O*-(D-1-carboxyethyl)-2-deoxy-D-glucose (muramic acid). J. Org. Chem. 30, 448-450.

Osawa, T. and Jeanloz, R.W. (1965 *b*) The synthesis of a 2-amino-2-deoxy-D-glucosyl-muramic acid disaccharide : 2-acetamido-6-*O*-(2-acetamido-3,4,6-tri-*O*-acetyl-2-deoxy-β-D-glucopyranosyl)-1, 4-di-*O*-acetyl-2-deoxy-3-*O*-[D-1-(methyl carboxylate)ethyl]-α-D-glucopyranose. Carbohydr. Res. 1, 181-186.

Park, J.T. (1952) Uridine-5′-pyrophosphate derivatives. I. Isolation from *Staphylococcus aureus*. J. Biol. Chem. 194, 877-884.

Pratviel-Sosa, F., Mengin-Lecreulx, D. and van Heijenoort, J. (1991) Over-production, purification and properties of the uridine diphosphate *N*-acetylmuramoyl-L-alanine : D-glutamate ligase from *Escherichia coli*. Eur. J. Biochem. 202, 1169-1176.

Roseman, S., Distler, J.J., Moffatt, J.G. and Khorana, H.G. (1961) Nucleoside polyphosphates. XI. An improved general method for the synthesis of nucleotide coenzymes. Syntheses of uridine-5′, cytidine-5′ and guanosine-5′ diphosphate derivatives. J. Amer. Chem. Soc. 83, 659-663.

THE MUROPEPTIDE COMPOSITION OF THE PEPTIDOGLYCAN OF STAPHYLOCOCCUS AUREUS DETERMINED WITH REVERSED-PHASE HIGH PERFORMANCE LIQUID CHROMATOGRAPHY

Boudewijn L.M. de Jonge, and Alexander Tomasz

Laboratory of Microbiology
The Rockefeller University
New York, NY 10021
U.S.A.

INTRODUCTION

While extensive early studies on the chemical composition of the cell walls of Staphylococcus aureus have been described in the literature (e.g., Tipper and Strominger, 1965), the high resolution technique of reversed-phase high performance liquid chromatography (HPLC), recently developed by Glauner et al. (1988), has not been applied to staphylococcal cell walls until now. In this communication, we report the results of studies in which the HPLC method, adapted to staphylococci, was used to determine the muropeptide composition of a highly methicillin resistant isolate of S. aureus and some of its transposon mutants with decreased antibiotic resistance.

CELL WALL MUROPEPTIDE COMPOSITION OF A HIGHLY METHICILLIN RESISTANT STRAIN

Peptidoglycan from purified cell walls was digested with a muramidase into disaccharide units (muropeptides) which were separated by HPLC, as described by de Jonge et al. (1992a). The method, which separates muropeptides on the basis of both chemical composition (hydrophobicity) and molecular size, has resolved the staphylococcal cell wall hydrolysate to a large number of reproducible components (Fig. 1A). The first part of the chromatogram showed good separation, but at the end separation was incomplete, resulting in a hump. The components of this hump must be multimeric structures since the hump could be resolved upon secondary digestion with lysostaphin, an endopeptidase that hydrolyzes glycine bridges.

The chemical structure of twenty-five components was elucidated by amino acid analysis, HPLC retention times and mass spectrometry. Fig. 2 shows the structural assignments to the muropeptides.

The main peaks (peak no. 5, 11, 15 through 21, Fig. 2A) were pentaglycine substituted muropeptides. Peak no. 5 was the pentapeptide

Bacterial Growth and Lysis, Edited by M.A. de Pedro
et al., Plenum Press, New York, 1993

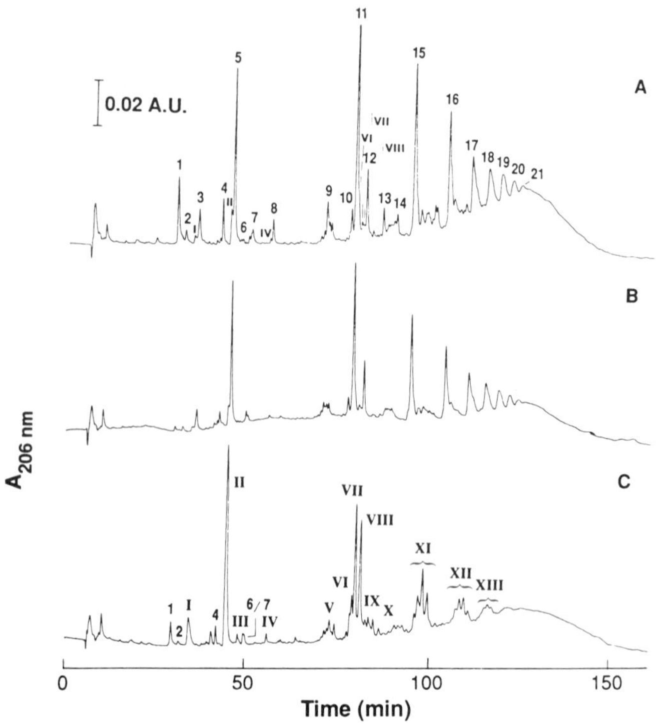

Fig. 1. Separation of muropeptides by reversed-phase HPLC. Muropeptides were isolated from the highly resistant strain COL (A) and two of its transposon mutants with reduced methicillin resistance. Adapted from de Jonge et al. (1992b) with the authors' permission.

monomer, peak no. 11 the dimer of 5 and peak no. 15 through 21 represented the trimer and higher oligomers of the pentaglycine substituted monomer. In addition to pentaglycine-substituted muropeptide species the epsilon amino group of lysine in some of the stem peptides was also found substituted with tetra-, tri- or monoglycine residues or was present without substitution (peaks 1, 4 and II, Fig. 2B). Interestingly, a diglycine substituted muropeptide was not detectable. One would expect such a compound to elute between the mono- and triglycine muropeptides (peak 4 and II).

Monomeric peptides can be crosslinked to one another and indeed most of the muropeptide monomers were also detected in dimeric forms (peaks 9, VI-VIII). Higher oligomeric forms of these muropeptides are likely to be present in the hump at the end of the chromatogram.

Besides the glycine substituted species, alanyl-glycine substituted muropeptides were also detected. The alanine was directly linked to the stem peptide lysine residue followed by glycine units in the side chain (peaks 7, 8, IV and 13, Fig. 2C).

Monomeric muropeptides were invariably present as pentapeptides, confirming the low DD-carboxypeptidase activity in S. aureus (Tipper and Strominger, 1968). The only compounds found with tetrapeptide stem peptides (peaks 3 and I, Fig. 2D) contained glycine peptides on the lysine as well as the terminal alanine residue. These were most likely degradation products of multimeric compounds produced by endopeptidase activity in S. aureus.

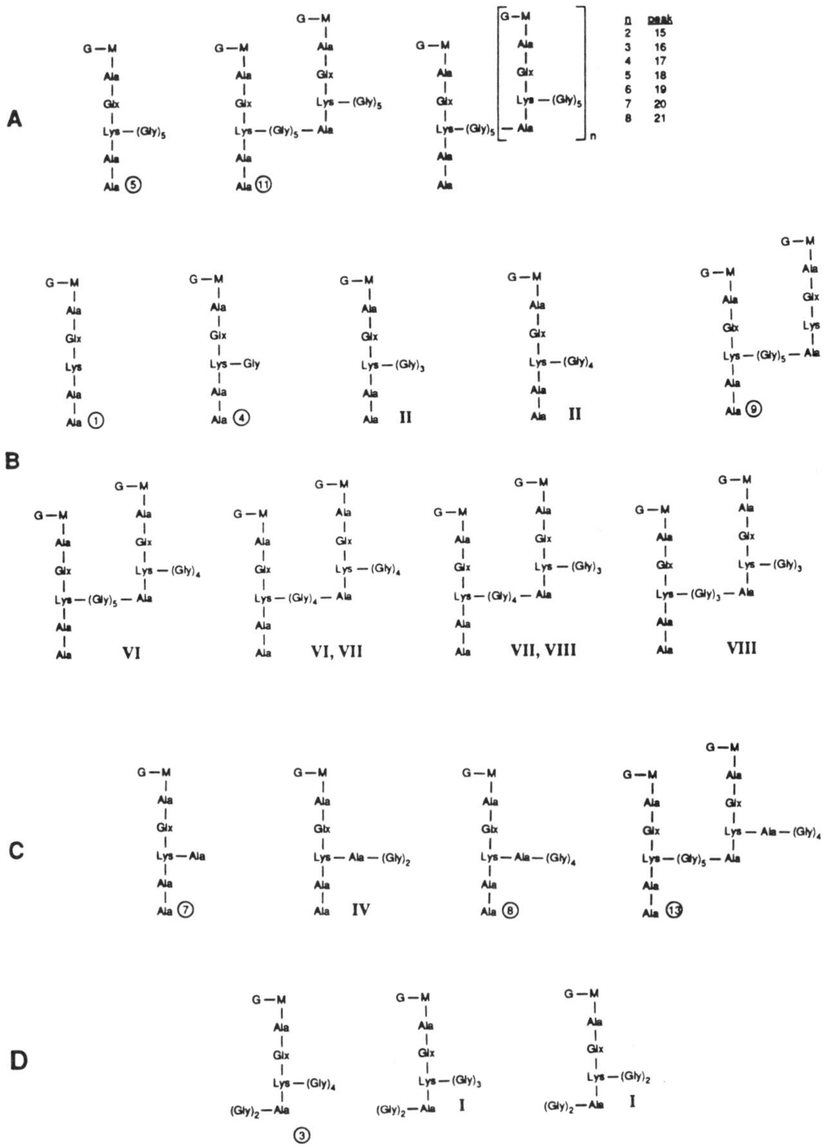

Fig. 2. Muropeptide structures of the peptidoglycan of S. aureus as separated with reversed-phase HPLC. Adapted with the authors' permission, from de Jonge et al. (1992 a,b).

Analysis of the muropeptide composition of several methicillin susceptible S. aureus isolates gave virtually identical HPLC patterns as that of the highly resistant strain shown in Fig. 1A. Thus, our conclusions concerning peptidoglycan composition appear to be generally valid for S. aureus.

ALTERED CELL WALL MUROPEPTIDE COMPOSITION IN TRANSPOSON MUTANTS WITH DECREASED METHICILLIN RESISTANCE

Tn551 insertion at several chromosomal sites can cause decrease in the level of methicillin resistance (Berger-Bachi and Kohler, 1983. Kornblum et al., 1986. Murakami and Tomasz, 1989). Recent analysis of the peptidoglycan of some of these mutants showed major alterations in muropeptide composition as compared to the isogenic parental strain (de Jonge et al., 1992b). One mutant (Fig. 1C) lacked all the pentaglycine muropeptides (peaks 5, 11 and 15-21), but retained the ability to form tetraglycine substituted compounds (Roman numerals in Fig. 1C). We concluded from this that the addition of the fifth glycine residue in the pentaglycine side chains may be specifically blocked. Another mutant (Fig. 1B) lacked the unsubstituted pentapeptide (peak 1) and the alanine-glycine substituted muropeptides (peaks 7 and 8) as well as crosslinked derivatives of these monomers (peaks 9 and 13). A possible explanation of this particular defect may be the shortage of lipid-linked pentapeptide caused by the Tn551 insertion (see Fig. 3).

PEPTIDOGLYCAN BIOSYNTHESIS

The chemical nature of muropeptides identified in the cell walls of the resistant <u>S. aureus</u> strain and its Tn551 mutant derivatives allows some speculations concerning peptidoglycan biosynthesis in this bacterium, and these are summarized in a model outlined in Fig. 3.

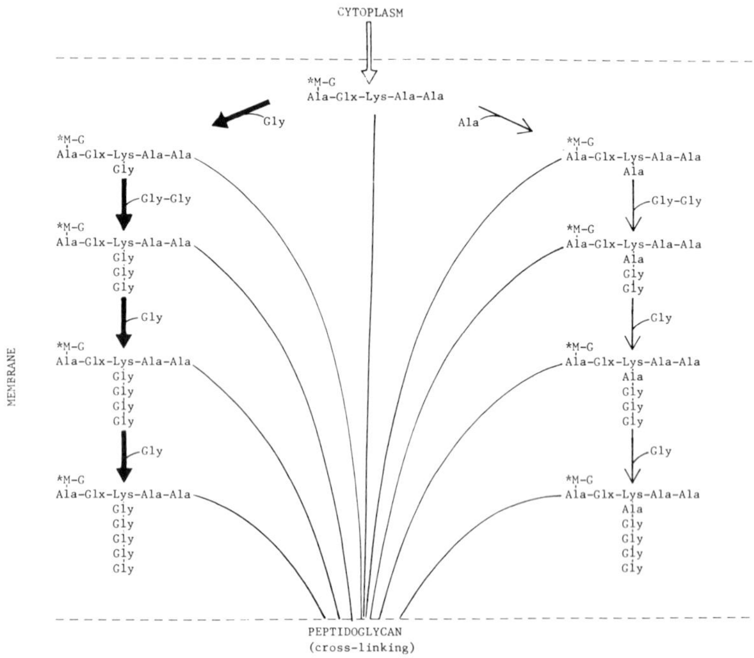

<u>Fig. 3</u>. Biosynthetic pathway of the peptidoglycan in <u>S</u>. <u>aureus</u>. For explanation see text.

We assume that after the formation of the lipid-linked disaccharide pentapeptide either alanine or glycine may be added to the stem peptide lysine residues. Since the glycine substituted compounds are quantitatively more abundant in the peptidoglycan of <u>S</u>. <u>aureus</u>, we assume that addition of glycine is more favored (see thick arrows in Fig. 3) and upon addition of the first glycine, the product is channeled into the pentaglycine pathway through several steps. It seems that the next step in this sequence is the addition of diglycine (leading to a triglycine substituted species) since diglycine substituted muropeptide species could not be detected in the peptidoglycan hydrolysates. The fourth and fifth glycines appear to be added in distinct steps. The tetraglycyl muropeptides are particularly evident in one of the Tn551 mutants which seems to be specifically blocked in the biosynthetic step involving the addition of the fifth residue. The presence of all these muropeptides in the peptidoglycan indicates that each muropeptide monomer may incorporate into the cell wall where they may also be crosslinked to one another. A similar sequence of reactions may be envisioned for the alanine-containing branch of muropeptides.

If one assumes that a limitation in the amount of lipid-linked pentapeptide will result in the preferential conversion of this intermediate into the quantitatively predominant pentaglycine substituted species, the disappearance of the alanine-containing muropeptides from the peptidoglycan of the second type of transposon mutant may be explained.

FUNCTION OF PBP2A IN PEPTIDOGLYCAN METABOLISM

One of the genetic determinants involved with methicillin resistance is the mec gene which is generally believed to have originated outside the species. The mec gene encodes for a penicillin binding protein (PBP2A) with extremely low affinity for most beta lactam antibiotics. PBP2A contains the structural motifs characteristic of serine transpeptidases (Song et al., 1987) and it is generally assumed that PBP2A serves as a surrogate enzyme for cell wall synthesis in the presence of high concentrations of antibiotic in the medium. Because of its unusual features, PBP2A might also exhibit unique transpeptidase activity that may leave its mark on the composition of the peptidoglycan. For this reason, we compared the peptidoglycan composition of the highly methicillin

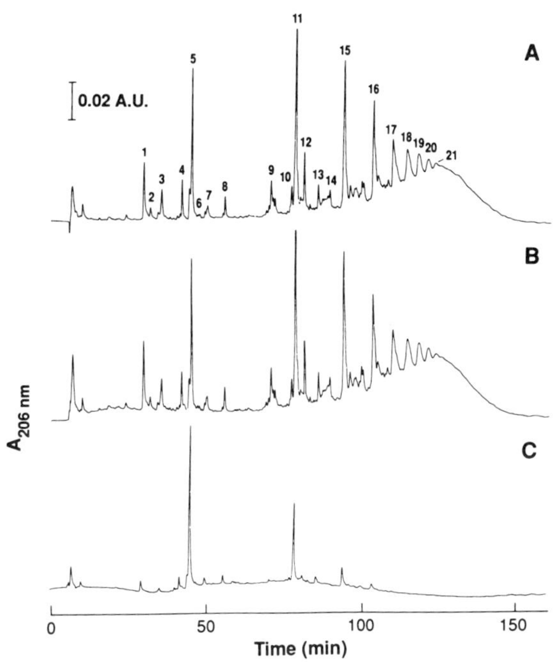

Fig. 4. Separation of muropeptides by reversed-phase HPLC. Muropeptides were isolated from the highly resistant strain COL (A), a Tn551 mutant of COL with the transposon insert within the mec gene (B) and COL grown in the presence of 50 µg per ml methicillin (C). Adapted, with the authors' permission, from de Jonge et al. (1992a).

resistant PBP2A containing strain COL and one of its Tn551 mutants in which the structural determinant of PBP2A was inactivated by transposon insertion within the gene (Murakami and Tomasz, 1989. Matthews and Tomasz, 1990). Peptidoglycan isolated from both strains showed no differences in the muropeptide composition (Fig. 4A and B).

It may be argued that PBP2A becomes only activated when the bacteria are exposed to methicillin. Therefore the cell wall analysis was repeated with the highly resistant staphylococcal strain grown with 50 µg per ml of methicillin in the medium, a concentration sufficient to saturate all PBPs except PBP2A. Peptidoglycan made under these conditions was found to be poorly crosslinked with the disappearance of the highly crosslinked oligomers from the HPLC chromatogram (Fig. 4C). Virtually the only crosslinked muropeptides detectable in the cell walls of these staphylococci were dimers with a few trimers and tetramers. No qualitatively new compounds could be detected. If one assumes that this unusual peptidoglycan was the product of PBP2A, then one may consider this protein as a transpeptidase unable to use muropeptides more complex than dimers as acceptors (or donors) in the transpeptidase reaction.

ACKNOWLEDGEMENTS

These investigations received partial support from a grant from the US Public Health Service (AI 16794).

REFERENCES

Berger-Bachi, B., and Kohler, M.L. (1983) A novel site on the chromosome of Staphylococcus aureus influencing the level of methicillin resistance: genetic mapping. FEMS Microbiol. Lett. 20, 305-309.

De Jonge, B.L.M., Chang, Y.-S., Gage, D., and Tomasz, A. (1992a) Peptidoglycan composition of a highly methicillin-resistant Staphylococcus aureus strain: the role of penicillin binding protein 2A. J. Biol. Chem. 267, in press.

De Jonge, B.L.M., Chang, Y.-S., Gage, D., and Tomasz, A. (1992b) Peptidoglycan composition in heterogeneous Tn551 mutants of a methicillin-resistant Staphylococcus aureus strain. J. Biol. Chem. 267, in press.

Glauner, B., Holtje, J.-V., and Schwarz, U. (1988) The composition of the murein of Escherichia coli. J. Biol. Chem. 263, 10888-10095.

Matthews, P., and Tomasz, A. (1990) Insertional inactivation of the mec gene in a transposon mutant of a methicillin-resistant clinical isolate of Staphylococcus aureus. Antimicrob. Agents Chemother. 34, 1777-1779.

Kornblum, J., Hartman, B.J., Novick, R.P., and Tomasz, A. (1986) Conversion of a homogeneously methicillin- resistant strain of Staphylococcus aureus to heterogeneous resistance by Tn551-mediated insertional inactivation. J. Clin. Microbiol. 5, 714-718.

Murakami, K., and Tomasz, A. (1989) Involvement of multiple genetic determinants in high-level methicillin resistance in Staphylococcus aureus. J. Bacteriol. 171, 874-879.

Song, M.D., Wachi, M., Doi, M., Ishino, F., and Matsuhashi, M. (1987) Evolution of an inducible penicillin-target protein in methicillin-resistant Staphylococcus aureus by gene fusion. FEBS Lett. 221, 167-171.

Tipper, D.J., and Strominger, J.L. (1965) Mechanism of action of penicillins: a proposal based on their structural similarity to acyl-D-alanyl-D-alanine. Proc. Natl. Acad. Sci. USA 54, 1133-1141.

Tipper, D.J., and Strominger, J.L. (1968) Biosynthesis of the peptidoglycan of bacterial cell walls. XII Inhibition of cross-linking by penicillins and cephalosporins: studies in Staphylococcus aureus in vivo. J. Biol. Chem. 243, 3169-3179.

DISTRIBUTION OF *O*-ACETYLATION IN THE PEPTIDOGLYCAN FROM *Proteus mirabilis* 19

Anthony J. Clarke, Steve Gyorffy, and James W. Chase

Department of Microbiology
University of Guelph
Guelph, Ontario N1G 2W1

INTRODUCTION

The detection of *O*-acetylated peptidoglycan has previously required the isolation and purification of peptidoglycan from the study microorganisms followed by the release (saponification) and quantitation of acetate. Since this methodology is laborious and in addition, the ester-linked acetyl moiety is highly labile, it is not surprising that the prevalence of *O*-acetylated peptidoglycan among the eubacteria is largely unknown. The only microorganisms confirmed to possess *O*-acetylated peptidoglycan include *Lactobacillus fermentum, Moraxella glucidolytica, Neisseria gonorrhoeae, N. perflava, Proteus mirabilis, P. vulgaris, Pseudomonas alcaligenes, Staphylococcus aureus* and *Streptococcus faecalis* (reviewed in Clarke and Dupont, 1992). Many of these microorganisms represent important human pathogens but a correlation, if any, between peptidoglycan *O*-acetylation and pathogenicity/ virulence has not been examined. Moreover, very little is known regarding both the physiological role and the mechanistic process of the modification (Clarke and Dupont, 1992).

Anti-peptidoglycan antibodies have been detected in the sera of patients with bacterial infections and in rabbit models immunized with peptidoglycan (reviewed in Heymer *et al.*, 1985; Seidl and Schleifer, 1985*a*). Five antigenic epitopes have been discerned: the glycan stand, the N-terminal and C-terminal sequences of the interpeptide bridges, non-crosslinked tetrapeptides, and non-crosslinked pentapeptides (Seidl and Schleifer, 1985*b*). With the aim to develop an efficient screen for peptidoglycan *O*-acetylation and further investigate the physiological role of the modification, we have produced a monoclonal antibody (MAb PmPG5-3) with specificity for the *O*-acetylated peptidoglycan of *P. mirabilis*. Characterization of MAb PmPG5-3 indicated that it recognizes neither the unmodified peptidoglycan of *Escherichia coli* nor de-*O*-acetylated *P. mirabilis* peptidoglycan but it is weakly inhibited by the chemical subunits of peptidoglycan, *viz. N*-acetylglucosamine (GlcNAc) and *N*-acetylmuramic acid-dipeptide (MurNAc-dipeptide). The MAb was used to localize the *O*-acetyl moieties on *P. mirabilis* peptidoglycan through immunogold electron microscopy.

Bacterial Growth and Lysis, Edited by M.A. de Pedro *et al.*, Plenum Press, New York, 1993

MATERIALS AND METHODS

Isolation and purification of peptidoglycan

Peptidoglycan and *O*-acetylated peptidoglycan was isolated and purified from cultures of *E. coli* (K12)CSH4 and *P. mirabilis* 19, respectively, by the method of Hoyle and Beveridge (1983) as modified and described in Dupont and Clarke (1991).

Production of MAb PmPG5-3

Female BALB/c mice were immunized intradermally with a 1:1 mixture of *O*-acetylated peptidoglycan (Dupont and Clarke, 1991) in vesicles of phosphatidyl choline (250 μg) and Freund's incomplete adjuvant. An intravenous/interperitoneal booster immunization with 150 μg soluble immunogen followed 14 days later. Mice demonstrating activity against *O*-acetylated peptidoglycan were selected for a second booster immunization on day 21 and spleen cell-murine myeloma cell fusion. Hybridomas were produced essentially by the method of Oi and Herzenberg (1979) as modified by Lam *et al.* (1987). Nineteen hybridoma cell lines secreting MAbs to peptidoglycan, as determined by ELISA, were obtained and expressed a broad range of reactivities. The hybridoma, PmPG5-3, expressing both the strongest response to *O*-acetylated peptidoglycan and weakest reactivity toward both de-*O*-acetylated peptidoglycan and the non-*O*-acetylated peptidoglycan of *E. coli* was selected for limiting dilution and ascites production to facilitate further study. MAb PmPG5-3 of the ascites fluid was purified to homogeneity, as judged by SDS-PAGE, by a combination of ammonium sulfate precipitation and affinity chromatographies. ELISA subtyping assay identified MAb PmPG5-3 as isotype IgM.

Immunoassay

Quantitative immunoassays were performed employing an enzyme-linked immunosorbent assay (ELISA) with non-pronase-treated, mutanolysin-solubilized *O*-acetylated peptidoglycan as antigen in 10 mM phosphate saline pH 7.4. Unbound antigen and antibodies were removed by washing three times with 0.1% Tween 20 in the phosphate buffer. Alkaline phosphatase-linked goat anti-mouse IgG (pre-absorbed with 1.0 mg of non-digested *O*-acetylated peptidoglycan overnight at 4 °C) and *p*-nitrophenyl phosphate in 10% diethanolamine buffer pH 9.6 were used as the second antibody and substrate, respectively.

Synchronization of *P. mirabilis* and isolation of murein sacculi

Cultures (400 ml) of *P. mirabilis* were synchronized by sucrose density gradient centrifugation as described by Gmeiner and Sarnow (1987). Samples of cells were collected at various stages during the cell cycle (20 min, 60 ml; 35 min, 48 ml; 50 min, 33 ml; 70 min, 22 ml; 90 min, 18 ml) and murein sacculi were prepared as described above.

Immunogold labelling of *O*-acetylated peptidoglycan

Purified sacculi were treated with homogeneous MAb PmPG5-3 at 4°C, collected by centrifugation, and washed twice with 10 mM sodium phosphate buffer, pH 5.5 containing 0.1% (v/v) Tween 20 and 1.0% (w/v) BSA. The sacculi were then labelled with Protein G-colloidal gold (10 nm) complex, washed three times with the PBS-Tween-BSA buffer and once with Milli-Q water. Labelled sacculi were mounted on carbon Formvar-coated nickel (200 mesh) grids and negative stained with 2% (wt/v) uranyl acetate before examination using a Phillips EM 300 transmission microscope operating at 60 kV.

RESULTS

As seen in Figure 1, MAb PmPG5-3 recognized the soluble form of the *O*-acetylated peptidoglycan of *P. mirabilis* but only weakly reacted with the chemically de-*O*-acetylated (incubation in 50 mM sodium carbonate buffer, pH 9.8, 15 h, RT) material. This decreased reactivity was comparable to that for the non-*O*-acetylated peptidoglycan of *E. coli*. Competitive inhibition assays were used to further characterize the specificity of the MAb for the *O*-acetyl epitope. Preincubation of PmPG5-3 with soluble and insoluble *O*-acetylated peptidoglycan inhibited the MAb from binding to the *O*-acetylated peptidoglycan antigen bound to the microtiter plate whereas the *O*-acetylated peptidoglycan subjected to alkali hydrolysis resulted in only partial inhibition (Figure 2). The naturally non-*O*-acetylated peptidoglycan of *E. coli* CSH4 was also only as effective as de-*O*-acetylated peptidoglycan in the inhibition of PmPG5-3. However, GlcNAc and MurNAc-dipeptide were equally more effective in partially inhibiting the ELISA assay while D-alanine was totally ineffective suggesting that the binding epitope of the antigen comprises only the glycan chain.

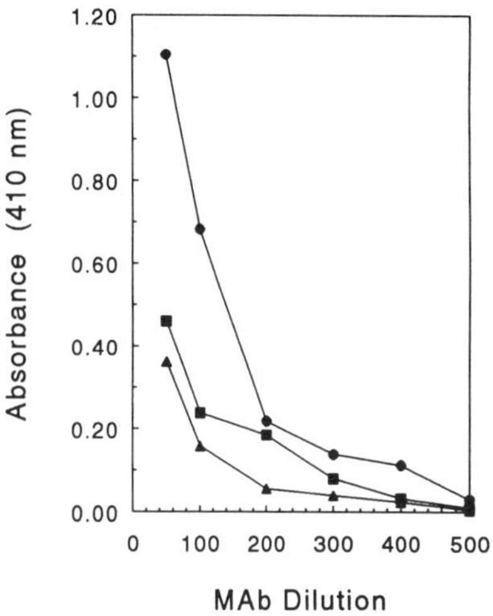

Figure 1. Characterization of the binding specificity of MAb PmPG5-3 by ELISA. Peptidoglycan isolated from: (■) *E. coli* CSH4; and *P. mirabilis* 19, (●) native and (▲) de-*O*-acetylated by mild base hydrolysis, were tested as antigens for MAb PmPG5-3 at the dilutions indicated.

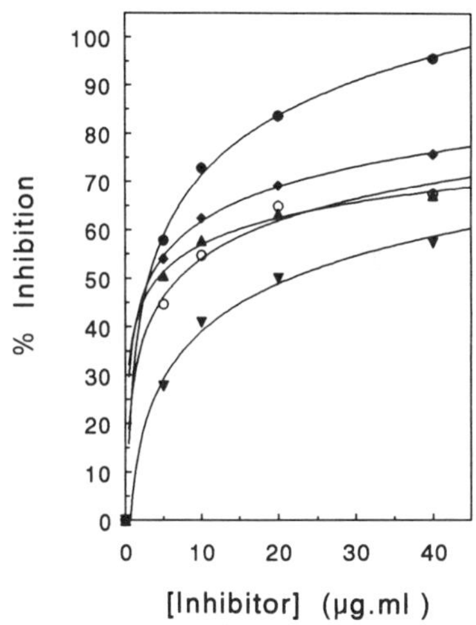

Figure 2. Competitive inhibition ELISA of MAb PmPG5-3. MAb (1:100 dilution) was incubated overnight at pH 6.8 with: (●) soluble *O*-acetylated, and (○) mild-base hydrolysed *P. mirabilis* peptidoglycan; (▲) *E. coli* peptidoglycan; (◆) MurNAc-dipeptide; and (▼) chitin, at the concentrations indicated, prior to ELISA using *O*-acetylated peptidoglycan as antigen.

Treatment of murein sacculi isolated from batch cultures of *P. mirabilis* 19 with MAb PmPG5-3 and Protein G-colloidal gold, followed by negative staining resulted in both heavy and specific labelling of the peptidoglycan when visualized by TEM (Figure 3). The distribution of the *O*-acetyl moieties appears to be random, with no indication of an accumulation in specific regions such as the polar caps or equatorial zones. An average labelling density of 417 gold particles (± 87, standard deviation) per μm^2 was determined from the examination of 10 murein sacculi. With the controls of de-*O*-acetylated *P. mirabilis* peptidoglycan and *E. coli* sacculi, immunogold labelling was reduced to approx. 15 and 0 gold particles per μm^2, respectively.

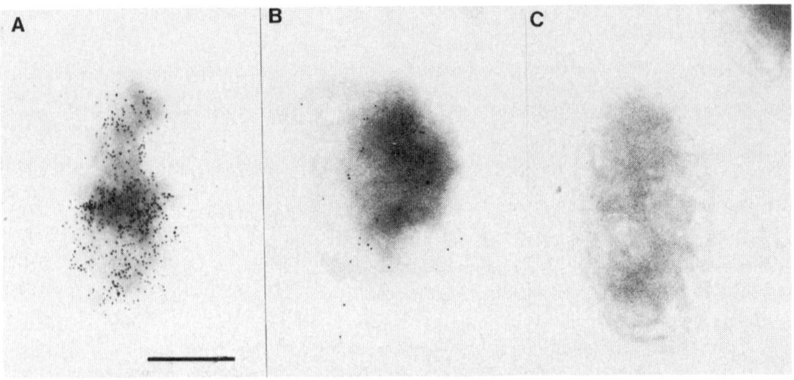

Figure 3. Immunogold labelling of murein sacculi with MAb PmPg5-3 and Protein G-gold. Murein sacculi of SDS-solubilized cells from the exponential growth phase of (A) *P. mirabilis* 19 and (C) *E. coli* CSH4 were isolated and incubated first with MAb PmPG5-3 and then with protein G-gold (10 nm) as described in Materials and Methods. (B), De-*O*-acetylated murein sacculi of *P. mirabilis* 19 prepared by mild-base hydrolysis prior to immunogold labelling. Bar, 0.5 μm.

Synchronized cells of *P. mirabilis* were obtained by sucrose density gradient centrifugation in diluted complex medium at 25°C. The doubling time under these conditions was between 30 and 40 minutes and synchrony of the cells persisted for three to four generations (Figure 4).

Decreasing volumes of the synchronized culture containing approximately equal numbers of cells were harvested at the indicated times and murein sacculi were prepared for immunogold labelling studies. Representative transmission electron micrographs of the results are presented in Figure 5. Regardless of the time of isolation, the sacculi were similarly labelled and again, no obvious distinction in the pattern of the gold particles was discerned.

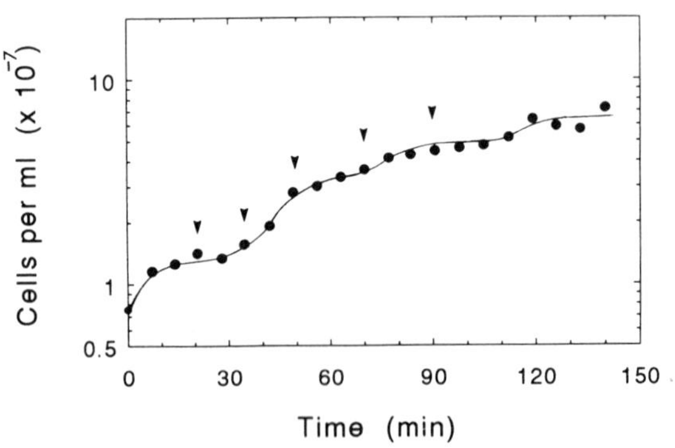

Figure 4. Growth of synchronized cells of *P. mirabilis* 19. Cells were cultivated in complex medium following synchronization by sucrose density gradient centrifugation and enumerated using a Coulter Counter ZM. At the indicated times, samples of the cells were harvested and murein sacculi were prepared for immunoelectron microscopy as described in Materials and Methods.

DISCUSSION

This study is the first to report the production of a MAb to peptidoglycan with the *O*-acetyl group being recognized as the major antigenic epitope. Shockman and coworkers (Jackson *et al.*, 1983) have previously prepared MAbs to the peptidoglycan of *Streptococcus mutans* and several, which were all of the IgM class, recognized the glycan portion of peptidoglycan while others recognized the peptide side chains. With MAb PmPG5-3, competition ELISA experiments suggest that the binding epitope of the *O*-acetyl-peptidoglycan antigen would comprise the glycan backbone and likely, the acetyl moieties of both the N^2 of *N*-acetylglucosaminyl and C^6 of muramyl residues. The specificity of this MAb for the *O*-acetylated glycan portion of peptidoglycan thus makes it valuable for use in a screening protocol since the binding epitope will be common to all *O*-acetylated peptidoglycan.

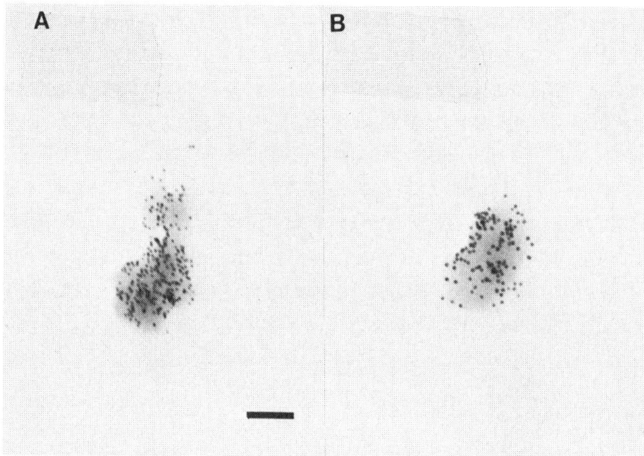

Figure 5. Immunogold labelling of murein sacculi isolated from synchronized cultures of *P. mirabilis* 19. Representative sacculi are presented and were obtained from the culture at (A) 20 min and (B) 50 min. Bar, 0.2 μm.

Immunoelectron microscopy of the *P. mirabilis* murein sacculus obtained from a batch culture using MAb PmPG5-3 revealed a heavy but random distribution of the colloidal gold. This likely reflects the pattern of *O*-acetylation present on the sacculi since little and no colloidal gold was observed by TEM on the murein sacculi of *E. coli* and de-*O*-acetylated *P. mirabilis*, respectively. It is unclear at present why the murein sacculi of *E. coli* were not labelled by the immunogold technique and yet weakly reacted with MAb PmPG5-3 in the ELISA. It is conceivable that the epitope of the *E. coli* peptidoglycan is more accessible in the solubilized antigen used in ELISA as compared to the murein sacculi prepared for immunogold labelling. This would thus imply that despite the similarities in the general chemical composition of the A1γ peptidoglycans, subtle conformational differences in overall architecture may exist between the sacculi of *E. coli* and *P. mirabilis* which affect the highly specific binding interactions of the antibody and antigen.

The physiological role of *O*-acetylation of peptidoglycan remains unknown. The modification may serve to protect the peptidoglycan strands from the hydrolytic activity of muramidase-type autolysins since it has been shown to block the action of lysozymes (reviewed in Clarke and Dupont, 1992). Conversely, the presence of acetyl moieties may indicate sites for autolytic action since Blundell and Perkins (1985) observed an endopeptidase activity in permeabilized *N. gonorrhoeae* with selectivity for *O*-acetylated peptidoglycan. Given the extensive and random *O*-acetylation observed by immunogold labelling of the sacculi, it is likely that the latter hypothesis can be excluded for *P. mirabilis*. The absence of any obvious zones, including the equatorial regions, of unmodified peptidoglycan in the immunogold labelled sacculi isolated during different stages of the cell cycle would tend to also discount the former postulate. It is possible, however, that very narrow zones lacking *O*-acetylation may exist on the sacculi which remained undetected by the crude experimental approach of immunoelectron microscopy of isolated murein sacculi. Further details may in the future be provided by the post-embedding labelling of thin sections of *P. mirabilis* employing the freeze substitution technique.

The *O*-acetylation of carbohydrate polymers as a means of preservation is not uncommon in nature. For example, the xylans of the hemicellulose component of many hard woods are extensively *O*-acetylated (up to 70%) and thereby inhibit the action of the extracellular hemicellulases and cellulases of plant pathogens (Timmell, 1964). Likewise, *O*-acetylation of peptidoglycan may represent a further strategy by bacteria to evade the host response and in the case of pathogens, constitute another virulence factor. This could be equally true for gram negative microorganisms as for gram positives since host muramidases that penetrate to the peptidoglycan layer after the outer membrane has been breached by the membrane attack complex of human complement (Esser, 1991) would be ineffective in hydrolysing the rigid cell wall. These cells would thus escape lysis and be afforded the time to regenerate their outer membrane. With the availability of MAb PmPG5-3, an effective screen for peptidoglycan *O*-acetylation may now be established and the question of a correlation between the modification and the virulence of strains of *P. mirabilis* and other pathogens with *O*-acetylated peptidoglycan could be addressed.

ACKNOWLEDGMENTS

We thank Joe Lam and Gord Southam for their helpful advice concerning the production of the MAbs, and Theresa Mok for her technical assistance with immunogold labelling and TEM. This work was supported by an operating grant (MT-10954) from the Medical Research Council of Canada.

REFERENCES

Blundell, J.K. and Perkins, H.R. (1985) Selectivity for *O*-acetylated peptidoglycan during endopeptidase action by permeabilized *Neisseria gonorrhoeae*. FEMS Microbiol. Lett. 30, 67-69.

Clarke, A.J. and Dupont, C. (1992) *O*-Acetylated peptidoglycan: its occurrence, pathobiological significance, and biosynthesis. Can J. Microbiol. In press.

Dupont, C. and Clarke, A.J. (1991) Dependence of lysozyme-catalysed solubilization of *Protues mirabilis* peptidoglycan on the extent of *O*-acetylation. Eur. J. Biochem. 195, 763-769.

Esser, A.F. (1991) Big MAC attack: complement proteins cause leaky patches. Immunol. Today 12, 316-318.

Gmeiner, J. and Sarnow, E. (1987) Murein biosynthesis in synchronized cells of *Proteus mirabilis*. Quantitative analysis of *O*-acetylated murein subunits and of chain terminators incorporated into the sacculus during the cell cycle. Eur. J. Biochem. 163, 389-395.

Heymer, J., Seidl,P. and Schleifer, K. (1985) Immunochemistry and biological activity of peptidoglycan, in "Immunology of the Bacterial Cell Envelope" (Stewart, D. and M. Davies, M., Eds.), pp. 11-46. John Wiley and Sons, New York.

Hoyle, B.D. and Beveridge, T. (1983) Metal binding by the peptidoglycan sacculus of *Escherichia coli* K-12. Can. J. Microbiol. 30, 204-211.

Jackson, D.E., Wong, W., Swavely, T.S. and Shockman, G.D. (1983) Monoclonal antibodies that recognize bacterial cell wall peptidoglycan. Abstract No. K151. 83rd Annual Meeting of the American Society for Microbiology, March, New Orleans

Lam, J.S., MacDonald, L.A., Lam, M.Y.C., Duchesne, L.G.M. and Southam, G.S. (1987) Production and characterization of monoclonal antibodies against serotype strains of *Pseudomonas aeruginosa.* Infect. Immun. 55, 1051-1057.

Oi, V.T. and Herzenberg, L.A. (1979) Immunoglobulin production hybrid cell lines, in "Selected Methods in Cellular Immunology" (Mishell, B.B. and S. M. Shiigi, S.M., Eds), pp. 351-371. W. H. Freeman & Co., San Francisco.

Seidl, P.H. and Schleifer, K.H. (1985*a*) "Biological Properties of Peptidoglycan," Walter de Gruyter, New York.

Seidl, P.H. and Schleifer, K.H. (1985*b*) Structure and immunochemistry of peptidoglycan, in "Biological Properties of Peptidoglycan" (Seidl, P.H. and Schleifer, K.H. Eds.), pp. 1-20. Walter de Gruyter, New York.

Timmell, T.E. (1964) Wood hemicelluloses: Part I. Adv. Carbohydr. Chem. 19, 247-302.

MUREIN STRUCTURE OF THREE DIFFERENT SPECIES OF CHEMOLITHOTROPHIC SULFUR BACTERIA: *Thiobacillus tepidarius, Thiobacillus neapolitanus and Thiobacillus versutus*

Zdzislaw Markiewicz*, Jadwiga Baj, Iwona Grabowska, and Zofia Sniezek

Institute of Microbiology
University of Warsaw
Nowy Swiat 67
Warszawa 64
00-046 Poland

INTRODUCTION

The genus *Thiobacillus* is an extremely heterogeneous group of bacteria. The only criterion for placing all the species in a single genus is that they all obtain energy for autotrophic growth from the oxidation of inorganic sulfur substrates. Some species are obligately chemolitotrophic and autotrophic (e.g. *Thiobacillus neapolitanus* and *Thiobacillus tepidarius*) whereas others are facultatively autotrophic and capable of heterotrophic growth using a wide range of organic substrates (e.g. *Thiobacillus versutus*). The facultative species are frequently indistinguishable from some of the bacteria grouped as *Pseudomonas*, except for their capacity for growth with reduced sulfur compounds (Kelly and Harrison, 1989). It has been suggested that the facultative or mixotrophic *Thiobacillus* species be reassigned to the appropriate genera of chemoorganotrophic bacteria (Friedrich and Mitrenga, 1981).

The cell envelope structure of the thiobacilli has so far received very little attention, Holt *et al.* (1974) demonstrated the three-layered structure of the envelopes of several thiobacilli, including *T. neapolitanus*. More recently Yokota *et al.* (1987) described the structure of *T. versutus*. lipopolysaccharide.There has been no mention of murein (peptidoglycan) structure of representatives of the thiobacilli and this, to the best of our knowledge, holds true for the whole large group of chemolitoautotrophic bacteria. In comparison, the mureins of several species of photosynthetic bacteria, the second large group of autotrophs, have recently been characterized (e.g. Schmelzer *et al.*, 1982; Jürgens *et al.*, 1987) and shown to contain modifications rather frequently encountered of the murein of Gram-positive bacteria.

In this paper we compare the mureins of two obligately chemolitotrophic thiobacillus species: *T. neapolitanus* (Baj *et al.*, 1991; Baj *et al.*, 1992) and the moderately thermophilic *T. tepidarius* with that of the facultative chemolithotroph *T. versutus* (Baj and Markiewicz, 1988). The mureins of the former two species are shown to contain a number of modifications not found in the latter. This finding supports the previously postulated lack of relatedness between *T. versutus* and the obligately autotrophic thiobacilli.

*Corresponding author

Bacterial Growth and Lysis, Edited by M.A. de Pedro *et al.*, Plenum Press, New York, 1993

MATERIALS AND METHODS

Bacterial Strains and Growth Conditions

T. neapolitanus and *T. versutus* were obtained from Dr.O.Tuovinen (University of Groningen, The Netherlands), *T. tepidarius* was from Dr.D.P.Kelly (University of Warwick, Great Britain). The strains were grown with vigorous shaking in mineral medium containing 0.5% Na_2SO_3 as described elsewhere (Baj *et al.*,1992). The growth temperatures for the strains were 28°, 32° and 44°C, respectively. Cells were harvested at $A_{578} = 0.25$.

Preparation of Murein

Isolation of murein was essentially as described by Glauner (1988), including α-amylase and pronase digestion For some experiments the cells of *T. versutus* were broken in French press prior to boiling in SDS. Alternatively, isolated murein sacculi were additionally treated with sodium hydroxide to remove contaminating proteins (Markiewicz *et al.*,1983). Murein from *Bacillus cereus* AHU 13567 was a kind gift from Dr E.Ito (Hokkaido University, Japan).

Chemical Modification of Murein

N-acetylation was performed according to the method of Hayashi et al. (1973). Dinitrophenylation of murein was as described by Schmelzer *et al.* (1982). Standards for thin layer chromatography were prepared as described by Schleifer (1985).

Biochemical Determinations

The presence of O-acetyl groups on the murein was assayed according to Hestrin (1949) with acetylcholine chloride as a standard. The presence of acetamide groups was identified with bis(1,1-trifluoroacetoxy)iodo-benzene (BTI) as described by Soby and Johnson. Amino acid and amino sugar analysis was carried out with Durrum D-500 autoanalyzer. Reducing groups released by the action of lysozyme or muramidase SG were quantitated according to Thompson and Shockman. Phosphate was determined as described by Lowry *et al.* (1954). The method used to isolate and identify muramic acid-6-phosphate was that of Liu and Gotschlich (1967). Muramic acid content was estimated by the method of Hadzija (1974) which makes use of the degradation of muramid acid to lactic acid and then acetic anhydride. Hexosamine was determined colorimetrically (Tsuji *et al.*, 1969) with glucosamine as a standard.

Other Methods

Thin-layer chromatography of acid hydrolysate of dinitrophenylated murein was in two dimensions on Cellulose F254 plates (Araki *et al.*, 1972). First dimension, 1-butanol : pyridine : water (6 : 4 : 3); second dimension, 1.5 M potassium phosphate buffer, pH 6.0.

RESULTS

Isolation of Murein

Murein was obtained by extraction of bacterial cells with boiling sodium dodecylsulfate, followed by alfa-amylase and pronase digestion of the rigid sacculi. The procedure yielded a pure murein fraction for both *T. neapolitanus* and *T. tepidarius*. For the former, pronase digestion could be omitted, pointing to the absence of covalently bound protein. *T. versutus* murein obtained in the described manner was heavily contaminated with protein. It was possible to remove most of the protein by breaking the cells open before two cycles of SDS extraction and pronase treatment. In some cases it was necessary, however, to treat the murein fraction with sodium hydroxide to remove all contaminating protein.

The crosslinkage of the mureins was assayed by paper chromatography of a muramidase SG digest of labelled murein (Primosigh *et al.*, 1961) and was found to strongly differ from species to species (Table 1). High performance liquid chromatography analysis of cellosyl digests of the three isolated mureins revealed strikingly dissimilar patterns, different also from that for the murein of *Escherichia coli* (data not shown). The individual muropeptides were not, however, identified.

The studied mureins also differed in their sensitivity to lysozyme. which could point to the presence of non-N-acetylated glucosamine residues in the glycan strands.

Chemical Analysis of Murein

The three mureins contained glutamic acid, diaminopimelic acid and alanine. The ratio of diaminopimelic acid to alanine in *T. neapolitanus* murein was 1:1.25, as opposed to 1:1.8 in the mureins from the other two species. Differences were also seen in amino sugar content. The molar equivalents of glucosamine and muramic acid were much lower in *T. neapolitanus* murein than in those from *T. tepidarius* and *T. versutus*. On the other hand, the hydrolysate contained two unknown constituents eluting before muramic acid (**A**) and just behind glucosamine (**B**). Compound **A**, but not **B**, was also present in *T. tepidarius* murein. Finally, the hydrolysate of *T. neapoiitanus* murein was found to contain a high amount of ammonia.

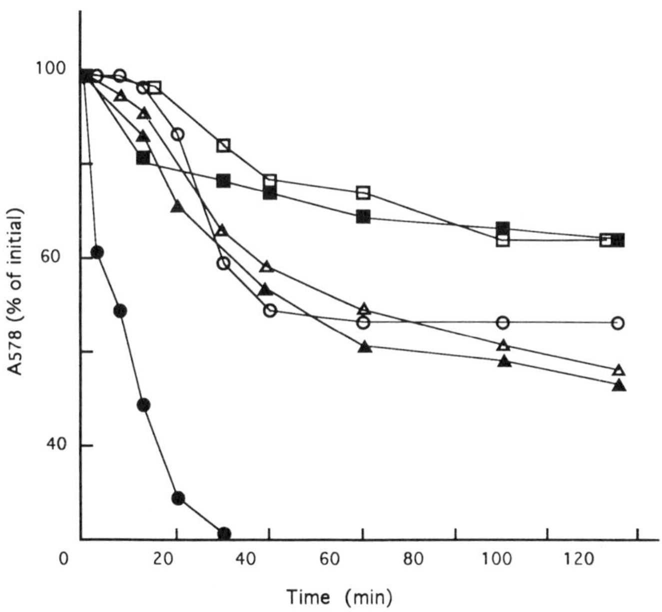

Fig.1. Sensitivity of mureins from *T. tepidarius* (triangles), *T. versutus* (squares) and *T. neapolitanus* (circles) to lysozyme as measured by a decrease in absorbance. Solid symbols represent N-acetylated murein.

Modifications of Sugar Components

The poor digestion of *T. neapolitanus* murein, mentioned above, greatly improved upon N-acetylation of the sacculi prior to treatment with the enzyme (Fig.1). There was no difference in decrease of optical density or release of reducing groups in the case of the mureins from *T. versutus* and *T. neapolitanus*, irrespective of whether they were N-acetylated prior to lysozyme digestion or not. The chemical analysis of an acid hydrolysate of N-acetylated *T. neapolitanus* murein showed an increase in the amounts of both muramic acid and glucosamine, accompanied by almost complete disappearance of compound **B**. The latter was shown to have an identical retention time on the amino acid analyser as authentic Glcβ(1-4)Mur from *Bacillus cereus* This compound is formed on hydrolysis of the murein from this species which lacks N-acetyl substitution of glucosamine (Araki *et al.*, 1972).

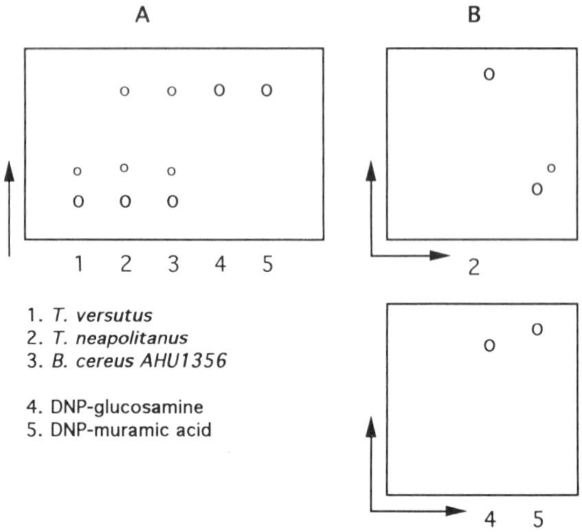

Fig.2. Thin layer chromatography of acid hydrolysates of dinitrophenylated mureins and dinitrophenylated standards.
A. one-dimensional separation, **B**. two-dimensional separation. DNP-compounds were visible as yellow spots.

The partial lack of N-acetyl substitution of glucosamine in the murein of *T.neapolitanus* was further confirmed by thin layer chromatography of an acid hydrolysate of dinitrophenylated murein, compared with authentic dinitrophenylated glucosamine and dinitrophenylated muramic acid standards (Fig.2).

Calculations based on both above methods revealed a 35% lack of N-acetyl substitution of glucosamine.

Compound **A** in acid hydrolysates of *T. neapolitanus* and *T. tepidarius* mureins was isolated by cation exchange chromatograpy on Dowex 50W-X8 column (Liu and Gotschlich, 1967). The fractions eluted with water were assayed for the presence of hexosamines, neutral sugars and organic phosphorus. Hexosamine, subsequently colorimetrically identified as muramic acid, was found in equimolar ratio with organic phosphorus. The oxidation of the muramic acid phosphate with sodium *meta*-phosphate, followed by anion exchange chromatography on Dowex 1-X8 column (Cl^- form), showed the phosphate to be in the C-6 position. The produced compound cochromatographed with glycoaldehyde phosphate, which is produced by periodate oxidation only when muramic acid has a phosphate group on C-6 (Liu and Gotschlich, 1967; Jürgens and Weckesser, 1986).

No O-acetyl was found in any of the three studied mureins.

Modification of Peptide Components

The high amount of ammonia in acid hydrolysates of *T. neapolitanus* murein pointed to the possibility of the presence of acetamide groups in the peptide side chains. Treatment with BTI prior to acid hydrolysis allowed to demonstrate the presence of amidation of part of the *meso*-diaminopimelic acid residues in native murein. This modification was not observed in the murein of *T. versutus*.

The results of our studies on the murein structure of the thiobacilli: *T.tepidarius, T. versutus* and *T. neapolitanus* are summarized in Table 1.

DISCUSSION

In this paper we present the results of our studies on the structure of murein isolated from different species of chemolithotrophic sulfur bacteria. The three thiobacillus species studied strongly differ from one another. The first, *T. versutus*, is a facultative chemolithoautotroph

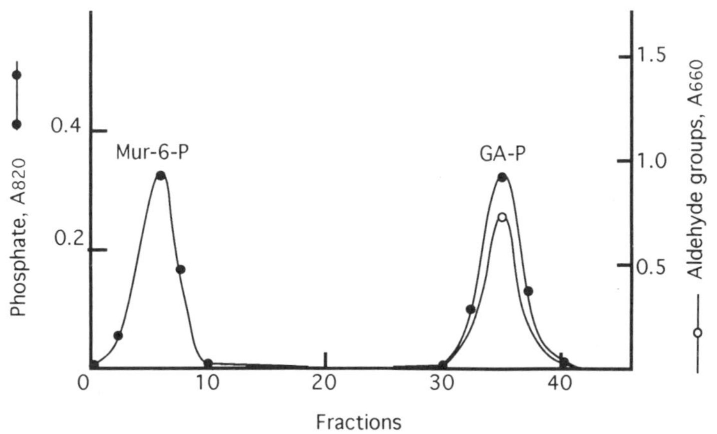

Fig.3. Chromatography of Mur-6-P before (fractions 0 - 20) and after (fractions 25 - 40) oxidation with sodium *meta*-periodate. Mur-6-P - position of muramic acid-6-phosphate, as estimated by presence of phosphorus, GA-P - position of glycolaldehyde phosphate as estimated by phosphorus and aldehyde groups.

which can grow heterotrophically, utilizing a wide variety of organic carbon sources (Kuenen and Tuovinen, 1981). The other two, *T. neapolitanus* and *T. tepidarius* are both obligately chemolithotrophic and autotrophic but differ with respect to optimal growth temperature, the latter being moderately thermophilic (Kelly and Harrison, 1989).

The differences between the species have in our studies been found to be accompanied by differences in murein structure. Although the primary structure of all three studied mureins is of the A1γ type (direct crosslinkage between *meso*-diaminopimelic acid and alanine) they were found to differ in degree of crosslinkage and in amount of murein-associated protein. Moreover, modifications of murein structure thought to be infrequent in murein from Gram-negative bacteria were identified in some of the mureins (Table.1). Lack of N-acetyl substitution in the glycan chains was observed only for the murein from *T. neapolitanus*.

Analysis of di-nitrophenylated murein revealed that 30 to 35% of the glucosamine residues lacked N-acetyl substitution. Incomplete N-acetylation of glucosamine residues explains the poor digestion of *T. neapolitanus* murein by lysozyme (Amano *et al.*, 1980).

The occurrence of non-N-acetyl substituted sugar moieties in the murein of Gram-negative bacteria has so far been described only for a few species belonging to the genera *Rhodopseudomonas* and *Rhodomicrobium* (Schmelzer *et al.*, 1982; Jürgens *et al.*, 1987).On the other hand, none of the studied thiobacillus mureins contained O-acetyl groups, a modification far more frequent in Gram-negative bacteria.

It is more difficult to explain the poor digestion of the other two mureins by lysozyme. The lack of N-acetyl substitution of a small percentage of glucosamine residues, though little likely, cannot be ruled out. The poor digestion of *T. versutus* murein may be due to

Table 1. Comparison of main structural features of mureins from *T. versutus, T. neapolitanus* and *T. tepidarius.*

Species	*T. versutus*	*T. neapolitanus*	*T. tepidarius*
Murein type	A1γ	A1γ	A1γ
Crosslinkage	38%	28%	26%
O-acetyl groups	-	-	-
Non-N-acetyl substitution of glucosamine	-	35%	-
Muramic acid-6-phosphate	-	+	+
Amidated amino acids	-	DAP	n.d.
Molar ratio, DAP:Ala	1:1.75	1:1.25	1:1.80

n.d. - not determined

its higher crosslinkage, uncommon for Gram-negative bacteria, and protein contamination of the sacculi.

A second modification of the glycan strands was found for *T. tepidarius* and *T. neapolitanus*. The mureins of both bacteria contained muramic acid-6-phosphate. This modification, which is frequent in Gram-positive bacteria, has so far been described for only a single Gram-negative species, *Chloroflexus aurantiacus* (Jürgens *et al.*, 1987b). Both *T. tepidarius* (unpublished) and *T. neapolitanus* (Baj *et al.*, 1992) have been shown to have neutral sugars presumably bound via a phosphodiester linkage to the cell wall murein. The nature of these sugar components is unknown.

A further modification found in these studies was the amidation of part of the diaminopimelic acid residues in the stem peptides of *T. neapolitanus* murein. Amidated DAP has been found in the murein of such species as *Bacillus subtilis* and *Corynebacterium diphtheriae* (Rogers *et al.*, 1980). but to our knowledge, *T. neapolitanus* is the first Gram-negative bacterium whose murein is thus modified.

The results of our studies, together with those described for several species of phototrophic bacteria (Schmelzer et al., 1982; Jürgens et al., 1987a) show that the murein of Gram-negative bacteria is subject to far greater structural variation than hitherto believed.(e.g. Rogers *et al.*, 1980). The structure of murein has been suggested as a useful chemotaxonomic marker (Schleifer and Seidl), 1985). In fact, the observed differences in murein structure between *T. versutus* and the other two species of thiobacilli studied are compatible with the results of physiological and genetic studies showing that the former should be withdrawn from the group of obligately autotrophic thiobacilli and classified to an appropriate genus of chemoorganotrophs (Kelly and Harrison, 1989).

Acknowledgments

The authors are grateful to Dr E. Ito for supplying us with GlcNβ(1-4)Mur and murein from *Bacillus cereus*. Support from grant G-MEN 73/91 is acknowledged.

REFERENCES

Araki, Y., Nakatani, T., Nakayama, K, and Ito, E. (1972) Occurrence of N-nonsubstituted glucosamine residues in peptidoglycan resistant cell wall from *Bacillus cereus*. J. Biol. Chem. 247, 6312-6322.

Baj, J. and Markiewicz Z. (1988) Characterization of cell wall murein of *Thiobacillus versutus*. Acta Microbiol. Polon. 37, 6-16.

Baj, J., Grabowska, I., Wolska, I. and Markiewicz, Z. (1991) The murein of *Thiobacillus versutus*. Acta Mic robiol. Polon. 40, 27-35.

Baj, J., Grabowska, I. and Markiewicz Z. (1992) N-nonsubstituted glucosamine residues and other modifications in murein of the obligatory chemolithotroph *Thiobacillus neapolitanus*. Res. Microbiol. 143, 47-54.

Friedrich, C.G. and Mitrenga, G. (1981) Oxidation of thiosulfate by *Paracoccus denitrificans* and other hydro gen bacteria. FEMS Microbiol. Lett. 10, 209-212.

Glauner, B. (1988) Separation and quantification of muropeptides with high-performance liquid chromatography. Analyt. Biochem. 172, 451-464.

Hadzija, O. (1974) A simple method for the quantitative determination of muramic acid. Analyt. Biochem.60 512-517.

Hestrin, S. (1949) The reaction of acetylcholine and other carboxylic acid derivatives with hydroxylamine and its analytical application. J. Biol. Chem. 180, 249-261.

Holt , S.C., Shively, J.M. and Greenawalt, J.W. (1974) Fine structure of selected species of the genus *Thiobacillus* as revealed by chemical fixation and freeze etching. Can. J. Microbiol. 20, 1347-1251.

Jürgens, U. and Weckesser, J. (1986) Polysaccharide covalently linked to the peptidoglycan of the cyanobacterium *Synechocystis* sp. strain PCC 6714, J. Bacteriol. 168, 568-573.

Jürgens, U.J., Rieth,B. and Weckesser,J. (1987a) Partial lack of N-acetyl substitution of glucosamine in the peptidoglycan of the budding, phototrophic *Rhodomicrobium vannieli*.. Z. Naturforsch. 42c, 1165-1170.

Jürgens, U.J., Meissner, J., Fischer, J., König, W. and Weckesser, J. (1987b) Ornithine as a constituent of the peptidoglycan of *Chloroflexus aurantiacus*, diaminopimelic acid in that of *Chlorobium vibrioforme f. thiosulfatophilum*. Arch. Microbiol. 148, 72-76.

Kelly, D. and Harrison, A.P. (1989) Genus *Thiobacillus*, in "Bergey's Manual of Systematics Bacteriology" (Folt, J.G., Ed.), pp. 1842-1858. Williams and Wilkins, Baltimore, Hong Kong, London.

Liu, T.Y. and Gotschlich, E.C. (1967) Muramic acid phosphate as a component of the mucopeptide of gram -positive bacteria. J. Biol. Chem. 242, 471-476.

Lowry, O.H., Roberts, N.R., Leiner, K.Y., Wu, M.-J. and Farr, L. (1954) The quantitative biochemistry of the brain. Chemical Methods. J. Biol. Chem. 207, 1-17.

Markiewicz, Z., Glauner, B. and Schwarz, U. (1983) Murein structure and lack of DD- and LD-carboxypeptidase activities in *Caulobacter crescentus*. J. Bacteriol. 156, 649-655.

Primosigh,J., Pelzer, H., Maass, D. and Weidel, W. (1961) Chemical characterization of muropeptides released from the *E. coli* B cell wall by enzymic action, Biochim. Biophys. Acta 46, 68-80.

Rogers, H.J., Perkins, H.R. and Ward, J.B. (1980) "Microbial Cell Walls and Membranes". Chapman and Hall, London, New York.

Schleifer, K.H. (1985) Analysis of the chemical composition and primary structure of murein, in "Methods in Microbiology, Vol. 18" , pp. 123-156. Academic Press, Inc., London.

Schleifer, K.H. and Seidl, P.H. (1985) Chemical composition and structure of murein, in "Methods in Microbiology, Vol. 18" , pp. 201-219. Academic Press, Inc., London.

Schmelzer, E., Weckesser, J., Warth, R. and Mayer, H. (1982) Peptidoglycan of *Rhodopseudomonas viridis*, partial lack of N-acetyl substitution of glucosamine. J. Bacteriol. 149, 151-155.

Soby, L.M. and Johnson, P. (1981) Determination of Asparagine and glutamine in polypeptides using bis(1. 1-trifluoroacetoxy)iodobenzene. Analyt. Biochem. 113, 149-153.

Thompson, J.S. and Shockman, P. (1968) A modification of the Park and Johnson reducing sugar determination suitable for the assay of insoluble material: its application to bacterial cell walls. Analyt. Biochem. 22, 260-268.

Tsuji , A., Kinoshita, T. and Hoshono, M. (1969) Microdetermination of hexosamines. Chem. Pharm. Bull 17, 217-218.

Yokota, A., Schlecht, S and Mayer, H. (1987) Lipopolysaccharides of chemolithotrophic bacteria *Thiobacillus versutus* and a related *Thiobacillus* species. FEMS Microbiol. Lett. 44, 197-201.

PHENOTYPIC, BIOCHEMICAL AND STRUCTURAL ANALYSIS OF "S-LAYER" MUTANTS FROM *Thermus thermophilus* HB8

Iñigo Lasa, José R. Castón, Luis A. Fernández-Herrero, Myrian de Grado, Garbiñe Olabarría, José L. Carrascosa, M.A. de Pedro, and J. Berenguer

Centro de Biología Molecular
Universidad Autónoma de Madrid
28049, Madrid

INTRODUCTION

S-layers are crystalline proteinaceous structures that surround a high number of wild type bacteria, being specially frequent among thermophiles (Sleytr and Messner, 1983). Their role and biochemical properties have been discussed extensively by a number of authors (Sàra and Sleytr, 1977; Baumeister and Engelhardt, 1987; Hövmoller *et al.*, 1988), and, from these analyses, it could be concluded that there exist, at least, two well differentiated classes of S-layers, depending on the degree of interaction with the cell envelope. The first group is represented by *bona fide* Gram positive organisms, such as *Bacillus brevis*, *B. stearothermophilus* etc.., in which the crystalline structure can be easily extracted from the envelope (Sleytr and Messner,1983). In these bacteria, defective mutants in the S-layer synthesis with no apparent phenotypic defects can be obtained by spontaneous mutations (Ishiguro *et al.*, 1981). Furthermore, it has been suggested that the presence of the S-layer is a counter-selected character under laboratory conditions for this group of organisms (Baumeister and Engelhardt, 1987).

A second class of S-layers are found in bacteria such as *Deinococcus*, and thermoacidophilic Achaebacteria, in which the interaction between the S-layer and the envelope is so strong that they could be separated only after very destructive procedures (Baumeister *et al.*, 1982). The genus *Thermus*, and in particular *T. thermophilus* HB8, belongs to this second group of microorganisms, as it contains an S-layer (Castón *et al.*, 1988) that can be solubilized in neutral detergents only after extensive digestion of the peptidoglycan (Faraldo *et al.*, 1988). During the last years, we have been studying the hexagonal S-layer of *T.thermophilus* HB8, from which we have purified (Faraldo *et al.*, 1988), cloned (Faraldo *et al.*, 1991) and sequenced (Faraldo, 1991) its main component, a protein with an apparent electrophoretic mobility of 100 kDa (P_{100}).

In order to obtain information about the role of the S-layer in this bacterium, we endeavored to obtain defective mutants by insertional inactivation of the gene responsible for its synthesis (*slpA*). For this purpose, we have developed a thermostable and selectable marker (*kat*) that, as a single copy, confers to *Thermus* a resistance level to kanamycin of 20 μg/ml (Lasa *et al.*, 1992). This marker was inserted *in vitro* into the structural sequence of the S-layer gene (*slpA*), and the construction was further used to transform *T.thermophilus* HB8. As the plasmid used for the construction (pUC) was unable to replicate in *Thermus*, selection of kanamycin resistance could only be achieved by insertion of the *kat* gene into the chromosome after homologous recombination.

In this report, we describe the most important properties of such S-layer mutants. We also describe here the surprising presence of crystalline structures in these mutants, apparently not attached to their envelopes.

Bacterial Growth and Lysis, Edited by M.A. de Pedro
et al., Plenum Press, New York, 1993

RESULTS AND DISCUSSION

Construction of S-layer Mutants

In order to ensure the expression in *Thermus* of the thermostable kanamycin adenyl transferase, we developed a fusion gene in which the resistance marker was controlled by the promoter and the ribosome binding site of *slpA*. Furthermore, we added to both ends of this fusion gene duplicate recognition sites for the restriction enzymes PstI, BamHI, and XmaI, in such a way that it could be purified after digestion with any of them.

Taking advantage of the presence in the S-layer gene of two close PstI sites, overlapping with amino acid positions 261 and 288 (Faraldo, 1991), we inserted *in vitro* the resistance cartridge between them. In Fig. 1 we show the construction obtained, pMF45KAT, that was used then to transform wild type cells of *T. thermophilus* HB8. After 48 hours of incubation at 70°C, most of the colonies, grown on plates containing 10 μg/ml of kanamycin, were proven to be defective for the P_{100} protein by SDS-PAGE (data not shown).

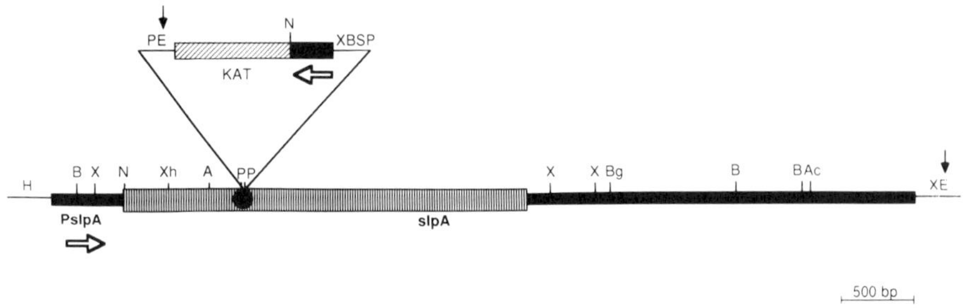

Figure 1. Restriction map of the insertional form of *slpA* used to obtain the S-layer mutants. The structural part and the position and transcription sense of the promoter of *slpA* are labeled with vertical bars and white arrows, respectively. Symbols: A) *AatII*; Ac) *AccI*; B) *BamHI*; Bg) *BglII*; E) *EcoRI*; H) *HindIII*; N) *NdeI*; P) *PstI*; S) *SalI*; X) *XmaI*; Xh) *XhoI*.

Biochemical Analysis of S-layer Mutants

All the P_{100}^- mutants obtained after the transformation step showed identical protein patterns when stained with Coomassie Blue (data not shown). In this pattern, the presence of a 50 kDa protein, not present in the wild type membranes, was as evident as the disappearance of the P_{100} protein was (Fig. 5). In order to know the possible relationships between these proteins and the P_{100}, we compared those from two of the mutants with those of the wild type for their labeling by specific antiserum against P_{100}. The results of this experiment is shown in Fig.2A.

A first important observation from Fig.2A was the labeling in the mutants of three bands of 50, 36 and 22 kDa, respectively. Their low level of detection suggested that, if any of these bands were related with native P_{100}, they should have lost very important epitopes with respect to this. In order to probe such hypothesis, we tried to microsequence the amino terminus of the 50 kDa protein after selective extraction with detergents, electrophoretic separation, and transference to a immobilon-P membrane. However, the results of this experiment demonstrated that, as in the case of native P_{100} (Faraldo *et al*., 1991), the amino terminus of the 50 kDa protein was blocked against the Edman degradation, impeding confirmation of its origin. Nevertheless, the idea of a relationship between the labeled bands of the mutants and the P_{100} was further supported by the results of an experiment, not shown here, in which total proteins from *E. coli* cells, transformed with pMF45KAT, were assayed by western blot with antiserum against P_{100}. Such results demonstrated that, also in the mesophile, low amounts of a 50 and a 15 kDa proteins could be detected, suggesting that these proteins were synthesized from the construction used to obtain the mutants.

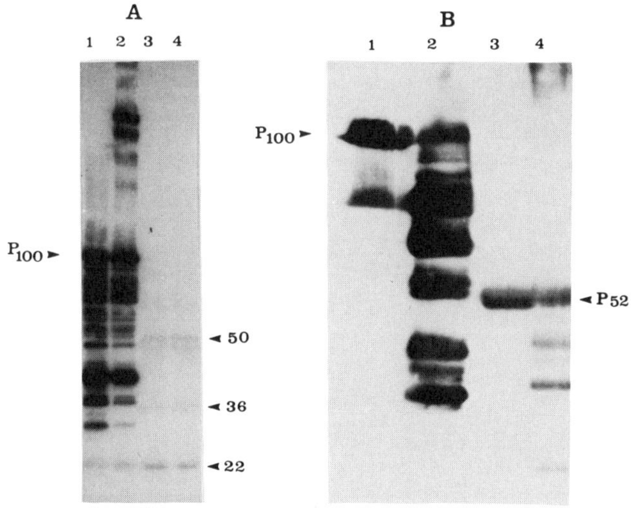

Figure 2. A) Membrane proteins from wild type cells, treated (1) and untreated with EDTA (2), and from two P_{100} mutants (3,4), were assayed by western blotting with antiserum against native P_{100} and iodinated protein A. B) Gel pieces from a SDS-PAGE, containing P_{100} (1,2) or a 52 kDa product (P_{52}) of *slpA* expressed in *E. coli* (3,4), were incubated for 1 hour in the presence (2,4) or absence (1,3) of 1% SDS, and subsequently run in a second SDS-PAGE. Products related to P_{100} were then detected by western blotting as in part A.

In addition to this, a deletion of the carboxi-terminal part of *slpA* from pMF45KAT did not prevent the expression in *E.coli* of low amounts of these proteins. From this, we deduced that they were synthesized from the amino terminal part of *slpA*. Furthermore, as mentioned before, the blockade to Edman degradation of both, the 50 kDa protein of the mutants, and the P_{100} of the wild type, reinforced our suspicion about the nature of, at least, the 50 kDa protein as a product of the amino terminus of *slpA*.

A second important observation from the results seen in Fig. 2A was the presence in the wild type of a high number of bands recognized by antiserum against P_{100}, and their disappearance in the mutants. This strongly suggested again the nature of such bands as *slpA* products, which has already been demonstrated for the high Mr bands, previously identified as oligomers of P_{100} stabilized by calcium (Berenguer *et al.*, 1988). In order to demonstrate the origin of the low Mr bands labeled with antibodies in the wild type, we developed an experiment whose results are shown in Fig. 2B. In this experiment, two gel pieces from a SDS-PAGE containing P_{100} (1, 2) or a 52 kDa amino terminal fragment of P_{100} (3,4), expressed in *E. coli* from plasmid pMF51 (Faraldo *et al.*, 1991), were incubated at 60°C in the presence (2,4) or absence of 1% (w/v) SDS, and run again in a second dimension SDS-PAGE. Proteolytic fragments were detected by western blot with the same antibodies used for the experiment of Fig. 2A. From these results, it could be concluded that P_{100} is self degraded when treated with the detergent, even when expressed in *E. coli*. However, these results do not imply that any of the fragments observed in wild type cells were produced by the uncontrolled activity of membrane proteases.

Phenotypic Analysis of S-layer Mutants

The microscopic analysis of the mutants, shown in Fig. 3B, revealed important morphological alterations. First, the cells appeared as filaments without apparent internal separations. Furthermore, very frequently these filaments were not straight as exponentially growing wild type cells, but presented an spectacular twisted aspect. These results suggest

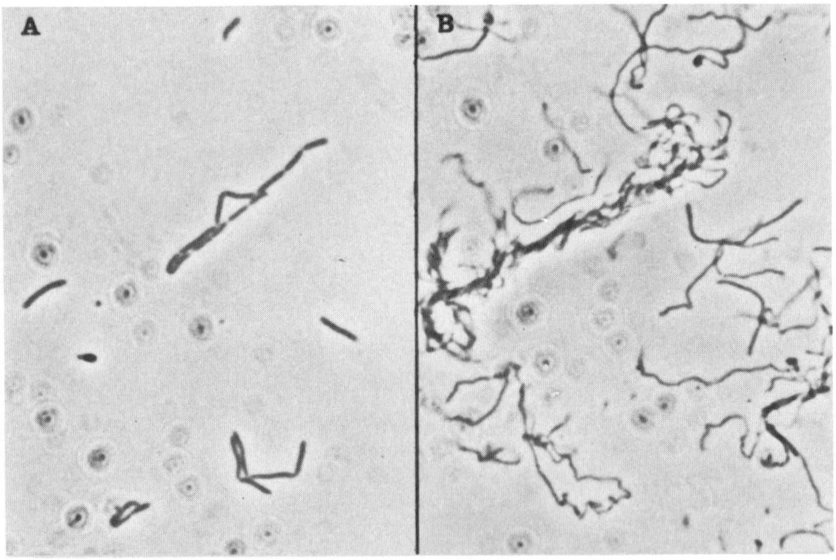

Figure 3. Phase contrast micrographs of wild type cells (A) and S-layer mutants (B) of *T. thermophilus* HB8.

that the S-layer of *T. thermophilus* is a morphogenetic element very important for the correct shape and division of this thermophile.

Another important morphological characteristic of these mutants was their frequent appearance in groups. Although very difficult to see by phase contrast microscopy, these groups are in most cases surrounded by a membrane-like structure, similar to the "rotund bodies" described by Brock (1978). In order to characterize such membranes, thin sections of a mutant culture were analyzed by electron microscopy. The results, shown in Fig. 4A, clearly demonstrated that a membrane-like layer effectively surround groups of cells.

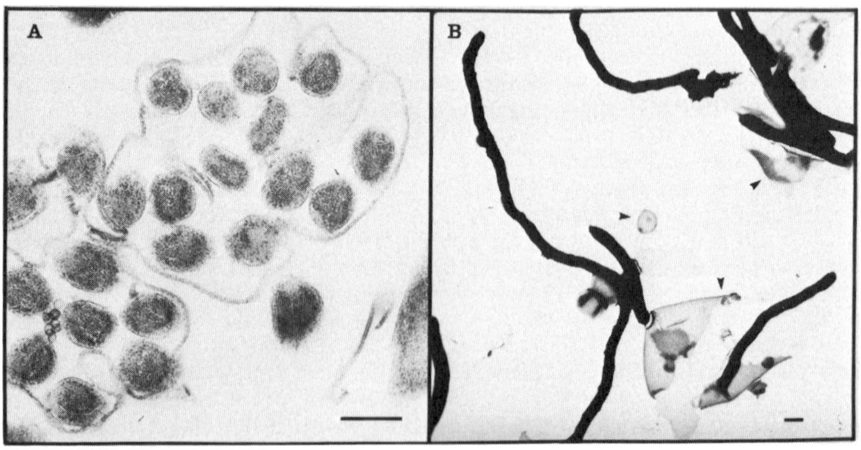

Figure 4. Thin section (A) and negatively stained whole cells preparation (B) of a S-layer mutant from *T. thermophilus* HB8. Bars represent 200 nm.

As Fig. 4A did not allow us to define the possible presence of any periodic structure, we decided to analyze whole cells directly by negative staining. The results of these assays, shown in Fig. 4B, demonstrated that the common membrane collapsed and broke apart (labeled with arrowheads), allowing the observation of planar projections. This permitted high resolution views as that shown in Fig. 5B. Surprisingly, this image demonstrated the presence of an hexagonal crystalline structure, practically indistinguishable from those obtained from P_{100}-rich membrane fractions of the wild type (Fig. 5C).

In order to determine if the 50 kDa protein of the mutants had something to do with the crystalline layer observed in Fig. 5B, we purified membranes from the mutants (Fig. 5A, lane 2), and extracted them with Triton X-100 (TX100) and EDTA, a method routinely followed to prepare native S-layers. The insoluble fraction (lane 3) was subsequently extracted with TX100, and the new insoluble fraction obtained was analyzed by electron microscopy (Fig. 5D) and SDS-PAGE (lane 6). As we expected, the fraction of the mutant membranes that showed a crystalline layer was essentially composed by the 50 kDa protein. The presence of a second band of slightly lower Mr could be due to degradation thereof, as the amount of this second band increased dramatically after prolonged treatment with hot detergent. When the same treatment was applied to membranes of the wild type, the presence of the P_{100} protein as the major component was evident (lane 7). These results clearly suggest that the 50 kDa protein, in addition to being recognized by antiserum against P_{100}, has biochemical properties very similar to those of native S-layer protein.

CONCLUSIONS

The analysis of the mutants described in this report demonstrated that the S-layer is extremely important for the correct growth and shape of *T. thermophilus* HB8. Nevertheless, we don't know if the low rate of growth of the mutants (75 min, about twice the doubling time of the wild type) is a consequence of their tendency to form aggregates, in which the availability of nutrients could be limited, or it really represents the intrinsic

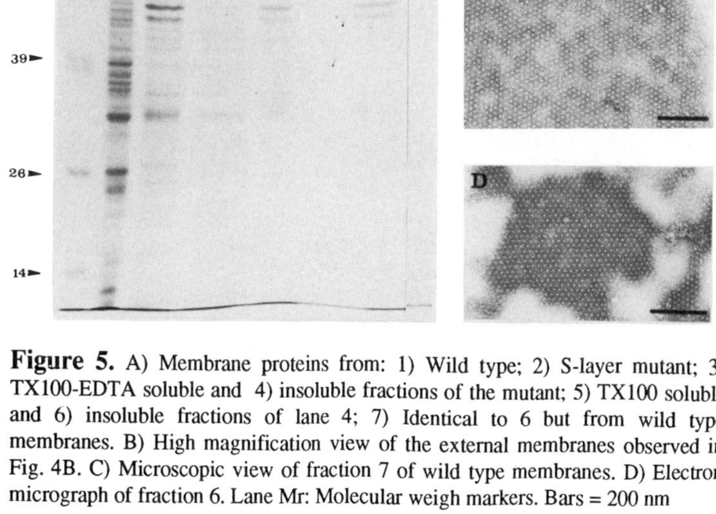

Figure 5. A) Membrane proteins from: 1) Wild type; 2) S-layer mutant; 3) TX100-EDTA soluble and 4) insoluble fractions of the mutant; 5) TX100 soluble and 6) insoluble fractions of lane 4; 7) Identical to 6 but from wild type membranes. B) High magnification view of the external membranes observed in Fig. 4B. C) Microscopic view of fraction 7 of wild type membranes. D) Electron micrograph of fraction 6. Lane Mr: Molecular weigh markers. Bars = 200 nm

necessity for the presence of the S-layer. In any case, the presence of twisted filaments outside the external membranes supports the second interpretation.

The nature of the external membrane observed in the mutants, remains difficult to explain. Its seems to be made by fragments of P_{100} that are expressed by the mutant. Tentatively, we have identified these fragments as an amino terminal part of the gene, although their protection against the Edman degradation still impedes to confirm this interpretation. Anyway, the reason by which a fragment that represent about a half of P_{100} could form a crystalline structure which projection image is indistinguishable from that of native S-layer, remains unknown. We hope that, in the future, high resolution images of thin sections of the mutants could definitively confirm our working hypothesis: The external membrane of the "rotund bodies" that appears in the mutants represents a defective S-layer, in which the deletion of its carboxyl terminus prevented the attachment to the envelope, thus separating it apart from the whole cell.

ACKNOWLEDGEMENTS

This work has been supported by grant BIO91-0523 from the CICYT and by an institutional grant from "Fundación Ramón Areces". I. Lasa, J. R. Castón, and M. de Grado are the holders of fellowships from F.P.I. L. A. Fernández-Herrero and G. Olabarría are supported by fellowships from the Comunidades Autónomas of Madrid and País Vasco, respectively.

REFERENCES

Baumeister, W., Karrenberg, F., Rachel,R., Engel,A., Ten Heggeler,B. and Saxton,W.O. (1982) The major cell envelope protein of *Micrococcus radiodurans* (R1). Structural and chemical characterization. Eur. J. Biochem. 125, 535-544.

Baumeister, W. and Engelhardt, H. (1987) Three-dimensional structure of bacterial surface layers. In "Electron microscopy of proteins" (Harris,J. R., and Horne R. W.,Eds.), pp. 110-154. Academic Press. London.

Brock, T. D. (1978) The genus *Thermus* In "Thermophilic microorganisms and life at high temperatures". (Brock,T.D., Ed). pp. 72-91. Springer-Verlag, New York.

Castón, J. R., Carrascosa, J. L., de Pedro, M. A. and Berenguer, J. (1988) Identification of a crystalline surface layer on the cell envelope of the thermophilic eubacterium *Thermus thermophilus* HB8. FEMS Microbiol. Lett. 51, 225-230.

Faraldo, M. M., de Pedro, M. A. and Berenguer, J. (1988) Purification, composition and Ca^{2+}-binding properties of the monomeric protein of the S-layer of *Thermus thermophilus* HB8. FEBS. Lett. 235, 117-121.

Faraldo, M. M. (1991) Ph Doctoral Thesis. Universidad Autónoma de Madrid.

Faraldo, M. M., de Pedro, M. A. and Berenguer, J. (1991) Cloning and expression in *E.coli* of the structural gene coding for the monomeric protein of the S-layer of *Thermus thermophilus* HB8. J. Bacteriol. 173, 5346-5351.

Hövmoller, S., Sjögren, A. and Wang, D. N. (1988) The structure of crystalline bacterial surface layers. Prog. on Biophys. and Mol. Biol. 51, 131-163.

Ishiguro, E. E., Kay, W., Ainsworth, T., Chamberlain, J. B., Austin, R. A., Buckley, T. J. and Trust, T. J. (1981) Loss of virulence during culture of *Aeromonas salmonicida* at high temperature. J. Bacteriol. 148, 333-340.

Lasa,I., Castón,J.R., Fernández-Herrero, L.A., de Pedro, M.A. and Berenguer, J. (1992) Insertional mutagenesis in the extreme thermophilic eubacteria *Thermus thermophilus* HB8. Mol. Microbiol. 6, 1555-1564.

Sára, M. and Sleytr, U. B. (1987) Molecular sieving through S-layer of *Bacillus stearothermophilus* strains. J. Bacteriol. 169, 4092-4098.

Sleytr, U. B. and Messner,P. (1983) Crystalline surface layers on bacteria. Ann. Rev. Microbiol. 37, 311-339.

CRYSTALLINE BACTERIAL CELL SURFACE LAYERS AND THEIR APPLICATION POTENTIALS

Paul Messner, Margit Sára, Dietmar Pum, and Uwe B. Sleytr

Zentrum für Ultrastrukturforschung und Ludwig-Boltzmann-Institut
für Molekulare Nanotechnologie
Universität für Bodenkultur
A-1180 Wien, Austria

INTRODUCTION

Crystalline cell surface layers (S-layers) have been identified as the outermost envelope component in hundreds of different species of every taxonomical group of walled eubacteria and archaebacteria (Fig. 1) by electron microscopy of either freeze-etched, freeze-dried and shadowed, negatively stained, or thin sectioned preparations (for reviews see Sleytr, 1978; Sleytr and Messner, 1983, 1988 *a*; Sleytr *et al.*, 1988; Messner and Sleytr, 1992).

ULTRASTRUCTURE, CHEMISTRY, AND GENETIC STUDIES OF S-LAYERS

The regular arrays are generally composed of a single homogeneous protein or glycoprotein subunit (Sleytr and Messner, 1988 *b*) although recently data have been presented showing that the S-layer lattice can consist of different subunits (Takeoka *et al.*, 1991; Yang *et al.*, 1992). Other studies have shown that several superimposed, immunologically different S-layers can be present on bacterial cells (Beveridge and Murray, 1975; Tsuboi *et al.*, 1982; Kist and Murray, 1984; Smith and Murray, 1990; Sára *et al.*, 1990). The molecular weights of the subunits as determined by sodium dodecylsulphate polyacrylamide gel electrophoresis range from approximately 35,000 to 200,000 (Fig. 2) and the lattices can exhibit oblique (p2), square (p4) or hexagonal (p6) symmetry (for reviews see Sleytr and Messner, 1988 *a*; Messner and Sleytr, 1992). Dependent on the lattice type, the morphological units in the crystalline arrays consist of two, four, or six identical monomers, respectively, with a centre-to-centre spacing ranging from 3 to 32 nm (Fig. 3). High resolution electron microscopical studies, in combination with digital image reconstruction procedures have revealed that S-layers form highly porous mesh works (for reviews see Baumeister and Engelhardt, 1987; Hovmöller *et al.*, 1988). Due to the regular arrangement of the constituent subunits, the pores of the lattice have both identical size and morphology and well defined physicochemical properties.

Comparison of amino acid analyses and genetic studies on numerous S-layers have

Bacterial Growth and Lysis, Edited by M.A. de Pedro *et al.*, Plenum Press, New York, 1993

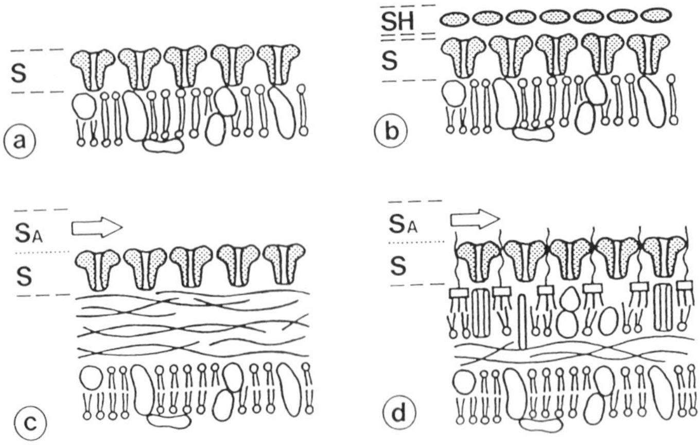

Figure 1. Schematic illustrations of the major classes of prokaryotic cell envelopes containing S-layers. (a) The cell envelope structure of Gram-negative archaebacteria with crystalline S-layers as an exclusive cell wall component. (b) S-layers with an additional regularly arranged sheath. (c) The cell envelope as observed in Gram-positive eubacteria and archaebacteria (containing peptidoglycan or pseudomurein, respectively). (d) S-layers in the cell envelope of Gram-negative eubacteria. S indicates the crystalline surface layer, S_A the location for possible additional S-layer, and SH a sheath. (From Messner and Sleytr, 1992.) With permission of Academic Press Inc., London.

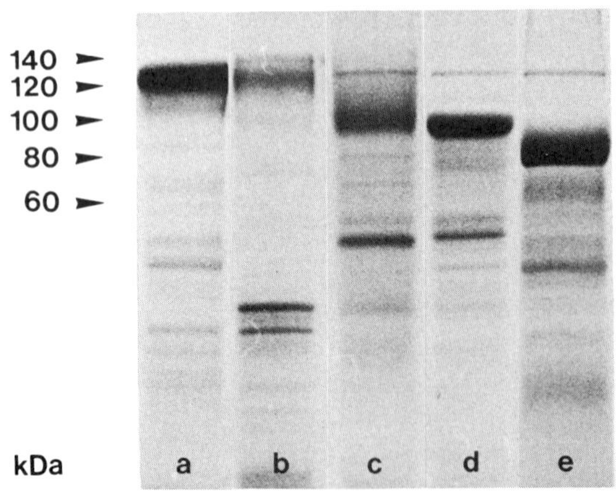

Figure 2. Sodium dodecylsulphate polyacrylamide gel electrophoresis of SDS-solubilized whole cell extracts of *Clostridium thermohydrosulfuricum* strains (Matteuzzi *et al.*, 1978) showing the considerable differences in the molecular mass of S-layer subunits (heavy high-molecular mass bands) even among closely related strains. Lane a, L111-69; b, L77-66; c, H101-69; d, S102-70; e, T15-66; molecular weights given x 10^3.

shown that the constituent subunits are usually composed of weakly acidic proteins (for reviews see Sleytr and Messner, 1983; Messner and Sleytr, 1992). The content of hydrophobic amino acids is generally high whereas the cysteine or methionine content is low. In glycosylated S-layers the glycan chains can be linked to the protein moiety both, by *N*- and *O*-glycosidic linkages (Messner and Sleytr, 1988 *a*; Lechner and Wieland, 1989; Messner *et al.*, 1992 *a*). Most of the glycan chains studied so far are composed of linear or branched repeat sequences of two to six monosaccharide units (Table 1) which include a wide range of hexoses, deoxy sugars and amino sugars, uronic acids and even sulfate or phosphate residues as constituents. The glycan chain length varies strain-specifically from a few sugars up to approximately 150 monosaccharide residues (for review see Messner and Sleytr, 1991).

Molecular cloning and characterization of the genes encoding the crystalline S-layer (glyco)proteins is important for the understanding of both, the precise properties of these macromolecules and the mechanisms involved in their biosynthesis, translocation across membranes, assembly at the external surface and, where present, glycosylation of S-layer proteins. Up to now nucleotide sequences of four archaebacterial and nine eubacterial S-layer genes are partly or completely known (for review see Messner and Sleytr, 1992). Homology comparisons have shown that, with the exception of closely related strains (Kay *et al.*, 1984, Peters *et al.*, 1989; Ebisu *et al.*, 1990), there is relatively little overall homology between the characterized S-layer proteins. These observations concur with the diversity of S-layers observed by ultrastructural and biochemical characterizations (for reviews see Sleytr and Messner, 1988 *a*; Messner and Sleytr, 1992). Together the data reflect the non-conservative character of these prokaryotic cell surface macromolecules.

SELF-ASSEMBLY IN THE ABSENCE AND PRESENCE OF SUPPORTING LAYERS

Isolation of S-layers subunits requires breaking of the bonding forces between the proteinaceous subunits and between subunits and the underlying cell wall components (for reviews see Sleytr, 1978; König, 1988). Binding can involve hydrophilic interactions, salt

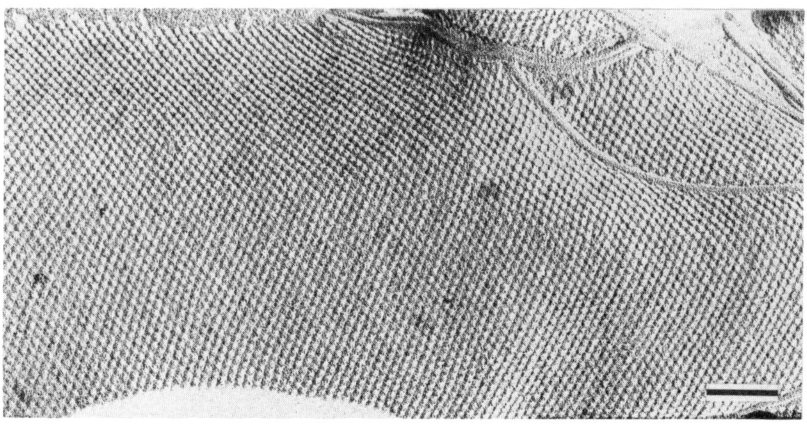

Figure 3. Electron micrograph of a freeze-etched preparation of intact cells of *Clostridium thermohydrosulfuricum* L111-69 showing the hexagonal S-layer lattice. Bar = 100 nm.

Table 1. Glycan structures of S-layer glycoproteins.[1]

● *Halobacterium halobium* R_1M_1

```
                 2 SO4(2-)                                  Ala-NH2
                                                               |
→3)-GalNAc-(1→4)-GlcNAc-(1→4)-GalA-(1→3)-GalNAc -(1→N)-Asn
                 6               3                             |
                 ↑               ↑                            Ala
                 1               1                             |
           3-OMe-GalA          Galf                  n        Ser
                                                               |

                                                  |
GlcA-(1→4)-IdA-(1→4)-GlcA-(1→4)-β-D-Glc-(1→N)-Asn
                                                  |
                                                  X
       3 SO4(2-)                                  |
                                               Thr/Ser
                                                  |

                            |
α-D-Glc-(1→2)-Gal-(1→O)-Thr
                            |
```

● *Haloferax volcanii* DS2

N-Glycans, Structure not analyzed

```
                        |
Glc-(1→2)-Gal-(1→O)-Thr
                        |
```

● *Bacillus stearothermophilus* NRS 2004/3a

```
                                                                                 |
→2)-α-L-Rhap-(1→2)-α-L-Rhap-(1→3)-β-L-Rhap-(1→ n~50                 Rha-(1→N)-Asn
                                                                                 |
and

→4)-β-ManpA-2,3-diNAc-(1→3)-α-GlcpNAc-(1→4)-β-ManpA-2,3-diNAc-(1→6)-α-Glcp-(1→ n~15
```

● *Bacillus alvei* CCM 2051

```
→3)-β-D-Galp-(1→4)-β-D-ManpNAc-(1→
                          6
                          ↑
                          1
                      α-D-Glcp          n~40
```

● *Clostridium symbiosum* HB25

→6)-α-D-ManpNAc-(1→4)-β-D-GalpNAc-(1→3)-α-D-BacpNAc-(1→4)-α-D-GalpNAc-(1→PO_3H)→ n~15

● *Clostridium thermohydrosulfuricum* L111-69, L110-69

→4)-α-D-Manp-(1→3)-α-*L*-Rhap-(1→ n~30 *O*-glycosidic bond →Tyr bridges, hydrogen bonds, or direct electrostatic interactions (Masuda and Kawata, 1985; Jaenicke *et al.*, 1985; Sára and Sleytr, 1987 *a*). Different isolation methods and regimes have been developed for archaebacteria (for review see König and Stetter, 1986), Gram-negative eubacteria (for reviews see Koval and Murray, 1984; Dubreuil *et al.*, 1988) and Gram-positive eubacteria (for review see Messner and Sleytr, 1988 *b*). Upon removal of the disrupting agent used for the isolation, the isolated S-layer subunits from different organisms have been shown to assemble merely by an entropy-driven process into two-dimensional arrays exhibiting the same lattice dimensions as those observed on intact cells (for review see Sleytr and Messner, 1989).

Studies on the self-assembly process of isolated subunits in suspension using S-layers from thermophilic Bacillaceae have shown that the formation of the crystalline arrays is initiated by a rapid nucleation of subunits into oligomeric precursors consisting of several unit cells, and a subsequent slow recrystallization into larger assembly products (Jaenicke *et al.*, 1985). The assembly products can form flat sheets, open ended cylinders or even

---

● *Clostridium thermohydrosulfuricum* L77-66, L92-71 (Altman *et al.*, 1992)

→3)-α-D-Gal*p*NAc-(1→3)-α-D-Gal*p*NAc-(1→ *O*-glycosidic bond →Tyr
4
↑
1
β-D-Man*p*
2
↑
1
α-D-Glc*p*NAc n~30

● *Clostridium thermohydrosulfuricum* S102-70 (Messner *et al.*, 1992 *a*)

β-D-Gal*f*-(1→3)-α-D-Gal*p*-(1→2)-α-L-Rha*p*-(1→3)-α-D-Man*p*-(1→3)-α-L-Rha*p*-(1→3)-β-D-Glc*p*-(1→*O*)-Tyr

● *Clostridium thermosaccharolyticum* D120-70

→3)-β-D-Man*p*-(1→4)-α-L-Rha*p*-(1→3)-α-D-Glc*p*-(1→4)-α-L-Rha*p*-(1→
6 (on β-D-Man*p*) ↑ 1 β-D-Glc*p*; 2 (on α-L-Rha*p*) ↑ 1 α-D-Gal*p* n

and

→4)-β-D-Glc*p*NAc-(1→3)-β-D-Man*p*NAc-(1→
4
↑
1
(α-D-Gal*p*)$_{0.5}$ n

Abbreviations: Glc*p*, glucose (pyranose form); Gal*f*, galactose (furanose form), Man, mannose; Rha, rhamnose; GlcNAc, GalNAc, ManNAc, *N*-acetylhexosamines; GlcA, GalA, ManA, hexuronic acids; IdA, iduronic acid; 3-*O*Me-GalA, 3-*O*-methyl-galacturonic acid; BacNAc, *N*-acetylbacillosamine (2-*N*-acetyl-4-amino-2,4,6-trideoxy glucose); SO_4, sulfate; PO_4, phosphate; Asn, asparagine; Ser, serine; Thr, threonine; Ala, alanine; Tyr, tyrosine; *X*, interchangeable amino acid.

[1]Modified from Messner and Sleytr (1992). With permission of Academic Press Inc., London.

closed vesicles (for review see Sleytr and Messner, 1989). Double layer assembly products can also be formed when the two constituent monolayers are either bound with their inner or outer faces (Kist and Murray, 1984; Messner *et al.*, 1986; Pum *et al.*, 1989 *a*).

S-layer subunits or S-layer fragments from both Gram-positive and Gram-negative eubacteria are able to reattach to those cell walls from which they have been isolated (homologeous reattachment), or to other cell walls (heterologeous reattachment) (Buckmire and Murray, 1976; Sleytr, 1978; Masuda and Kawata, 1985). Recently, reattachment experiments were performed on *Campylobacter fetus* cells. Different to the reattachment experiments on Gram-positive *Clostridium* strains (Sleytr, 1975; Sleytr and Plohberger, 1980), binding of S-layer protein turned out to be highly serotype-specific in Gram-negative *Campylobacter* strains (Yang *et al.*, 1992). Since these S-layers require divalent cations for binding to lipopolysaccharides, reattachment was only possible on templets containing the correct serotype-specific lipopolysaccharide.

S-layer subunits from different Bacillaceae have shown the ability to recrystallize on solid supports such as silica, glass, mica, carbon or synthetic polymers upon removal of the disintegrating agents used for their isolation (Pum *et al.*, 1989 *b*; Sleytr *et al.*, 1989, 1992). Environmental conditions similar to those chosen for self-assembly experiments in suspensions were applied. Crystal growth is initiated at several distant nucleation points. The individual monocrystalline patches grow isotropically in all directions until the edge of the neighbouring crystal is reached. The capability of S-layers to recrystallize on solid supports is essential for their use as patterning elements in the molecular nanotechnologies (Pum *et al.*, 1992 *a*; Sleytr *et al.*, 1992).

S-LAYER FORMATION ON GROWING CELLS

Labelling experiments on many Gram-positive and Gram-negative bacteria have demonstrated that once the protomeric subunits are incorporated into the S-layer lattice they virtually undergo no turnover (Thorne *et al.*, 1976; Smit and Agabian, 1982; Howard *et al.*, 1982). An exception is *Bacillus brevis* 47 which overproduces S-layer protein. This material is subsequently shed into the surrounding medium (Yamada *et al.*, 1981). By labelling techniques such as fluorescent antibodies and protein A/colloidal gold, the insertion of newly synthesized S-layers can be followed. Using the indirect fluorescent antibody technique in *Bacillus sphaericus* several zones of insertion of newly synthesized S-layer material were observed along the cylindrical part of the cell (Howard *et al.*, 1982). Investigation of two *Bacillus stearothermophilus* strains possessing different S-layer lattices revealed that on both strains the new S-layer was inserted in helically arranged bands over the cylindrical surface of the cell at a pitch angle related to the orientation of the lattice vectors of the crystalline array (Fig. 4). Minimal insertion was observed at the cell poles of the mother cell but at septal regions and, subsequently, at the newly formed cell poles new S-layer protein was found (Gruber and Sleytr, 1988). Recent investigations on *Lactobacillus helveticus* have shown that after stripping the S-layer from intact cells by lithium chloride treatment (5 M LiCl) new S-layer material appeared all over the cell surface in a random fashion during exponential phase growth (Lortal *et al.*, 1992). The difference to the localized insertion of new S-layer material as observed in Bacillaceae shows that excreted S-layer subunits must be available in the entire peptidoglycan network and not only at the areas of peptidoglycan growth. Different to the lattice growth of Gram-positive organisms in *Caulobacter crescentus,* random incorporation of new S-layers was found on cells which are completely covered with the regular arrays (Smit and Agabian, 1982).

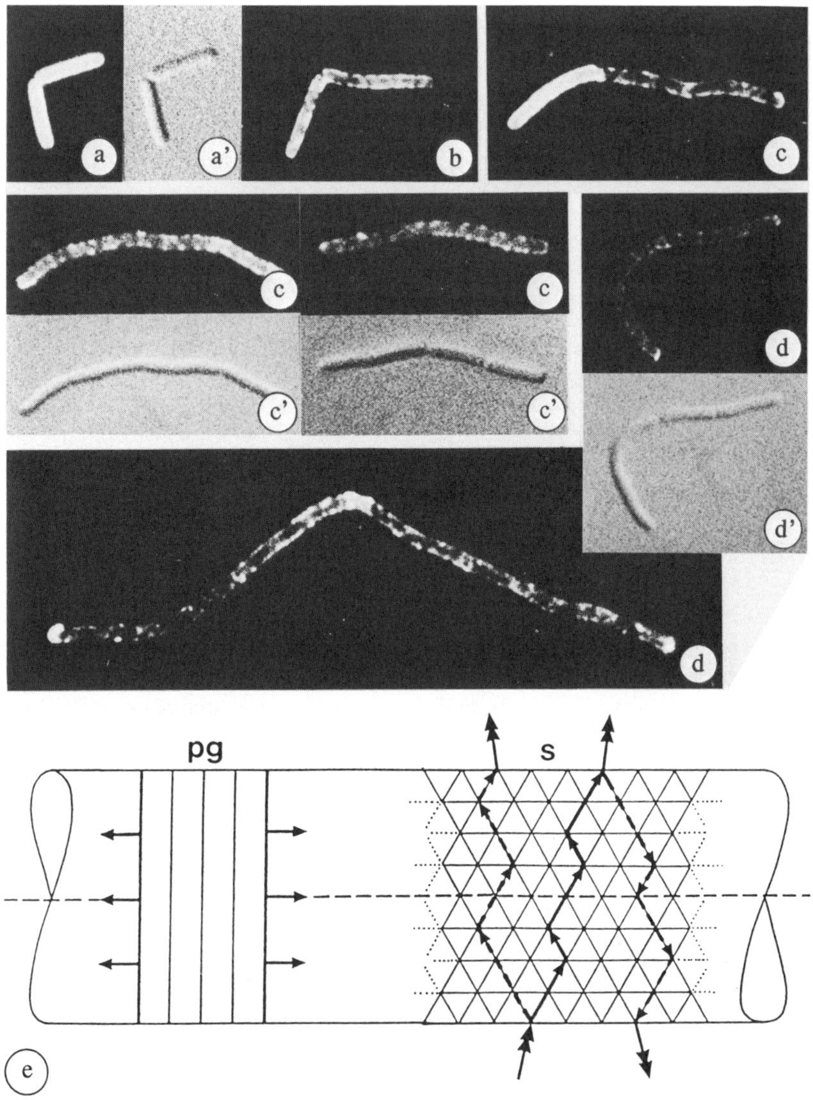

Figure 4. Labelling of the old S-layer of *Bacillus stearothermophilus* PV72 with indirect fluorescent antibodies (a, a') at 0 time; (b) after 15 min; (c, c') after 30 min; (d, d') after 60 min of growth. (a, b, c, d) fluorescence illumination; (a', c', d') phase interference contrast. Magnification x 2,135. Generation time approximately 20 min. (e) Diagramatic representation of the expansion of the peptidoglycan (pg) and the S-layer (s) of *B. stearothermophilus* PV72. Localized insertion of new peptidoglycan material can be expected to occur in bands perpendicular to the longitudinal axis of the rod. Incorporation of new S-layer subunits can take place at sites of sliding dislocations (single-headed arrows). The pitch of the helical bands of new S-layer at growth zones (doubleheaded arrows) is determined by the stress generated in the S-layer lattice by the expanding peptidoglycan. (From Gruber and Sleytr, 1988). With permission of Springer-Verlag, Heidelberg.

SURFACE AND PERMEABILITY PROPERTIES OF S-LAYERS

Two- and three-dimensional image reconstructions of the mass distribution of S-layers have shown a remarkable asymmetry in the topography of the inner and outer face of S-layers (for reviews see Hovmöller *et al.*, 1988; Baumeister *et al.*, 1989). Usually, the inner face is more corrugated than the outer face, which can be demonstrated by unidirectional and rotary shadowing of freeze-dried specimens and three-dimensional computer reconstruction procedures (for reviews see Baumeister and Engelhardt, 1987; Hovmöller *et al.*, 1988; Sleytr *et al.*, 1988). A comparable anisotropy has also been observed with the net surface charge of S-layers. Due to an equal number of amino and carboxyl groups the outer face reveals a neutral surface charge. The inner face, however, is net negatively charged as a consequence of an excess of free carboxyl groups (Sára and Sleytr, 1987 *a*; Sára *et al.*, 1992 *a*).

The adhesive properties of bacteria to all kinds of surfaces are influenced by several parameters including van der Waals forces, electrostatic forces, and hydrophobic interactions (for reviews see Wicken, 1985; Marshall, 1991). Comparison of the cell surface charge and hydrophobicity of an S-layer carrying strain of *Bacillus stearothermophilus* and of an S-layer deficient mutant strain by electrostatic interaction and hydrophobic interaction chromatography has shown significant differences (Gruber and Sleytr, 1991). Both strains possess a hydrophilic surface but the hydrophilicity is more pronounced on the S-layer lacking organism.

Since S-layers represent the outermost cell envelope component and completely cover the cell surface they must possess pores in the lattice large enough for the passage of nutrients, enzymes, and metabolic degradation products. Results from high resolution electron microscopical studies indicate that dependent on the type of the S-layer lattice the pore dimensions can range from 2 to 8 nm with a porosity of the lattice between 30% and 70% (Sjögren *et al.*, 1985; Wildhaber *et al.*, 1987; Sára and Sleytr, 1988; Pum *et al.*, 1992 *a*). Due to the regular arrangement of the constituent subunits all functional groups exposed on the S-layer lattice are in an identical position and orientation (for review see Pum *et al.*, 1992 *a*). Muramidases were frequently used for determining the functional pore diameters (Messner *et al.*, 1984; Sára *et al.*, 1990). For example, lysis of intact cells by the action of lysozyme and mutanolysine indicated pore diameters of the S-layer larger than 3.5 nm. On the other hand, resistance of S-layer carrying Bacillaceae towards muramidases was shown to be caused by muramidase resistant peptidoglycan (Sára *et al.*, 1990). The functional pore diameters were determined in permeability studies using a set of globular test proteins. These studies showed that most S-layers of Bacillaceae possess pores ranging from 3.5 to 4 nm (Sára and Sleytr, 1987 *b*, 1988).

FUNCTION

From the data available it appears that S-layers as porous crystalline arrays completely covering the cell surface have the potential to function as (i) protective coats, molecular sieves, and molecule and ion traps, (ii) structures involved in cell adhesion and surface recognition, and (iii) as frameworks which determine and maintain cell shape (for reviews see Beveridge, 1981, 1991; Sleytr and Messner, 1983, 1988 *b*; Smit, 1987; Koval, 1988; Baumeister *et al.*, 1989; Koval and Hynes, 1991; Messner and Sleytr, 1992).

APPLICATION POTENTIAL OF S-LAYERS

The considerable body of knowledge accumulated on ultrastructure, chemistry, assembly, biosynthesis, genetics, physicochemical properties and pathogenicity of S-layers has led to several applications particularly in the field of biotechnology and biomedicine.

S-layers as Isoporous Ultrafiltration Membranes

Analysis of the mass distribution in S-layer lattices and the results from permeability studies have shown that S-layers are suitable for the production of isoporous ultrafiltration membranes with well-defined molecular weight cut offs (Sára and Sleytr, 1987 *b*, *c*, 1988). For this purpose S-layer fragments are deposited on microfiltration membranes and stabilized by cross-linking (e.g. with glutaraldehyde). By chemical modifications the physicochemical surface properties of the membranes such as surface charge and hydrophobicity can be changed. Immobilized molecules in the pore area of the S-layer lattice cause specific shifts in the molecular sieving properties and the cut off values obtained (Küpcü *et al.*, 1991).

S-layers as Support for the Covalent Attachment of Macromolecules

Since the crystalline arrays of S-layers are composed of identical subunits it is evident that functional groups such as carboxyl, amino and hydroxyl groups must be present on the polypeptide or carbohydrate moiety of each protomer in an identical position and orientation. Crosslinking with glutaraldehyde involves reaction with free amino groups. This leads to an excess of free carboxyl groups and a net negative charge of the protein lattice. Activated carboxyl groups (e.g. by carbodiimide) can subsequently react with free amino groups of functional macromolecules such as enzymes (Sára and Sleytr, 1989; Küpcü *et al.*, 1991). S-layers composed of glycoprotein subunits have an increased potential as immobilization matrix. Since the carbohydrate chains protrude on the surface of the protein lattice at well defined positions (Sára *et al.*, 1989), ligand binding is possible in the third dimension (Sára and Sleytr, 1989; Messner et al., 1991).

High resolution electron microscopical and chemical studies have shown that depending on their molecular dimensions foreign molecules can be bound to S-layer-lattices in a defined number and geometrical distribution. This characteristic feature has led to applications of S-layers as patterning elements and immobilization matrices for functional molecules as required for affinity and enzyme membranes and biosensors (Pum *et al.*, 1989 *b*, 1992 *a*, *b*; Sára *et al.*, 1992 *b*).

S-layers for Vaccine Development

In the last years S-layers have been found in an increasing number of pathogenic microorganisms (for review see Messner and Sleytr, 1992). To fight the respective diseases the use of S-layers as peptide subunit vaccines has been considered. Particular interest is being taken in the development of fish vaccines against *Aeromonas salmonicida* in salmon nurseries (Evenberg *et al.*, 1988; Dubreuil *et al.*, 1990; Kay and Trust, 1991).

Another line of development is directed to the use of S-layers as carrier for conjugate vaccines (Malcolm *et al.*, 1992). Both protein and glycoprotein S-layers were used for the immobilization of selected weakly immunogenic carbohydrate antigens and haptens (Messner *et al.*, 1992 *b*). Immunological tests of S-layer conjugates in mice have shown that S-layers are effective carrier/adjuvants for eliciting T- and B-cell responses to specific antigens (Malcolm *et al.*, 1992).

Future Outlook

The observation that many archaebacteria which grow under extreme environmental conditions possess a simple cell envelope structure consisting exclusively of a cytoplasmic membrane and an S-layer (Fig. 1a), led to the idea of using reconstituted S-layers as carrier for Langmuir-Blodgett lipid film. Such lipid-S-layer composites, in which functional molecules such as pore-forming proteins, proton pumps, receptor molecules or other biologically active molecules can be incorporated, could lead to open a broad spectrum of new applications in the development of biosensors or even more complex molecular machines (Pum *et al*,. 1989b, 1992; Sleytr *et al.*, 1992).

ACKNOWLEDGMENTS

This work was supported by the Fonds zu Förderung der wissenschaftlichen Forschung in Österreich, projects P8922-MOB, S50/02, and S5705-PHY, and the Österreichisches Bundesministerium für Wissenschaft und Forschung.

REFERENCES

Altman, E., Brisson, J.-R., Gagné, S.M., Kolbe, J., Messner, P. and Sleytr, U.B. (1992) Structure of the glycan chain from the surface layer glycoprotein of *Clostridium thermohydrosulfuricum* L77-66. Biochim. Biophys. Acta, in press.

Baumeister, W. and Engelhardt, H. (1987) Three-dimensional structure of bacterial surface layers, in "Electron Microscopy of Proteins" vol. 6, "Membraneous Structure" (Harris, J.R. and Horne, R.W., Eds.), pp. 109-154. Academic Press, London.

Baumeister, W., Wildhaber, I. and Phipps, B.M. (1989) Principles of organization in eubacterial and archaebacterial surface proteins. Can. J. Microbiol. 35, 215-227.

Beveridge, T.J. (1981) Ultrastructure, chemistry, and function of the bacterial wall. Int. Rev. Cytol. 72, 229-317.

Beveridge, T.J. and Graham, L.L. (1991) Surface layers of bacteria. Microbiol. Rev. 55, 684-705.

Beveridge, T.J. and Murray, R.G.E. (1975) Surface arrays on the cell wall of *Spirillum metamorphum*. J. Bacteriol. 124, 1529-1544.

Buckmire, F.L.A. and Murray, R.G.E. (1976) Substructure and in vitro assembly of the outer, structured layer of *Spirillum serpens*. J. Bacteriol. 125, 290-299.

Dubreuil, J.D., Logan, S.M., Cubbage, S., Ni Eidhin, D., McCubbin, W.D., Kay, C.M., Beveridge, T.J., Ferris, F.G. and Trust, T.J. (1988) Structural and biochemical analyses of a surface array protein of *Campylobacter fetus*. J. Bacteriol. 170, 4165-4173.

Dubreuil, J.D., Kostrzynska, M., Austin, J.W. and Trust, T.J. (1990) Antigenic differences among *Campylobacter fetus* S-layer proteins. J. Bacteriol. 172, 5035-5043.

Ebisu, S., Tsuboi, A., Takagi, H., Naruse, Y., Yamagata, H., Tsukagoshi, N. and Udaka, S. (1990) Conserved structures of cell wall protein genes among protein-producing *Bacillus brevis* strains. J. Bacteriol. 172, 1312-1320.

Evenberg, D., de Graaff, P., Lugtenberg, B. and van Muiswinkel, W.B. (1988) Vaccine-induced protective immunity against *Aeromonas salmonicida* tested in experimental carp erythrodermatits. J. Fish Disease 11, 337-350.

Gruber, K. and Sleytr, U.B. (1988) Localized insertion of new S-layer during growth of *Bacillus stearothermophilus* strains. Arch. Microbiol. 149, 485-491.

Gruber, K. and Sleytr, U.B. (1991) Influence of an S-layer on surface properties of *Bacillus stearothermophilus*. Arch. Microbiol. 156, 181-185.

Hovmöller, S., Sjögren, A. and Wang, D.N. (1988) The structure of crystalline bacterial surface layers. Prog. Biophys. Mol. Biol. 51, 131-163.

Howard, L.V., Dalton, D.D. and McCoubrey, Jr., W.K. (1982) Expansion of the tetragonally arrayed cell wall protein layer during growth of *Bacillus sphaericus*. J. Bacteriol. 149, 748-757.

Jaenicke, R., Welsch, R., Sára, M. and Sleytr, U.B. (1985) Stability and self-assembly of the S-layer protein of the cell wall of *Bacillus stearothermophilus*. Biol. Chem. Hoppe-Seyler 366, 663-670.

Kay, W.W. and Trust, T.J. (1991) Form and functions of the regular surface array (S-layer) of *Aeromonas salmonicida*. Experientia 47, 412-414.

Kay, W.W., Phipps, B.M., Ishiguro, E.E., Olafson, R.W. and Trust, T.J. (1984) Surface layer virulence A-protein from *Aeromonas salmonicida* strains. Can. J. Biochem. Cell Biol. 62, 1064-1071.

Kist, M.L. and Murray, R.G.E. (1984) Components of the regular surface array of *Aquaspirillum serpens* MW5 and their assembly in vitro. J. Bacteriol. 157, 599-606.

König, H. (1988) Archaeobacteria, in "Biotechnology" vol. 6b (Rehm, H.-J., Ed.), pp. 697-728. VCH Publishers, Weinheim.

König, H. and Stetter, K.O. (1986) Studies on archaebacterial S-layers. System. Appl. Microbiol. 7, 300-309.

Koval, S.F. (1988) Paracrystalline protein surface arrays on bacteria. Can. J. Microbiol. 34, 407-414.

Koval, S.F. and Hynes, S.H. (1991) Effect of paracrystalline protein surface layers on predation by *Bdellovibrio bacteriovorus*. J. Bacteriol. 173, 2244-2249.

Koval, S.F. and Murray, R.G.E. (1984) The isolation of surface array proteins from bacteria. Can. J. Biochem. Cell Biol. 62, 1181-1189.

Küpcü, S., Sára, M. and Sleytr, U.B. (1991) Chemical modification of crystalline ultrafiltration membranes and immobilization of macromolecules. J. Membrane Sci. 61, 167-175.

Lechner, J. and Wieland, F. (1989) Structure and biosynthesis of prokaryotic glycoproteins. Annu. Rev. Biochem. 58, 173-194.

Lortal, S., van Heijenoort, J., Gruber, K. and Sleytr, U.B. (1992) S-layer of *Lactobacillus helveticus* ATCC 12046: isolation, chemical characterization and re-formation after extraction with lithium chloride. J. Gen. Microbiol. 138, 611-618.

Malcolm, A.J., Messner, P., Sleytr, U.B., Smith, R.H. and Unger, F.M. (1992) Crystalline bacterial cell surface layers (S-layers) as combined carrier/adjuvants for conjugate vaccines, in "Immobilised Macromolecules. Application Potentials in Biotechnology and Biosensors" (Sleytr, U.B., Messner, P., Pum, D. and Sára, M., Eds.), Springer-Verlag, London, in press.

Marshall, K.C. (1991) The importance of studying microbial cell surfaces, in "Microbial Cell Surface Analysis. Structural and Physicochemical Methods" (Mozes, N., Handley, P.S., Busscher, H.J. and Rouxhet, P.G., Eds.), pp. 3-19. VCH Publishers, New York.

Masuda, K. and Kawata, T. (1985) Reassembly of a regularly arranged protein in the cell wall of *Lactobacillus buchneri* and its reattachment to cell walls: chemical modification studies. Microbiol. Immunol. 29, 927-938.

Matteuzzi, D., Hollaus, F. and Biavati, B. (1978) Proposal of neotype for *Clostridium thermohydrosulfuricum* and the merging of *Clostridium tartarivorum* with *Clostridium thermosaccharolyticum*. Int. J. Syst. Bacteriol. 28, 528-531.

Messner, P. and Sleytr, U.B. (1988 *a*) Asparaginyl-rhamnose: a novel type of protein-carbohydrate linkage in a eubacterial surface-layer glycoprotein. FEBS Lett. 228, 317-320.

Messner, P. and Sleytr, U.B. (1988 *b*) Separation and purification of S-layers from gram-positive and gram-negative bacteria, in "Bacterial Cell Surface Techniques" (Hancock, I.C. and Poxton, I.R., Eds.), pp. 97-104. John Wiley & Sons, Chichester.

Messner, P. and Sleytr, U.B. (1991) Bacterial surface layer glycoproteins. Glycobiology 1, 545-551.

Messner, P. and Sleytr, U.B. (1992) Crystalline bacterial cell-surface layers, in "Advances in Microbial Physiology" vol. 33 (Rose, A.H. and Tempest, D.W., Eds.), pp. 213-275. Academic Press, London.

Messner, P., Hollaus, F. and Sleytr, U.B. (1984) Paracrystalline cell wall surface layers of different *Bacillus stearothermophilus* strains. Int. J. Syst. Bacteriol. 34, 202-210.

Messner, P., Pum, D. and Sleytr, U.B. (1986) Characterization of the ultrastructure and the self-assembly of the surface layer of *Bacillus stearothermophilus* strain NRS 2004/3a. J. Ultrastruct. Mol. Struct. Res. 97, 73-88.

Messner, P., Küpcü, S., Sára, M., Pum, D. and Sleytr, U.B. (1991) Characterization and biotechnological application of eubacterial glycoproteins, in "Protein Glycosylation: Cellular, Biotechnological and Analytical Aspects" (Conradt, H.S., Ed.), pp. 111-116. GBF Monographs, vol. 15. VHC-Verlagsgesellschaft, Weinheim.

Messner, P., Christian, R., Kolbe, J., Schulz. G. and Sleytr, U.B. (1992 *a*) Analysis of a novel linkage unit of O-linked carbohydrates from crystalline surface layer glycoprotein of *Clostridium thermohydrosulfuricum* S102-70. J. Bacteriol. 174, 2236-2240.

Messner, P., Mazid, M.A., Unger, F.M. and Sleytr, U.B. (1992 *b*) Artificial antigens. Synthetic carbohydrate haptens immobilized on crystalline bacterial surface layer glycoproteins. Carbohydr. Res., in press.

Peters, J., Peters, M., Lottspeich, F. and Baumeister, W. (1989) S-layer protein gene of *Acetogenium kivui*: cloning and expression in *Escherichia coli* and determination of the nucleotide sequence. J. Bacteriol. 171, 6307-6315.

Pum, D., Sára, M. and Sleytr, U.B. (1989 *a*) Structure, surface charge, and self-assembly of the S-layer lattice from *Bacillus coagulans* E38-66. J. Bacteriol. 171, 5296-5303.

Pum, D., Sára, M. and Sleytr, U.B. (1989 *b*) Use of two-dimensional protein crystals from bacteria for nonbiological applications. J. Vac. Sci. Technol. B 7, 1391-1397.

Pum, D., Sára, M. and Sleytr, U.B. (1992 *a*) Two-dimensional (glyco)protein crystals as patterning elements and immobilisation matrices for the development of biosensors, in "Immobilised Macromolecules. Application Potentials in Biotechnology and Biosensors" (Sleytr, U.B., Messner, P., Pum, D. and Sára, M., Eds.), Springer-Verlag, London, in press.
Pum, D., Sára, M., Messner, P. and Sleytr, U.B. (1992 *b*) Two-dimensional (glyco)protein crystals as patterning elements for the controlled immobilisation of functional molecules. Nanotechnology 4, 196-202.
Sára, M. and Sleytr, U.B. (1987 *a*) Charge distribution on the S-layer of *Bacillus stearothermophilus* NRS 1536/3c and importance of charged groups for morphogenesis and function. J. Bacteriol. 169, 2804-2809.
Sára, M. and Sleytr, U.B. (1987 *b*) Molecular sieving through S-layers of *Bacillus stearothermophilus* strains. J. Bacteriol. 169, 4092-4098.
Sára, M. and Sleytr, U.B. (1987 *c*) Production and characteristics of ultrafiltration membranes with uniform pores from two-dimensional arrays of proteins. J. Membrane Sci. 33, 27-49.
Sára, M. and Sleytr, U.B. (1988) Membrane biotechnology: Two-dimensional protein crystals for ultrafiltration purposes, in "Biotechnology" vol. 6b (Rehm, H.-J., Ed.), pp. 615-636. VCH Publishers, Weinheim.
Sára, M. and Sleytr, U.B. (1989) Use of regularly structured bacterial cell envelope layers as matrix for the immobilization of macromolecules. Appl. Microbiol. Biotechnol. 30, 184-189.
Sára, M., Küpcü, S. and Sleytr, U.B. (1989) Localization of the carbohydrate residue of the S-layer glycoprotein from *Clostridium thermohydrosulfuricum* L111-69. Arch. Microbiol. 151, 416-420.
Sára, M., Moser-Thier, K., Kainz, U. and Sleytr, U.B. (1990) Characterization of S-layers from mesophilic bacillaceae and studies on their protective role towards muramidases. Arch. Microbiol. 153, 209-214.
Sára, M., Pum, D. and Sleytr, U.B. (1992 *a*) Permeability and charge dependent adsorption properties of the S-layer lattice from *Bacillus coagulans* E38-66. J. Bacteriol. 174, 3487-3493.
Sára, M., Küpcü, S., Weiner, C., Weigert, S. and Sleytr, U.B. (1992 *b*) Crystalline protein layers as isoporous molecular sieves and immobilisation and affinity matrices, in "Immobilised Macromolecules. Application Potentials in Biotechnology and Biosensors" (Sleytr, U.B., Messner, P., Pum, D. and Sára, M., Eds.), Springer-Verlag, London, in press.
Sjögren, A., Hovmöller, S., Farrants, G., Ranta, H., Haapsalo, M., Ranta, K. and Lounatmaa, K. (1985) Structures of two different surface layers found in six *Bacteroides* strains. J. Bacteriol. 164, 1278-1282.
Sleytr, U.B. (1975) Heterologous reattachment of regular arrays of glycoproteins on bacterial surfaces. Nature 257, 400-402.
Sleytr, U.B. (1978) Regular arrays of macromolecules on bacterial cell walls: structure, chemistry, assembly, and function. Int. Rev. Cytol. 53, 1-64.
Sleytr, U.B. and Messner, P. (1983) Crystalline surface layers on bacteria. Annu. Rev. Microbiol. 37, 311-339.
Sleytr, U.B. and Messner, P. (1988 *a*) Crystalline surface layers on bacteria, in "Crystalline Bacterial Cell Surface Layers" (Sleytr, U.B., Messner, P., Pum, D. and Sára, M., Eds.), pp. 160-186. Springer-Verlag, Berlin.
Sleytr, U.B. and Messner, P. (1988 *b*) Crystalline surface layers in procaryotes. J. Bacteriol. 170, 2891-2897.
Sleytr, U.B. and Messner, P. (1989) Self-assemblies of crystalline bacterial cell surface layers, in "Electron Microscopy of Subcellular Dynamics" (Plattner, H., Ed.), pp. 13-31. CRC Press, Boca Raton.
Sleytr, U.B. and Plohberger, R. (1980) The dynamic process of assembly of two-dimensional arrays of macromolecules on bacterial cell walls, in "Electron Microscopy at Molecular Dimensions" (Baumeister, W. and Vogell, W., Eds.), pp. 36-47. Springer-Verlag, Berlin.
Sleytr, U.B., Messner, P. and Pum, D. (1988) Analysis of crystalline bacterial surface layers by freeze-etching, metal shadowing, negative staining and ultrathin sectioning, in "Methods in Microbiology" vol. 20 (Mayer, F., Ed.), pp. 29-60. Academic Press, London.
Sleytr, U.B., Sára, M. and Pum, D. (1989) Application potential of two dimensional protein crystals. Microelectronic Engineering 9, 13-20.
Sleytr, U.B., Pum, D., Sára, M. and Messner, P. (1992) Two-dimensional protein crystals as patterning elements in molecular nanotechnology, in "Proc. 2nd Int. Conference on Molecular Electronics - Science and Technology" (Aviram, A., Ed.). The Engineering Foundation, New York, in press.
Smit, J. (1987) Protein surface layers of bacteria, in "Bacterial Outer Membranes as Model Systems" (Inouye, M., Ed.), pp. 343-376. John Wiley & Sons, New York.

Smit, J. and Agabian, N. (1982) Cell surface patterning and morphogenesis: biogenesis of a periodic surface array during *Caulobacter* development. J. Cell Biol. 95, 41-49.
Smith, S.H. and Murray, R.G.E. (1990) The structure and associations of the double S-layer on the cell wall of *Aquaspirillum sinuosum*. Can. J. Microbiol. 36, 327-335.
Takeoka, A., Takumi, K., Koga, T. and Kawata, T. (1991) Purification and characterization of S layer proteins from *Clostridium difficile* GAI 0714. J. Gen. Microbiol. 137, 261-267.
Thorne, K.J.I., Oliver, R.C. and Glauert, A.M. (1976) Synthesis and turnover of the regularly arranged surface protein of *Acinetobacter* sp. relative to the other components of the cell envelope. J. Bacteriol. 127, 440-450.
Tsuboi, A., Tsukagoshi, N. and Udaka, S. (1982) Reassembly *in vitro* of hexagonal surface arrays in a protein-producing bacterium, *Bacillus brevis* 47. J. Bacteriol. 151, 1485-1497.
Wicken, A.J. (1985) Bacterial cell walls and surfaces, in "Bacterial Adhesion. Mechanisms and Physiological Significance" (Savage, D.C. and Fletcher, M., Eds.), pp. 45-70. Plenum Press, New York.
Wildhaber, I., Santarius, U. and Baumeister, W. (1987) Three-dimensional structure of the surface protein of *Desulfurococcus mobilis*. J. Bacteriol. 169, 5563-5568.
Yamada, H., Tsukagoshi, N. and Udaka, S. (1981) Morphological alterations of cell wall concomitant with protein release in a protein-producing bacterium, *Bacillus brevis* 47. J. Bacteriol. 148, 322-332.
Yang, L., Pei, Z., Fujimoto, S. and Blaser, M.J. (1992) Reattachment of surface array proteins to *Campylobacter fetus* cells. J. Bacteriol. 174, 1258-1267.

AN OVERVIEW OF THE ASSEMBLY, TURNOVER AND RECYCLING OF THE MUREIN SACCULUS

James T. Park

Department of Molecular Biology and Microbiology
Tufts University School of Medicine
Boston, Massachusetts 02111, U.S.A.

In presenting an overview of this subject, I will limit my remarks to the situation in *E. coli*. The discussion is divided into two parts.

The first part is concerned with the roles of the various penicillin-binding proteins. Since only PBP-3 is absolutely essential and PBP-2 only under most circumstances, it is obviously difficult to surmise what each of the PBPs of interest may be doing under normal circumstances. Although there have been no dramatic developments in the past ten years, I want to point out how our view of the normal functions of the PBPs has gradually shifted by pointing out a few observations which seem to be telling us something about the roles of individual PBPs.

In the second part, the phenomena of turnover of peptidoglycan and recycling of muropeptides will be discussed. This bears on the question of whether *E. coli* has a multilayered or monolayered wall. I will summarize some new data on turnover and recycling obtained in our laboratory but unfortunately it doesn't lead to a firm conclusion on the wall thickness question.

Returning to the role of the various PBPs in the assembly of the murein sacculus, consider the situation eight or ten years ago. Table 1 lists the properties of the various PBPs. From the pioneer work of Brian Spratt (1975), we know that *E. coli* has at least 7 PBPs and his estimates of their approximate molecular weights are shown. We are concerned here primarily with the four largest molecular weight PBPs which have known biosynthetic activities. However, I list three additional PBPs (PBP-1C, PBP-7 and PBP-8) which deserve more attention than they've received in the past. Also shown is Spratt's estimate of the number of molecules of each PBP present in a cell and so far as I am aware, no one has refined these numbers. Note that most of the PBPs are present in very small amounts. It is interesting that PBP-2 and PBP-3 are the only PBPs that are essential under most conditions.

The known enzymatic activities of the PBPs have been elucidated mainly through the monumental work of Michio Matsuhashi and his co-workers on PBPs-1A, 1B, 2 and 3 (Nakagawa et al, 1984) and of Jack Strominger's laboratory on PBPs-4, 5, and 6. Unfortunately, little if any new information has appeared in the last 10 years with regard to properties of the so-called large molecular weight PBPs, *i.e.* those with transglycosylase-transpeptidase activities. Instead, only studies with whole cells have been forthcoming that throw some light on the role of individual PBPs under normal conditions. As shown in Table 1, all of the large PBPs appear to have the same enzymatic activities, namely the ability to polymerize the glycan strands and to link them together by transpeptidation, and yet

Bacterial Growth and Lysis, Edited by M.A. de Pedro
et al., Plenum Press, New York, 1993

Table 1. Properties of the penicillin-sensitive enzymes of *E. coli.*

PBP	Mol. wt.[1]	Molecules/cell[1]	Known enzymatic activities
1A	92,000	100	Transglycoslase-transpeptidase
1B	90,000	120	Transglycosylase-transpeptidase
2	66,000	20	Transpeptidase
3	60,000	50	Transglycoslase-transpeptidase
4	49,000	110	DD-Endopeptidase, DD-carboxpeptidase
5	42,000	1,800	DD-Carboxypeptidase
6	40,000	600	DD-Carboxypeptidase

[1]Modified from Spratt (1975).

presumably they perform different functions under normal conditions rather than simply providing backup.

If we look at suggested functions of the large PBPs as listed in Table 2, bearing in mind that supportive evidence is largely lacking, one sees that three separate functions were envisioned: PBP-1A, 1B and 2 were thought to participate in cell elongation, PBP-3 more certainly was involved in cell septation, and PBP-2 was somehow necessary to preserve the rod-shape of the cell. Let's now look at key observations, new and old, relavent to the function of each of the large molecular weight PBPs. Table 3 lists several milestones concerning PBP-2. Twenty years ago, Lund and Tybring (1972) reported that a new antibiotic, FL-1060, now called mecillinam, caused *E. coli* to change to a spherical form suggesting a unique mode of action. Soon thereafter, Spratt (1975) showed that the target for mecillinam was PBP-2. Also in these ancient times Lars Burman and I (1973) reported that mecillinam inhibited murein synthesis by 50% which implied that PBP-2 was a major factor in net synthesis of peptidoglycan. However, the idea that PBP-2 was somehow responsible for maintenance of the rod-shape obscured its role in net synthesis for a dozen years or so. de la Rosa et al (1985) and Pisabarro et al (1987) recently demonstrated that PBP-2 was needed for initiation of growth from stationary phase cells and elongation of these cells. This was shown by inhibiting all other PBPs with cefmetazole and showing that PBP-2 alone permitted elongation of the sacculus. Finally, James, Haga and Pardee (1975)

Table 2. Predicted functions of the PBPs, 1982.

PBP	Function
1A	Murein synthesis during cell elongation
1B	Murein synthesis during cell elongation
1C	Unknown
2	Growth in rod shape, cell elongation
3	Murein synthesis during cell septation
4	Cross-link hydrolysis during cell elongation
5	Destruction of unutilized pentapeptide
6	Destruction of unutilized pentapeptide
7	Unknown

Table 3. Processes involving PBP-2.

PBP-2 is required to maintain rod-shape.	Spratt (1975), Lund and Tybring (1972)
PBP-2 is required for 50% of net synthesis of murein.	Park and Burman (1973)
PBP-2 is required to initiate growth and elongate cells.	de la Rosa et al (1985), Pisabarro et al. (1987)
PBP-2 affects the septation process.	James et al (1975) Rodriguez and de Pedro (1990)

Table 4. Known activities of PBP-3.

PBP-3 is required for septation.	Spratt (1975)
PBP-3 makes septal murein.	Botta and Park (1981)
PBP-3 and PBP-2 act alternately.	Woldringh et al. (1987)

Table 5. Evidence for different functions for PBP-1A and PBP-1Bs.

Either PBP-1A or PBP-1B must be functional for growth to occur.	Suzuki *et al.* (1978)
PBP-1A and PBP-1B have different functions.	Schmidt *et al.* (1981), del Portillo and de Pedro (1990)
PBP-1A and/or PBP-1Bs required for septation.	del Portillo *et al.* (1989).

demonstrated that mecillinam blocked cell division, and Rodriguez and de Pedro (1990) confirmed that PBP-2 was needed for normal cell division; since without it aberrant divisions and mini-cell like structures formed. Thus the weight of current evidence suggests PBP-2 is responsible for net synthesis of murein during elongation, producing over 70% of the lateral wall and lastly the division process cannot proceed normally unless PBP-2 remains active until septation is initiated.

Our ideas covering PBP-3 are summarized in Table 4. These have changed little in the past ten years. Spratt, in his classic 1975 paper showed that PBP-3 was required for cell division. Some years later Botta and I (1978) were able to show that PBP-3 was needed to make the murein for the septum, since antibiotics that specifically bound to PBP-3, namely furazlocillin and piperacillin, inhibited murein synthesis during septation but not during elongation. Woldringh *et al.* (1987) analyzed pulse-labeled cells by autoradiography and showed that during septation the rate of incorporation into the sidewall was reduced and hence the activity of PBP-3 and PBP-2 tend to act alternately. Obviously the regulation of these activities is a crucial part of normal cell growth and remains to be investigated in detail.

The few clues concerning the functions of PBP-1A and PBP-1B are listed in Table 5. We know from the classic paper on mutants of PBPs by Suzuki, Nishimura and Hirota (1978) that either PBP-1A or PBP-1Bs must be functional in order for *E. coli.* to grow. Schmidt *et al.* (1981) showed, however, that these PBPs served different functions, since PBP-1B mutants when prevented from dividing by the presence of furazlocillin, grew as filaments for an extended period and then lysed at a division site. PBP-1A mutants, on the other hand, under similar conditions barely doubled in length and then ceased growth without lysing. One could interpret this as indicating that PBP-2 does not become activated after division in the absence of PBP-1A. Incidentally, PBP-1A deletion mutants do not follow this pattern (unpublished). del Portillo and de Pedro (1990) also demonstrated through a different approach employing cell division mutants that PBPs-1A and 1B play distinct roles in cell wall synthesis and further concluded that PBP-1B is involved at the time of septation. del Portillo *et al.* (1989) had previously observed, by studying the lytic behavior of cells aligned for cell division, that one of these PBPs must be functional for septation to occur (which one couldn't be determined).

Thus, the present view, still largely hypothetical, of the role of the biosynthetic PBPs during normal growth is as follows (Table 6). PBP-2 is largely responsible for elongation of the cell rather than PBP-1B as was initially inferred from its superior *in vitro* enzymatic activity. PBP-3 synthesizes septal murein once the process has been initiated, and PBP-1B, as well as PBP-2 is needed during the initiation process. Finally, all of the biosynthetic PBPs may undergo activation and deactivatin each cell cycle.

I want to mention again PBP-1C and PBP-7 which have received little attentionin the past. Some years ago when Guiseppe Botta screened a large number of β-lactam antibiotics for ability to bind to PBPs, it was found that many β-lactams bound to PBP-1C and PBP-7 at low concentration (Table 7). This indicates to us that they have an active site that recognizes D-ala-D-ala and therefore may be playing a role in murein metabolism. PBP-8 displayed the same sensitivity spectrum as PBP-7. It has been shown by Barbas *et al.* (1986) to be a break down product of PBP-7. Tuomanen and Schwartz (1987) have shown that only β-lactams which bind to PBP-7 cause lysis of amino acid-starved *E. coli* and hence have suggested that PBP-7 may be physiologically significant. A search for mutants of PBP-1C and PBP-7 or biochemical investigations could prove rewarding.

Table 6. Activities of the PBPs as surmised from the new data.

PBP-2 is the major enzyme for elongation of the cell.
PBP-3 forms the septal murein following initiation.
PBP-1B is needed for initiation of septation.
All of the biosynthetic PBPs may require activation and deactivation during the cell cycle.

Table 7. Binding of selected β-lactam antibiotics to the PBPs of *E. coli.*[1] d·

PBP	Penicillin G	Cephaloridine	Furazlocillin	Cefsulodin	Moxalactam
1A	1.5	0.4	3	1.5	1.
1B	3	6	6	6	6
1C	-	0.4	0.05	-	1.5
2	3	40	0.5	>100	>100
3	1.5	6	0.05	>100	0.4
4	0.4	3	6	>100	1.5
5	12	>100	6	>100	6
6	6	>100	12	>100	6
7	1.5	1.5	0.8	12	1.5

[1]Concentration (μg/ml) to half-saturate the PBP in 10 min at 30°C. Results are very approximate.

Now let's consider turnover and recycling and its significance. Turnover has come to mean the actual loss of wall material during normal growth. Over the years it has been seen as a relatively minor occurrence in *E. coli* whereas in Gram-positive bacteria the turnover rate can be as high as 50% per generation. Recycling, which will be defined as reutilization of muropeptides released by turnover, was discovered by Goodell in 1985, and his data suggested that, and I quote "About 50% of the murein in the sacculus is degraded to peptides and recycled each generation." Goodell and Higgins (1987) subsequently reported that *E. coli* would take up the tripeptide, L-ala-D-glu-γ-*meso*-diaminopimelic acid (DAP) from the medium and reincorporate it into the sacculus (as measured by incorporation of the radiolabeled DAP). They further showed that for uptake, oligopeptide permease (*opp*) was specifically required. This remarkable result led us to investigate turnover in *opp* negative strains,expecting to observe 50% turnover. Let me summarize our results, by saying that in *opp* strains, in which recycling was presumably blocked, turnover remained as low as in wild type cells, rather than 50% per generation as predicted. The rate of loss from the sacculus varied from 0 to 7% per generation depending on the strain. We confirmed Goodell's result that wild type *E. coli* W-7 had a turnover rate of about 7% per generation. We also found that turnover was greater from uniformly-labeled cells than from cells labeled only in the lateral wall suggesting that cell separation caused a significant loss of murein. This could be viewed as supporting the idea that the poles may be multilayered.

When we repeated and extended certain pulse-chase experiments, employing the acceptor-donor radioactivity ratio to determine the stability of cross-links in the lateral wall of *opp* strains, we were surprised to find essentially no change from the earlier result (Park, 1988) which had led us to suggest that half of the cross-links in the lateral wall were stable.

These two results made us suspicious that the lack of oligopeptide permease may not prevent recycling. Hence we investigated recycling in an *opp* strain. As judged by the radioactivity remaining in the UDP-N-acetylmuramyl pentapeptide pool following a three generation chase of cells uniformly-labeled with ^{14}C-DAP, the turnover rate had to be at least 40% per generation and if the poles are assumed to be stable the lateral walls may be turned over and recycled at a rate of well over 50%.

One caveat, we have identified the UDP-N-acetylmuramyl pentapeptide by ion exchange and paper chromatography but have not yet tested it for biological activity. This was done by Goodell and Schwarz (1985) in their initial report so presumably the precursor in the cytoplasm is authentic.

Assuming that a very high rate of turnover and recycling does exist in *opp* cells, it has several implications:

1. There must be a second permease, presumably one with a lower affinity for the substrate since the experiments of Goodell and Higgins must have involved a high affinity permease.
2. Recycling requires that L-ala-D-glu-γ-*meso*-DAP be used preferentially to L-ala for the synthesis of UDP-N-acetylmuramyl pentapeptide. This must have a significant effect on experiments employing the acceptor-donor radioactivity ratio especially during a chase when radioactive peptide from the murein would be recycled into the donor position.

This leads me to say a few words about a model for elongation of the murein sacculus that Lars Burman and I (1984) proposed some years ago. The model contained three ideas and was based on experiments which followed the fate of new muropeptides as they were cross-linked to the sacculus. The data indicated that insertion of new strands into the sacculus was a multisite process; that each site or membrane-bound enzyme complex actually traversed the circumference of the cell following the strands of the existing sacculus in a truly processive manner; and that two strands seemed to be inserted together. This latter suggestion, was based on the assumption that DAP-starved cells did not immediately resume septation and that Bug 6 at the restrictive temperature formed no septal murein so that we would be measuring only lateral wall synthesis. Both of these assumptions proved to be wrong. Bug 6 was recently shown to be an *ftsA* mutant (Begg et al , 1986) and fts A mutants are known to initiate septation at the restrictive temperature. Though we had shown that W-7 cells starved for DAP for 25 minutes were delayed in septation, we starved cells only 20 minutes so as not to stress the cells and obviously septation was not delayed. Thus in both these situations septal murein formed entirely of new strands was being made and this raised the acceptor-donor radioactivity ratio such that it suggested two strands might go in together. When we pulse-labeled *ftsZ* strains at the restrictive temperature so as to prevent septation, we found that over 95% of the label was in the donor position of cross-linked dimers which clearly indicates that single strands are being inserted during elongation. This is in agreement with the results of Cooper et al (1988) in *Salmonella* and of de Jonge et al (1989).

3. A final implication of a high recycling rate. It makes the question of whether the murein sacculus is multilayered or not even more difficult to assess. My prejudice is that the lateral wall is monolayered and the poles are initially triple layered with considerable material turning over during cell separation.. This is one possibility suggested by the results of Labischinsky *et al.* (1991) and further supported by the recent chemical analyses of Wientjes *et al.* (1991).

Goodell (1985) proposed a pathway for recycling the tripeptide and figure 1 shows a slightly modified version of his proposal. Our result simply adds an additional uptake pathway for transport of tripeptide into the cell. Thus murein is degraded by the combined action of endopeptidases (Tamura *et all.*, 1976; Keck and Schwarz, 1968), transglycosylases (Mett et al,1980), amidase (Parquet *et al.*, 1983) and carboxypeptidase II (Izaki and Strominger, 1968) to yield the tripeptide. The tripeptide is transported into the cytoplasm either by oligopeptide permease or a hypothetical low affinity transport system. There it is added to UDP-N-acetylmuramic acid in place of L-ala and hence is recycled extremely efficiently. As mentioned earlier, well over 50% of the side wall murein is recycled per generation.

To summarize this overview of the assembly, turnover and recycling of the murein sacculus, no dramatic break-throughs have occurred in the past ten years concerning specific functions for the biosynthetic PBPs. We see more clearly that PBP-2 is responsible, in large part, for elongation, that both PBP-2 and PBP-1B are required for initiation of septation and that PBP-3 completes the septum following initiation.

The principal new message concerns the fact that recycling (and hence turnover), as observed by Goodell (1985), occurs at a tremendous rate even in the absence of oligopeptide permease. Obviously, recycling must be taken into account in all work that involves pulse-labelling of the sacculus.

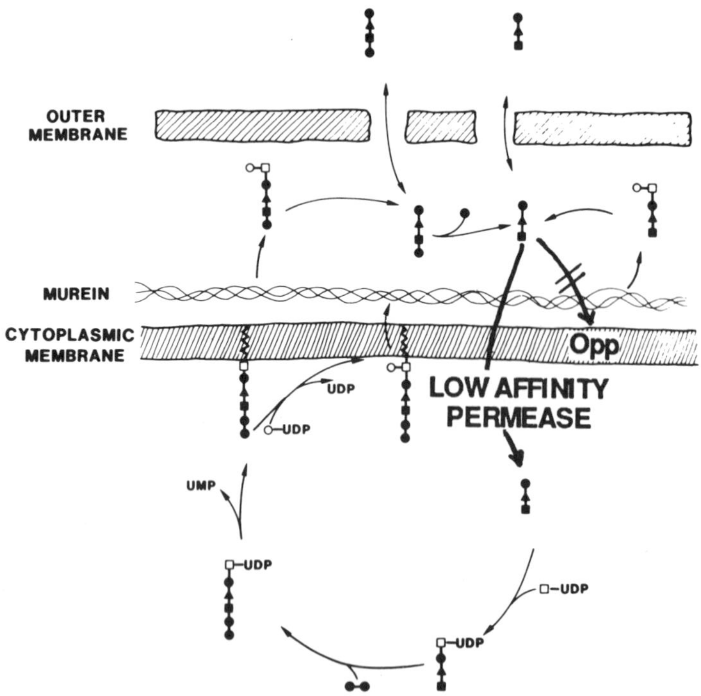

Figure 1. Murein recycling in *E. coli.* Modified from Goodell (1985)

ACKNOWLEDGMENTS

This work was supported, in part, by grant AI05090 from the National Institutes of Health.

REFERENCES

Barbas, J. A., Diaz. J., Rodriguez-Tebar, A. and Vazquez, D. (1986) Specific location of penicillin-binding proteins within the cell envelope of *Escherichia coli*. J. Bacteriol. 165, 269-275.

Begg, K. J., Spratt, B. G. and Donachie, W. D. (1986) Interaction between membrane proteins PBP 3 and RodA is required for normal cell shape and division in *Escherichia coli*. J. Bacteriol. 167, 1004-1008.

Botta, G. A. and Park, J. T. (1981) Evidence for involvement of penicllin-binding protein 3 in murein synthesis during septation but not during cell elongation. J. Bacteriol. 145, 333-340.

Burman, L. and Park, J. T. (1984) Molecular model for elongation of the murein sacculus of *Escherichia coli*. Proc. Natl. Acad. Sci. U.S.A. 81, 1844-1848.

Cooper, S., Hsieh, M.-L. and Guenther, B. (1988) Mode of peptidoglycan synthesis in *Salmonella typhimurium*: single-strand insertion. J. Bacteriol. 170, 3509-3512.

de la Rosa, E. J., de Pedro, M. A. and Vazquez, D. (1985) Penicillin binding proteins: role in initiation of murein synthesis in *Escherichia coli*. Proc. Natl. Acad. Sci. USA 82, 5632-5635.

de Jonge, B. L. M., Wientjes, F. B., Jurida, I., Driehuis, F., Wouters, J. T. M. and Nanninga, N. (1989) Peptidoglycan synthesis during the cell cycle of *Escherichia coli*: composition and mode of insertion. J. Bacteriol. 171, 5783-5792.

del Portillo, F. G. and de Pedro, M. A. (1990) Differential effect of mutational impairment of penicillin-binding proteins 1A and 1B on *Escherichia coli* strains harboring thermosensitive mutations in the celll division genes *ftsA*, *ftsQ*, *ftsZ*, and *pbpB*. J. Bacteriol. 172, 5863-5870.

del Portillo, F. G., de Pedro, M. A., Joseleau-Petit, D. and D'Ari, R. (1989) Lytic response of *Escherichia coli* cells to inhibitors of penicillin-binding proteins 1a and 1b as a timed event related to cell division. J. Bacteriol. 171, 4217-4221.

Goodell, E. W. (1985) Recycling of murein by *Escherichis coli*. J. Bacteriol. 163, 305-310.

Goodell, E. W. and Schwarz, U. (1985) Release of cell wall peptides into culture medium by exponentially growing *Escherichia coli*. J. Bacteriol. 162, 391-397.

Goodell, E. W. and Higgins, C. F. (1987) Uptake of cell wall peptides by *Salmonella typhimurium* and *Escherichia coli*. J. Bacteriol. 169, 3861-3865.

Izaki, K. and Strominger, J. L.(1968) Biosynthesis of the peptidoglycan of bacterial cell walls. XIV. purification and properties of two D-alanine carboxypeptidases from *Escherichia coli*. J. Biol. Chem. 243, 3193-3201.

James, R., Haga, J. Y. and Pardee, A. B. (1975) Inhibition of an early event in the cell division cycle of *Escerichia coli* by FL1060, an amidinopenicillanic acid. J. Bacteriol. 122, 1283-1292.

Keck, W. and Schwarz, U. (1979) *Escherichia coli* murein DD-endopeptidase insensitive to β-lactam antibiotics. J. Bacteriol. 139, 770-774.

Labischinski, H., Goodell, E. W., Goodell, A. and Hochberg, M. L. (1991) Direct proof of a "more-than-single-layered" peptidoglycan architecture of *Escherichia coli* W7: a neutron small angle scattering study. J. Bacteriol. 173, 751-756.

Lund, F. and Tybring, L. (1972) 6β-amidinopenicillanic acids—a new group of antibiotics. Nature New Biol. 236, 135-137.

Mett, H., Keck, W., Funk, A. and Schwarz, U. (1980) Two differing species of murein transglycosylase in *Escherichia coli*. J. Bacteriol. 144, 45-52.

Nakagawa, J., Tamaki, S., Tomioka, S. and Matsuhashi, M. (1984) Functional biosynthesis of cell wall peptidoglycan by polymorphic bifunctional polypeptides. J. Biol. Chem. 259, 13937-13946.

Park, J. T. (1988) Does *Escherichia coli* use a left-handed murein endopeptidase and transpeptidase to assemble its sacculus? in "Antibiotic inhibition of bacterial cell surface assembly and function" (Actor, P., Daneo-Moore, L., Higgins, M. L., Salton, M. R. J. and Shockman, G. D., Eds), pp.60-65. Soc. for Microbiol., Washington, D.C.

Park, J. T. and Burman, L. (1973) FL1060: a new penicillin with a unique mode of action. Biochem. Biophys. Res. Commun. 51, 863-868.

Parquet, C., Fouret, B., Leduc, M., Hirota, Y. and van Heijenoort, J. (1983) N-acetylmuramoyl-L-alanine amidase of *Escherichia coli* K12. Possible physiological functions. Eur. J. Biochem. 133, 371-377.

Pisabarro, A. G., del Portillo, F. G., de la Rosa, E. J. and de Pedro, M.A. (1987) Initiation of murein biosynthesis in *Escherichia coli* cells treated with β-lactam antibiotics. FEMS Microbiol. Lett. 42, 81-84.

Rodriguez, M. C. and de Pedro, M. A. (1990) Initiation of growth in $pbpA^{ts}$ and $rodA^{ts}$ mutants of *Escherichia coli*. FEMS Microbiol. Lett. 72,235-240.

Schmidt, L. S., Botta, G. and Park, J. T. (1981) Effects of furazlocillin, a β-lactam antibiotic which binds selectively to penicillin-binding protein 3, on *Escherichia coli* mutants deficient in other penicillin-binding proteilns. J. Bacteriol. 145, 632-637.

Spratt, B. G. (1975) Distinct penicillin binding proteins involved in the division, elongation, and shape of *Escherichia coli* K12. Proc. Natl. Acad. Sci. USA 72, 2999-3003.
Suzuki, H., Nishimura, Y. and Hirota,Y. (1978) On the process of cellular division in *Escherichia cloi* a series of mutants of *E. coli* altered in the penicillin-binding proteins. Proc. Natl. Acad. Sci. USA 75, 664-668.
Tamura, T., Imae,Y. and Strominger, J. L. (1976) Purification to homogeniety and properties of two D-alanine carboxypeptidases from *Escherichia coli* J. Biol. Chem. 251, 414-423.
Tuomanen, E. and Schwartz, J. (1987) Penicillin-binding protein 7 and its relationship to lysis of nongrowing *Escherichia coli.* J. Bacteriol. 169, 4912-4915.
Wientjes, F. B., Woldringh, C. L. and Nanninga, N. (1991) Amount of peptidoglycan in cell walls of gram-negative bacteria. J. Bacteriiol. 173, 7684-7691.
Woldringh, C. L., Huls, P., Pas, E., Brakenhoff, G. J. and Nanninga, N. (1987) Topography of peptidoglycan synthesis during elongation and polar cap formation in a cell division mutant of *Escherichia coli* MC4100. J. Gen. Microbiol. 133, 575-586.

VARIATIONS IN THE METABOLISM OF PEPTIDOGLYCAN PRIOR TO POLYMERIZATION

Jean van Heijenoort[1], Dominique Mengin-Lecreulx[1], Yveline van Heijenoort[1], Didier Blanot[1], Bernard Flouret[1], Catherine Michaud[1], Claudine Parquet[1], Flore Pratviel-Sosa[1], Manolo Gomez[2] and Juan A. Ayala[2]

[1]) URA 1131 du C.N.R.S.,
Université Paris-Sud, Bât. 432,
91405 Orsay Cedex, France
[2]) Centro de Biologia Molecular, Universidad Autonoma de Madrid, Canto Blanco, 28049 Madrid, Spain

INTRODUCTION

The biosynthesis of bacterial cell wall peptidoglycan is a complex process which comprises two major stages (Rogers *et al.*, 1980). The first one concerns the sequential formation of a series of cytoplasmic and membrane precursors. The second one concerns the polymerisation reactions with the insertion of the new peptidoglycan material and its maturation. The first stage of peptidoglycan synthesis is considered to begin with UDP-*N*-acetylglucosamine (UDP-GlcNAc) and to end with *N*-acetylglucosaminyl-*N*-acetylmuramyl(pentapeptide)-pyrophosphate undecaprenol (lipid II in Fig.1) which is the substrate for the polymerization reactions. The different steps prior to polymerization have long been identified and certain studied to some extent (Rogers *et al.*, 1980). However little attention had been paid to their *in vivo* functioning.

Over the past few years, in *Escherichia coli* efforts have been devoted to the genetic, biochemical and physiological study of the steps of this pathway and many of their features investigated. Pratically all the genes coding for the enzymes involved have now been identified and sequenced. Most are located at 2 min in a large cluster of genes extending from *pbpB* to *envA* and coding for proteins all involved in peptidoglycan synthesis and cell division (Mengin-Lecreulx *et al.*, 1989 *a* ; Ikeda *et al.*, 1991 ; Mengin-Lecreulx *et al.*, 1991). It is also possible to evaluate the specific activity of all the enzymes

Bacterial Growth and Lysis, Edited by M.A. de Pedro *et al.*, Plenum Press, New York, 1993

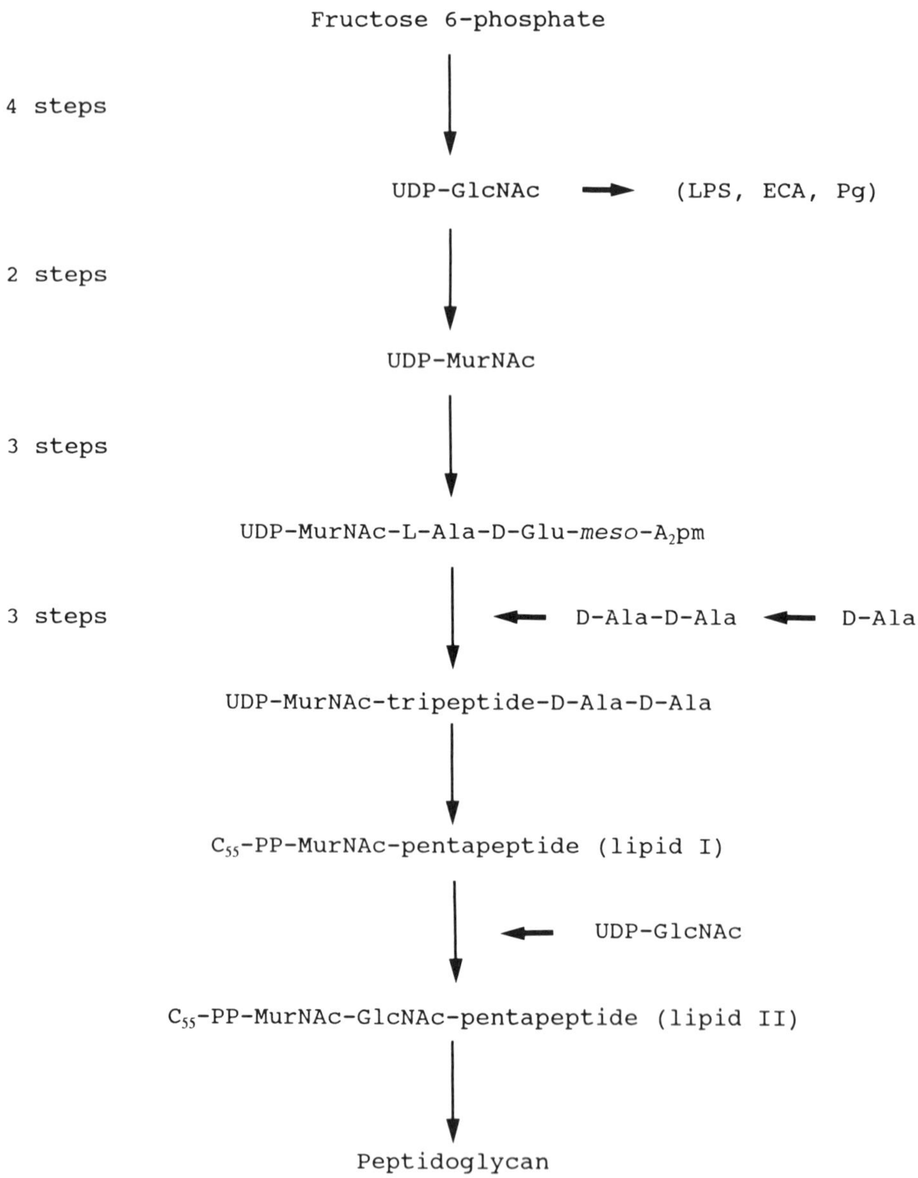

Fig. 1. Peptidoglycan synthesis in *E. coli* prior to polymerization.

(Mengin-Lecreulx *et al.*, 1982 ; Mengin-Lecreulx *et al.*, 1983 ; Geis and Plapp, 1978). Moreover, a whole set of analytical methods enables to determine the pool levels of the cytoplasmic precursors (Flouret *et al.*, 1981 ; Mengin-Lecreulx *et al.*, 1982; Mengin-Lecreulx *et al.*, 1983) and lipid intermediates (Ishiguro and Ramey, 1978; Kohlrausch *et al.*, 1989 ; Mengin-Lecreulx *et al.*, 1991 ; van Heijenoort *et al.*, 1992).

In the past, variations in the pool levels of the peptidoglycan nucleotide precursors have been observed either in conditional mutants altered in a specific step of the pathway (Lugtenberg *et al.*, 1972 ; Rogers *et al.*, 1980) or in cells treated with an antibiotic having a specific step for target (Gale *et al.*, 1972 ; Rogers *et al.*, 1980). The systematic use of the analytical methods now available has enabled to analyze in detail in *E. coli* the pool levels of all peptidoglycan precursors and the enzymatic activities of the pathway under many different conditions. This approach has led to a better understanding of the first stage of peptidoglycan metabolism, in particular, of its flexibility, its correlation with other cell functions and of the possible regulatory mechanisms involved. It also provided explanations for certain variations in peptidoglycan structure and effects of antibiotics. The aim of the present report is to sum up some of the recent data and to illustrate their usefulness by a few examples.

PEPTIDOGLYCAN BIOSYNTHESIS PRIOR TO POLYMERIZATION IN *E.COLI*

In *E. coli* UDP-GlcNAc is an important branch point for the synthesis of three distinct cell wall polymers : peptidoglycan, lipopolysaccharide and the enterobacterial common antigen. This is perhaps the reason why it is at a fairly high pool level of 10^{-4} M (Mengin-Lecreulx *et al.*, 1983). At the first step of the peptidoglycan pathway a transferase catalyzes the formation of UDP-GlcNAc-enolpyruvate from UDP-GlcNAc and PEP. UDP-GlcNAc is also involved in the formation of lipid II (Fig.1). UDP-GlcNAc-enolpyruvate was only recently isolated as such and its very low pool level (10^{-6} M) quantified (Mengin-Lecreulx *et al.*, 1983). It is the substrate of the reductase catalyzing the formation of UDP-*N*-acetylmuramic acid (Anwar and Vlaovic, 1979). The gene *murB* for this reductase has just been identified at 90 min as the sequenced ORF next to *birA* (Doublet *et al.*, 1992 ; Pucci *et al.*, 1992).

The sequential addition of Ala, D-Glu, *meso*-A_2pm and D-alanyl-D-alanine to UDP-MurNAc leads to the formation of UDP-MurNAc-pentapeptide (Rogers *et al.*, 1980 ; Mengin-Lecreulx *et al.*, 1982). Synthetases MurD, MurE and MurF have been overproduced and purified (Pratviel-Sosa *et al.*,1991 ; Michaud *et al.*,1990; Duncan *et al.*, 1990). Certain of their properties were investigated (substrate specificity, kinetic parameters, reversibility, etc.). All four synthetases catalyze a very similar type of reaction :

$$\text{UDP-MurNAc} + \text{L-Ala} + \text{ATP} \rightarrow \text{UDP-MurNAc-L-Ala} + \text{ADP} + P_i \quad (\textit{murC})$$

$$\text{UDP-MurNAc-L-Ala} + \text{D-Glu} + \text{ATP} \rightarrow \text{UDP-MurNAc-L-Ala-D-Glu} + \text{ADP} + P_i \quad (\textit{murD})$$

$$\text{UDP-MurNAc-L-Ala-D-Glu} + \textit{meso}\text{-DAP} + \text{ATP} \rightarrow \text{UDP-MurNAc-tripeptide} + \text{ADP} + P_i \quad (\textit{murE})$$

$$\text{UDP-MurNAc-tripeptide} + \text{D-Ala-D-Ala} + \text{ATP} \rightarrow \text{UDP-MurNAc-pentapeptide} + \text{ADP} + P_i \quad (\textit{murF})$$

The mechanisms involved in these reactions remain to be investigated. Formation of acyl phosphate intermediates and of tetrahedral transition states has been postulated (Abo-Ghalia *et al.*, 1988). The specific activities of the four synthetases were found to not vary considerably with growth rate and growth phase (Lugtenberg, 1972 ; Mengin-Lecreulx and van Heijenoort, 1985). They are therefore presumably in excess under most growth conditions and this part of the pathway appears to function as a constitutive machinery.

The pool levels of the various precursors involved in these four steps have also been determined for different growth conditions (Mengin-Lecreulx and van Heijenoort, 1985). In exponential phase cells the pool levels of nucleotide substrates appear adjusted to the Km of the synthetases or lower, whereas those of the other substrates (ATP, Ala, D-Glu, *meso*-A_2pm, D-alanyl-D-alanine) are more or less at saturating concentrations. In stationary phase cells important variations have been observed mainly with cells grown in glucose minimum medium (de Roubin *et al.*, 1992). Excepting for UDP-MurNAc-pentapeptide pool, which is at 10^{-4} M, the pools of the other nucleotide precursors in growing cells are at *ca.* 10^{-5} M. Considering the rate of peptidoglycan synthesis to be *ca.* 1000 units per second, the turnover rate of these pools must be quite rapid. *In vitro* inhibitory effects of the nucleotide precursors on various synthetases of the pathway have been observed (Mengin-Lecreulx *et al.*, 1982, 1983 ; Abo-Ghalia *et al.*, 1985 ; Michaud *et al.*, 1987). However, excepting for certain extreme cases, these effects are probably of no physiological significance since they were observed at concentrations 100-fold higher than the normal pool values.

Lipid II is formed from UDP-MurNAc-pentapeptide by a two-step process (Rogers *et al.*, 1980). First a translocase catalyzes the transfer of the phospho-MurNAc-pentapeptide moiety to the membrane acceptor, undecaprenyl phosphate, yielding MurNAc(pentapeptide)-pyrophosphate-undecaprenol (lipid I).

Thereafter a transferase catalyzes the addition of *N*-acetylglucosamine yielding lipid II. The translocase also catalyzes the reverse reaction and the equilibrum has been found to be greatly in favor of UDP-MurNAc-pentapeptide both *in vitro* (Geis and Plapp, 1978) and *in vivo* (Mengin-Lecreulx *et al.*, 1991). The irreversibility of the transferase reaction allows for the *in vivo* formation of lipid II. Recently the translocase and the transferase have been identified as the products of gene *mraY* and gene *murG*, respectively, located in the 2 min region (Mengin-Lecreulx *et al.*, 1991 ; Ikeda *et al.*, 1991). The ratios of UDP-MurNAc pentapeptide : lipid I : lipid II have been estimated at *ca.* 300:1:3 (Kohlrausch *et al.*,1989) or 140:1:2.7 (Mengin-Lecreulx *et al.*, 1991). Considering that there are *ca.* 10^5 molecules of UDP-MurNAc-pentapeptide per cell (Mengin-Lecreulx *et al.*, 1982), there would thus be only 1000 to 2000 molecules of lipid II and less than 700 of lipid I (van Heijenoort *et al.*, 1992). These extremely low values imply an extremely rapid turnover rate. Moreover, a high UDP-MurNAc pentapeptide : lipid I ratio is presumably an efficient way of avoiding a too low level of the undecaprenyl pool, which is also necessary for the synthesis of other cell wall heteropolymers (LPS, ECA).

Several specific side pathways (D-Glu, A_2pm, D-alanyl-D-alanine, and undecaprenyl phosphate) are associated with the sequential formation of the UDP-MurNAc-peptides and lipid I. The A_2pm pathway has been extensively investigated (Patte, 1983 ; Cohen and Saint-Girons, 1987). The gene responsible for the formation of D-Glu has now been identified (Doublet *et al.*, 1992). Interesting results concerning the formation of D-alanyl-D-alanine have recently been published (Zawadzke *et al.*, 1991; de Roubin *et al.*, 1992). In particular, it has been established that there are in *E. coli* two distinct genes corresponding to two D-alanyl-D-alanine ligases possessing very similar kinetical characteristics. Undecaprenyl phosphate is involved not only in the synthesis of lipid I but also in that of other heteropolymers such as the LPS O-antigen. Presently little is known about its metabolism in *E. coli*.

EFFECT OF AN INHIBITION OF PROTEIN SYNTHESIS

When the pool levels of the peptidoglycan nucleotide precursors were analyzed after inhibition of protein synthesis in *E. coli*, manifold increases of the UDP-GlcNAc, UDP-MurNAc-pentapeptide and D-alanyl-D-alanine pool levels were observed (Mengin-Lecreulx *et al.*, 1989 *b* ; de Roubin *et al.*, 1992). The increase of the UDP-MurNAc-pentapeptide pool appeared as a consequence of that of the UDP-GlcNAc pool by the unrestricted functioning of the intermediate steps of the pathway. The increase of the UDP-GlcNAc pool after inhibition of protein synthesis could be due to an effect on some upstream step by a yet unknown mechanism. The increase of the D-alanyl-D-alanine pool was presumably a consequence of that of the alanine pool.

These effects provide an interesting explanation for a correlation observed between inhibition of peptidoglycan synthesis by fosfomycin, cell lysis and inhibition of protein synthesis (Mengin-Lecreulx and van Heijenoort, 1990). Inhibition of 70% of the transferase activity catalyzing the formation of UDP-GlcNAc-enolpyruvate by fosfomycin leads to lysis within 40 min. This fosfomycin-induced lysis was suppressed by chloramphenicol. Analysis of the precursor pool levels clearly showed that fosfomycin alone led to a depletion of the precursors downstream from UDP-GlcNAc and that the addition of chloramphenicol 20 min after fosfomycin led to their partial restoration. This was explained by the fact that UDP-GlcNAc is normally not at a saturating concentration for the transferase (Mengin-Lecreulx *et al.*, 1989 *b*) and that the residual transferase activity can therefore be modulated by the greatly increased UDP-GlcNAc pool. In turn this will insure the formation of enough UDP-MurNAc-pentapeptide to sustain peptidoglycan synthesis at a rate that will antagonize fosfomycin-induced lysis. A similar explanation could perhaps be given for the effect of chloramphenicol on D-cycloserine-induced lysis (de Roubin *et al.*, 1992).

VARIATIONS IN DIAMINOPIMELIC ACID METABOLISM

We report here on two aspects of variations in the metabolism of diaminopimelic acid (A_2pm). One deals with the effects of aminopimelic acid (Apm) and the other one with mutants altered in the *dapF* gene.

Tetrahydrodipicolinate succinylase is an enzyme of the A_2pm pathway : it catalyzes the conversion of tetrahydrodipicolinic acid to 2-succinylamino-6-oxo-L-pimelic acid. L-Apm is an alternate substrate for this enzyme leading to unproductive L-2-succinylaminopimelic acid (Berges *et al.*, 1986). Since Apm is impermeant, peptides containing it must be used for its transport into the bacterial cell, where it is released by action of peptidases (Berges *et al.*, 1986 ; Le Roux *et al.*, 1991). When exponential-phase cells of *E. coli* were treated with DL-Apm-L-Ala a strong depletion of the UDP-MurNAc-pentapeptide pool and considerable increase of the UDP-MurNAc-dipeptide were observed (Table 1). The decrease of peptidoglycan content paralleled the increase of UDP-MurNAc-dipeptide. This indicates that, at least during the laps of time considered, inhibition of the A_2pm pathway has no effect on the upstream steps of the peptidoglycan pathway and in particular the accumulation of UDP-MurNAc-dipeptide has little effect on the upstream precursors.

A mutant of *E. coli* lacking the diaminopimelate epimerase coded by *dapF* was found to have an unusually large LL-A_2pm pool as compared with that of *meso*-A_2pm (Richaud *et al.*, 1987). The consequences of high cellular LL-A_2pm/*meso*-A_2pm ratios on the structure and metabolism of peptidoglycan were investigated (Mengin-Lecreulx *et al.*, 1988). For this purpose new efficient high-pressure liquid chromatography techniques for the separation of the A_2pm isomers were developed. Sacculi from *dapF* mutants contained a high proportion of LL-A_2pm that varied greatly with growth conditions. The same was observed with the two A_2pm-containing precursors, UDP-*N*-acetylmuramyl-tripeptide

Table 1. Effects of DL-Apm-L-Ala on peptidoglycan and its precursors in *E. coli*.

DL-Apm-L-Ala (100 μg x ml^{-1}) was added to exponential phase cells of *E. coli* K12 HfrH at a concentration promoting cell lysis within 40 min. Analysis of precursor pools and peptidoglycan contents were performed 10 min prior to the onset of lysis.

	nmoles per g dry bacterial weight	
	Control	DL-Apm-Ala
UDP-GlcNAc	870	900
UDP-MurNAc	235	255
UDP-MurNAc-Ala	5	15
UDP-MurNAc-dipeptide	6	2080
UDP-MurNAc-tripeptide	60	10
UDP-MurNAc-pentapeptide	225	30
Peptidoglycan	9110	7190

and UDP-*N*-acetylmuramyl-pentapeptide. The limiting steps for the incorporation of LL-A_2pm into peptidoglycan were found to be its addition to UDP-*N*-acetylmuramyl-L-alanyl-D-glutamate and the formation of the D-alanyl-A_2pm cross-bridges. The K_m value of the A_2pm-adding enzyme for LL-A_2pm was 3.6 x 10^{-2} M as compared with 1.1 x 10^{-5} M for *meso*-A_2pm. When isolated sacculi were treated with *Chalaropsis N*-acetylmuramidase and the resulting soluble products were analyzed by high-pressure liquid chromatography, the proportion of the main dimer was lower in the *dapF* mutant than in the parental strain. Moreover, the proportion of LL-A_2pm was higher in the main monomer than in the main dimer, where it was almost exclusively located in the donor unit. There are thus very few D-alanyl-LL-A_2pm cross-bridges, if any. We also observed that large amounts of LL-A_2pm and *N*-succinyl-LL-A_2pm were excreted in the growth medium by the *dapF* mutant. These results show that the bacterial cell can adjust to the blocking of an apparently essential step by accepting important modification (high LL-A_2pm pool, low extent of peptidoglycan cross-linkage, etc.) and it is thereby doubtful that the A_2pm epimerase can be an efficient target for new antibacterial agents.

FORMATION OF TRIPEPTIDE LIPIDS I AND II

It was recently proposed that in the cell cycle of *E. coli* the alternating phases of cell elongation and septation depend on a regularly shifting balance between activities of two competing morphogenetic systems (Begg *et al.*, 1990). In growth by elongation of the cylindrical cell transpeptidation would be carried out preferentially with pentapeptide side chains, whereas tripeptide subunits would be the preferred acceptors for transpeptidation carried out by PBP3 during septum formation. Variations in the tripeptide subunit content would determine the shift from cell elongation to septation.

In peptidoglycan of normally growing cells mainly tetrapeptides are present and the tripeptide subunit content is fairly low and it does not seem to vary significantly with elongation or septation (Glauner *et al.*, 1988 ; Nanninga, 1991). Whatever the case, the presence of tripeptide subunits in peptidoglycan can *a priori* be explained in two ways. Either they arise by polymerization of tripeptide lipid II after formation of tripeptide lipids from UDP-MurNAc tripeptide or they are formed by action of a DL-carboxypeptidase on peptidoglycan tetrapeptide subunits (Beck and Park, 1976 ; Izaki and Strominger, 1968).

The possible presence of tripeptide lipids I and II was investigated (van Heijenoort *et al.*, 1992). First, it was established that tripeptide lipid II could be made from UDP-MurNAc-tripeptide (^{14}C-A_2pm labelled) with a cell free system. Thereafter it was isolated on a preparative scale, purified and shown to be used as substrate in the *in vitro* polymerization assay catalyzed by PBP1b as efficiently as pentapeptide lipid II. However, the formation of tripeptide lipid II itself in the cell-free system was 3 to 4 times less efficient than with the pentapeptide nucleotide and it was therefrom estimated that, if present in cells, the tripeptide lipids I and II were at no more than 100 copies per cell. A method was developed for analyzing tripeptide lipid cell contents (use of strain FB8 *lysA* labelled with ^{3}H-A_2pm and of HPLC detection of reduced muropeptides). Reduced murotripeptide from the membrane fraction was clearly identified in D-cycloserine treated cells but with more difficulty in untreated cells where it accounted for less than 100 molecules of tripeptide lipids per cell. However, to truly discriminate between the *in vivo* use of the tripeptide lipid in polymerization reactions and *in vivo* modifications of peptidoglycan, control experiments with a DL-carboxypeptidase deficient strain will have to be carried out.

ACKNOWLEDGMENTS

This work was supported by grants from the Centre National de la Recherche Scientifique (URA 1131), the Institut National de la Santé et de la Recherche Médicale (n° 90 0314), and the Actions Intégrées Franco-Espagnoles (1989-1990-1991).

REFERENCES

Abo-Ghalia, M., Michaud, C., Blanot, D. and van Heijenoort, J. (1985) Specificity of the uridine-diphosphate-*N*-acetylmura myl-L-alanyl-D-glutamate : *meso*-2,6-diaminopimelate synthetase from *Escherichia coli*. Eur. J. Biochem. 153, 81-87.

Abo-Ghalia, M., Flegel, M., Blanot, D., and van Heijenoort, J. (1988) Synthesis of inhibitors of the *meso*-diaminopimelate-adding enzyme from *Escherichia coli*. Int. J. Peptide Protein Res. 32, 208-222.

Anwar, R.A. and Vlaovic, M. (1978) Purification of UDP-*N*-acetylenolpyruvoylglucosamine reductase from *Escherichia coli* by affinity chromatography, its subunit structure and the absence of flavin as the prosthetic group. Can. J. Biochem. 57, 188-196.

Beck, B.D. and Park, J.T. (1976) Activity of three murein hydrolases during the cell division cycle of *Escherichia coli* K12 as measured in toluene-treated cells. J. Bacte riol. 126, 1250-1260.

Begg, K.J., Takasuga, A., Edwards, D.H., Dewar, S.J., Spratt, B.G., Adachi, H., Ohta, T., Matsuzawa, H. and Donachie, W.D. (1990) The balance between different peptidoglycan precursors determines whether *Escherichia coli* cells will elongate or divide. J. Bacteriol. 172, 6697-6703.

Berges, D.A., Dewolf, W.E., Jr, Dunn, G.L., Grappel, S.F., Newman, D.J., Taggart, J.J. and Gilvarg, C. (1986) Peptides of 2-aminopimelic acid : antibacterial agents that inhibit diaminopimelic acid biosynthesis. J. Med. Chem. 29, 89-95.

Cohen, G. and Saint-Girons, I. (1987) Biosynthesis of threonine, lysine and methionine, in "*Escherichia coli* and *Salmonella typhimurium* : cellular and molecular biology", pp. 429-444" ASM, Washington, D.C.

De Roubin, M.R., Mengin-Lecreulx, D. and van Heijenoort, J. (1992) Peptidoglycan biosynthesis in *Escherichia coli* : variations in the metabolism of alanine and D-alanyl-D-alanine. J. Gen. Microbiol. (in press).

Doublet, P., van Heijenoort, J. and Mengin-Lecreulx, D. (1992) Identification of the *murI* and *murB* genes coding for cytoplasmic peptidoglycan synthetases in the 90-min region of the *E. coli* chromosome. This issue.

Duncan, K., van Heijenoort, J. and Walsh, C.T. (1990) Purifica tion and characterization of the D-alanyl-D-alanine-adding enzyme from *Escherichia coli*. Biochem. 29, 2379-2386.

Flouret, B., Mengin-Lecreulx, D. and van Heijenoort, J. (1981) Reverse-phase high pressure liquid chromatography of uridine diphosphate *N*-acetylmuramylpeptide precursors of bacterial cell wall peptidoglycan.Anal.Biochem. 114, 59-63.

Gale, E.F., Cundliffe, E., Reynolds, P.E., Richmond, M.H. and Waring, M.J. (1972) Inhibitors of bacterial cell wall synthesis, in "The molecular basis of antibiotic action". pp. 49-174 John Wiley & Sons, London.

Geis, A. and Plapp, R. (1978) Phospho-*N*-acetylmuramoyl-pentapep tide-transferase of *Escherichia coli* K12. Properties of the membrane bound and the extracted and partially purified enzyme. Biochim. Biophys. Acta. 527, 414-424.

Glauner, B., Höltje, J.V. and Schwarz, U. (1988) Composition of the murein of *Escherichia coli*. J. Biol. Chem. 263, 10088-10095.

Ikeda, M., Wachi, M., Jung, H.K., Ishino, F. and Matsuhashi, M. (1991) The *Escherichia coli mraY* gene encoding UDP-*N*-acetylmuramoyl-pentapeptide:undecaprenyl-phosphate phospho-*N*-acetylmuramoyl-pentapeptide transferase. J. Bacteriol. 173, 1021-1026.

Ishiguro, E.E. and Ramey, W. (1978) Involvement of the *relA* gene product and feedback inhibition in the regulation of UDP-*N*-acetylmuramyl-peptide synthesis in *Escherichia coli*. J. Bacteriol. 135, 766-774.

Izaki, K. and Strominger, J.L. (1968) Biosynthesis of the peptidoglycan of bacterial cell walls. XIV. Purification and properties of two D-alanine carboxypeptidases from *Escherichia coli*. J. Biol. Chem. 245, 3660-3666.

Kohlrausch, U., Wientjes, F.B. and Höltje, J.V. (1989) Determi nation of murein precursors during the cell cycle of *Escherichia coli*. J. Gen. Microbiol. 135, 1499-1506.

Le Roux, P., Blanot, D., Mengin-Lecreulx, D. and van Heijenoort, J. (1991) Peptides containing 2-aminopimelic acid. Synthesis and study of *in vitro* effects on bacterial cells. Int. J. Peptide Protein Res. 37, 103-111.

Lugtenberg, E.J.J., de Hass-Menger, L. and Ruyters, W.H.M. (1972) Murein synthesis and identification of cell wall precursors of temperature sensitive lysis mutants of *Escherichia coli*. J. Bacteriol. 109, 326-335.

Lugtenberg, E.J.J. (1972) Studies on *Escherichia coli* enzymes involved in the synthesis of uridine diphosphate-*N*-acetylmuramyl-pentapeptide. J. Bacteriol. 110, 26-34.

Mengin-Lecreulx, D., Flouret, B. and van Heijenoort, J. (1982) Cytoplasmic steps of peptidoglycan synthesis in *Escherichia coli*. J. Bacteriol. 151, 1109-1117.

Mengin-Lecreulx, D., Flouret, B. and van Heijenoort, J. (1983) Pool levels of UDP-*N*-acetylglucosamine and UDP-*N*-acetylglucosamine-enolpyruvate in *Escherichia coli* and correlation with peptidoglycan synthesis. J. Bacteriol. 154, 1284-1290.
Mengin-Lecreulx, D. and van Heijenoort, J. (1985) Effect of growth conditions on peptidoglycan content and cytoplasmic steps of its biosynthesis in *Escherichia coli*. J. Bacteriol. 163, 208-212.
Mengin-Lecreulx, D., Michaud, C., Richaud, C., Blanot, D. and van Heijenoort J. (1988) Incorporation of LL-diaminopimelic acid into peptidoglycan of *Escherichia coli* mutants lacking diaminopimelate epimerase encoded by *dapF*. J. Bacteriol. 170, 2031-2039.
Mengin-Lecreulx, D., Parquet, C. Desviat, L.R., Pla, J., Flouret, B., Ayala, J.A. and van Heijenoort, J. (1989 *a*) Organization of the *murE-murG* region of *Escherichia coli*: identification of the *murD* gene encoding the D-glutamic acid-adding enzyme. J. Bacteriol. 171, 6126-6134.
Mengin-Lecreulx, D., Siegel, E. and van Heijenoort, J. (1989 *b*) Variations in UDP-*N*-acetylglucosamine and UDP-*N*-acetylmuramyl-pentapeptide pools in *Escherichia coli* after inhibition of protein synthsis. J. Bacteriol. 171, 3282-3287.
Mengin-Lecreulx, D. and van Heijenoort, J. (1990) Correlation between the effects of fosfomycin and chloramphenicol on *Escherichia coli*. FEMS Microbiol. Lett. 66, 129-134.
Mengin-Lecreulx, D., Texier, L., Rousseau, M. and van Heijenoort, J. (1991) The *murG* gene of *Escherichia coli* codes for the UDP-*N*-acetylglucosamine : *N*-acetylmuramyl-(pentapeptide) pyrophosphoryl-undecaprenol *N*-acetylglucosamine transferase involved in the membrane steps of peptidoglycan synthesis. J. Bacteriol. 173, 4625-4636.
Michaud, C., Blanot, D., Flouret, B. and van Heijenoort, J. (1987) Partial purification and specificity studies of the D-glutamate-adding and D-alanyl-D-alanine-adding enzymes from *Escherichia coli*. Eur. J. Biochem. 166, 631-637.
Michaud, C., Mengin-Lecreulx, D., van Heijenoort, J. and Blanot, D. (1990) Over-production, purification and properties of the uridine-diphosphate-*N*-acetylmuramyl-L-alanyl-D-glutamate : *meso*-2,6-diaminopimelate ligase from *Escherichia coli*. Eur. J. Biochem. 194, 853-861.
Nanninga, N. (1991) Cell division and peptidoglycan assembly in *Escherichia coli*. Mol. Microbiol. 5, 791-795.
Park, J.T. (1987) Murein synthesis, in "*Escherichia coli* and *Salmonella typhimurium* : cellular and molecular biology" pp. 663-671. ASM, Washington, D.C.
Patte, J.C. (1983) Diaminopimelate and lysine, in "Amino Acids: biosynthesis and genetic regulation", pp. 213-228. Addison-Wesley Publishing Company. Reading, Mass.
Pratviel-Sosa, F., Mengin-Lecreulx, D. and van Heijenoort, J. (1991) Over-production, purification and properties of the uridine diphosphate *N*-acetylmuramyl-L-alanine : D-glutamate ligase from *Escherichia coli*. Eur. J. Biochem. 202, 1169-1176.

Pucci, M.J., Discotto, L.F. and Dougherty, T.J. (1992) Cloning and identification of the *Escherichia coli murB* DNA sequence, which encodes UDP-*N*-acetylenolpyruvoylglucosamine reductase. J. Bacteriol. 174, 1690-1693.

Richaud, C., Higgins, W., Mengin-Lecreulx, D. and Stragier, P. (1987) Molecular cloning, characterization, and chromosomal localization of *dapF*, the *Escherichia coli* gene for diaminopimelate epimerase. J. Bacteriol. 169, 1454-1459.

Rogers, H.J., Perkins, H.R. and Ward, J.B. (1980) Biosynthesis of peptidoglycan, in "Microbial cell walls and membranes", pp. 239-297. Chapman & Hall, Ltd., London.

van Heijenoort, Y., Gomez, M., Derrien, M., Ayala, J.A. and van Heijenoort, J. (1992) Membrane intermediates in the peptidoglycan metabolism of *Escherichia coli* : possible roles of PBP1b and PBP3. J. Bacteriol. (in press).

Zawadzke, L.E., Bugg, T.D.H. and Walsh, C.T. (1991) Existence of two D-alanine : D-alanine ligases in *Escherichia coli:* cloning and sequencing of the *ddlA* gene and purification and characterization of the DdlA and DdlB enzymes. Biochemistry 30, 1673-1682.

IDENTIFICATION OF THE *murI* and *murB* GENES CODING FOR CYTOPLASMIC PEPTIDOGLYCAN SYNTHETASES IN THE 90-MIN REGION OF THE *E. COLI* CHROMOSOME

Patricia Doublet, Jean van Heijenoort
and Dominique Mengin-Lecreulx*

URA 1131 du C.N.R.S.
Enveloppes bactériennes et peptides
Université Paris-Sud, Bât. 432
91405 Orsay Cedex, France

INTRODUCTION

The main cytoplasmic precursors of peptidoglycan synthesis are a series of uridine nucleotides which have been previously characterized in *Escherichia coli* as UDP-*N*-acetylglucosamine (UDP-GlcNAc), UDP-GlcNAc-enolpyruvate, UDP-*N*-acetylmuramic acid (UDP-MurNAc) and its four peptide derivatives from UDP-MurNAc-L-Ala to UDP-MurNAc-L-Ala-γ-D-Glu-*meso*-DAP-D-Ala-D-Ala. Their sequential formation is catalyzed by a set of eight highly specific enzymes which were earlier studied in detail (Rogers *et al.*,1980; Mengin-Lecreulx and van Heijenoort, 1985).

Conditional lethal mutants of *E. coli* with certain of these enzymes altered have been described and most of the mutations have been mapped in several regions of the chromosome (Fletcher *et al.*, 1978; Lugtenberg *et al.*, 1973; Miyakawa *et al.*, 1972; Wijsman, 1972). One of these regions at 2 min was studied in detail since it contained a large cluster of genes from *pbpB* to *envA* that code for proteins all involved in different aspects of peptidoglycan synthesis and cell division. The *murC*, *murD*, *murE*, *murF* and *ddl* genes coding for the L-alanine, D-glutamic acid, *meso*-DAP and D-alanyl-D-alanine adding enzymes and the D-alanine:D-alanine ligase, respectively, as well as the two genes *mraY* and *murG* involved in the synthesis of lipid intermediates from the nucleotide precursors were identified in this particular region (Mengin-Lecreulx *et al.*, 1989; Ikeda *et al.*, 1991; Mengin-Lecreulx *et al.*, 1991).

Bacterial Growth and Lysis, Edited by M.A. de Pedro
et al., Plenum Press, New York, 1993

However, not all the genes for cytoplasmic peptidoglycan synthetases are found in the 2-min region. The *murA* and *murB* genes coding respectively for the transferase and reductase activities which catalyze the two-step formation of UDP-MurNAc, and the gene we named *murI* which is required for the biosynthesis of D-glutamic acid, were mapped previously in a completely separate region of the chromosome (Venkateswaran and Wu, 1972; Miyakawa *et al.*, 1972; Lugtenberg *et al.*, 1973).

Here we report the identification of the *murI* and *murB* DNA sequences within the same chromosomal region at 90 min on the *E. coli* map.

IDENTIFICATION OF THE *murI* GENE REQUIRED FOR D-GLUTAMIC ACID SYNTHESIS IN *E. COLI*

Properties of a D-glutamic acid requiring mutant

A mutant WM335 of *E. coli* B/r auxotrophic for D-glutamic acid (Lugtenberg *et al.*, 1973) was used as a tool to clone the *murI* gene of *E. coli* K12. It resembles other previously described mutants with defects in peptidoglycan synthesis in terms of the lytic phenotype it exhibits under restrictive growth conditions.

When deprived of D-glutamic acid, mutant cells accumulate UDP-MurNAc-L-Ala while pools of precursors located downstream in the pathway are rapidly depleted, resulting in a decreased rate of peptidoglycan synthesis and finally cell lysis.

The biosynthesis of D-glutamic acid was most probably altered in this mutant but attempts to identify the enzymatic defect were up to now unsuccessful (Lugtenberg *et al.*, 1973).

Mapping of the mutation

A preliminary genetic location of the mutation was earlier determined between the *rpsL* marker at 73-min and the origin of transfer of the HfrH strain at 98 min (Lugtenberg *et al.*, 1973).

We localized the mutation more precisely in conjugation and transduction experiments using donor strains with selectable transposon Tn*10* inserted at different sites of the chromosome. We observed that D-glutamic acid prototrophy was linked only to markers from the 90 min region of the *E. coli* map: 36%, 71% and 25% of linkage were estimated with *argE* (89.8 min), *thi*-39 (90.5 min) and *metA* (91 min), respectively.

Complementation of the mutation

A prerequisite to complementation experiments using *E. coli* genomic libraries constructed in phage or phasmid vectors was that recipient cells were sensitive and immune to lambda. Since WM335 was lambda resistant (*lamB* mutation), we thus constructed plasmid pPDM3 carrying both *lamB* (receptor to lambda) and *cI857* (thermosensitive form of the lambda repressor) genes.

WM335(pPDM3) cells were first infected by phages from the collection of Kohara *et al.* (1987) that carry chromosomal inserts from the 87-92-min region of the *E. coli* map. Only phage clone λE11C11 (clone number 534) was able to complement the auxotrophy for D-glutamic acid of strain WM335 (figure 1).

We also assayed for complementation the whole *Sau*3A genomic library of Elledge and Walker (1985) prepared in vector λSE6 which confers kanamycin resistance and is capable of growing lysogenically as a phasmid in an immune host. Five different Ap^r Km^r clones that could grow at 30°C in the absence of D-glutamic acid were isolated by this procedure. The characterization of

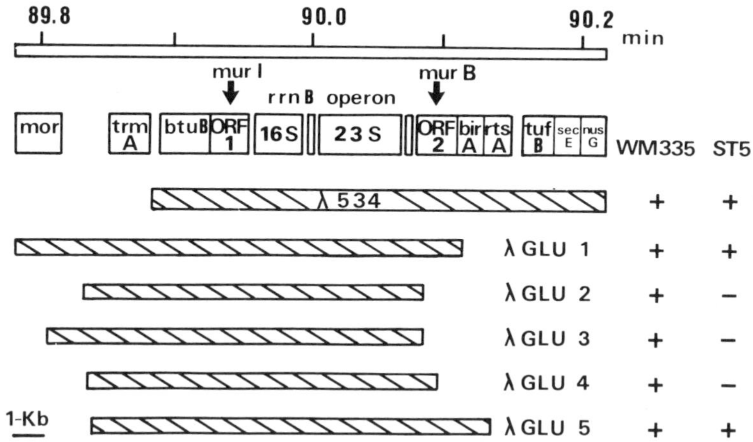

Figure 1. Complementation of the *murI* and *murB* mutations by phage and phasmid clones from different *E. coli* genomic libraries. The top of the figure shows the location of the genes from the 90-min region in relation to the physical map of the *E. coli* chromosome (Kohara *et al.*, 1987; Medigue *et al.*, 1990; Kröger *et al.*, 1991). Bacterial DNA present in λ534 and in the five λGLU phasmids that confer D-glutamic acid prototrophy to WM335 cells is represented (hatched regions). The ability of these different DNA sources to complement the specific defect of strain ST5 is also indicated on the right: + = complementation; - = no complementation.

the chromosomal DNA carried by the corresponding phasmids λGLU1 to λGLU5 confirmed the localization of the *murI* gene at 90 min, within a 15 kilobase-pair chromosomal fragment located between the *argECBH* cluster and the *rpoBC* genes (figure 1). DNA in this particular chromosomal region has been previously almost completely sequenced (for references, see Kröger *et al.*, 1991). The size of the chromosomal fragment present in each phasmid was precisely determined by DNA sequencing of its extremities using as primers 5'-CCGGGGCCGGGGACCTTG-3' and 5'-GGTTCTCTCCAGAGGTTC-3' oligonucleotides corresponding to sequences chosen adjacent to the insert in the left and right arms of the λSE6 vector.

Phage λ534 and λGLU phasmids carried in common 9-kb of chromosomal DNA that contained only one open reading frame of unknown function: ORF1 (Brosius *et al.*, 1981), flanked on the left by the *btuB* gene coding for the vitamin B_{12} receptor protein (Heller and Kadner, 1985) and on the right by the *rrnB* operon (Brosius *et al.*, 1981) that contains the genes for 16S, 23S, and 5S ribosomal RNAs (figure 1). ORF1 was thus the only candidate for being the *murI* gene required for D-glutamic acid synthesis in *E. coli*.

Cloning of the *murI* gene

The 2.5-kb *Eco*RI-*Eco*RI fragment that contains only ORF1 was inserted into the corresponding site of pUC19 vector. WM335 was transformed by this ligation mixture and transformants were directly selected for acquisition of both ampicillin resistance and D-glutamic acid prototrophy. All the Ap^r transformants chosen that grew in the absence of added D-glutamic acid were shown to contain the plasmid pPDM25 with the insert in the orientation dependent on *lac* promoter control (figure 2).
A 0.6-kbp *Pst*I deletion of the left of the insert generated pPDM26 which no longer complemented the mutation (figure 2). This result was consistent with the identification of the *murI* gene as the ORF1 which begins within the end of the preceding *btuB* gene at a putative initiation codon located upstream from the *Pst*I site (Brosius *et al.*, 1981; Heller and Kadner, 1985).

IDENTIFICATION OF THE *murB* GENE OF *ESCHERICHIA COLI* ENCODING UDP-*N*-ACETYLGLUCOSAMINE-ENOLPYRUVATE REDUCTASE

It was previously shown by Miyakawa *et al.* (1972) that the *murB* gene was located together with other genes related to peptidoglycan metabolism in the same chromosomal region close to *argH* they named *mrb* (for murein region b). It was therefore tempting to speculate that the *murB* gene might be located in the vicinity of the *murI* gene.

We thus assayed our five λGLU phasmids for their ability to complement the specific defect of thermosensitive mutant strain ST5 of *E. coli* altered in the *murB* gene (Miyakawa *et al.*, 1972). We observed that growth at 42°C was effectively restored for ST5(λ^+) cells harboring two of these phasmids : λGLU1 and λGLU5. When comparing the chromosomal fragments carried by the five phasmids (figure 1), it was clear that complementation occurred only when the whole sequence of open reading frame ORF2 (Brosius *et al.*, 1981) of unknown function was present . This ORF2 was located between the *rrnB* operon and the *birA* gene required for biotin synthesis (Brosius *et al.*, 1981; Howard *et al.*, 1985), 6-kb clockwise with respect to the *murI* gene.

The identification of the *murB* gene as this ORF2 was further substantiated by subcloning different fragments

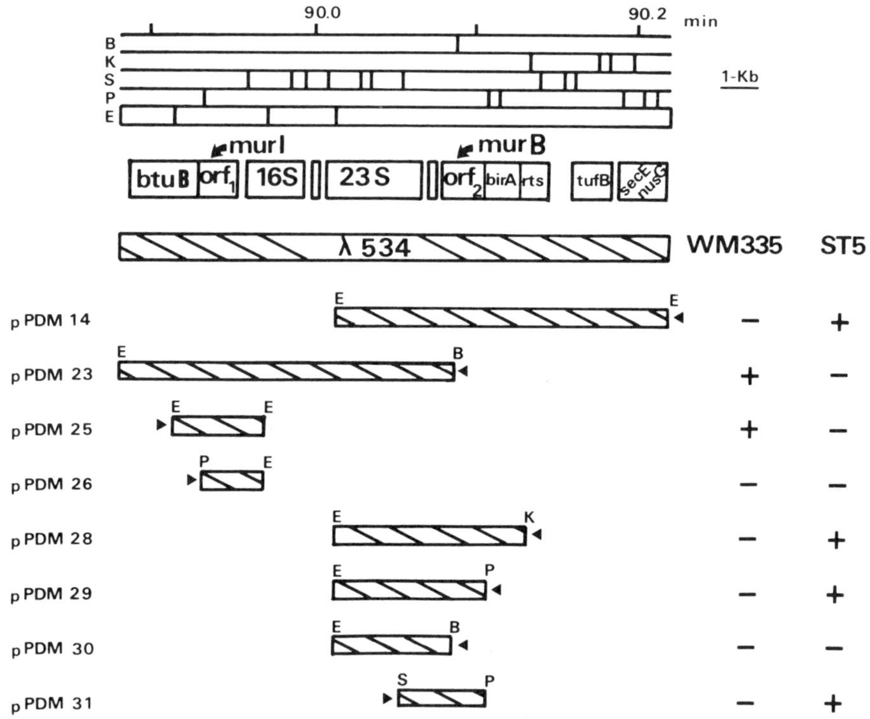

Figure 2. Identification of the *murI* and *murB* DNA sequences. The top of the figure shows the location of the genes from the 90-min region in relation to the physical map of the *E. coli* chromosome. The position of cleavage sites is indicated for the following enzymes: B, *Bam*HI; K, *Kpn*I; S, *Sty*I; P, *Pst*I; E, *Eco*RI. Plasmid inserts derived from λ534 DNA are shown below and the position of the *lac* vector promoter relative to the inserts is indicated by an arrow. The ability of plasmids to complement the specific defect of WM335 and ST5 strains is indicated on the right : + = complementation ; - = no complementation.

originating from λ534 DNA into the pUC19 vector (figure 2). We observed that plasmid pPDM31 in which the chromosomal DNA was reduced to a 2.4-kb *Sty*I-*Pst*I fragment (containing only ORF2) was still able to complement the specific defect of strain ST5.

DISCUSSION

Here we report the identification of the gene we named *murI* which is required for the synthesis of D-glutamic acid and the *murB* gene encoding UDP-GlcNAc-enolpyruvate reductase in the 90-min region of the *E. coli* chromosome. The *murI* gene corresponds to a previously sequenced open reading frame of 867-bp (Genbank accession number J01695, Brosius *et al.*, 1981) calculated to encode a 289-aminoacid polypeptide of 31,500 Da. The *murB* gene corresponds to another sequenced open reading frame located 6-kb clockwise of *murI* (Genbank accession number M10123, Howard *et al.*, 1985) which theoretically encodes a 342-aminoacid polypeptide of 37,827 Da. This latter value is in agreement with the molecular weight of the purified reductase earlier estimated to be 35,500 by Anwar and Vlaovic (1978). In a recent paper, Pucci *et al.* (1992) also report the identification and DNA sequence of the *murB* gene of *E. coli* B.

With the recent identification of the *rtsA* gene product as the pantothenate kinase (Song and Jackowski, 1992), the cellular functions of all the genes in this particular region have now been clearly established. The precise map position of the *murA* gene encoding the PEP:UDP-GlcNAc enolpyruvyl transferase remains to be identified. However, it is clear that the *murA*, *murB*, and *murI* genes do not belong to a cluster of tightly packed genes of related functions as is the case for the other *mur* genes in the 2-min region (Mengin-Lecreulx *et al.*, 1989).

ACKNOWLEDGMENTS

We thank B. Bachmann for supplying bacterial strains and Y. Kohara and J.P. Bouché for providing phage clones. We are grateful to J. Caillet for the generous gift of the phasmid library. This work was supported by grants from the Centre National de la Recherche Scientifique (URA 1131) and the Institut National de la Santé et de la Recherche Médicale (900314). P. Doublet was supported by a scholarship from the Ministère de la Recherche et de la Technologie.

REFERENCES

Anwar, R.A. and Vlaovic, M. (1978) Purification of UDP-*N*-acetylenolpyruvoylglucosamine reductase from *Escherichia coli* by affinity chromatography, its subunit structure and the absence of flavin as the prosthetic group. Can. J. Biochem. 57, 188-196.

Brosius, J., Dull, T.J., Sleeter, D.D. and Noller, H.F. (1981) Gene organization and primary structure of a ribosomal RNA operon from *Escherichia coli*. J. Mol. Biol. 148, 107-127.

Elledge, S.J. and Walker, G.C. (1985) Phasmid vectors for identification of genes by comlementation of *Escherichia coli* mutants. J. Bacteriol. 162, 777-783.

Fletcher, G., Irwin, C.A., Henson, J.M., Fillingim, C., Malone, M. and Walker, J.R. (1978) Identification of the *Escherichia coli* cell division gene *sep* and organization of the cell division-cell envelope genes in the *sep-mur-ftsA-envA* cluster as determined with specialized transducing lambda bacteriophages. J. Bacteriol. 133, 91-100.

Heller, K. and Kadner, R.J. (1985) Nucleotide sequence of the gene for the vitamin B_{12} receptor protein in the outer membrane of *Escherichia coli*. J. Bacteriol. 161, 904-908.

Howard, P.K., Shaw, J. and Otsuka, A.J. (1985) Nucleotide sequence of the *birA* gene encoding the biotin operon repressor and biotin holoenzyme synthetase functions of *Escherichia coli*. Gene 35, 321-331.

Ikeda, M., Wachi, M., Jung, H.K., Ishino, F. and Matsuhashi, M. (1991) The *Escherichia coli mraY* gene encoding UDP-*N*-acetylmuramoyl-pentapeptide:undecaprenyl-phosphate phospho-*N*-acetylmuramoyl-pentapeptide transferase. J. Bacteriol. 173, 1021-1026.

Kohara, Y., Akiyama, K. and Isono, K. (1987) The physical map of the whole *E. coli* chromosome: application of a new strategy for rapid analysis and sorting of a large genomic library. Cell 50, 495-508.

Kröger, M., Wahl, R. and Rice, P. (1991) Compilation of DNA sequences of *Escherichia coli* (update 1991). Nucleic Acids Res. 19 Supplement, 2023-2043.

Lugtenberg, E.J.J., Wijsman, H.J.W. and van Zaane, D. (1973) Properties of a D-glutamic acid-requiring mutant of *Escherichia coli*. J. Bacteriol. 114, 499-506.

Medigue, C., Bouché, J.P., Henaut, A. and Danchin, A. (1990) Mapping of sequenced genes (700 kbp) in the restriction map of the *Escherichia coli* chromosome. Mol. Microbiol. 4, 169-187.

Mengin-Lecreulx, D. and van Heijenoort, J. (1985) Effect of growth conditions on peptidoglycan content and cytoplasmic steps of its biosynthesis in *Escherichia coli*. J. Bacteriol. 163, 208-212.

Mengin-Lecreulx, D., Parquet, C., Desviat, L.R., Pla, J., Flouret, B., Ayala, J. and van Heijenoort, J. (1989) Organization of the *murE-murG* region of *Escherichia coli*: identification of the *murD* gene encoding the D-glutamic acid-adding enzyme. J. Bacteriol. 171, 6126-6134.

Mengin-Lecreulx, D., Texier, L., Rousseau, M. and van Heijenoort, J. (1991) The *murG* gene of *Escherichia coli* codes for the UDP-*N*-acetylglucosamine:*N*-acetylmuramyl-(pentapeptide) pyrophosphoryl-undecaprenol *N*-acetylglucosamine transferase involved in the membrane steps of peptidoglycan synthesis. J. Bacteriol. 173, 4625-4636.

Miyakawa, T., Matsuzawa, H., Matsuhashi, M. and Sugino, Y. (1972) Cell wall peptidoglycan mutants of *Escherichia coli* K-12: existence of two clusters of genes, *mra* and *mrb*, for cell wall peptidoglycan biosynthesis. J. Bacteriol. 112, 950-958.

Pucci, M.J., Discotto, L.F. and Dougherty, T.J. (1992) Cloning and identification of the *Escherichia coli murB* DNA sequence, which encodes UDP-*N*-acetylenolpyruvoylglucosamine reductase. J. Bacteriol. 174, 1690-1693.

Rogers, H.J., Perkins, H.R. and Ward, J.B. (1980) Biosynthesis of peptidoglycan, in "Microbial cell walls and membranes", pp. 239-297. Chapman & Hall, Ltd., London.

Song, W.J. and Jackowski, S. (1992) *coaA* and *rts* are allelic and located at kilobase 3532 on the *Escherichia coli* physical map. J. Bacteriol. 174, 1705-1706.

Venkateswaran, P.S. and Wu, H.C. (1972) Isolation and characterization of a phosphonomycin-resistant mutant of *Escherichia coli* K-12. J. Bacteriol. 110, 935-944.
Wijsman, H.J.W. (1972) A genetic map of several mutations affecting the mucopeptide layer of *Escherichia coli*. Genet. Res. 20, 65-74.

AMINO ACIDS AS USEFUL TOOLS IN THE STUDY OF MUREIN METABOLISM IN *Escherichia coli*

Marta Caparrós, José C. Quintela, José I. Leguina, and
Miguel A. de Pedro

Centro de Bilogía Molecular C.S.I.C.-U.A.M.
Facultad de Ciencias Universidad Autónoma de Madrid
Campus de Cantoblance
28049 Madrid, Spain

INTRODUCTION

The murein sacculus can be envisaged as a bacterial exoskeleton fulfilling mechanical and morphogenetic functions. Metabolism of the murein sacculus is an essential process that compares in complexity with chromosome replication and protein synthesis in bacteria (Höltje and Schwarz, 1985; Höltje and Tuomanen, 1991).

The chemical nature and layout of the sacculus impose very severe constraints on its biosynthesis. Cell growth is strictly dependent on the enlargement of the murein layer and cell division requires concomitant synthesis of a transverse wall, the septum, at a precise place and time. To complicate the situation further, continuity of the sacculus must be warranted throughout the process. Otherwise mechanical stability of the cell wall would be insufficient to stand the turgor pressure of the cell and lysis would follow (Koch, 1988).

Impressive advances took place during the last few years in the investigation of the biochemistry and genetics of murein biosynthesis. However, a large number of crucial questions ranging from the mode of insertion of new precursors to the very same structure of the murein sacculus remain open.

Traditionally murein synthesis has been divided into two different stages: synthesis of the membrane-bound activated precursor disaccharide pentapeptide-P-P-bactoprenol, and integration of the activated precursor into macromolecular murein by means of tansglycosylation/transpeptidation reactions (Höltje and Schwarz, 1985; Waxman and Strominger, 1983). These stages differ basically in location and nature. Whilst, the former is cytoplasmic and occurs basically in a liquid phase, in the later membrane associated enzymes catalyze reactions among membrane anchored (disaccharide pentapeptide-P-P-bactoprenol) and immobilized (murein) substrates in the periplasmic space.

A point of particular relevance has been the realization that the murein sacculus of Gram-negative bacteria should not be considered a static, metabolically inert structure. On the contrary, evidence indicates that it is subjected to a dynamic and complex metabolism, throughout the life span of the cell. Therefore, metabolism of the sacculus does not end with the insertion of new precursors into the preexisting structure, but rather includes an elaborate post-insertional processing of macromolecular murein. At least three different post-insertional processs are known at present: Maturation of newly inserted murein (de Pedro and Schwarz, 1981), adaptation of murein to resting conditions (Pisabarro *et al.,* 1985), and turnover (recycling) (Goodell and Schwarz, 1983; Godell, 1985).

The physiological meaning and detailed biochemistry of these processes is still rather obscure. Turnover is perhaps the best understood. Models for growth of the murein sacculus based on "patched structures" postulate solubilization of that fraction of the old murein which is displaced out of the stress bearing layer concomitantly with the incorporation of new precursors. Utilization (turnover) of the solubilized materials would obviously mean a significant saving in metabolic energy. The biochemistry of turnover could be basically explained on the basis of already known enzymes, at least for the periplasmic stage. Nevertheless, some gaps at the level of transport and cytoplasmic reactions still have to be filled.

Maturation and adaptive changes of murein differ widely in timing, and likely in function. Maturation specifically affects the newly made murein, whereas adaptation occurs concomitantly with cessation of active growth and represents a global structural

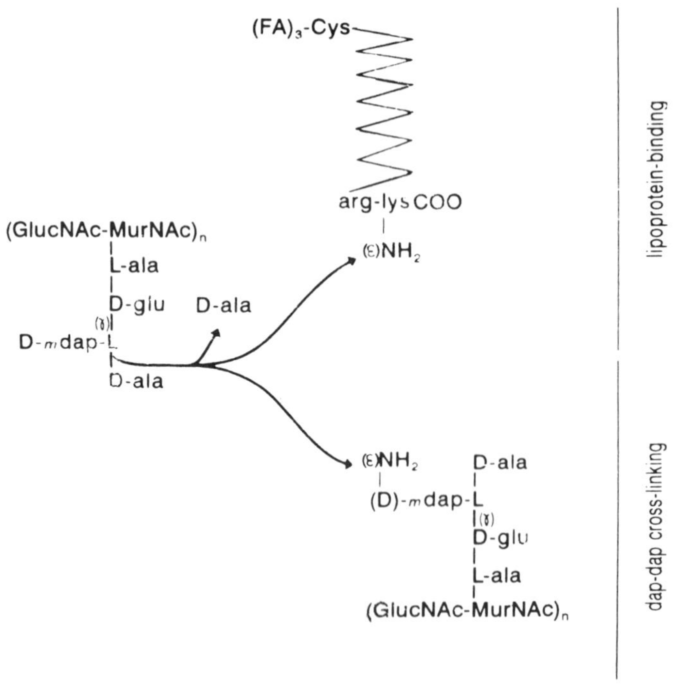

Figure 1. Schematic representation of lipoprotein binding and dap-dap crosslinking reactings.

change. The structural modifications generated by these two processes in the murein are quite similar in nature; in both cases the degree of cross-linkage and the amount of murein-bound lipoprotein increase drastically. In the case of adaptation there is also a large increase (two- to three-fold), in the proportion of the glycan chain terminating (1-6)anhydro-muramyl muropeptides (de Pedre and Schwarz, 1981; Pisabarro *et al.* 1985). The rise in cross-linkage is mainly caused by the augmentation in the proportion of diaminopimelyl-diaminopimelate cross-linked muropeptides (dap-dap muropeptides) (Glauner *et al.* 1988). Both reactions are apparently catalyzed by penicillin-insensitive enzymes, and are esentially post-insertional in the sense that newly made murein is very poor in dap-dap cross-linked muropeptides and in bound lipoprotein (Pisabarro *et al.* 1985). Neither the enzymes nor the reaction mechanisms behind these post-insertional processes are known. Nevertheless, synthesis of dap-dap muropeptides are binding of lipoprotein to murein are thought to occur by means of similar transpeptidation reactions catalyzed by specific enzymes. In both cases the reaction would imply the transfer of the LD peptide bond between *meso*-diaminopimelic acid and D-alanine at position 4 in the peptide side chain of a muropeptide, to an ε-NH_2 group in the acceptor molecule (Figure 1) which would be achiral or have a D configuration in the lipoprotein binding and dap-dap cross-linking reactions, respectively (Braun, 1975).

Investigation of the enzymatic systems involved in the metabolism of macromolecular murein, and specially on biosynthetic reactions, has been often hampered by lack of appropriate *in vitro* assay methods. The extreme complexity of topological relations among enzymes and substrates as well as the impossibility to maintain *in vitro* the physico-chemical state of the sacculus as it is *in vivo,* severly limit development of fast, reliable assay methods. A high molecular weight PBP, for instance, is a membrane bound enzyme, catalyzing a chemical reaction between a membrane anchored (C_{55}-disaccharide pentapeptide) and an insoluble (sacculus), periplasmic substrate whose geometry and energetic state are partially determined by the turgor pressure of the cell (Koch, 1988). As a matter of fact 17 years after the identification of the high Mr PBPs as crucial enzymes in murein synthesis (Spratt, 1975), no really satisfactory *in vitro* assay system has been developed yet.

Application of indirect methods offers a complementary approach. Methods based on the exploitation of side reactions catalyzed by important enzymes, and in the study of the response of the biosynthetic machinery to alterations in the normal state of the cells have provided important information. Perhaps the most dramatic demonstration for the validity of indirect approaches has been characterization of PBPs by means of β-lactam-binding assays (Spratt, 1975; Spratt, 1983; Waxman and Strominger, 1983).

In this context, we have been investigating in *Escherichia coli* the possibilities offered by two processes described years ago but not yet studied in depth; Incorporation of exogenous D-amino acids into macromolecular murein and substitution by an analog (*meso*-lanthionine) of the normal diamino acid at position three (*meso*-diaminopimelic acid). In this report we summarize our finding on these topics and comment on potential applications of these methods to study some aspects of murein metabolism, in particular the post-insertional stages.

INCORPORATION OF D-AMINO ACIDS INTO MACROMOLECULAR MUREIN OF *Escherichia coli*

Early work by Lark and Lark (1959) on the lethal effect of D-methionine against

Alcaligenes faecalis, lead to the discovery of the ability of this Gram negative organism to incorporate externally added D-amino acids in to the "R-layer," the murein sacculus nowadays (Lark *et al.* 1963). The same phenomenon was subsequently described in *E. coli* by Tsuruoka *et al.* (1984) who observed the incorporation of radioactive D-methionine into the murein sacculus of cells grown in the presence of this compound. These authors suggested that incorporation of D-methionine was an enzymatic reaction catalyzed by a β-lactam resistant enzyme. Based on the observation of a reduction in the bound to free form ration of lipoprotein, the possible involvement of the still unknown lipoprotein-binding enzyme was postulated.

The possible connection between incorporation of D-amino acids and binding of lipoprotein, one of the key reactions in post-insertional murein metabolism, moved us to re-examine this process making use of the powerful analytical techniques developed meanwhile.

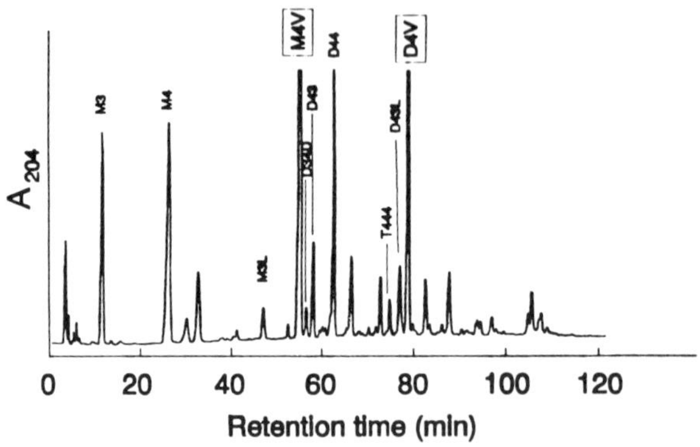

Figure 2. Muropeptide composition of D-valine modified-murein. Murein from cell grown in 20 mM D-valine for 8 generations was purified and analyzed by H.P.L.C. The graphic shows the A_{204} absorption profile of the eluate. The D-valine modified-muropeptides are indicated in boxes: M4V, D-valine modified monomer; D44V, D-valine modified cross-linked dimer. The chemical structure is described in figure 3. The more abundant normal muropeptides are labeled: M3, disaccharide tripeptide; M4, disaccharide tetrapeptide; M3L, disaccharide tripeptide bound to lipoprotein; D34D, disaccharide tripeptide-disaccharide tetrapeptide dap-dap cross-linked dimer; D43, disaccharide tetrapeptide-disaccharide tripeptide cross-linked dimer; D44, *bis*-disaccharide tetrapeptide, T444, *tris*-disaccharide tetrapeptide; D-43L, disaccharide tetrapeptide-disaccharide tripeptide cross-linked dimer bound to lipoprotein.

Some D-amino acids can be lethal to *E. coli* but only at high concentrations, normally above 6 mg/ml. D-cysteine is the exception with a minimal inhibitory concentration (M.I.C.) around 2 mg/ml (Caparró *et al.* 1991). Most of the work has been performed using D-methionine, S-methyl-D-cysteine, D-tryptophan and D-phenylalanine as representative amino acid. M.I.C.s were 6 mg/ml and 10 mg/ml for the first two and last two amino acids, respectively. At sub-lethal concentrations (below 2/3rds of the M.I.C.) D-amino acids have no detectable effects on the growth rate, morphology and viability of growing cells which are able of indefinite growth (>16 generations) (Capparrós *et al.* 1991; Caparrós *et al.*, 1992b).

Growth of *E. coli* cells in media supplemented with D-amino acids results in the accumulation of new muropeptides with a residue of the added D-amino acid into the murein sacculus. This reflected in a characteristic alteration of the normal chromatographic pattern obtained upon HPLC analysis of muramidase digests of D-amino acid modified murein, as described by Glauner *et al.* (1988).

In all instances, the new components appear as two major peaks plus a small number (2 to 5) of minor elements. The major components account for more than 95% of total new material. The retention times (Rt) of the new peaks depend on the nature of the D-amino acid, in particular on its hydrophobicity as would be expected for a reverse-phase based chromatographic analysis of series of structurally similar compounds differing in the nature of a single modifier, the D-amino acid residue (Caparrós *et al*, 1992*b*). As an example figure 2 shows the muropeptide profile obtained by HPLC analysis of murein extracted from D-valine treated cells.

Chemical and structural analyses of the major modified muropeptides accumulated in the presence of a number of different D-amino acids revealed that the structures of the early and late eluting new muropeptides were those of a disaccharide tetrapeptide and a bis-disaccharide tetrapeptide with a residue of the foreign D-amino acid at the position of the terminal amino acid instead of the normally found D-alanine residue, in all instances (Figure 3) (Caparrós *et al.* 1992*b:* Caparrós *et al.* in preparation).

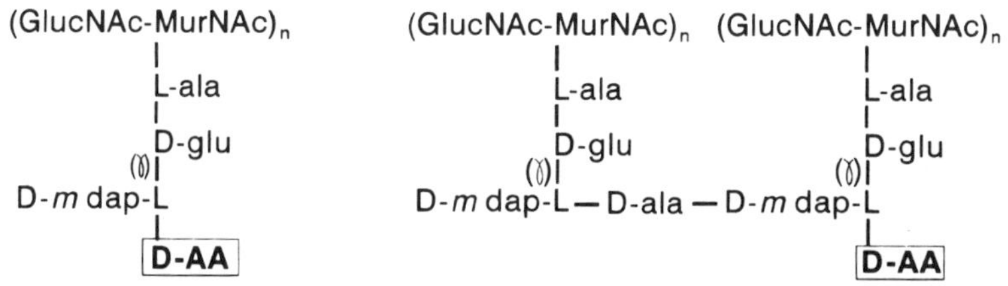

Figure 3. Structure of the major D-amino-acid modified muropeptides.

Accumulation of D-amino acid-modified muropeptides was dependent on the nature of the D-amino acid, as well as on the time of incubation and concentration. Incorporation of D-amino acids is a quantitatively important process. The proportion of modified muropeptides can account for as much as 45% of total murein, under conditions still permitting normal growth in terms of rate, morphology, and viability. Following addition of a D-amino acid the proportion of modified muropeptides increases quickly until an apparently constant level is reached. Time necessary to reach the final steady conditions seems to be around one mass doubling time, however, information is still restricted to two strains (*E. coli* MC6RP3 and W7) and a small number of D-amino acids. Incorporation of D-amino acids follows saturation kinetics as expected for an enzymatic reaction. A detailed kinetic analysis has not been possible yet. However an estimation of the apparent half saturating concentrations estimated for D-methionine, D-tryptophan and S-methyl-D-cysteine gave values in the low mmolar range (3, 6.5 and 1 mM, respectively) (Caparrós *et al.* 1992*a*; Caparrós *et al.* 1992*b*).

Interestingly incorporation of D-amino acids is not an exclusive property of growing cells. Resting cells of *E. coli* incorporate D-amino acids efficiently. For instance, stationary phase cells of *E. coli* MC6RP3 incubated in LB medium supplemented with 20 mM D-tryptophan for 12h accumulated modified muropeptides to an extent of 19%

of total murein. This value compares with the 26% found in cells growing exponentially in the same medium for 4 h (6 mass doubling times) (Caparrós *et al.* 1992*b*).

Incorporation of D-amino acids into murein is independent of the murein biosynthetic pathway. Insensitivity of the process to β-lactams discarded the involvement of penicillin-binding proteins (PBPs) in the incorporation reaction (Caparrós *et al.* 1992*b*; Tsuruoka *et al.* 1984). Furthermore, no D-amino acid-modified soluble murein precursor (modified UDP-muramyl-pentapeptide) could be found in the cytoplasmic fraction of cells grown in 20 mM D-phenylalanine, one of the better substrates for the incorporation reaction (Work done in collaboration with Dr. J. van Heijenoort. Université Paris Sud, Orsay. France).

The incorporation reaction does not require the integrity of the cell to take place. In fact, the reaction proceeds efficiently in ether permeabilized cells (Caparrós *et al.* 1992*a*). Furthermore, very recently a fully *in vitro* assay method in which purified murein is incubated with a cell extract and a radioactively labeled D-amino acid (S-[^{3}H]methyl-D-cysteine) has been developed in our laboratory (manuscript in preparation). In both cases the products of the reaction were identical to those detected *in vivo*. This indicates that incorporation of D-amino acids is a direct reaction between macromolecular murein and the D-amino acid catalyzed by an enzyme which apparently does not require an additional energy source.

Analysis of the specificity of the reaction for the soluble substrate, the D-amino acid, revealed a strict requirement for a free α-D-NH_2 group. The reaction is highly stereospecific. Achiral NH_2 groups are not recognized. Interestingly, it has been shown that the presence of additional free NH_2 groups in the molecule impede the reaction (Caparrós *et al.* 1992*b*).

Based on the information presented above, we proposed that D-amino acids were incorporated into the macromolecular murein by means of an amino acid exchange reaction catalyzed by an enzyme endowed with LD-transpeptidase activity. The reaction schematized in figure 4, would consist on the transfer of the LD-peptide bond between the L-center of dap and the terminal D-alanine in the peptide side chain of a

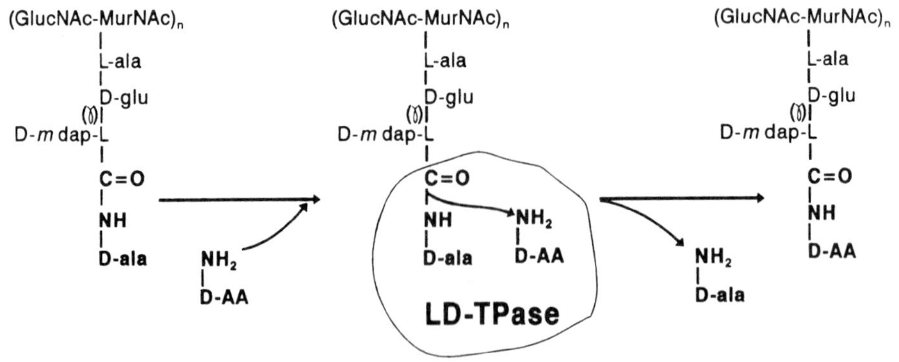

Figure 4. Schematic representation of the D-amino acid exchange reaction

muropeptide, to the α-D NH_2 group of the D-amino acid which would act as the acceptor substrate. Therefore the enzyme can be described as a murein:D-amino acid LD-transpeptidase. This reaction is formally similar to that proposed for the binding of lipoprotein to murein and almost identical to the LD-transpeptidation reaction postulated to explain the synthesis of dap-dap cross-linked muropeptides (Figure 1).

As commented before, Tsuruoka *et al.* (1984) postulated the involvement of the lipoprotein binding enzyme in the incorporation of D-amino acids. This proposal was mainly based on the apparent reduction in the ratio between the bound and free forms of lipoprotein induced by some D-amino acids. However, this hypothesis does not fit well with the high stereo-selectivity exhibited by the D-amino acid exchange reaction which does not recognize the achiral ε-NH_2 of lysine and acylated derivatives as an acceptor substrate (Caparrós *et al.* 1992*b*). Furthermore, not all incorporable D-amino acids inhibited binding of lipoprotein to murein. HPLC analysis of murein purified from cells of *E. coli* MC6RP3 grown in the presence of 20 mM D-tryptophan showed that in spite of the high modification of murein (27% of D-tryptophan modified muropeptides) the proportion of lipoprotein bound muropeptides was as in the control (9.3 and 10 % for

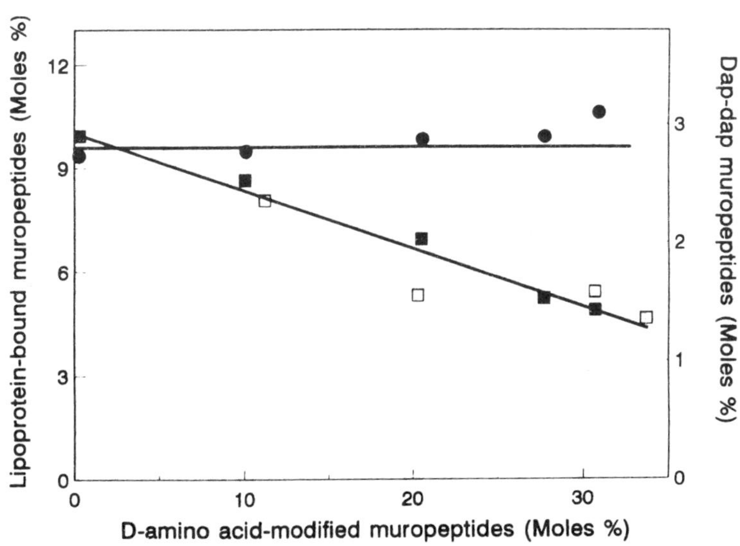

Figure 5. Effect of the accumulation of D-amino acid-modified muropeptides on the proportions of lipoprotein-bound and dap-dap cross-linked muropeptides. Murein purified from samples taken at different times from cultures of *E. coli* MC6RP3 after addition to the growth medium of 20 mM D-tryptophan (closed symbols) or D-methionine (open symbol) was analyzed by H.P.L.C. to determine the muropeptide composition. Symbols: (●) Lipoprotein-bound muropeptides in D-tryptophan modified murein; (■) dap-dap cross linked muropeptides in D-tryptophan-modified murein; (□) dap-dap cross linked muropeptides in D-methionine-modified murein

modified and control murein respectively) (Figure 5). This lack of correlation seriously questions the proposed involvement of the lipoprotein binding enzyme in the D-amino acid exchange reaction.

The differential effects of at least some D-amino acids on the abundance of dap-dap cross-linked and lipoprotein-bound muropeptides rules out the theoretical possibility of a single enzyme catalyzing both reactions (Figure 5).

In this context, the characteristics expected for the putative LD-transpeptidase responsible for dap-dap cross-linking fit better with the properties deduced for the murein:D-amino-acid LD-transpeptidase. The stereochemistry of both reactions would be identical (compare figures 1 and 4), and would be consistent with the drastic inhibition of dap-dap cross-linking induced by the incorporable D-amino acids studied at present (Figure 5). Furthermore, similar levels of murein modification by different D-amino acids lead to equivalent reductions in the proportions of dap-dap cross linked muropeptides (Figure 5).

A related aspect deserving discussion is the synthesis of the D-amino acid-modified cross-linked dimers, the second major product of the D-amino acid exchange reaction. In principle modified dimers could derive from modified monomers used as the acceptor substrate in a regular DD-transpeptidation reaction catalyzed by a PBP, or/and directly by the murein:D-amino acid LD-transpeptidase taking a normal cross-linked dimer (*bis*-disaccharide tetrapeptide) as the donor substrate. The solution of this alternative would be important to define the specificity of the later enzyme with respect to the donor substrate in the exchange reaction. As has been stated before, incorporation of D-amino acids is insensitive to β-lactam antibiotics. This allowed the examination of the products accumulated in the murein of stationary phase cells treated with D-amino acids (20 mM D-methionine or D-tryptophan) in the presence of a β-lactam able to inhibit the activity of the biosynthetic DD-transpeptidases (1 mg/ml ampicillin). Under these conditions, the relative proportion of cross-linked D-amino acid-modified muropeptides accumulated in the murein of β-lactam treated cells should be much lower than in the control if synthesis were dependent on the activity of the DD-transpeptidases. This was not observed. On the opposite, our results clearly indicated that the presence of ampicillin had no significant effect on the relative proportion of D-amino acid-modified monomers and dimers. In the case of cultures grown in 20 mM D-methionine the proportions of modified dimers were 6.1 and 5.4 % for the control and β-lactam treated samples, respectively. This result does not rule out synthesis of D-amino acid-modified dimers from modified monomers by DD-transpeptidation, but strongly indicates that the murein:D-amino acid LD-transpeptidase is able to recognize the muropeptide *bis*-disaccharide tetrapeptide as donor substrate in the D-amino acid exchange reaction. Assuming an hypothetical involvement of this enzyme in dap-dap cross-linking, its ability to recognize *bis*-disaccharide tetrapeptide as substrate would explain in a straightforward way the presence in the murein of trimeric muropeptides cross-linked by one dap-dap and one dap D-ala bridge (Glauner *et al.* 1988).

It is also interesting to notice that the ratios between the dimeric and monomeric modified muropeptides accumulated in the murein of cells treated with different D-amino acids were not significantly influenced by the nature of the D-amino acid, although the level of modification differed widely. Results obtained from cells grown in nine D-amino acids gave a mean value for the relative abundance of dimers *vs* total modified muropeptides of 20.2 $\pm$ 2 %, whilst accumulation of modified material varied from 4 (aspartic acid) to 45 % (phenylalanine) of total murein. It is also worth to point out that the relative abundance of dimers in the modified muropeptides matches well with the relative proportions of the equivalent muropeptides (disaccharide tetrapeptide and *bis*-disaccharide tetrapeptide) in non-modified murein (about 25%). These observations suggest that the murein:D-amino acid LD-transpeptidase has similar affinities for monomeric and dimeric cross-linked substrates. The difference in the abundance of the two major products would simply reflect the difference in the relative concentrations of the substrates.

Incorporation of D-amino acids into the murein sacculus is apparently well tolerated by the cells, and has no detectable effects on growth rate and cell morphology. However, a closer study of the effect of D-amino acids on murein synthesis suggests a direct inhibitory effect of D-amino acids on the activity of (some) high Mr PBPs leading to drastic reductions in the rate of synthesis and cross-linkage of murein in D-amino acid treated cells (Caparrós *et al.* 1992*b*). The ability of at least some D-amino acids to severely reduce the rate of murein synthesis (by 60 and 40% in the cases of D-methionine and D-tryptophan at 20 mM, respectively), maintaining constant the rate of growth and morphology in a stable form (>16 generations) rises a most interesting topic. In fact, such a behavior can only be explained assuming a drastic reduction in the amount of murein per unit of cell surface. In other words, *E. coli* cells are apparently able to grow and divide normally with a reduced amount of murein. Experiments aimed to the direct determination of the effect of D-amino acids on the cellular murein content are still in progress. Nevertheless, the first results confirmed our hypothesis. The amount of murein per unit of cell surface in cells grown in the presence of 20 mM D-methionine for a prolonged period of time (8 generations) was only 47% the amount in the control cells. Therefore according to our results *E. coli* is able of sustained growth with about one half the normal murein content. The ability of *E. coli* to grow with a reduced amount of murein was indicated before by our group based on a completely different experimental approach (Prats and de Pedro, 1989).

This observation has a number of important implications: i) A large proportion of the normal murein content of the cell is dispensable for growth; ii) In *E. coli* apparently there is no direct modulation of the growth rate by the murein biosynthetic capacity of the cell; iii) The murein content of the cell is likely to be variable depending on growth conditions; and iv) D-amino acids are a good experimental tool to manipulate the murein content of *E. coli* in a well defined and stable system.

D-amino acids also offer a considerable potential to study how the bacterial sacculus recovers from structural alterations *in vivo.* In fact, D-amino acids induce a number of structural alterations in the murein which are, nevertheless, compatible with an apparently normal cell growth. Therefore the system looks ideally suited to study how the murein layer is "repaired" to regain its normal structure.

At present we are studying how D-amino acids are eliminated from the murein of cells grown in their presence once the D-amino acid is removed from the medium. As for the studies on the incorporation reaction, D-methionine was selected as the representative amino acid, although others were often used.

Analysis of murein samples taken at different times from a culture of *E. coli* MC6RP1 grown in LB plus 20 mM D-methionine for 6 h, and transferred to D-methionine-free, preconditioned medium 2h after initiation of the stationary phase, showed that the amount of murein-bound D-methionine decreased during the stationary phase at a considerable rate. The proportion of modified muropeptides dropped from 43 % immediately after dilution in D-methionine-free medium, to 31 % in 3h, and to 23 % 15 h later. Interestingly, removal of D-methionine was highly preferential for the modified-monomeric muropeptide. Whilst the proportion of D-methionine modified monomer dropped from 35 to 15 % in 18h, the proportion of the modified dimer remained constant (8.2%) throughout the experimental period. Similar experiments performed in the presence of ampicillin (1 mg/ml) indicated that removal of D-methionine was insensitive to this β-lactam and therefore a PBP independent process.

Experiments performed with exponentially growing cells are rather difficult to interpret. The diluting effect of ongoing murein synthesis must be accounted for and, as indicated before, D-amino acids interfere with the biosynthetic process itself making difficult the precise estimation of the rate of accumulation of new murein. Nevertheless, our results indicate that the proportion of D-amino acid in the murein decreases considerably faster (1.5 to 2 fold) than expected from simple dilution with new material, supporting the action of an active removal mechanism. As in the case of stationary phase cells elimination seems to be consistently slower for modified dimers than for monomers (Quintela and de Pedro, in preparation).

Data available at present suggest that elimination of D-amino acids from modified murein occurs essentially by removal of the entire modified muropeptides, both in resting and growing cells. Such a mechanism would require the implication of murein hydrolases. It should be noted that D-amino acid (D-tryptophan) modified murein seems to be a better substrate for one of the lytic transglycosylases (Slt35, the former Mlt)(Höltje and Tuomanen, 1991) than control murein (Caparrós *et al.* 1991). A preferential activity *in vivo* of such an enzyme could explain the rapid elimination of modified muropeptides. Experiments performed with ether permeabilized cells indicate that the presence of D-amino acid-modified muropeptides in the murein leads to fast solubilization of a large fraction of total murein (about 50% in 2h) upon permeabilization. Murein of control cells remained stable (<10% solubilized material). This result indicates that the structural alterations of murein induced by D-amino acids have a clear influence on the regulation of murein hydrolase activities, an observation which in turn suggests that activation of murein hydrolases might be modulated by local alterations in the structure of murein. Such an ability would perfectly suit a system with "repair" potential. It would permit location and preferential release of altered material from the sacculus. Work aimed to investigate the likelihood of this speculation, and to identify the active mechanism involved in the elimination of D-amino acids is in progress at present.

SUBSTITUTION OF *m*-LANTHIONINE FOR *m*-DIAMINOPIMELIC ACID IN THE MACROMOLECULAR MUREIN OF *Escherichia coli*

A second approach to investigate the influence of a precise alteration in the structure of murein on the metabolism and physiology of the sacculus is offered by the possibility of substituting the di-amino acid (*m*-diaminopimelic acid), a key element in the determination of the overall structure of the sacculus. The study of the effects triggered by the substitution could give important information on the specificity and physiological meaning of those reactions in which the di-amino acid participates as a substrate such as DD and LD transpeptidations and binding of lipoprotein.

It is known since long that the requirement for exogenous *m*-dap of *E. coli* strains auxotrophic for this amino acid can be relieved by addition to the medium of certain analogs as *m*-lanthionine (Pelzer, 1969, Chaloupka *et al.* 1974). Lanthionine is a very close structural analog for dap, both molecules are almost identical, however the differences in the length and angle of the S-C bond relative to the C-C bond result in a slightly longer distance between the α and ε asymmetric carbons. In spite of the high structural similarity between dap and lanthionine *E. coli* is extremely selective for the former, and only upon complete depletion of dap starts the cell to use lanthionine. Interestingly the murein of some species of the Gram negative genus *Fusobacterium* contain lanthionine as the natural di-amino acid (Kato *et al.* 1979; Vaastrad *et al.* 1982).

Incorporation of lanthionine into murein apparently follows the normal biosynthetic pathway. Substitution proceeds without any apparent alterations in growth rate or morphology. When cells grown in dap-supplemented medium are transferred to lanthionine (20 μg/ml final concentration) supplemented medium, the cells keep growing exponentially at the same growth rate as in dap-medium, for as long as the culture is periodically diluted with new medium (>12 generations in our hands). However, when the culture approaches stationary phase an autolytic process is triggered, leading to the rapid loss of OD_{550} in the cultures (Figure 6). This behavior might explain the difficulty in recovering overnight stationary phase cultures grown in lathionine supplemented medium. Apparently autolysis at the cessation of growth is not linked to a progressive weakening of the cell wall. Sensitivity to a number of β-lactams were similar in lanthionine and dap grown cells (Leguina et al., manuscript in preparation).

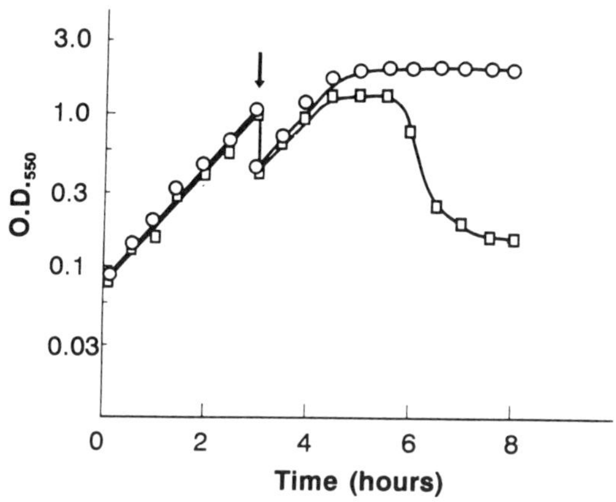

Figure 6. Growth of *E. coli* cells in medium supplemented with dap or lanthionine. Samples from an overnight culture of *E. coli* W7 (*dapA*, *lysA*)(Hartman *et al.* 1972) grown in mineral citrate medium (Vogel and Bonner, 1956)supplemented with 4 μg/ml of dap and 0.2 % glucose, were centrifuged, washed once in dap-free medium, and cells were resuspended in fresh medium supplemented either with 10 μg/ml dap or 20 μg/ml lanthionine. The subcultures were incubated at 37°C till both reached an OD_{550} of 1. At this moment both cultures were diluted to an OD_{550} of 0.07 (time 0) and growth was monitored by measurement of the OD_{550}. Symbols: (O) dap supplemented culture; (□) lanthionine supplemented culture. At the time indicated by the arrow both cultures were diluted 1:1 to keep cultures in exponential phase for one more generation.

Analysis by H.P.L.C. of murein from lanthionine and dap grown cells produced, as expected, different elution patterns (Figure 7). In general it looks as if the presence of lanthionine instead of dap would increase the Rt of the different muropeptides in a proportional way. Synthesis and utilization of radioactive lanthionine in a parallel experiment indicated the presence of the radioactive precursor in all the significant peaks (> 0.5 % of integrated area). Obviously, the change in the elution pattern of the muropeptides impedes at present any further calculation on the main parameters of the murein. However, direct estimation of the amount of murein-bound lipoprotein indicated that lanthionine has no major effect on this parameter.

An intriguing point is that comparison of the muropeptide patterns for lanthionine and dap containing mureins suggest that some muropeptides are missing in the former.

The expected positions for some of the missing peaks suggest that lanthionine-containing murein could be very poor in lanthionine-lanthionine cross-linked muropeptides, the equivalent to dap-dap muropeptides. If it were so it could be argued that dap-dap cross-linked muropeptides might be important for cell integrity when cell growth stops. In fact, it should be pointed out that dap-dap cross-linked muropeptides increase sharply at the initiation of the stationary phase (Pisabarro *et al.* 1985), precisely when lanthionine grown cells become prone to autolysis. At present this argument is purely speculative,

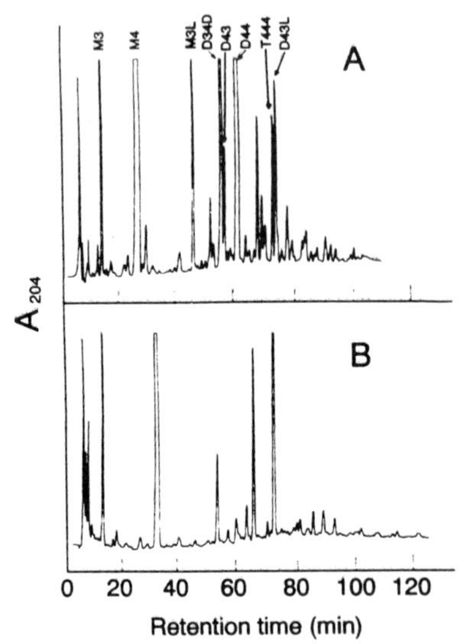

Figure 7. HPLC elution pattern of muramidase digests of murein purified from lanthionine and dap grown cells of *E. coli* W7. Murein from cells grown in the presence of 20 μg/ml lanthionine or 10 μg/ml dap for 10 generations was purified and analyzed by HPLC (Glauner *et al.* 1988). The figure shows the A_{204} registers for dap (**A**) and lanthionine (**B**) containing mureins. The position of the major muropeptides in the dap grown sample is indicated. M3, disaccharide tripeptide; M4, disaccharide tetrapeptide; M3L, disaccharide tripeptide bound to lipoprotein; D34D, disaccharide tripeptide-disaccharide tetrapeptide dap-dap cross-linked dimer; D43, disaccharide tetrapeptide-disaccharide tripeptide cross-linked dimer; D44, *bis*-disaccharide tetrapeptide, T444, *tris*-disaccharide tetrapeptide; D-43L, disaccharide tetrapeptide-disaccharide tripeptide cross-linked dimer bound to lipoprotein.

however, most of the lanthionine muropeptides have been purified and are being subjected to structural analysis, therefore we expect to get definitive information in the immediate future.

Completion of the project briefly summarized here should contribute to a better understanding of murein metabolism, and should also give important information on the flexibility of the mechanisms responsible for cell growth, morphology and division with respect to variations in the amount and properties of the murein constitutive of the cell sacculus.

ACKNOWLEDGMENTS

The technical assistance of J. de la Rosa is greatly appreciated. We wish to express our gratitude to Drs. J. van Heijenoort, U. Schwarz and J.-V. Höltje for their invaluable help and warm hospitality. This work was supported by grant BIO91-0523 from the CICYT and an institutional grant from "Fundación Ramón Areces". JIL was supported by a grant from the Basque Country Government. Parts of this work have been performed during short stays of M.C. and M.A.P. at the Max-Planck Institut für Entwicklungsbiologie (Tübingen, RFA) funded by short term fellowships from the D.A.A.D. and the M.P.G.-C.S.I.C. agreement, respectively, and at the University of Paris-Sud (Orsay, France) funded by the C.N.R.S.-C.S.I.C. bilateral agreement.

REFERENCES

Braun, V. (1975) Covalent lipoprotein from the outer membrane of *Escherichia coli.* Biochim. Biophys. Acta 415, 335-377.

Caparrós, M., Arán, V., and de Pedro, M.A. (1992*a*) Incorporation of S-[^{3}H]methyl-D-cysteine into the peptidoglycan of ether-treated cells of *Escherichia coli.* FEMS Microbiol. Lett. 93, 139-146.

Caparrós, M., . Torrecuadrada, J.L.M. and de Pedro, M.A. (1991) Effect of D-amino acids on *Escherichia coli* strains with impaired penicillin-binding proteins. Res Microbiol. 142, 345-350.

Caparrós, M., Pisabarro, A.G. and de Pedro, M.A. (1992*b*) Effect of D-amino acids in the structure and synthesis of peptidoglycan in *Escherichia coli.* J. Bacteriol. (In the press).

Chaloupka, J., Strnadová, M., Cáslavská, J., and Veres, K. (1974) Growth and cell division of *Escherichia coli* 173-25 in the presence of some analogs of diaminopimelic acid. Zeitschrift für Allg. Mikrobiologie 14, 283-296.

de Pedro, M.A., and Schwarz, U. (1981) Heterogeneity of newly inserted and preexisting murein in the sacculus of *Escherichia coli.* Proc. Natl. Acad. Sci. US 78, 5856-5860.

Glauner, B., Höltje, J.-V. and Schwarz, U. (1988) The composition of the murein of *Escherichia coli.* J. Biol. Chem. 263, 10088-10095.

Goodell, E. W. (1985) Recycling of murein by *Escherichia coli.* J. Bacteriol. 163, 305-310.

Goodell, E. W., and Schwarz, U. (1983) Cleavage and resynthesis of peptide cross-bridges in *Escherichia coli.* J. Bacteriol. 156, 136-140.

Hartman, R., Höltje, J.-V. and Schwarz, U. (1972) Target of penicillin action in *Escherichia coli.* Nature 235, 426-429.

Höltje, J.-V., and Tuomanen, E.I. (1991) The murein hydrolases of *Escherichia coli*: Properties, functions and impact on the course of infections *in vivo.* J. Gen. Microbiol. 137, 441-454.

Höltje, J.-V., and Schwarz, U. (1985) Biosynthesis and growth of the murein sacculus, pp. 77-119. *In* N. Nanninga (ed), Molecular cytology of *E. coli.* Academic Press, London.

Kato, K., Umemoto, T., Sagawa, H. and Kotani, S. (1979) Lanthionine as an essential constituent of cell wall peptidoglycan of *Fusobacterium nucleatum.* Current Microbiol. 3, 147-151.

Koch, A. (1988), Biophysics of bacterial wall viewed as stress-bearing fabric. Microbiol. Rev. 52, 337-353.

Lark, C., and Lark, K.G. (1959), The effect of D-amino acids on *Alcaligenes fecalis*. Can. J. Microbiol 5, 369-379.

Lark, C., Bradley, D. and Lark, K.G. (1963), The incorporation of D-methionine into bacterial cell wall - its incorporation into the R-layer and the structural consequences. Biochem. Biophys. Acta 78, 278-288.

Pelzer, H. (1969), Incorporation of diaminopimelic acid analogs into cell wall murein of *Escherichia coli.* VIth Meeting of FEBS. Abstract nº 143. Madrid.

Pisabarro, A.G., de Pedro, M.A. and Vázquez, D. (1985), Structural modifications in the peptidoglycan of *Escherichia coli* associated with changes in the state of growth of the culture. J. Bacteriol. 161, 238-242.

Sprall, B.G. (1975), Distinct penicellin-binding proteins involved in the division, elongation, and shape of *Escherichia coli* K12. Proc. Natl. Acad. Sci. US 72, 2999-3003.

Spratt, B.G. (1983), Penicillin-binding proteins and the future of β-lactam antibiotics. J. Gen. Microbiol. 129, 1247-1260.

Tsuroka, T., Tamura, A., Miyata, A., Takei, T., Iwamatsu, K., Inouye, S. and Matsuhashi, M. (1984), Penicillin insensitive incorporation of D-amino acids into cell wall peptidoglycan influences the amount of bound lipoprotein in *Escherichia coli.* J. Bacteriol. 160, 889-894.

Vasstrand, E.N., Jensen, H.B., Miron, T. and Hopstad, T. (1982), Composition of peptidoglycan in *Bacteriodaceae*: Determination and distribution of lanthionine. Infect. Immun. 36, 114-122.

Waxman, D.J. and Strominger, J.L. (1983), Penicillin-binding proteins and the mechanisim of action of β-lactam antibiotics. Ann. Rev. Biochem. 52, 825-969.

PEPTIDOGLYCAN SYNTHESIS IN *SALMONELLA TYPHIMURIUM*

Stephen Cooper, David Gally, Yuko Suneoka, Melissa Penwell, Kelly Caldwell, and Kelvin Bray

Department of Microbiology and Immunology
University of Michigan Medical School
Ann Arbor, MI 48109-0620, USA

INTRODUCTION

The rate and topography of peptidoglycan synthesis during the division cycle of *Salmonella typhimurium* strain 2616 was determined by comparing incorporation of diaminopimelic acid (DAP) into peptidoglycan to incorporation of radioactive amino acids into protein (Cooper, 1988). The observed ratio of DAP incorporation to amino acid incorporation during the division cycle led to the development of a consistent description of peptidoglycan synthesis. It was proposed that (1) the cytoplasm increases exponentially during the division cycle; (2) the surface of the gram-negative cell grows in response to the increase in cytoplasm; (3) the surface grows in such a way as to produce an exponential increase in cell volume; (4) the exponential increase in cell volume accounts for constant cell density during the division cycle; and (5) the synthesis of peptidoglycan is smoothly partitioned between the poles and the cylindrical side walls, allowing exponential increase in cell volume. Similar experimental results were found with *Escherichia coli* (Cooper and Hsieh, 1988) suggesting that the rate and topography of peptidoglycan synthesis is similar in all rod-shaped, gram-negative bacteria.

This view of peptidoglycan synthesis during the division cycle accounted for the observed constant cell density during the division cycle (Kubitschek, 1987). It also explained the observed decrease in side wall synthesis in cells undergoing invagination (Woldringh, *et al.*, 1987). The increase in the rate of wall synthesis compared to mass synthesis toward the end of the division cycle may be the experimental correlate of the variable T model (Koch and Burdett, 1984). As T—the amount of work required to make a unit of surface area—decreases, a fixed amount of mass synthesis leads to an increased amount of surface synthesis. The synthesis of peptidoglycan during the division cycle displays a pattern that is unique to the cell surface, and therefore differs from cytoplasm and nucleic acid synthesis (Cooper, 1991a,b).

INCORPORATION OF DIAMINOPIMELIC ACID INTO GRAM-NEGATIVE BACTERIA

The original description of *Salmonella typhimurium* 2616 reported that the incorporation of DAP was 30-50 fold higher than other *E. coli* strains when short labeling times were considered (Cooper and Metzger, 1986). This is due partially to a lag in the incorporation of DAP in *E. coli*. The high incorporation efficiency of strain 2616 is not representative of all *Salmonella* strains, as other lysine requiring strains reveal incorporation kinetics similar to *E. coli* (Figure 1). The differences may be related to the fact that *Salmonella typhimurium* 2616

Bacterial Growth and Lysis, Edited by M.A. de Pedro *et al.*, Plenum Press, New York, 1993

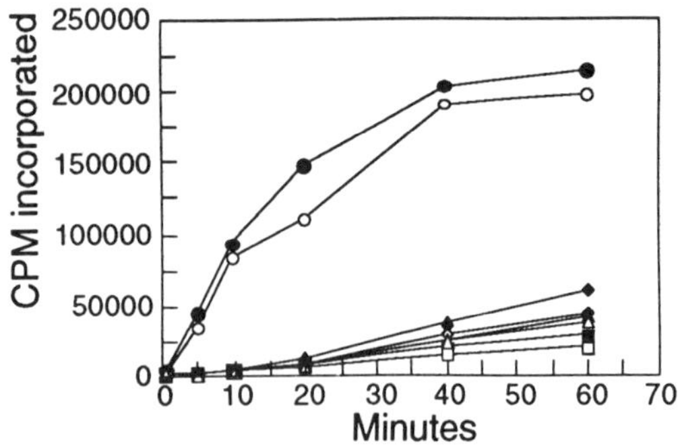

Figure 1. Comparison of incorporation of DAP into different bacterial strains. All strains were adjusted to the same absorbance at zero time and had essentially the same doubling time in glucose-minimal medium. Thus, the incorporation observed is a direct measure of the efficiency of peptidoglycan labeling. The open points are SDS treated material and the closed points indicate TCA treated material. Therefore the data refers to incorporation into macromolecular material and not into cellular pools. *Salmonella typhimurium* 2616, ●; *Escherichia coli* 4100, ■; *Salmonella typhimurium* 2801, ▲; *Salmonella typhimurium* 215, ◆.

is an LT7 strain, whereas the other *Salmonella typhimurium* strains are LT2 derivatives. Because of the incorporation lags, the relative efficiency of strain 2616 is highest at short labeling times, and grows less with extended labeling periods (Wientjes, *et al.* 1991; Cooper and Metzger, 1986).

LABELING OF *SALMONELLA TYPHIMURIUM 2616* WITH DIAMINOPIMELIC ACID

Wientjes, *et al.* (1991) have suggested that the efficient labeling of *Salmonella typhimurium* strain 2616 is due to the conversion of DAP to lysine which labels protein. Their supporting experiments were demonstrations that (1) strain 2616 grew well in the absence of lysine, indicating that the *lysA-* mutation was leaky, (2) there was a difference in TCA precipitable and SDS insoluble material after DAP labeling, and (3) there was considerable labeling of protein with DAP as shown by PAGE-autoradiographic analysis of material extracted from DAP labeled cells.

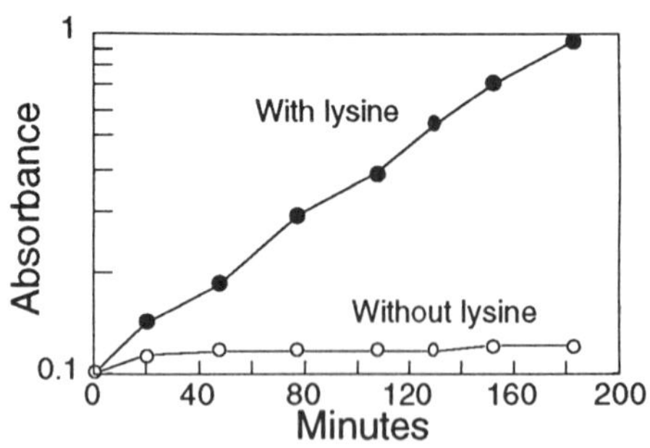

Figure 2. Growth of *Salmonella typhimurium* 2616 (*lys*⁻) in the presence and absence of lysine. Cells growing with lysine (50μg/ml) were filtered onto a membrane filter and washed with fresh medium without lysine. The cells were immediately taken up in fresh media with and without lysine. Growth was monitored by absorbance. Lysine added, ●; lysine omitted, ○.

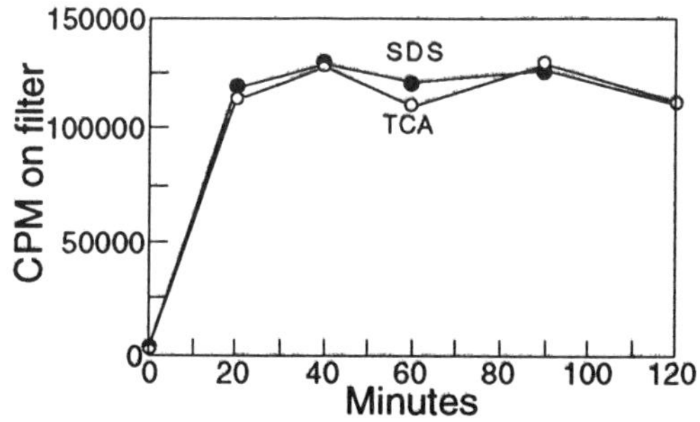

Figure 3. Comparison of SDS resistant material and TCA precipitable material. Tritiated DAP labeled cells were treated with boiling SDS or cold TCA and filtered onto membrane filters. SDS treatment, (10 minutes, boiling 4% SDS), ●; Cold TCA treatment, ○.

We have reinvestigated the high efficiency labeling of *Salmonella typhimurium* strain 2616 (Cooper and Metzger, 1986) and find that DAP is an excellent and specific label for peptidoglycan in these bacteria. Four sets of experiments showed that DAP is a specific label for peptidoglycan in strain 2616. We find that starvation for lysine leads to cessation of growth (Figure 2). This indicates that the *lysA* mutation, which prevents the conversion of DAP to lysine, is not leaky in strain 2616.

We find no difference in DAP incorporation when labeled bacteria are treated with either cold TCA or hot SDS prior to filtration (Figure 3). This experiment differs from that of Wientjes *et al.* (1991) in that we assayed the SDS resistant material by filtration onto a membrane filter. Under these conditions (10 minutes in boiling 4% SDS), over 95% of amino acid labeled protein passes through the filter.

The finding that different penicillins, when added to cells in increasing concentrations, can completely inhibit peptidoglycan synthesis (Figure 4) is additional support for labeling only the peptidoglycan with DAP.

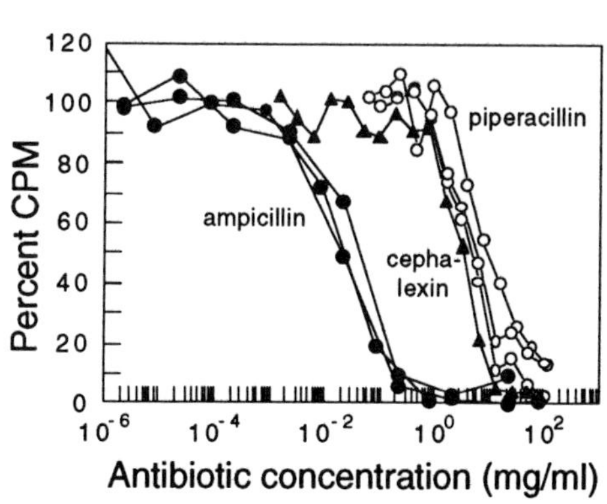

Figure 4. Inhibition of diaminopimelic acid incorporation by three antibiotics affecting peptidoglycan synthesis. *Salmonella typhimurium* 2616 was added to a dilution series of ampicillin, cephalexin, or piperacillin and a constant amount of tritiated DAP. After 8 minutes, the cells were treated with boiling SDS and the sacculi collected by filtration. The radioactive counts were normalized to 100% synthesis with no added antibiotic. Ampicillin, ●; piperacillin, ○; cephalexin, ▲.

Finally, for the most conclusive demonstration that DAP specifically labels peptidoglycan in *Salmonella typhimurium* strain 2616, we have looked directly, by reverse phase HPLC, for radioactive lysine in DAP labeled cells. Cells labeled with DAP were hydrolyzed and their radioactive lysine determined (Figure 5). Although a small peak was observed at the time of lysine elution, it is negligible compared to total incorporation (<3%). The lysine may have arisen during hydrolysis, since radioactive DAP subjected to the same hydrolysis procedure gave a similar labeling pattern with a small amount of lysine present. We conclude that DAP specifically labels the peptidoglycan of *Salmonella typhimurium* 2616 with a high efficiency and with no measurable lag in incorporation.

We did not assay material released by SDS treatment by electrophoresis and autoradiography as the radioactivity released was negligible. There was essentially no difference between the TCA precipitable and the SDS insoluble material. Since no counts were obtained in the SDS supernatant, no appearance of bands in an autoradiogram were expected.

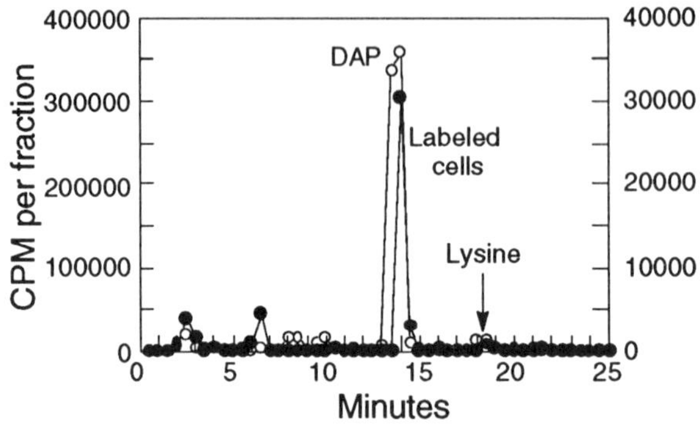

Figure 5. Assay of lysine in DAP labeled cells. Cells labeled with tritiated DAP were subjected to a standard hydrolysis and HPLC assay for amino acids (kindly performed by the Amino Acid Analysis Core Unit of the University of Michigan Medical School). The DAP labeled cells (●) refers to the left ordinate, while the pure radioactive DAP subjected to the same treatment and analysis (○) refers to the right ordinate. The time when lysine is eluted is indicated by the arrow.

AMOUNT OF PEPTIDOGLYCAN IN *SALMONELLA*

The observation that DAP label in *Salmonella typhimurium* strain 2616 is only in peptidoglycan means that peptidoglycan labeling in *E. coli* strain 4100 is less efficient (see Figure 1). This raises a question regarding the amount of peptidoglycan present in gram-negative cells. Wientjes *et al.* (1991) found that the chemically determined amount of peptidoglycan per unit surface area in *E. coli* and *Salmonella* were equivalent. They also found that the radiochemically determined peptidoglycan in *E. coli* was the same as the chemically determined peptidoglycan. If the radiochemical determination of peptidoglycan in *E. coli* is correct, then the amount of radiochemically determined peptidoglycan in *Salmonella* is higher, per unit area, than that of *E. coli*.

Another possibility, and one we consider more likely, is that the chemical determinations of peptidoglycan in *E. coli* and *Salmonella typhimurium* strain 2616 are correct, while the radiochemical assay of peptidoglycan in *E. coli* must be corrected by a factor of 3.5. We propose that chemical measurements are preferable to radiochemical measurements (due to problems with permeability, pools, endogenous synthesis, etc.), and therefore the peptidoglycan per unit surface area in *Salmonella typhimurium* is similar to that in *E. coli*. It is probable that the incorporation of DAP into *Salmonella typhimurium* 2616 peptidoglycan is essentially quantitative (Cooper and Metzger, 1986).

ANTIBIOTIC INHIBITION DURING THE DIVISION CYCLE

An analysis of different antibiotics and their affect on peptidoglycan synthesis reveals a similar patterns of inhibition as a function of antibiotic concentration. We find no shoulder or variation in the inhibition curve (Figure 4) that would suggest any differentiation in the site of action. This initial evidence has been supported by experiments on the affect of ampicillin and cephalexin on *Salmonella typhimurium* strain 2616 during the division cycle. We find no variation in the initial incorporation kinetics of DAP for either of these antibiotics as a function of the cell age between divisions.

STABILITY AND TURNOVER OF PEPTIDOGLYCAN

The question of peptidoglycan stability in gram-negative bacteria has been accorded significance by the proposal of an inside-to-outside peptidoglycan growth pattern (Glauner and Höltje, 1990). There have been reports of release of material into the medium during growth of labeled cells (Goodell and Schwarz, 1985). Our results on the membrane-elution apparatus ("baby machine") did not show any measurable turnover (Cooper, 1988). We have now investigated the turnover question more directly, and find a small but measurable amount of release of peptidoglycan label from cells. With labeled leucine as a control, we find a turnover rate of 7% per generation (Figure 6). There is no observable lag in release of label from the membrane, arguing against a simple pattern of inside-to-outside peptidoglycan growth.

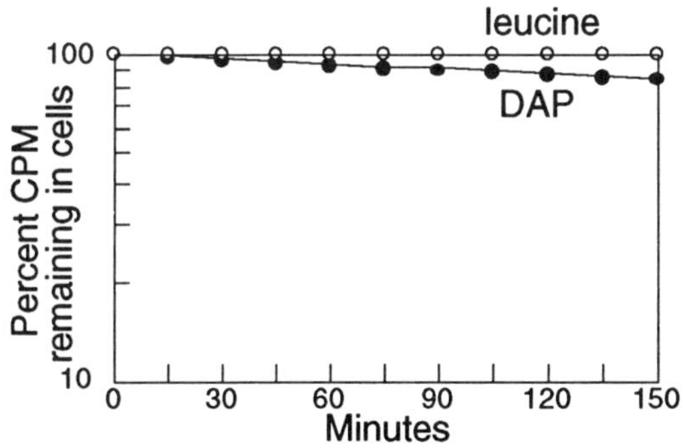

Figure 6. Release of peptidoglycan from DAP labeled cells. *Salmonella typhimurium* 2616 was labeled for 5 minutes with tritiated DAP and C-14 leucine, washed by filtration on a membrane filter, and taken up in fresh medium. The release of material from the cells was determined by filtration and measurement of the radioactivity in the filtrate. DAP label ●; leucine label ○.

At different growth rates, the rate of turnover per generation was consistently 7% per generation (Figure 7). While this constancy of turnover suggests some relationship of turnover to the division cycle, at this time we cannot relate turnover to any particular aspect of wall growth.

The release of label reported here may be considered the minimum level of turnover of peptidoglycan. There may be turnover occurring at a greater rate with a compensating reutilization. This possibility cannot be ruled out at this time.

PATTERN OF PEPTIDOGLYCAN SYNTHESIS IN *SALMONELLA*

We have studied the structure and pattern of synthesis of peptidoglycan in *Salmonella typhimurium* 2616 using reverse-phase HPLC (Glauner, 1988). The overall pattern of peptidoglycan structure and synthesis is similar to that of *E. coli* (Glauner, 1988; Glauner,

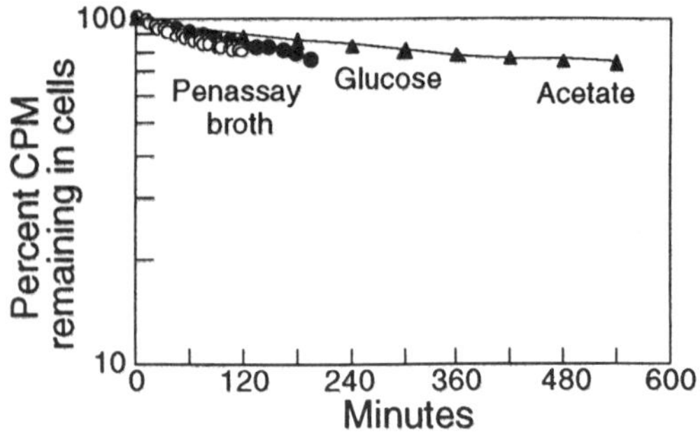

Figure 7. Release of peptidoglycan label in cells growing at different rates. Cells growing in different media were labeled with DAP. The label was removed by filtration and the release of label was measured during subsequent growth in the original medium. Glucose grown cells, ●; Penassay broth grown cells, ○; acetate grown cells, ▲.

Höltje, and Schwarz, 1988; Höltje and Schwarz, 1985; Glauner and Höltje, 1990). The detailed results will be reported at a later date (Gally and Cooper, manuscript in preparation).

ON THE PROBLEM OF CROSS-LINKING AND CROSS-LINKAGE

One parameter used to characterize peptidoglycan is the degree of cross-linkage. The way cross-linkage is calculated has been the subject of some discussion. For example, De Jonge, *et al.* (1989) proposed that the measured degree of cross-linkage, as calculated from the frequency of monomers, dimers, and trimers, should be adjusted for the acceptor-donor radioactivity ratio (ADRR). A rigorous analysis of the relationship of the ADRR to cross-linkage indicated that such a correction is not required (Cooper, 1990). There is no relationship of the ADRR to the calculation of cross-linkage.

It has been proposed that a distinction be made between cross-linkage and cross-linking (Driehuis, *et al.* 1992). Cross-linkage is the measured amount of linkage observed in a unit amount of peptidoglycan. For a given peptidoglycan, labeled or unlabeled, one can measure the number of monomers, dimers, trimers, and multimers, and calculate, without ambiguity, the degree of cross-linkage in that sample. Cross-linking, on the other hand, takes into account the mode of peptidoglycan synthesis. A calculated cross-linking value reflects the total number of cross-links formed in the process of making a unit amount of cross-linked peptidoglycan. For two pieces of peptidoglycan with the same number of cross-linkages, one may have used more total cross-linkings (i.e., connections between a D-alanine and a DAP). For example, if a cell makes peptidoglycan at a leading edge with no hydrolysis of preexisting cross-links, then every added cross-link results in a final cross-linkage. There would be no difference between the measured cross-linkages and the number of cross-links formed to make that peptidoglycan. However, if new strands are inserted between existing strands, cross-linkages formed by the new strands replace cross-linkages removed by hydrolysis. Compared to the leading edge case, more total cross-links would be required to make the same amount of peptidoglycan.

There is no distinction between the calculation of cross-linkage and cross-linking for single strand insertion with an ADRR of zero. Now consider peptidoglycan with an ADRR greater than zero. Consider two cases: two single strands are inserted sequentially, or two strands are inserted together (i.e., linked to each other and inserted as a pair) between two extant strands (rows 1 and 2 in Table 1). In both cases, the measured ADRR is the same (Cooper, 1990), and the cross-linkage at the end of the labeling period is the same. Our main concern with the cross-linking/cross-linkage distinction is that one does not know, for any non-zero ADRR, whether the ADRR value is due to rapid and localized strand insertion or to the insertion of multimers. Additionally, it can be seen (rows 2-6, Table 1), that different multimeric modes of insertion, coupled with the insertion of single strands, may give similar ADRR values. Thus, the ADRR and cross-linking values do not yield any information regarding the actual mode of strand insertion. The cross-linking value for non-zero ADRR insertion is the same for different modes of insertion (i.e., the cross-linking value is the same in rows 2-6 of Table 1). The cross-linking value does not really indicate how strands are inserted. All of the information related to strand insertion is contained in the ADRR. Unfortunately, there is no way at present to distinguish non-zero ADRRs due to two-strand insertion from the adjacent rapid insertion of two single strands.

Table 1. ADRR values derived for different proportions of single strand and multi-strand insertion.

Number of strands in a multimer	Calculated ADRR of multimer	ADRR of single-strand insertions	Proportion of multimer and single strand	ADRR of mixture
1+1	4/12	0/8	1:1	4/20 = 0.2
2	4/12	0/8	1:1	4/20 = 0.2
3	8/16	0/8	1:3	8/40 = 0.2
4	12/20	0/8	1:5	12/60 = 0.2
5	16/24	0/8	1:7	16/80 = 0.2
6	20/28	0/8	1:9	20/100 = 0.2

It is assumed that there are 8 cross-links formed when a single strand is inserted. For insertion of two strands adjacent to each other, whether by simultaneous insertion or rapid sequential insertion, the ADRR is 4/12 or 0.33. If one pair and one single strand are inserted during a labeling period, the resulting ADRR is 0.2. For higher multimers, one can always have a proportion of multimer and single strands that yield a given ADRR. The first column is the number of strands inserted as a sheet, except for the first value which indicates two strands inserted sequentially next to one another. The ADRR of the multimer (or pair) is given in column 2. A single strand ADRR (always zero) is given in column 3. There are different proportions of single and multimeric insertions that give a particular ADRR. It is demonstrated (column 5) that different mixtures (column 4) can give ADRRs of 0.2.

We suggest that the information content related to the mode of strand insertion is contained in the ADRR itself, and no added information is obtained by calculating a cross-linking value. Rather, the distinction between cross-linking and cross-linkage obscures the fact that we do not know the mode of strand insertion that produces a non-zero ADRR. A non-zero ADRR means that there is either multimeric insertion or rapid sequential insertion. The ADRR cannot yield the actual number of cross-links formed for the production of a given amount of peptidoglycan without further knowledge of the pattern of insertion.

THINKING ABOUT BACTERIAL SURFACE SYNTHESIS DURING THE DIVISION CYCLE

Our thinking about cell wall synthesis during the division cycle of rod-shaped, gram-negative bacteria has been influenced by the need to make the pattern of wall synthesis consistent with the overall pattern of cell growth (Cooper, 1991b). The model proposed takes into account the exponential synthesis of cell cytoplasm and the constant density of the cell. If wall synthesis responded to other controls (e.g., the stringent-relaxed system), rather than only to mass increase, one would expect variations in cell density or internal cell pressure during the division cycle. Such variations have not been found (Cooper, 1991a).

Another theme arising from our analysis of wall growth is that peptidoglycan synthesis is probably the result of a single biochemical synthetic system. We do not postulate any difference between the biochemistry of side wall or septum synthesis. Rather than a switch between these two synthesis locations, there is a steady partitioning in response to the cell's need to divide and to increase cell volume in a simple, exponential manner. At the moment, the trigger of septum formation is not known. We propose that once septum formation begins, triggered by whatever means, there is no need to invoke new or different systems for septal and side wall peptidoglycan synthesis. We are currently studying the affect of different antibiotics on peptidoglycan synthesis in a critical test of this model. Our preliminary results on the pattern of inhibition at different concentrations tend to support the unitary model. Further work on inhibition at different times during the division cycle is currently under way.

ACKNOWLEDGEMENTS

The preparation of this article and the experiments within were supported by National Institutes of Health Grant R01 GM44022-01A1. We thank the UM Protein Structure Core Unit for help with the amino acid analyses. Again, Sandi Cooper's editing helped make this a better paper.

REFERENCES

Cooper, S. (1988) Rate and topography of cell wall synthesis during the division cycle of *Salmonella typhimurium*. J. Bacteriol. 170, 422-430.

Cooper, S. (1990) Relationship between the acceptor/donor radioactivity ratio and cross-linking in bacterial peptidoglycan: application to surface synthesis during the division cycle. J. Bacteriol. 172, 5506-5510.

Cooper, S. (1991a) "Bacterial Growth and Division: Biochemistry and Regulation of the Division Cycle of Prokaryotes and Eukaryotes" Academic Press, Inc., San Diego, Calif.

Cooper, S. (1991b) Synthesis of the cell surface during the division cycle of rod-shaped, gram-negative bacteria. Microbiol. Rev. 55, 1649-674.

Cooper, S., and Hsieh, M.-L. (1988) The rate and topography of cell wall synthesis during the division cycle of *Escherichia coli* using N-acetylglucosamine as a peptidoglycan label. J. Gen. Microbiol. 134, 1717-1721.

Cooper, S., and Metzger, N. (1986) Efficient and quantitative incorporation of diaminopimelic acid into the peptidoglycan of *Salmonella typhimurium*. FEMS Microbiol. Lett. 36, 191-194.

De Jonge, B.L., Wientjes, F.B., Jurida, I., Driehuis, F., Wouters, J.T.M., and Nanninga, N. (1989) Peptidoglycan synthesis during the cell cycle of *Escherichia coli*: composition and mode of insertion. J. Bacteriol. 171, 5783-5794.

Driehuis, F., De Jonge, B.L.M., and Nanninga, N. (1992) Cross-linkage and cross-linking of peptidoglycan in *Escherichia coli*: definition, determination, and implications. J. Bacteriol. 174, 2028-2031.

Glauner, B. (1988) Separation and quantification of muropeptides with high-performance liquid chromatography. Anal. Biochem. 172, 451-464.

Glauner, B., and Höltje, J.-V. (1990) Growth pattern of the murein sacculus of *Escherichia coli*. J. Biol. Chem. 265, 18988-18996.

Glauner, B., Höltje, J.-V., and Schwarz, U., (1988) The composition of the murein of *Escherichia coli*. J. Biol. Chem. 263, 10088-10095.

Goodell, E.W., and Schwarz, U. (1985) Release of cell wall peptides into culture medium by exponentially growing *Escherichia coli*. J. Bacteriol. 162, 391-397.

Höltje, J.-V. and Schwarz, U. 1985. Biosynthesis and growth of the murein sacculus, p. 77-119. In N. Nanninga (ed.), Molecular cytology of *Escherichia coli*. Academic Press, Inc., New York.

Koch, A.L., and Burdett., I.D.J. (1984) The variable T model for gram-negative morphology. J. Gen. Microbiol. 130,2325-2338.

Kubitschek, H.E. (1987) Buoyant density variation during the cell cycle in microorganisms. Crit. Rev. Microbiol. 14, 73-97.

Wientjes, F.B., Woldringh, C.L., and Nanninga, N. (1991) Amount of peptidoglycan in cell walls of gram-negative bacteria. J. Bacteriol. 173, 7684-7691.

Woldringh, C.L., Huls, P., Pas, E., Brakenhoff, G.J. and Nanninga, N. (1987). Topography of peptidoglycan synthesis during elongation and polar cap formation in a cell division mutant of *Escherichia coli* MC4100. J. Gen. Microbiol. 133, 575-586.

BIOSYNTHESIS OF PEPTIDOGLYCAN IN GAFFKYA HOMARI: REGULATION OF TRANSPEPTIDATION BY ACCEPTOR PEPTIDE

Francis C. Neuhaus and Rabindra K. Sinha

Department of Biochemistry, Molecular Biology
and Cell Biology
Northwestern University
Evanston, IL 60208

INTRODUCTION

Gaffkya homari (*Aerococcus viridans*) contains a system for the biosynthesis of peptidoglycan(PG) that utilizes penicillin-sensitive targets which are not N^{ϵ}-(DAla) acceptor transpeptidases. In contrast to cross-linking enzymes from other bacteria, the transpeptidase from this gram-positive organism is essentially insensitive to inhibition by penicillin (Hammes, 1976). Thus, the target of penicillin in *G. homari* is different from that identified by Tipper and Strominger (1965) for *Staphylococcus aureus* and *Escherichia coli*. Because of this difference, an understanding of ß-lactam action in this unique organism is of primary interest.

The processing of nascent PG requires the regulated actions of at least three membrane-associated enzymes: DD-carboxypeptidase, LD-carboxypeptidase and the transpeptidase. The actions of these carboxypeptidases provide a population of tetrapeptide and tripeptide subunits which serve as acceptors in the reaction catalyzed by the transpeptidase. *G. homari* is one of a variety of bacteria that contain significant amounts of *iso*-D-glutaminyl residues in their wall PG (Nakel *et al.*, 1971). It was proposed that at least one of the peptide stems is required to be amidated for cross-linking to occur. The function of these amide substituents has not been established. It is suggested that these *iso*-D-Gln residues may either (1) facilitate the intramembranal translocation of nascent PG, (2) impart a unique conformation to the PG, or (3) enhance the effectiveness of the acceptor peptide in the cross-linking reaction. Thus, the actions of these enzymes, together with the amidation of the γ-D-glutamic acid residue, would appear to play a role in either determining or enhancing the activity of the acceptor peptide for the cross-linking reaction.

Hammes (1976) proposed that the DD-carboxypeptidase (PBP 9) is the primary *in vitro* target of penicillin in *G. homari*, an observation that has been confirmed in our laboratory (Wrezel *et al.*, 1986). However, penicillin binding experiments with intact cells failed to support this proposal in the *in vivo* system. The *in vivo* target is PBP 6, a protein of 112 kDa that accounts for 50% of the penicillin binding capacity (Wrezel *et al.*, 1986). Thus, while the *in vitro* target of ß-lactam action is the DD-carboxypeptidase, this enzyme does not appear to be the primary *in vivo* target even though its action is required for the processing of nascent PG.

Understanding the target of ß-lactam action in *G. homari* requires a detailed structural and quantitative analysis of the monomers and dimers from the PG of control and antibiotic inhibited

Bacterial Growth and Lysis, Edited by M.A. de Pedro
et al., Plenum Press, New York, 1993

cultures. From these analyses, we are able to assess the activities of those enzymes involved in the processing and cross-linking of PG. The results provide the basis for formulating a hypothesis which implicates the *iso*-D-Gln residue of the acceptor peptide as an important feature in the reaction catalyzed by the transpeptidase. It is our goal to test the hypothesis that hydrolysis of this amide substituent indirectly inhibits cross-linking by controlling the concentration of the preferred acceptor in the reaction catalyzed by the transpeptidase.

STRUCTURAL ANALYSIS OF PG

A characterization of the PG from *G. homari* by Nakel *et al.* (1971) revealed a complex mixture of muropeptides. For the present studies, walls, which were extracted with 4% SDS at 100 °C, were digested with muramidase. Using the HPLC procedure patterned after that described by Glauner (1988) for the analysis of PG from *E. coli*, we resolved and characterized 14 muropeptides (6 monomers and 8 dimers). There are two groups of dimers, heptapeptides and octapeptides, each containing 4 muropeptides which have been so far identified. The schematic structures of the octapeptide dimers are shown in Fig 1. Two dimers (compounds 12 and 13) with

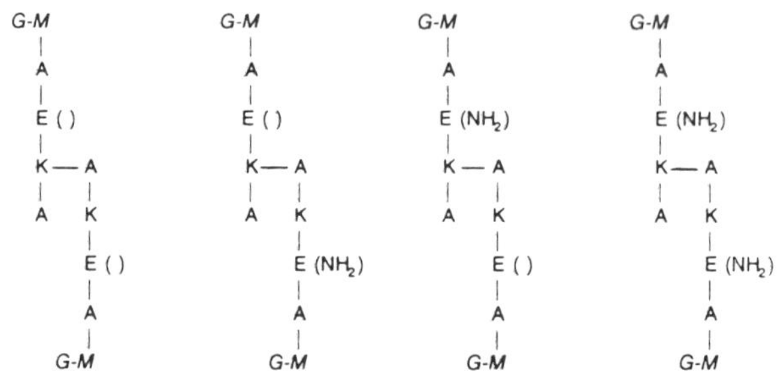

11 12 13 14

Figure 1. Schematic structures of octapeptide dimers (compounds 11, 12, 13, and 14). The assignment of the amide function in compounds 12 and 13 is arbitrary (Sinha and Neuhaus, 1991).

identical MH^+ mass ion numbers yielded nonamidated *bis*-DS-octapeptide (compound 11) after deamidation. Since the monoamidated octapeptide does not have an axis of symmetry, one stem peptide is not superimposable on the other. Thus, two isomers of this octapeptide are observed. Deamidation of compound 14 also yields the same nonamidated octapeptide as that formed from compounds 12 and 13. Under the growth conditions which we reported (Sinha and Neuhaus, 1991), the diamidated *bis*-DS-octapeptide is always the predominant dimer and the octapeptide group of dimers predominates over the heptapeptide group.

COMPOSITION OF PG SYNTHESIZED IN THE PRESENCE OF PENICILLIN

The role of the amidated tetrapeptide as the preferred acceptor in the cross-linking reaction is based on the *in vivo* effects of penicillin on the structure of PG. *G. homari* was grown in the presence of 3 concentrations of penicillin: 0.05 μg/ml (the MGIC), 0.20 μg/ml (the MIC) and 200 μg/ml (Fig. 2). To maximize the inhibition of cross-linking, the effect of 200 μg/ml, a concentration that is 1000 fold in excess of the MIC, was tested. Since benzylpenicillin does not cause lysis in *G. homari*, it was possible to study the compositional analysis of PG at all levels of ß-lactam. At the MIC, 100% of the cells survived the 5.5 hr of penicillin treatment, while at 200 μg of benzylpenicillin per ml, 33 % of the cells survived. *G. homari* is not lysed by penicillin and shows excellent tolerance to this ß-lactam. Thus, inhibition by penicillin is characterized by the arrest of cell growth without any of the secondary events observed in other organisms.

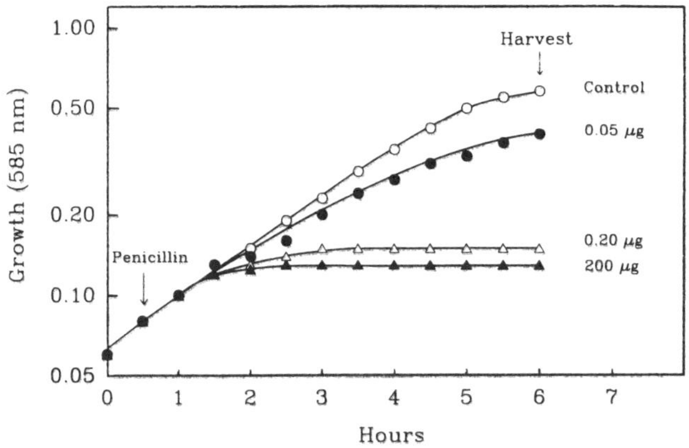

Figure 2. Growth of *G. homari* in the presence of benzylpenicillin. Symbols: O, control: ●, 0.05 μg/ml;Δ, 0.2 μg/ml; ▲ 200 μg/ml. At the first arrow, the β-lactam and radiolabeled glucosamine (20 μCi) were added. At the second arrow, the cells were harvested and used for the experiments in Fig. 3.

The muropeptide compositional analyses are shown in the three dimensional bargraph shown in Fig. 3. To monitor the PG synthesized during the application of the ß-lactam, D-[^{3}H]-glucosamine was added concurrently with the ß-lactam to the culture medium. While the A_{208} was used to monitor the monomers and dimers in the HPLC separation, the radiolabeled glucosamine was essential for monitoring the newly synthesized monomers and dimers that are formed during the application of the ß-lactam. The inhibition of bacterial growth by penicillin requires at least one generation for maximal expression. This lag may reflect the time-dependent acylation of the target proteins. Thus, with simultaneous addition of ß-lactam and [^{3}H]glucosamine, measurements of labeled muropeptides may reflect assembly in both the inhibited and uninhibited systems. To assess the contribution of the uninhibited system, we added the [^{3}H]glucosamine to the culture 1 hr after the application of the ß-lactam (0.2*). The observations under these conditions are essentially identical to those observed in cultures with the concurrent addition of penicillin.

Two conclusions may be drawn from these data. The percentage cross-linking decreased from 31% in the control culture to 24% for 0.05 μg/ml and 21% for 0.20 μg penicillin. The value for the control culture is similar to that (27%) reported by Nakel *et al.* (1971). The percent cross-linkage does not appear to decrease in a dose specific manner and, thus, supports the conclusion of Wilson *et al.*, (1985) that there is not a dose-dependent decrease in cross-link formation in this organism. An analysis of the octapeptide group of dimers (Fig. 4) is consistent with these data. The syntheses of three of the dimers, two isomers of monoamidated and a non-amidated dimer are relatively insensitive to the action of the ß-lactam whereas the synthesis of the diamidated octapeptide (compound 14) is inhibited by 58%. Under these conditions the synthesis of compounds 11, 12, and 13 is inhibited by 18%. If the transpeptidase were the target of penicillin action, one might expect the inhibition of each octapeptide dimer to be similar. These data support the conclusion that the transpeptidase is relatively insensitive to the action of the penicillin and that another enzyme or feature governs the amount of diamidated octapeptide synthesized.

One of the features which appears to determine the dimer profile is the degree of amidation of the acceptor peptide. It is suggested that a decrease in the amount of amidated DS-tetra will lead to a decrease in the amount of diamidated *bis*-DS-octapeptide. In the control culture, 26% of DS-tetra and 80% of DS-tri are not amidated. In contrast, in the culture treated with penicillin at 0.2 μg/ml, 50% of the DS-tetra and 90% of DS-tri are not amidated. The ratios of amidated DS-tetra to nonamidated DS-tetra are 2.8 and 1.0 in the control and penicillin-treated cultures, respectively. Thus, with an amount of benzylpenicillin sufficient for growth inhibition, a shift from more amidated to less amidated DS-tetra was observed. This increase in the amount of nonamidated DS-tetra represents a previously unrecognized feature to the action of penicillin and may be correlated with the inhibition of the diamidated octapeptide.

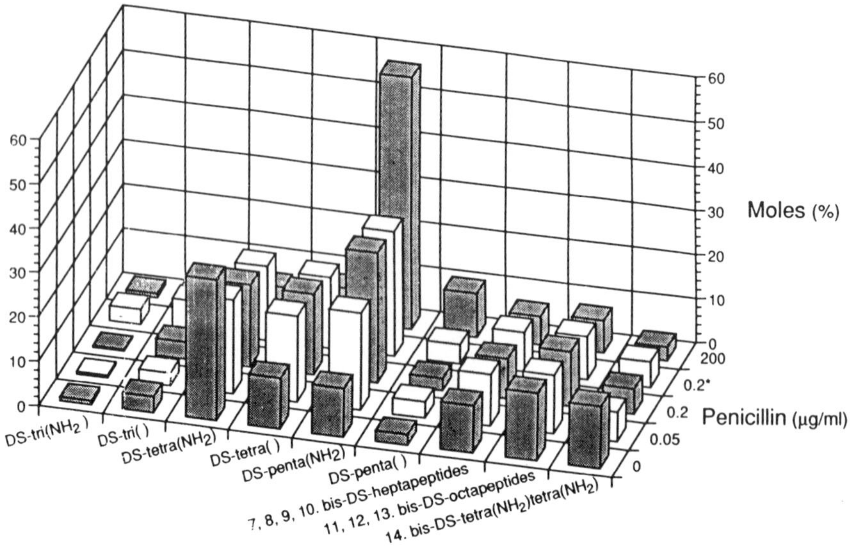

Figure 3. Muropeptide composition of PG from *G. homari* grown in the presence of benzylpenicillin. Cells labeled with [^{3}H]glucosamine were grown as described in Fig. 2. The reduced, radiolabeled muropeptides were separated and quantified as described by Sinha and Neuhaus (1991).

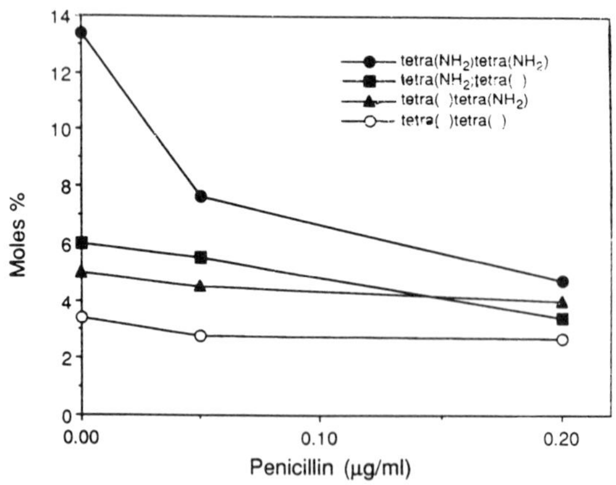

Figure 4. Inhibition of the octapeptide group of dimers by growth in the presence of penicillin.

COMPOSITION OF PG SYNTHESIZED IN THE PRESENCE OF OTHER ß-LACTAMS

One of our goals was to establish whether the observations with benzylpenicillin are characteristic of other ß-lactams. To address this question, we have studied the effects of methicillin, cephalothin, and cefoxitin on the composition of PG at their respective MICs. As in the case of benzylpenicillin, the ratio of amidated DS-tetra/nonamidated DS-tetra decreases to approximately 50% or less of that found in the control (Table 1). Under these conditions the ratio of diamidated *bis*-DS-octapeptide/nonamidated *bis*-DS-octapeptide decreased to approximately 50% of that found in the control. The results with each of these ß-lactams indicate the importance of amidated tetrapeptide as an acceptor in the cross-linking reaction. From these *in vivo* experiments with different ß-lactams it is suggested that the preferred acceptor in the cross-linking reaction is the amidated DS-tetrapeptide.

PROPOSED PATHWAY FOR THE PROCESSING OF ACCEPTOR IN G. HOMARI

From these data, a pathway for the processing of acceptor may be proposed. In addition to the roles of DD- and LD-carboxypeptidases, processing may also include cleavage of the amide function by an amidohydrolase. The regulated action of this proposed enzyme may play an indirect role in controlling cross-linking in *G. homari*. For example, increased activity would result in a decrease of the preferred acceptor (Fig. 5). Two mechanisms for regulating the action of this deamidation may be considered. In the first, penicillin stimulates the activity of the

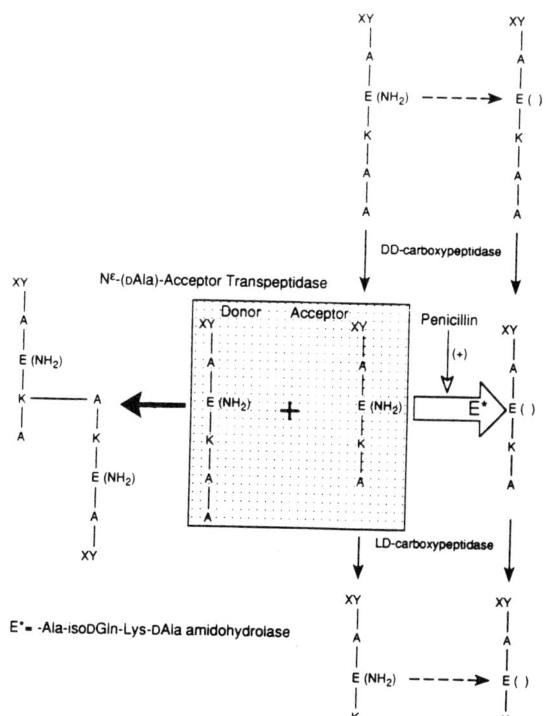

Figure 5. Proposed pathway for the processing of acceptor in *G. homari*. According to the hypothesis, the amidohydrolase primarily catalyzes the deamidation of the tetrapeptide stem i.e. Ala-*iso*DGln-Lys-DAla. The (+) with penicillin refers to the increased activity of the amidohydrolase.

Table 1. Tetrapeptide and octapeptide composition of *G. homari* PG synthesized in the presence of ß-lactams

	MIC ß-Lactam	DS-tetra(NH_2)/ DS-tetra( )	*bis*-DS-tetra(NH_2)tetra(NH_2)/ *bis*-DS-tetra( )tetra( )
	(μg/ml)	(Moles %/moles %)	
Control	0	2.8	3.7
Penicillin	0.2	1.0	1.7
Methicillin	10.	1.1	1.7
Cephalothin	5.	1.6	0.72
Cefoxitin	25.	1.5	2.0

amidohydrolase, and in the second, penicillin induces the synthesis of this enzyme. While the end result of each of these is the same, i.e. less of the preferred acceptor, the means for accomplishing the decrease of diamidated octapeptide are different. Thus, the stimulation (+) suggested in the model (Fig. 5) may result from either enzyme stimulation or an increased amount of enzyme.

It is our hypothesis that an enzyme, *iso*-D-glutaminyl amidohydrolase, plays an indirect role in the control of cross-linking by determining the amount of preferred acceptor for the transpeptidase. Although the proposal for a penicillin-stimulated or a penicillin-induced amidohydrolase is consistent with the observations, alternative models may be considered. Some of these are discussed by Sinha and Neuhaus (1991).

TESTING THE HYPOTHESIS

The ability to synthesize cross-linked PG by membranes of *G. homari* provided the system for testing this hypothesis. The synthesis of dimers by the transpeptidase(s) of this system utilizes amidated tetrapeptides in preference to nonamidated tetrapeptide (Bardin *et al.*, 1984). The synthesis of this amidated tetrapeptide is accomplished by the DD-carboxypeptidase, the primary *in*

Table 2. Composition of muropeptide monomers in *vitro* synthesized PG

	In vitro PG	
	Membranes (control)	Membranes (penicillin)[1]
	(Moles %)	
DS-Tri ( )	1.3	0
DS-Tri (NH_2)	3.9	0.7
DS-Tetra ( )	18.4	45.5
DS-Tetra (NH_2)	22.5	8.0
DS-Penta ( )	25.1	29.6
DS-Penta (NH_2)	18.9	12.3

[1] *G. homari* was grown in the presence of 0.2 μg/ml of penicillin and harvested 2 h after the application of the ß-lactam. Membranes were prepared by the method of Bardin *et al.* (1984). Sinha and Neuhaus, unpublished observations.

vitro target of penicillin (Hammes, 1976; Wrezel *et al.*, 1986). Inhibition of this enzyme by penicillin prevents the synthesis of the preferred acceptor and, hence, leads to the inhibition of cross-linking. The addition of ß-lactam to this system did *not* increase the level of nonamidated tetrapeptide, and thus, it was concluded that the putative amidohydrolase is not stimulated by penicillin. It would appear that the *in vitro* system used for our studies does not mimic the *in vivo* deamidation of the acceptor peptide.

A second approach for testing the above hypothesis involves a novel procedure for the preparation of membranes. If the amidohydrolase is induced in response to penicillin, membranes for the *in vitro* system should be prepared from organisms grown in the presence of the ß-lactam. In *Lactobacillus casei*, Pollack *et al.* (1992) observed that the profile of membrane proteins of the organism grown in the presence of benzylpenicillin differed from that found in membranes prepared from cells grown in the absence of ß-lactam. All of the *in vitro* systems used for the study of PG processing have utilized membranes prepared from organisms grown in the absence of ß-lactam. Because *G. homari* is tolerant to the action of penicillin (Sinha and Neuhaus, 1991), studies with membranes from penicillin-grown cells are feasible and would allow us to monitor the activities of those enzymes that have been induced or repressed by penicillin. If specific proteins involved in the processing of nascent PG, e.g. amidohydrolase, are induced by penicillin, their activities can be monitored with the HPLC analysis of the muramidase digest of the *in vitro* synthesized PG.

In Table 2, the composition of the muropeptide monomers is summarized for *in vitro* SDS-insoluble PG synthesized by membranes prepared from cells grown in the presence and absence of penicillin. A major difference in the ratio of amidated tetrapeptide/nonamidated tetrapeptide is observed. In the membranes from non-penicillin grown cells, the ratio is 1.22. In the PG synthesized by membranes from penicillin grown cells, the ratio is 0.17. This striking difference suggests that the amidohydrolase has been induced in response to penicillin. While these results require additional documentation, they nevertheless provide support to the proposal that an enzyme is induced in response to penicillin which hydrolyzes the *iso*-DGlutaminyl residue of the amidated tetrapeptide.

DISCUSSION

Our immediate goal is to establish the role of amidated tetrapeptide as the preferred acceptor in the formation of cross-bridges in *G. homari*. The *in vivo* experiments described in this paper and by Sinha and Neuhaus (1991) support the proposal by Nakel *et al.* (1971) that amidation of peptide stems is important for cross-linking. Thus, the metabolism of this amide function is an important feature of PG processing in this organism. The function of amidated tetrapeptide as a preferred acceptor in cross-bridge formation in other bacteria requires further investigation. Growth of *S. aureus* in the presence of 0.08 μg of penicillin per ml yielded monomer and polymer fractions of PG that are 30% less amidated than that of monomer from PG of control cells (Tipper and Strominger, 1965). It would appear that one of the effects of penicillin action in *S. aureus* may be similar to that described for *G. homari*. In Streptomyces strain R39 amidation of the α-carboxyl group of the γ-glutamyl residue of the peptide acceptor exerted strong influences on both the hydrolysis and transpeptidation activities of the DD-carboxypeptidase/transpeptidase (Ghuysen *et al.*, 1973). Thus, the metabolism of the *iso*-D-glutaminyl residues may represent an important aspect of cross-bridge formation in those organisms that have this type of residue in the stem peptide. ß-Lactams have provided a unique tool for modifying this metabolism and, hence, establishing the importance of this residue.

Establishing the mechanism of penicillin action in *G. homari* has also been a goal of this research. The results summarized in this paper support the suggestion that the amidohydrolase may play a role in the action of ß-lactams. Our experiments indicate that this enzyme is not stimulated but rather it is induced in response to growth in the presence of penicillin. As a result of increased levels of enzyme, the concentration of preferred acceptor for the transpeptidase is decreased. This mechanism of penicillin action must be correlated with that proposed by Wrezel *et al.* (1986) who suggested that PBP 6 is the primary target of growth inhibition by benzylpenicillin, cefmenoxime, cephalothin, and cefoxitin. The function of this PBP has not been established and one can only speculate about its relationship to the amidohydrolase. Since the amidohydrolase appears to be induced by penicillin, it is proposed that the PBP 6 is the receptor for the system that results in the elevation of the hydrolase. One might compare this system to the induction of ß-lactamase. In the induction of this enzyme, a penicillin sensory-transducer protein (BLaR) transmits the signal for induction of this enzyme (Ghuysen, 1991). Blar is a PBP isolated from *Bacillus licheniformis* (Zhu *et al.*, 1990). Whether PBP 6 from *G. homari* is a sensory protein and whether it is involved in the induction of the amidohydrolase remains to be investigated.

Acknowledgments

This study was supported by Public Health Service grant AI-04615 from the National Institute of Allergy and Infectious Diseases. The authors thank Stephen Beres for graphics.

REFERENCES

Bardin, C., Sinha, R.K., Kalomiris, E. and Neuhaus, F.C. (1984) Biosynthesis of peptidoglycan in *Gaffkya homari*: processing of nascent glycan by reactivated membranes. J. Bacteriol. 157, 398-404.

Ghuysen, J.-M., Leyh-Bouille, M., Campbell, J.N., Moreno, R., Frére, J.-M., Duez, C., Nieto, M. and Perkins, H.R. (1973) Structure of the wall peptidoglycan of Streptomyces R39 and the specificity profile of its exocellular DD-carboxypeptidase-transpeptidase for peptide acceptors. Biochemistry 12, 1243-1251.

Glauner, B. (1988) Separation and quantification of muropeptides with high-performance liquid chromatography. Anal. Biochem. 172, 451-464.

Ghuysen, J.-M. (1991) Serine ß-lactamases and penicillin-binding proteins. Annu. Rev. Microbiol. 45, 37-67.

Hammes, W.P. (1976) Biosynthesis of peptidoglycan in *Gaffkya homari*: the mode of action of penicillin G and mecillinam. Eur. J. Biochem. 70, 107-113.

Nakel, M., Ghuysen, J.-M. and Kandler, O. (1971) Wall peptidoglycan in *Aerococcus viridans* strains 201 Evans and ATCC 11563 and *Gaffkya homari* strain ATCC 10400. Biochemistry 10, 2170-2175.

Pollack, J.P., Ntamere, A.S. and Neuhaus, F.C. (1992) D-Alanyl-lipoteichoic acid in *Lactobacillus casei*: secretion of vesicles in response to benzylpenicillin J. Gen. Microbiol. in press.

Sinha, R.K. and Neuhaus, F.C. (1991) Biosynthesis of peptidoglycan in *Gaffkya homari*: on the target(s) of benzylpenicillin. Antimicrob. Agents Chemother. 35, 1753-1759.

Tipper, D.J. and Strominger, J.L. (1965) Mechanism of action of penicillins: a proposal based on the their structural similarity to acyl-D-alanyl-D-alanine. Proc. Natl. Acad. Sci. USA 54, 1133-1141.

Wilson, G.E., Jr., Jacob, G.S. and Schaefer, J. (1985) Solid-state ^{15}N NMR studies of the effects of penicillin on cell-wall metabolism of *Aerococcus viridans* (*Gaffkya homari*). Biochem. Biophys. Res. Commun. 126, 1006-1012.

Wrezel, P.W., Ellis, L.F. and Neuhaus, F.C. (1986) In vivo target of benzylpenicillin in *Gaffkya homari*. Antimicrob. Agents Chemother. 29, 432-439.

Zhu, Y.F., Curran, I.H.A., Joris, B., Ghuysen, J.-M. and Lampen, J.O. (1990) Identification of BlaR, the signal transducer for ß-lactamase production in *Bacillus licheniformis*, as a penicillin-binding protein with strong homology to the OXA-2, ß-lactamase (class D) of *Salmonella typhimurium*.

APPARENT OBLIGATORY DEPENDENCE OF PEPTIDOGLYCAN SYNTHESIS ON PHOSPHOLIPID SYNTHESIS STUDIED IN ETHER-TREATED *ESCHERICHIA COLI*

Edward E. Ishiguro

Department of Biochemistry and Microbiology
University of Victoria
P.O. Box 3055
Victoria, BC, Canada V8W 3P6

INTRODUCTION

Amino acid deprivation of *Escherichia coli* activates ppGpp synthetase I, a ribosome-associated enzyme encoded by the *relA* gene (reviewed by Cashel and Rudd, 1987). This results in the rapid accumulation of guanosine 5'-diphosphate 3'-diphosphate (ppGpp) in *relA*$^+$ bacteria but not in *relA* mutants; ppGpp is a putative stress signal compound specifying amino acid deficiency which apparently shuts down a variety of energy-consuming metabolic processes in starved bacteria in what has been termed the stringent response (Cashel and Rudd, 1987). The activities of both the peptidoglycan biosynthetic and the hydrolase systems are among the metabolic processes which are inhibited during the stringent response, and this may account for the well known penicillin tolerance exhibited by amino acid-deprived *relA*$^+$ bacteria (Pisabarro *et al.*, 1991). The fact that amino-acid deprived *relA* mutant strains lyse when treated with penicillin is consistent with this proposal. Furthermore, lysis of amino acid-deprived *relA*$^+$ cells by inhibitors of peptidoglycan synthesis can be induced by the addition of agents, such as chloramphenicol, which interfere with the stringent response. These stringent response antagonists block the synthesis of ppGpp during amino acid deprivation (Cashel and Rudd, 1987).

Although the exact mechanism by which the stringent response regulates peptidoglycan synthesis is not known, two findings are noteworthy. First, the regulatory phenomenon can be demonstrated *in vitro* using ether-permeablized bacteria (Harkness *et al.*, 1981). Ether-treated bacteria (ETB) are permeable to peptidoglycan nucleotide-activated precursors and can be used for *in vitro* studies on the terminal stages of peptidoglycan synthesis (Mirelman *et al.*, 1976). The normal cellular regulatory mechanism on peptidoglycan synthesis is apparently retained by ETB whereas such regulation is lost in cell-free systems, probably as a result of mechanical disruption. The second finding was that the relaxed synthesis of peptidoglycan in amino acid-deprived *relA* mutants was dependent on ongoing phospholipid synthesis (Ishiguro, 1983), and this result is repeated and expanded below. The main objective of this study was to demonstrate a direct correlation between these two phenomena.

MATERIALS AND METHODS

Strains VC7 (*thi-1 lysA*) and VC8 (*thi-1 lysA relA2*), are isogenic derivatives of

E. coli K-12. The bacteria were grown at 37°C as described previously (Harkness *et al.*, 1981) in glucose-M9 minimal medium in all experiments except those involving [^{32}P] orthophosphate ($^{32}P_i$) labeling, in which cases the modified M56LP medium of Bell (1974) was used. Isoleucine deprivation was achieved by adding L-valine at 500 µg per ml to the medium.

Peptidoglycan synthesis was determined by following the incorporation of [^{3}H] diaminopimelic acid ([^{3}H]DAP; Amersham Canada) into cold 5% trichloracetic acid-insoluble fractions (Harkness *et al.*, 1981). The incorporation of $^{32}P_i$ into phospholipids was determined after extraction of cell samples with chloroform-methanol (1:2) as described by Bligh and Dyer (1959).

The preparation of ETB and assays for peptidoglycan transpeptidase (TPase) and carboxypeptidase (CPase) were exactly as described by Harkness *et al.* (1981). Specific activities are expressed as pmol D-[^{14}C] alanine released per mg protein per hour from ^{14}C-labeled UDP-N-acetylmuramyl-pentapeptide.

RESULTS

Figures 1A and 1B confirm that the stringent response in strain VC7 (*relA*$^+$) resulted in the inhibition of phospholipid and peptidoglycan synthesis and that blockage of the stringent response by treatment with chloramphenicol relaxed both synthetic processes (curves a, b, and c). Furthermore, the addition of cerulenin resulted in the inhibition of relaxed phospholipid synthesis as expected; however, it is notable that this resulted in the concomitant inhibition of peptidoglycan synthesis (curves d). The addition of exogenous palmitate and oleate negated the inhibitory effect of cerulenin on phospholipid synthesis, and it is significant that it also restored and relaxed peptidoglycan synthesis (curves e). As reported previously (Ishiguro, 1983), these results suggest that peptidoglycan synthesis is obligatorily dependent on phospholipid synthesis, and the inhibition of peptidoglycan synthesis during the stringent response may therefore be coupled to the inhibition of phospholipid synthesis.

These conclusions were reinforced by complementary experiments with the isogenic *relA* mutant strain, VC8 (Figures 1C and 1D). In this case, phospholipid synthesis and peptidoglycan synthesis are relaxed in amino acid-deprived bacteria (curves a and b). In accordance with the above results, the inhibition of relaxed phospholipid

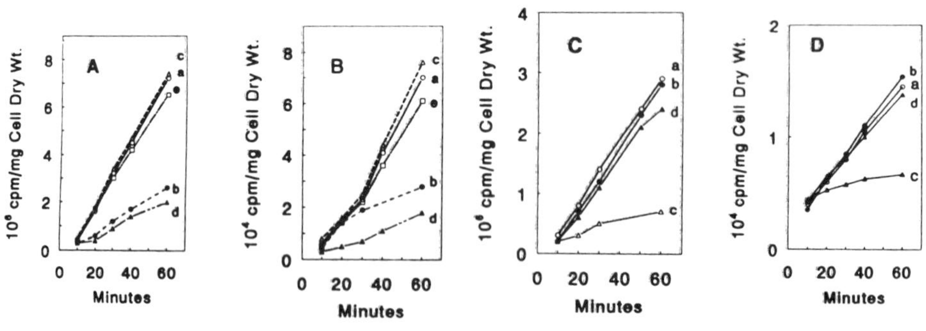

Figure 1. Dependence of peptidoglycan synthesis on phospholipid synthesis during amino acid deprivation. The synthesis of phospholipid (A) and peptidoglycan (B) in strain VC7 (*relA*$^+$). a) untreated control; b) isoleucine-deprived; c) isoleucine-deprived and chloramphenicol-treated; d) isoleucine-deprived with chloramphenicol and cerulenin; e) isoleucine-deprived with chloramphenicol, cerulenin, and a mixture of palmitate and oleate. Phospholipid (C) and peptidoglycan (D) synthesis in a complementary experiment with strain VC8 (*relA*). a) untreated control; b) isoleucine-deprived; c) isoleucine-deprived with cerulenin; d) isoleucine-deprived with cerulenin and a mixture of palmitate and oleate.

synthesis by cerulenin was accompanied by the inhibition of peptidoglycan synthesis (curves c). Both inhibitory effects of cerulenin were eliminated by exogenously supplied oleate and palmitate (curves d). Thus, the observed inhibition is not caused indirectly by the mere presence of cerulenin because the inhibition was abolished by palmitate and oleate even though cerulenin was still present.

ETB were prepared from samples taken at various times from the cultures described in Figure 1. Figure 2 shows the specific activities of CPase and TPase in these ETB preparations. The CPase and TPase activities in control ETB preparations of both strains VC7 and VC8 remained relatively constant over the course of the experiment (curves a). As reported previously (Ishiguro, 1983), amino acid deprivation caused a marked decline in both activities within 40 min in the *relA*+ strain (panels A and B, curves b) but had no effect on the *relA* mutant (panels C and D, curves b). Chloramphenicol prevented loss of these activities in amino acid-deprived *relA*+ bacteria (panels A and B, curves c). Collectively, the results of *in vitro* assays employing ETB were therefore consistent with the observed effects of the stringent response on peptidoglycan synthesis. The main purpose of these experiments was to determine the effects of cerulenin on TPase and CPase activities. Cerulenin caused the loss of both activities when combined with chloramphenicol in the *relA*+ strain (panels A and B, curves d) or by itself in the case of the *relA* mutant (panels C and D, curves c). Furthermore, the TPase and CPase activities in both strains were retained at control levels when the effects of cerulenin were negated by exogenous fatty acid supply (panels A and B, curves e; panels C and D, curves d). Thus, the apparent requirement for ongoing phospholipid synthesis could be correlated with TPase and CPase activities assayed in ETB.

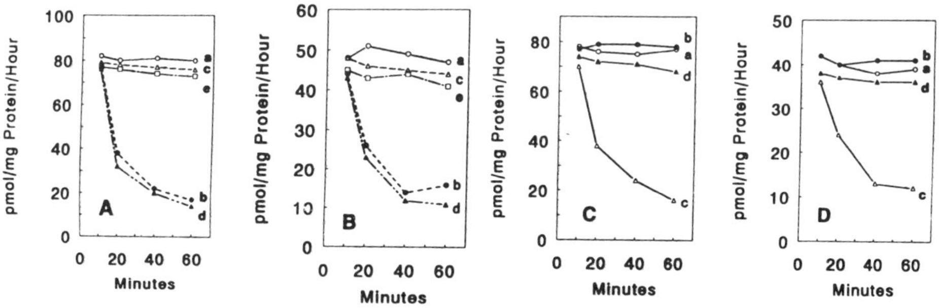

Figure 2. Effects of inhibition of phospholipid synthesis on the specific activities of CPase (A and C) and TPase (B and D). ETB of strain VC7 (A and B) were prepared from the following: a) untreated control; b) isoleucine-deprived; c) isoleucine-deprived and chloramphenicol-treated; d) isoleucine-deprived with chloramphenicol and cerulenin; e) isoleucine-deprived with chloramphenicol, cerulenin and a mixture of palmitate and oleate. ETB of strain VC8 (C and D) were prepared from the following: a) untreated control; b) isoleucine-deprived; c) isoleucine-deprived with cerulenin; d) isoleucine-deprived with chloramphenicol, cerulenin and a mixture of palmitate and oleate.

In agreement with a previous report (Ishiguro, 1983), CPase specific activities in control ETB of either strain VC7 or VC8 was at least doubled by brief (1 sec) sonic treatment (Figures 3A and 3B). Moreover, the activity which was apparently lost as a result of amino acid deprivation of VC7 was recovered by sonication (data not shown; see Harkness *et al.*, 1981). Figure 3 also shows that the CPase activities lost as a consequence of the inhibition of relaxed phospholipid synthesis during amino acid deprivation in strains of VC7 and VC8 were recovered upon sonication of the ETB. These results suggest that the loss of CPase activity reported here is probably not due to perma-

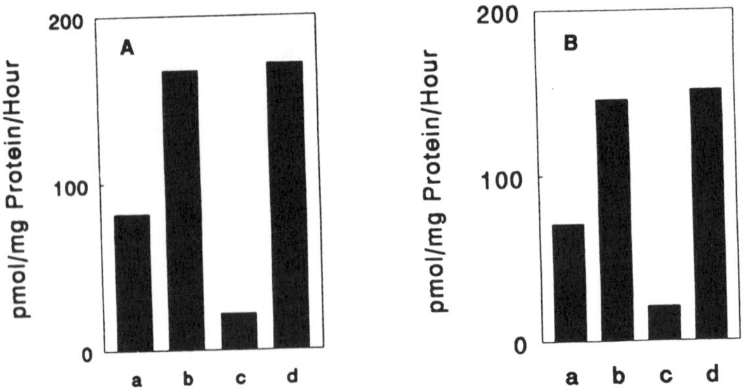

Figure 3. Effects of sonification on specific activities of CPase. ETB prepared at the 60 min point in Figure 1 were used. (A) Strain VC7: untreated control ETB assayed before (a) and after (b) sonication; isoleucine-deprived, chloramphenicol/cerulenin-treated ETB assayed before (c) and after (d) sonication. (B) Strain VC8: untreated control ETB assayed before (a) and after (b) sonication; isoleucine-deprived, cerulenin-treated ETB assayed before (c) and after (d) sonication.

nent inactivitation or degradation of the enzyme and may represent a transition to a cryptic form. It has not been possible to test TPase activity in this way because sonication completely destroyed the activity in control ETB.

DISCUSSION AND CONCLUSIONS

We have previously proposed that the loss of peptidoglycan synthetic activity observed in ETB prepared from amino acid-deprived *relA*+ cultures probably reflected the effects of the stringent response (Harkness *et al.*, 1981). A direct effect of ppGpp was ruled out because ether permeabilization resulted in the release of such low molecular weight compounds. Alternatively, it was suggested that the observed inhibition of peptidoglycan synthesis may be a manifestation of some inherent change in cell membranes brought about by the stringent response. We think the observed dependence of peptidoglycan synthesis and phospholipid synthesis during the stringent response as demonstrated here with cerulenin provides further support for this proposal.

The coupling between peptidoglycan synthesis and phospholipid synthesis was also demonstrated previously with a glycerol auxotroph (Ishiguro, 1983). The relaxed syntheses of both phospholipids and peptidoglycan were inhibited when a *relA* glycerol auxotroph was simultaneously deprived of glycerol and a required amino acid. Moreover, both inhibitory effects were reversed upon restoration of the glycerol supply. We have subsequently shown that peptidoglycan synthesis is also inhibited by cerulenin during normal bacterial growth (unpublished data).

A direct correlation between the inhibition of phospholipid synthesis and the loss of *in vitro* peptidoglycan synthetic activities has been demonstrated here. We have so far been unable to identify the exact enzyme activities measured in these experiments by assaying penicillin-binding proteins either in vivo or in ETB. Despite the inability to identify these enzyme activities, it is significant, in my opinion, that there is a clearcut correlation between their activities and the stringent response. We hypothesize that some functional or physical change in cell membranes accompanies the inhibition of phospholipid synthesis in amino acid-deprived *relA*+ bacteria and this in turn is responsible for the observed inhibition of peptidoglycan synthesis. The pleiotropic nature of the stringent response is noteworthy, and it is possible that several other membrane-associated metabolic phenomena known to be controlled by the stringent response are similarly dependent on phospholipid synthesis, *e.g.*, lipopolysaccharide synthesis (Ishiguro *et al.*, 1986), initiation of DNA synthesis (Cashel and Rudd, 1987), and various transport systems (Cashel and Rudd, 1987). This may be an efficient global means of regulating such processes. The underlying mechanisms involved in regulating peptidoglycan synthesis are currently unknown. It is interesting that CPase activities lost during amino acid deprivation can be recovered by sonication. It may be that sonication merely enhances accessibility of CPase to its substrate. If so, we speculate that the changes induced by the stringent response and the inhibition of phospholipid synthesis result in a significant loss of substrate accessibility.

ACKNOWLEDGMENTS

This work was supported by grants from the Natural Sciences and Engineering Research Council of Canada. I thank Desiree Vanderwel for excellent technical assistance.

REFERENCES

Cashel, M. and Rudd, K.E. (1987) The stringent response, in "*Escherichia coli* and *Salmonella typhimurium*. Cellular and Molecular Biology" (F.C. Neidhardt, J.L. Ingraham, K.B. Low, B. Magasanik, J. Schaecter, and H.E. Umbarger, Ed.), pp. 1410–1438. American Society for Microbiology, Washington, D.C.

Bell, R.M. (1974) Mutants of *Escherichia coli* defective in membrane phospholipid synthesis: macromolecular synthesis in an sn-glycerol 3-phosphate acyltransferase K_m mutant. J. Bacteriol. 117, 1065–1076.

Bligh, E.G. and Dyer, W.J. (1959). A rapid method of total lipid extraction and purification. Can. J. Biochem. Physiol. 37, 911–917.

Harkness, R.E., Mirelman, D. and Ishiguro, E.E. (1983) Regulation of D-alanine carboxypeptidase and peptidoglycan cross-linkage in amino acid-deprived *Escherichia coli*. J. Bacteriol. 145, 845–849.

Ishiguro, E.E. (1983) Mechanism of stringent control of peptidoglycan synthesis in *Escherichia coli*, in "The Target of Penicillin" (R. Hakenbeck, J.-V. Höltje and H. Labischinski, Ed.), pp. 631–636, Walter de Gruyter & Co., Berlin.

Ishiguro, E.E., Vanderwel, D. and Kusser, W. (1986) Control of lipopolysaccharide biosynthesis and release by *Escherichia coli* and *Salmonella typhimurium*. J. Bacteriol. 168, 328–333 .

Mirelman, D., Yashouv-Gan, Y. and Schwarz, U. (1976) Peptidoglycan synthesis in a thermosensitive division mutant of *Escherichia coli. Biochemistry*. 15, 1781–1790.

Pisabarro, A.G., De Pedro, M.A. and Ishiguro, E.E. (1990) Dissociation of the ampicillin-induced lysis of amino acid-deprived *Escherichia coli* into two stages. J. Bacteriol. 172, 2187–2190.

DOES PBP2 REGULATE CELL DIVISION IN *E. coli*?

Philippe Bouloc, Daniel Vinella, Danièle Joseleau-Petit and Richard D'Ari

Institut Jacques Monod
C.N.R.S., Université Paris 7
2, place Jussieu
F-75251 Paris Cedex 05, France

Although penicillin has not made war humane, it probably saved over a million lives during World War II. The action of penicillin and related β-lactam antibiotics was early recognized to be extrmely broad, affecting the vast majority of bacterial species, and at the same time highly specific, generally producing little effect on eukaryotic cells. As the complex structure of the bacterial cell wall became known, the mechanism of action of β-lactams was also revealed: they bind covalently to the PBPs ('penicillin binding proteins'), a set of integral membrane proteins which catalyse the terminal steps in the synthesis of the rigid peptoglycan wall (for review see Waxman and Strominger, 1983; Ghuysen, 1991).

The PBPs of *Escherichia coli* have been shown to catalyse specific types of chemical reactions. The lower molecular weight enzymes (PBPs 4, 5 and 6) have been reported to have carboxypeptidase activity, which removes the fifth amino acid, D-alanine, from N-acetyl muramic acid pentapeptide in newly synthesized peptidoglycan (Spratt, 1976; Matsuhashi *et al.,* 1977, 1979), whereas the higher molecular weight enzymes (PBPs 1a, 1b, 2 and 3) participate in two activities, transglycosylation—adding new units to growing glycan chains— and transpeptidation—crosslinking parallel glycans chains to form a highly reticulated network (Tamaki *et al.*, 1977; Ishino *et al.*, 1980; Ishino and Matsuhashi, 1981; Ishino *et al.*, 1986). Penicillin inhibits the carboxypeptidase and transpeptidase activities (Izaki *et al.*, 1968). This essentially prevents crosslinking of newly synthesized peptidoglycan, resulting in lysis of growing cells.

The apparent redundancy in PBPs has been partially confirmed genetically. Mutants lacking PBP4 (Matsuhashi *et al.*, 1977), PBP5 (Spratt, 1980) or PBP6 (Broome-Smith and Spratt, 1982) are perfectly viable, as is the double mutant lacking PBPs 5 and 6 (Broome-Smith, 1985) and, indeed, these mutants have esentially no phenotype. It would be satisfying to have a genetic demonstration that at least one low molecular weight PBP is necessary for viability. Similarly, PBPs 1a and 1b can be eliminated by separate mutations, but the double mutant is lethal (Suzuki *et al.*, 1978; Kato *et al.*, 1985). On the other hand, there are indications that the low molecular weight PBPs are not, in fact, equivalent (Buchanan and Sowell, 1982; Korat *et al.*, 1991), and there is considerable evidence that the high molecular weight PBPs are not all interchangeable. For example, when PBP3 or

Bacterial Growth and Lysis, Edited by M.A. de Pedro
et al., Plenum Press, New York, 1993

PBP2 is inhibited, PBVP1b activity prevents cell lysis, whereas PBP1a does not (Schmidt *et al.*, 1981; del Portillo and de Pedro, 1991). It is not known whether this reflects a quantitative difference in the synthetic capacity of the two enzymes or whether one of them also affects autolysin activity (Höltje and Tuomanen, 1991). Similarly, the two principal modes of cell wall growth, pole formation (or septation) and lateral elongation (Lleo *et al.*, 1990), seem to require specialised enzymes: PBP3 has been shown to be an essential septation enzyme (Spratt and Pardee, 1975; Botta and Park, 1981), whereas PBP2 is essential for lateral elongation of the cylindrical cell wall and maintenance of rod shape (Spratt and Pardee, 1975; Ogura *et al.*, 1989). Furthermore, the proper balance between these two modes of wall growth may be a target of cell cycle regulation, possibly by the gene products of the *mre* locus (Wachi *et al.*, 1987).

The distinctions between PBPs may well reflect true enzymatic differences. Over the past decade, elegant chemical analysis has revealed that the crosslinks between glycan chains are of several different types (Glauner *et al.*, 1988), and it is possible that the transpeptidase activity of each high molecular weight PBP is specialised to form a single specific type of crosslink. There is evidence, for example, that the septation enzyme PBP3 forms links preferentially (or exclusively) with muramic acid tripeptides (Pisabarro *et al.*, 1986), and it is known that an excess of carboxypeptidase activity, increasing the amount of tetrapeptides and tripeptides, makes wild type cells spherical (Markiewicz *et al.*, 1982) and permits cells with a mutant PBP3 to function with lower levels of activity, presumably because of the higher substrate concentration (Begg *et al.*, 1990). We wish to discuss here the possibility that PBP2 may have additional activities.

PBP2 has been at the centre of a paradox. Its study has been greatly facilitated by the discovery of mecillinam (Lund and Tybring, 1972), a ß-lactam that binds specifically to PBP2. The addition of mecillinam to cultures of *E. coli* induces a morphological change, with all cells becoming spherical within about one generation time. After this transition, cell division gradually stops and there is a loss of viability (Park and Burman, 1973; Tybring and Mechior, 1975; Matsuhashi *et al.*, 1979), suggesting that PBP2 is an essential protein. However, Spratt isolated a mecillinam resistant mutant, SP45, affected in the PBP2 structural gene, *pbpA* (Spratt, 1978). This mutant has no detectable PBP2, yet it is perfectly viable and grows as a coccus, suggesting that PBP2 is dispensable (Donachie and Begg, 1989). On the other hand, we isolated a *pbpA* mutant which at high temperature becomes spherical and dies, again suggesting that PBP2 is essential (Ogura *et al.*, 1989).

We became interested in this question through the study of mecillinam resistant mutants in which PBP2 activity and cell shape remained normal in the absence of the drug. Such strains include *cya* and *crp* mutants, lacking a functional CAP–cyclic AMP complex (Aono *et al.*, 1978; Jaffé *et al.*, 1983; D'Ari *et al.*, 1988). They also comprise mutants in which one or another tRNA synthetase is partially defective, resulting in partial starvation for the corresponding aminoacyl-tRNA species and high levels of the nucleotide ppGpp, effector of the stringent response. We showed that the high ppGpp level in these mutants is necessary for their mecillinam resistance, since a *relA* mutation, affecting ppGpp synthesis, restores sensitivity, and, in fact, high ppGpp seems to be sufficient, since excess RelA activity confers mecillinam resistance on wild type strains (Vinella *et al.*, 1992).

To resolve the question of the essentiality of PBP2, we cloned the *pbpA* gene and constructed an insertion-deletion *in vitro*. The *ΔpbpA* allele is lethal in our wild type strain (Ogura *et al.*, 1989). However, it can be introduced into mecillinam resistant mutants — *cya*, *argS*, *alaS*, *leuS*(Ts) or p*relA*′-bearing strains (Ogura *et al.*, 1989; Vinella *et al.*, 1992). Furthermore, when we introduced a *pbpA*$^+$ allele into Spratt's mutant SP45, it remained mecillinam resistant, suggesting that the loss of PBP2 in the original strain was associated with a Mec^R mutation, as in our *ΔpbpA* strains. These observations show clearly that PBP2 is an essential protein and suggest that it plays two roles in the cell. Its transglycosylase and transpeptidase activities are required for lateral cell wall elongation. This role is not essential for cell survival, as demonstrated by the existence of viable spherical mutants of *E. coli*. The second role of PBP2, however, is indispensable in wild type cells, in which the inactivation of PBP2 leads to cell death.

How can the inactivation of PBP2 block a vital process other than peptidoglycan synthesis? The simplest way would be by direct interaction with another enzyme, presumably a membrane protein, like PBP2. A cell division protein would be an attractive candidate. This possibility stimulated a search for mutations conferring mecillinam resistance and located in the 2 min cluster, which includes the cell division genes *ftsI*, *ftsW*, *ftsQ*, *ftsA* and *ftsZ*; despite considerable effort, none were found (K. Begg, personal communication). Secretory proteins such as SecA and SecY are also potential targets of PBP2 inactivation. In the case of SecA, at least, inactivation of the protein leads to early division arrest, followed considerably later by growth arrest (Oliver and Beckwith, 1982). Similarly, mutations in the heat shock genes *groEL*, *groES* and *dnaK* cause filamentation followed by growth arrest (Georgopoulos and Eisen, 1974; Paek and Walker, 1987), and thermal shock in wild type strains produces a temporary division block, suggesting the heat shock regulon as another a potential target of PBP2 inactivation (Powell and Young, 1991). One could also imagine that inactive PBP2 interferes with some vital membrane-bound sensor protein, leading to misinterpretation of environmental signals. The inactivation of PBP2 might produce its effect indirectly, *via* a cascade of interactions, ultimately blocking some vital process.

What is the second, essential role of PBP2? Mecillinam clearly interferes with both roles, since it prevents cell wall elongation and results in death. Early studies pointed to an arrest of cell division in the presence of mecillinam before the final cessation of growth, suggesting that the essential function of PBP2 may concern septation (James *et al.*, 1975). In the light of evidence for a dual role of PBP2, it seemed advisable to study the essential function under conditions that did not at the same time alter cell shape, thereby avoiding complications due to the morphological change from rod to sphere. Ideally, one could use a conditional, rod-shaped *pbpA* mutant in which only the vital function is affected under nonpermissive conditions. However, it is possible that the vital process cannot be dissociated from peptidoglycan synthetase activity. In any case, to date no such *pbpA* allele has been described and there is no obvious selection for one. Alternatively, it is possible to use a spherical *ΔpbpA* strain completely lacking PBP2, in which viability is maintained thanks to a conditional Mec^R mutation; shifting the cells to Mec^S conditions should cause cell death without restoring rod shape.

We have used two conditional systems of this type. The first is a *ΔpbpA Δcya* strain, which loses viability in the presence of exogenous cyclic AMP (Ogura *et al.*, 1989). Under these conditions, cell division is indeed affected, as evidenced by the formation of large spherical cells. There is also some degree of cell lysis, which makes it exceedingly difficult to use the Coulter counter to determine cell counts and cell volume distribution and thus to establish the precise kinetics of cell division arrest. We recently set up a second conditional system, taking advantage of the fact that a *leuS*(Ts) mutant, in which the leucyl-tRNA synthetase is temperature sensitive, is resistant to mecillinam at 37°C but sensitive at 25°C (and Ts for growth at 42°C) (Vinella *et al.*, 1992). A *ΔpbpA leuS*(Ts) double mutant is viable at 37°C but not at 25°C (and it is Ts for growth at 42°C). There is essentially no lysis in this strain at 25°C; the partial lysis observed in the *ΔpbpA Δcya* strain may reflect the pleiotropic phenotype due to the absence of CAP-cAMP, possibly a regulatory effect of this complex on autolysin expression. After a shift to 25°C, cell division slows down immediately; over a 3 h period, cell number doubles while average cell mass increases more than 7-fold. In the phase contrast microscope, very large spheres are observed, and average cell volume increases in parallel with the optical density of the culture.

The viability of the *ΔpbpA leuS*(Ts) strain at 37°C is presumably due to a high ppGpp pool (Vinella *et al.*, 1992). Since the turnover of ppGpp is rapid, it is safe to assume that the new, lower equilibrium level is reached within minutes after the shift from 37°C to 25°C. The residual division observed suggests that protection afforded by the high pool at 37°C remains operative for a certain time after the pool collapses. It is thus unlikely that the protection reflects, for example, direct activation of some vital enzyme by ppGpp. It is more likely to be a transcriptional effect, although we cannot rule out the interesting possibility that a high ppGpp pool induces a special event exactly once per cycle.

How can we account for the ability of cells to do without PBP2 when the ppGpp pool is high or when the cAMP pool is nil? If we assume a direct interaction between PBP2 and some vital enzyme, expression of the latter might be increased under these conditions, compensating for the negative effect of PBP2 inactivation. One would therefore expect such an enzyme to be positively regulated by ppGpp and negatively by CAP-cAMP. Knowledge of the regulation of cell division, secretion, heat shock and sensor proteins is sufficiently scanty that most of them remain candidates for the PBP2 target. For the case of a chain of interactions transmitting information on PBP2 activity to some vital process, one can imagine dispensable intermediates, regulated by ppGpp and CAP-cAMP. It is interesting to note that *ΔrelA ΔspoT* mutants, entirely lacking ppGpp, tend to filament (Xiao *et al.*, 1991), compatible with the division-promoting role we postulate for ppGpp.

Alterations of ribosomal proteins have been described which can affect cell division. Certain *rpsD* and *rpsE* mutants (proteins S4 and S5), isolated for their resistance to spectinomycin in media of high osmolarity, have been shown to filament and die when the osmolarity is lowered (Mizuno *et al.*, 1976; Dombou *et al.*, 1981). Streptomycin-dependent *rpsL* mutants (protein S12) also filament and die when the drug is removed (Spotts and Stanier, 1961). Evidence for a connection between the ribosomes and PBP2 comes from our observation that streptomycin resistant *rpsL* alleles which increase ribosomal fidelity suppress the mecillinam resistance due to an *argS* mutation (Bouloc *et al.*, 1988, 1989; Vinella *et al.*, 1992). We have also found that certain *rpsD*, *rpsE* and *rpsL* alleles can confer mecillinam resistance on otherwise wild type strains (Vinella *et al.*, 1992; unpublished observations); we are currently investigating the role of the ppGpp pool in these phenomena. It is entirely possible that there exists some additional connection which transmits information on the state of ribosome activity directly to the PBPs. Considering that PBP2 inactivation ultimately results in growth arrest, it is conceivable — albeit unlikely — that this hypothetical connection works in both directions, permitting a direct effect of the PBPs on ribosome activity.

In conclusion, we feel convinced that PBP2, at least, plays a vital role in *E. coli* different from its activity as peptidoglycan synthetase. The same question can be asked about the other high molecular weight PBPs. Positive answers to these questions and the identification of vital PBP-dependent processes should shed considerable light on the coupling mechanisms responsible for the harmonious growth of *E. coli.*

ACKNOWLEDGEMENT

This work was supported in part by the Association pour la Recherche sur le Cancer (contract N° 6696).

REFERENCES

Aono, R., Yamasaki, M. and Tamura, G. (1978) Changes in composition of envelope proteins in adenylate cyclase- or cyclic AMP receptor protein-deficient mutants of *Escherichia coli*. J. Bacteriol. 136, 812-814.

Begg, K.J., Takasuga, A., Edwards, D.H., J., D.S., Spratt, B.G., Adachi, H., Ohta, T., Matsuzawa, H. and Donachie, W.D. (1990) The balance between different peptidoglycan precursors determines whether *Escherichia coli* cells will elongate or divide. J. Bacteriol. 172, 6697-6708.

Botta, G.A. and Park, J.T. (1981) Evidence of penicillin-binding protein 3 in murein synthesis during septation but not during cell elongation. J. Bacteriol. 145, 333-340.

Bouloc, P., Jaffé, A. and D'Ari, R. (1988) Preliminary physiologic characterization and genetic analysis of a new *Escherichia coli* mutant, *lov*, resistant to mecillinam. Rev. Infect. Dis. 10, 905-910.

Bouloc, P., Jaffé, A. and D'Ari, R. (1989) The *Escherichia coli lov* gene product connects peptidoglycan synthesis, ribosomes and growth rate. EMBO J. 8, 317-323.

Broome-Smith, J.K. (1985) Construction of a mutant of *Escherichia coli* that has deletions of both the penicillin-binding protein 5 and 6 genes. J. Gen. Microbiol. 331, 2115-2118.

Broome-Smith, J.K. and Spratt, B.G. (1982) Deletion of the penicillin-binding protein 6 gene of *Escherichia coli*. J. Bacteriol. 152, 904-906.

Buchanan, C.E. and Sowell, M.O. (1982) Synthesis of penicillin-binding protein 6 by stationary-phase of *Escherichia coli*. J. Bacteriol. 151, 491-494.

D'Ari, R., Jaffé, A., Bouloc, P. and Robin, A. (1988) Cyclic AMP and cell division in *Escherichia coli*. J. Bacteriol. 170, 65-70.

del Portillo, F.G. and de Pedro, M.A. (1991) Penicillin-Binding Protein 2 is essential for the integrity of growing cells of *Escherichia coli ponB* strains. J. Bacteriol. 173, 4530-4532.

Dombou, M., Bhide, S.V. and Mizushima, S. (1981) Appearance of elongation factor Tu in the outer membrane of sucrose-dependent spectinomycin-resistant mutants of *Escherichia coli*. Eur. J. Biochem. 113, 397-403.

Donachie, W.D. and Begg, K.J. (1989) Cell length, nucleoid separation, and cell division of rod-shaped and spherical cells of *Escherichia coli*. J. Bacteriol. 171, 4633-4639.

Georgopoulos, C.P. and Eisen, H. (1974) Bacterial mutants which block phage assembly. J. Supramolec. Struct. 2, 349-359.

Ghuysen, J.-M. (1991) Serine ß-lactamases and penicillin-binding proteins. Annu. Rev. Microbiol. 45, 37-67.

Glauner, B., Höltje, J.-V. and Schwarz, U. (1988) The composition of the murein of *Escherichia coli*. J. Biol. Chem. 263, 10088-10095.

Höltje, J.-V. and Tuomanen, E.I. (1991) The murein hydrolases of *Escherichia coli*: properties, functions and impact on the course of infections *in vivo*. J. Gen. Microbiol. 137, 441-454.

Ishino, F. and Matsuhashi, M. (1981) Peptidoglycan synthetic enzyme activities of highly purified penicillin-binding protein 3 in *Escherichia coli*: a septum-forming reaction sequence. Biochem. Biophys. Res. Comm. 101, 905-911.

Ishino, F., Mitsui, K., Tanaki, S. and Matsuhashi, M. (1980) Dual enzyme activities of cell wall peptidoglycan synthesis, peptidoglycan transglycosylase and penicillin sensitive transpeptidase, in purified preparations of *Escherichia coli* penicillin-binding protein 1A. Biochem. Biophys. Res. Comm. 97, 287-293.

Ishino, F., Park, W., Tomioka, S., Tamaki, S., Takase, I., Kunugita, K., Matsuzawa, H., Asoh, S., Ohta, T., Spratt, B.G. and Matsuhashi, M. (1986) Peptidoglycan synthetic activities in membranes of *Escherichia coli* caused by overproduction of penicillin-binding protein 2 and RodA protein. J. Biol. Chem. 261, 7024-7031.

Izaki, K., Matsuhashi, M. and Strominger, J.L. (1968) Biosynthesis of the bacterial cell walls. XIII. Peptidoglycan transpeptidase and D-alanine carboxypeptidase: penicillin-sensitive enzymatic reaction in strains of *Escherichia coli*. J. Biol. Chem. 243, 3180-3192.

Jaffé, A., Chabbert, Y.A. and Derlot, E. (1983) Selection and characterization of ß-lactam-resistant *Escherichia coli* K12 mutants. Antimicrob. Agents Chemother. 23, 622-625.

James, R., Haga, J.Y. and Pardee, A.B. (1975) Inhibition of an early event in the cell division cycle of *Escherichia coli* by FL 1060, an amidinopenicillanic acid. J. Bacteriol. 122, 1283-1292.

Kato, J., Suzuki, H. and Hirota, Y. (1985) Dispensability of either penicillin binding protein-1a or -1b involved in the essential process for cell elongation in *Escherichia coli*. Mol. Gen. Genet. 200, 272-277.

Korat, B., Mottl, H. and Kech, W. (1991) Penicillin-binding protein 4 of *Escherichia coli*: molecular cloning of the *dacB* gene, controlled overexpression, and alterations in murein composition. Mol. Microbiol. 5, 675-684.

Lleo, M.M., Canepari, P. and Satta, G. (1990) Bacterial cell shape regulation: testing of additional predictions unique to the two-competing-sites model for peptidoglycan assembly and isolation of conditional rod-shaped mutants from some wild-type cocci. J. Bacteriol. 172, 3758-3771.

Lund, F. and Tybring, L. (1972) 6ß-amidinopenicillanic acid — a new group of antibiotics. Nature New Biol. 236, 135-137.

Markiewicz, Z., Broome-Smith, J.K., Schwarz, U. and Spratt, B.G. (1982) Spherical *E. coli* due to elevated levels of D-alanine carboxypeptidase. Nature 297, 702-704.

Matsuhashi, M., Takagaki, Y., Maruyama, I.N., Tanaki, S., Nishimura, Y., Suzuki, H., Ogino, U. and Hirota, Y. (1977) Mutants of *Echerichia coli* lacking in highly penicillin-sensitive D-alanine carboxypeptidase activity. Proc. Natl. Acad. Sci. USA. 74, 2976-2979.

Matsuhashi, M., Tamaki, S., Curtis, T.S. and Strominger, J.L. (1979) Mutational evidence for identity of penicillin-binding protein 5 in *Escherichia coli* with major D-alanine carboxypeptidase IA activity. J. Bacteriol. 137, 644-647.

Mizuno, T., Yamada, H., Yamagata, H. and Mizushima, S. (1976) Coordinated alteration in ribosomes and cytoplasmic membrane in sucrose-dependent, spectinomycin-resistant mutants of *Escherichia coli.* J. Bacteriol. 125, 524-530.

Ogura, T., Bouloc, P., Niki, H., D'Ari, R., Hiraga, S. and Jaffé, A. (1989) Penicillin-Binding Protein 2 is essential in wild-type *Escherichia coli* but not in *lov* or *cya* mutants. J. Bacteriol. 171, 3025-3030.

Oliver, D. and Beckwith, J. (1982) Identification of a new gene (*secA*) and gene product involved in the secretion of the envelope proteins. J. Bacteriol. 150, 686-691.

Paek, K.-H. and Walker, G.C. (1987) *Escherichia coli dnaK* null mutants are inviable at high temperature. J. Bacteriol. 169, 283-290.

Park, J.T. and Burman, L. (1973) FL1060 — A new penicillin with a unique mode of action. Biochem. Biophys. Res. Comm. 51, 863-868.

Pisabarro, A.G., Prats, R., Vázquez, D. and Rodríguez-Tébar, A. (1986) Activity of penicillin-binding protein 3 from *Escherichia coli.* J. Bacteriol. 168, 199-206.

Powell, J.K. and Young, K.D. (1991) Lysis of *Escherichia coli* by beta-Lactams which bind penicillin-binding proteins 1a and 1b - Inhibition by Heat Shock Proteins. J. Bacteriol 173, 4021-4026.

Schmidt, L., Botta, G. and Park, J.T. (1981) Effects of furazlocillin, a ß-lactam antibiotic which binds selectively to penicillin-binding protein 3, on *Escherichia coli* mutant deficient in other penicillin-binding proteins. J. Bacteriol. 145, 632-637.

Spotts, C.R. and Stanier, R.Y. (1961) Mechanism of streptomycin action on bacteria: a unitary hypothesis. Nature 192, 633-637.

Spratt, B.G. (1976) Identification of the major penicillin binding proteins of *Escherichia coli* as D-alanine carboxypeptidase IA. J. Bacteriol. 127, 660-663.

Spratt, B.G. (1978) *Escherichia coli* resistance to ß-lactam antibiotics throught a decrease in the affinity of a target for lethality. Nature 274, 713-715.

Spratt, B.G. (1980) Deletion of the penicillin-binding protein 5 gene of *Escherichia coli.* J. Bacteriol. 144, 1190-1192.

Spratt, B.G. and Pardee, A.B. (1975) Penicillin-binding protein and cell shape in *E. coli.* Nature 254, 515-517.

Suzuki, H., Nishimura, Y. and Hirota, Y. (1978) On the process of cellular division in *Escherichia coli*: a series of mutants of *E. coli* altered in the penicillin-binding proteins. Proc. Natl. Acad. Sci. U.S.A. 75, 664-668.

Tamaki, S., Nakajima, S. and Matsuhashi, M. (1977) Thermosensitive mutation in *Escherichia coli* simultaneously causing defects in penicillin-binding protein-1Bs and in enzyme activity for peptidoglycan synthesis *in vitro*. Proc. Natl. Acad. Sci. USA. 74, 5472-5476.

Tybring, L. and Mechior, N.H. (1975) Mecillinam (FL 1060), a 6ß-amidinopencillanic acid derivative: bacterial action and synergy in vitro. Antimicrobiol. Agents Chemother. 8, 271-276.

Vinella, D., D'Ari, R. and Bouloc, P. (1992) Penicillin-binding protein 2 is dispensable in *Escherichia coli* when ppGpp synthesis is induced. EMBO 11, in press.

Wachi, M., Doi, M., Tamaki, S., Park, W., Nakajima-Iijima, S. and Matsuhashi, M. (1987) Mutant isolation and molecular cloning of *mre* genes, which determine cell shape, sensitivity to mecillinam, and amount of penicillin-binding proteins in *Escherichia coli.* J. Bacteriol. 169, 4935-4940.

Waxman, D.J. and Strominger, J.L. (1983) Penicillin-binding proteins and the mechanism of action of ß-lactam antibiotics. Annu. Rev. Biochem. 52, 825-869.

Xiao, H., Kalman, M., Ikehara, K., Zemel, S., Glaser, G. and Cashel, M. (1991) Residual guanosine 3´,5´-bispyrophosphate synthetic activity of *relA* null mutants can be eliminated by *spoT* null mutations. J. Biol. Chem. 266, 5980-5990.

PEPTIDOGLYCAN SYNTHESIS DURING HYPHAL ELONGATION IN **Streptomyces antibioticus**

Elisa M. Miguélez, Carlos Hardisson and Manuel B. Manzanal

Departamento de Biología Funcional
Laboratorio de Microbiología
University of Oviedo
33006 Oviedo, Spain

INTRODUCTION

The streptomycetes are gram-positive bacteria with a complex and specialized cycle of cellular differentiation. Initially, the spores germinate forming multinuclear hyphae which branch repeatedly, originating a substrate mycelium that penetrates into the medium. Later, the colony differentiates and forms a "reproductive" aerial mycelium that grows into the air and sporulates.

In regard to mycelial growth it is generally assumed that the streptomycetes elongate by apical extension, that is, by means of a single envelope-growing site located at the tip of the hyphae (Kretschmer, 1982). Most data on the growth of the streptomycetes have been obtained from microscopic observations (Kretschmer, 1982; Allan and Prosser, 1983) or from autoradiographic studies (Braña **et al.**, 1982; Gray **et al.**, 1990). However, biochemical studies on the synthesis of peptidoglycan during hyphal growth have not been made.

In this report we describe an investigation designed to provide more information on the process of hyphal growth in **Streptomyces**. With this purpose synchronous cultures were used to investigate the kinetics of hyphal elongation, the synthesis, incorporation and fate of the peptidoglycan during growth, as well as the possible mechanisms involved in its regulation.

MATERIALS AND METHODS

Spores of **Streptomyces antibioticus** (ATCC 11891) were obtained as described previously (Hardisson **et al.**, 1978). Suspensions of spores were ultrasonicated (1 min, 150 W), and suspended in GYA medium containing: 0.2% yeast extract, 50 mM glucose, 15 mM ammonium sulphate, 3 mM HK_2PO_4, 1 mM Mg_2SO_4, 0.4 mM $MnSO_4$, 5 µM $FeSO_4$, 3.4 µM $ZnSO_4$, 100 mM MOPS buffer (pH 7.0) and an amino acid mixture formed by L-lysine, L-alanine, L-arginine, L-valine, L-threonine, L-leucine, L-isoleucine, L-proline and L-serine (100 µg/ml each, final concentration). Incubation was carried out at 35°C in glass-tubes aerated by bubbling air through the culture.

Bacterial Growth and Lysis, Edited by M.A. de Pedro *et al.*, Plenum Press, New York, 1993

For continuous labeling experiments the hyphae were grown in GYA medium containing 0.1 μCi/ml of N-acetyl-D-[^{14}C]glucosamine (GlcNAc) (58.7 mCi/mmol; Amersham) and 0.015% of cold GlcNAc. For pulse labeling experiments, hyphae growing in GYA medium were filtered and suspended in GYA medium containing 0.7 μCi/ml of [^{14}C]GlcNAc. After 3 min of incubation at 35°C incorporation was stopped by adding cold GlcNAc (1.5%, final concentration). Samples were filtered, treated with trichloroacetic acid (TCA) and washed.

The incorporation of [^{14}C]GlcNAc into different cell fractions was determined as follows. Hyphae continuously or pulse labeled were treated sequentially with: 5% TCA (1 min at 0°C); RNase (1 mg/ml) in 50 mM sodium phosphate buffer (pH 8.0) (45 min at 37°C); trypsin (1 mg/ml) in the above phosphate buffer (90 min at 37°C); and lysozyme (100 μg/ml) in 50 mM phosphate buffer containing 5 mM sodium citrate (16 h at 37°C). In all cases the radioactivity solubilized and that remaining insoluble after treatments were measured in triplicate. Filters containing labeled hyphae or digested samples were dried and placed into vials containing standard toluene-based scintillation fluid. Liquid samples were added to vials and mixed with a toluene-triton X 100-based scintillation fluid. Radioactivity was measured in a Beckman LS 100C liquid scintillation counter.

For autoradiographic studies young hyphae growing in GYA medium were pulse-labeled for 1 min in GYA medium containing 50 μCi/ml of [^{3}H]GlcNAc (2.84 Ci/mmol; Amersham). The hyphae were then treated with cold-TCA, RNase and trypsin as above. Samples were spread on collodion-coated slides and covered with half-strength Ilford L4 nuclear emulsion. After 4 to 7 days of exposure, the slides were developed at 20°C during 1 min in the physical developer of Caro and van Tubergen (1962) for observation in the electron microscope.

RESULTS

Labeling of the peptidoglycan

The study of cell wall synthesis and its regulation during hyphal growth requires specifical labeling of the peptidoglycan. GlcNAc has been widely used for peptidoglycan labeling in gram-positive bacteria. However, in some species and under certain conditions of growth, this compound can be partially degraded and used for protein synthesis, originating an unspecific labeling of the cell (Mobley **et al.**, 1982). Therefore, it was necessary to investigate the effect of some culture media on the incorporation and fate of the GlcNAc in **S. antibioticus**.

For these experiments, hyphae of **S. antibioticus** growing in different media were either continuously or pulse-labeled with [^{14}C]GlcNAc. After labeling, samples were treated sequentially with TCA and different lytic enzymes. For all treatments, the radioactivity liberated and the radioactivity that remained insoluble after digestion were measured. The best results for specifical labeling of the peptidoglycan were obtained in hyphae growing in a medium containing glucose, yeast extract and a mixture of nine amino acids (GYA medium). As Table 1 shows, under these conditions of incubation and in pulse labeling experiments, only 8% of the radioactivity precipitated by TCA was incorporated into proteins. Moreover, most of the radioactivity (about 86.5%) that remained after TCA-RNase-trypsin treatments was solubilized by lysozyme.

Studies on cell wall turnover

To investigate the possible existence of turnover in the

TABLE 1. Incorporation of [^{14}C]GlcNAc into cell fractions of **S. antibioticus**

Treatment	Fraction	[^{14}C]GlcNAc incorporation [10^3 cpm (% of total)] Continuously labeled cells	Pulse labeled cells
Whole washed cells		54.6(100)	642.1(100)
Trichloroacetic acid	Soluble	8.7(17.2)	152.3(25)
	Insoluble	42.3(82.8)	458.8(75)
RNase-Trypsin	Soluble	6.3(16.0)	39.7(8.2)
	Insoluble	33.3(84.0)	446.9(91.8)
Lysozyme	Soluble	24.8(79.9)	401.9(86.5)
	Insoluble	6.3(20.1)	62.6(13.5)

Spores of **S. antibioticus** were germinated in GYA medium and the young hyphae were continuously or pulse labeled with [^{14}C]GlcNAc. Labeled hyphae were collected by filtration, washed and treated sequentially with trichloroacetic acid (TCA) and different lytic enzymes.

peptidoglycan of streptomycetes, hyphae of **S. antibioticus** growing in GYA medium were continuously labeled with [^{14}C]GlcNAc as described in METHODS. After labeling, the hyphae were filtered and suspended into fresh GYA medium containing cold GlcNAc. Once in the chase medium, hyphae resumed growth immediately but the radioactivity in the wall remained constant throughout the experiment (data not shown). This result seems to exclude the existence of a process of cell wall turnover during hyphal growth in **S. antibioticus**.

Autoradiographic localization of the sites of peptidoglycan incorporation

In the electron microscope the hyphae, pulse-labeled with [^{3}H]GlcNAc, appeared heavily labeled at the tip, but many silver grains also appeared randomly distributed on a broader region that, in some hyphae, extends until about 15 μm from the tip. The distribution of the radioactivity obtained by counting the number of silver grains per μm over the first 10 μm from the tip is shown in Fig. 1. As can be seen, about 30% of the grains were concentrated on the first μm (apical tip) and the remainder 70% appeared uniformly distributed over a 9 μm-length segment of the hyphae.

Kinetics of hyphal growth and branching

In liquid cultures, the streptomycetes do not differentiate into aerial mycelium, but form aggregates containing hyphae in different developmental stages. Therefore, to investigate the process of growth in **Streptomyces** it was necessary to obtain homogeneous populations of hyphae growing synchronously in liquid medium.

Synchronous cultures were obtained by ultrasonic treatment of the spores. After this treatment, most of the spores germinated at the same time, yielding hyphae very similar in length (Fig. 2). Moreover,

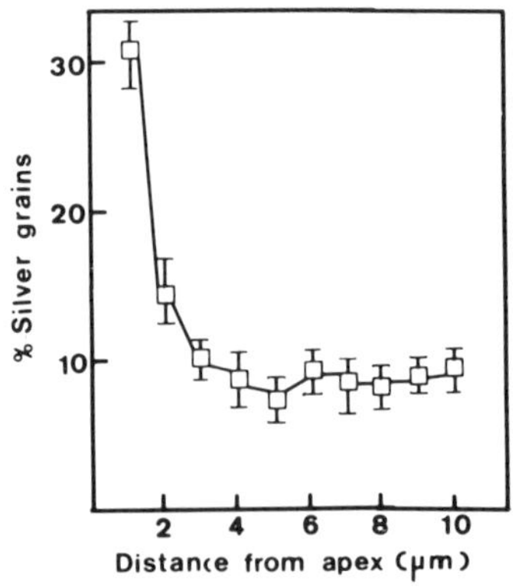

Fig. 1. Distribution of silver grains along the hyphae of **S. antibioticus**. Hyphae were labeled with [^{3}H]GlcNAc for 1 min, processed for autoradiography and observed with the electron microscope. Quantity of silver grains per μm is expressed as a percentage of the total number of silver grains contained over 10 μm from the apex.

when these populations of young hyphae were cultured at 35°C in well aerated glass-tubes containing GYA medium (in which the concentrations of Fe and Mg were carefully controlled), the hyphal aggregation was lower and the growth could be maintained during several hours in dispersed form.

Synchronous cultures of young hyphae were used to investigate the kinetics of growth and branching during hyphal development. At different times of incubation the OD of the cultures was measured and samples were taken and photographed in the phase-contrast microscope. Average lengths were calculated on enlarged images of about 50 hyphae. For each sample, the number of tips was counted on at least 100 hyphae.

As Fig. 3A shows, the OD of the cultures increased linearly, showing regular stepwise increases in growth rate. A similar pattern of growth was obtained when the increases in length were measured in the microscope (Fig. 3B). As Fig. 3B shows, hyphae with only one tip (leading

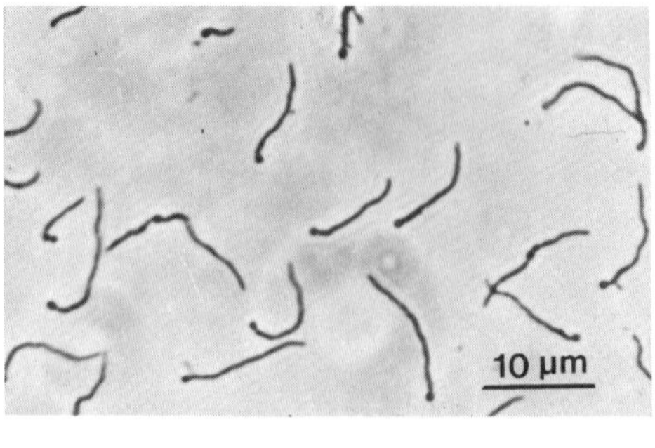

Fig. 2. Synchronous cultures of **S. antibioticus** hyphae growing in GYA medium.

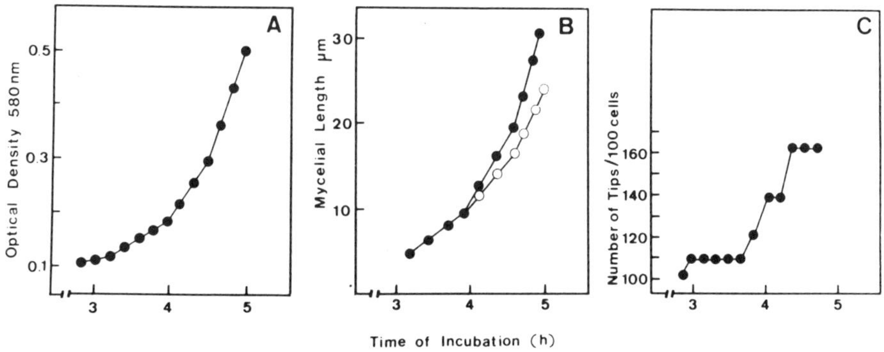

Fig. 3. Kinetics of growth in synchronous cultures of **S. antibioticus**. A, Changes in optical density. B, Total mycelial length (●) and length of the leading hypha (○). C, Time course of branch formation.

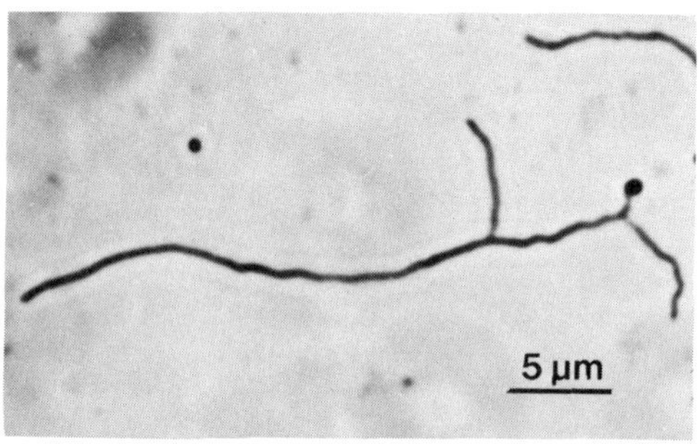

Fig. 4. Young hypha of **S. antibioticus** after 5 h of growth, showing two lateral branches.

hyphae) also exhibited changes in the rate of elongation. The time course of branch formation is shown in Fig. 3C. Initially, the number of tips was constant, but after about 3 h of incubation their number increased slightly since some spores developed a second germ-tube. Lateral branches started to be observed after about 4 h of incubation. The branches appeared close to the spore and in no case were seen on the apical segment of the hyphae (Fig. 4).

Synthesis of peptidoglycan during hyphal growth

Synchronous cultures of young hyphae growing in GYA medium were labeled with [^{14}C]GlcNAc. Total peptidoglycan synthesis was followed in continuous labeling experiments. The rate of peptidoglycan synthesis was determined by means of pulse-labeling experiments. The results obtained revealed that synthesis of peptidoglycan follows a multiphasic pattern, with successive periods in which the rate of synthesis remained constant. Thus, as Fig. 5A shows, total peptidoglycan content increased linearly, varying the slope at intervals. Pulse-labeling experiments confirmed that peptidoglycan was synthesized at a constant rate and exhibited changes in rate at specific times (Fig. 5B).

On the other hand, since no information on the regulation of peptidoglycan synthesis in **Streptomyces** is available, its possible coupling with the replication of DNA was investigated. For these

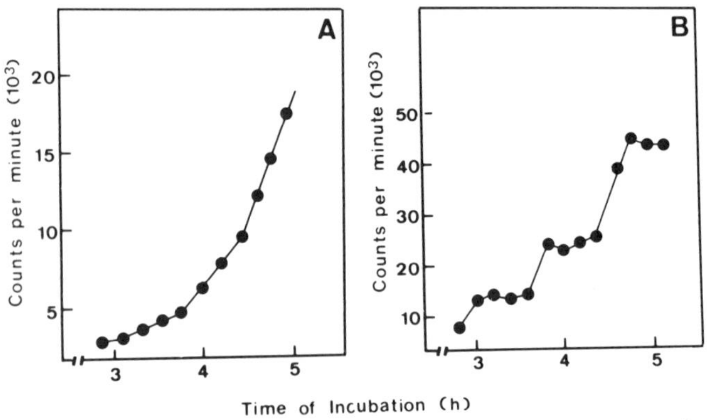

Fig. 5. Synthesis of peptidoglycan during hyphal growth. Peptidoglycan was labeled with [^{14}C]GlcNAc. A, Total peptidoglycan accumulation determined by continuous labeling. B, Rate of peptidoglycan synthesis determined by pulse-labeling.

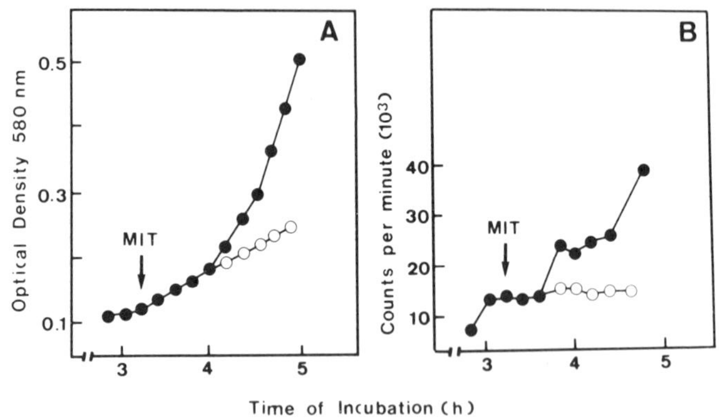

Fig. 6. Effect of mitomycin C (MIT) on growth and peptidoglycan synthesis in **S. antibioticus**. A, Changes in optical density in untreated cultures (●) and after addition of MIT (5 μg/ml) (○). B, Rates of peptidoglycan synthesis in the absence (●) and in the presence of MIT (5μg/ml) (○).

experiments we used mitomycin C (MIT) because it inhibits DNA synthesis with a minimal effect on the synthesis of other macromolecules (Hardisson **et al.**, 1978).

As Fig. 6A shows, after addition of MIT (5 μg/ml; final concentration), linear growth was maintained for about 120 min and no changes in slope were seen. On the other hand, in the presence of MIT the rate of peptidoglycan synthesis (determined by pulse labeling) remained constant throughout the experiment (Fig. 6B). These results clearly indicate that in the absence of DNA synthesis the biosynthetic activity in the wall is maintained, whithout increases in the rate of peptidoglycan synthesis.

DISCUSSION

The results obtained in this work seem to indicate that, in streptomycetes, the incorporation of peptidoglycan that takes place during hyphal growth is not restricted to a single point located on the very tip. Autoradiographic analysis of hyphae pulse-labeled with [^{3}H]GlcNAc demonstrated that 30% of the radioactivity was concentrated on the first μm of the hyphae but the remainder 70% appeared uniformly distributed along a broad segment that extends about 10 µm from the tip. Similar studies carried out by Gray **et al.** (1990) in **S. coelicolor** also demonstrated incorporation of the label behind the tip.

The possibility that this pattern of labeling could have been originated by mechanisms different to those of a true process of wall growth should be excluded since: (1) the peptidoglycan was specifically labeled and the low levels of radioactivity incorporated into protein were removed by enzymic treatment before autoradiography; (2) growth occurred in the absence of cell wall turnover, which excludes the possibility that a turnover-induced wall synthesis might be the explanation for the incorporation of peptidoglycan observed behind the tip; and (3) this labeling was not due to other process of wall synthesis, since branches were not observed in this region of the hyphae.

On the other hand, studies on the time course of length extension demonstrated that hyphal length increases linearly, showing changes in slope at particular times. Biochemical data corroborated this observation since, during periods of linear growth, peptidoglycan was synthesized at constant rate. Moreover, changes in slope also coincided with changes in the rate of peptidoglycan synthesis. These changes in the rate of peptidoglycan synthesis were not originated by a process of hyphal branching since: (1) studies on branch formation showed no correlation between increases in the growth rate and tip formation and (2) hyphae with only one tip also exhibited changes in the rate of elongation. However, experiments carried out with DNA-inhibited hyphae clearly demonstrated that changes in growth rate were dependent on DNA synthesis.

All these data seem to be consistent with a model of hyphal growth in which peptidoglycan is added to the wall at a constant rate in a well defined number of sites, the number of which depends on, or is coupled to the replication cycle of the DNA. However our data do not indicate how these growth sites are arranged in the wall of a growing hypha. Probably, the most satisfactory explanation for the autoradiographic data could be a model of hyphal growth in which growth sites are initially formed as a cluster at the tip. Later, and as a consequence of their biosynthetic activity, the wall enlarges at the tip and the sites, that still remain active, separate from each other, being dispersed along the growing portion of the hypha. A model in which surface extension is achieved by enlargement and segregation of wall growth zones has been proposed for **Bacillus** (Schlaeppi **et al.**, 1982). However, this model, although quite attractive, has been recently questioned in favour of a diffuse mode for cell wall growth (Koch, 1988). Obviously, more experimental data are needed before the model proposed here for hyphal growth could be accepted for **Streptomyces**.

ACKNOWLEDGEMENTS

This work was supported by grant 0865-84 from the Comisión Interministerial de Ciencia y Tecnología (Spain).

REFERENCES

Allan, E. J. and Prosser, J. I. (1983) Mycelial growth and branching of **Streptomyces coelicolor** A3(2) on solid medium. J. Gen. Microbiol. 129, 2029-2532.

Braña, A. F., Manzanal, M. B. and Hardisson, C. (1982) Mode of cell wall growth of **Streptomyces antibioticus**. FEMS Microbiol. Letters 13, 231-235.

Caro, L. G. and van Tubergen, R. P. (1962) High resolution autoradiography. J. of Cell Biol. 15, 173-188.

Gray, D. I., Gooday, G. W. and Prosser, J. I. (1990) Apical hyphal extension in **Streptomyces coelicolor** A3(2). J. Gen. Microbiol. 136, 1077-1084.

Hardisson, C., Manzanal, M. B., Salas, J. A. and Suárez, J. E. (1978) Fine structure and biochemistry of arthrospore germination in **Streptomyces antibioticus**. J. Gen. Microbiol. 105, 203-214.

Koch, A. L. (1988) Biophysics of bacterial walls viewed as stress-bearing fabric. Microbiol. Rev. 52, 337-353.

Kretschmer, S. (1982) Dependence of the mycelial growth pattern on the individually regulated cell cycle in **Streptomyces granaticolor**. Z. Allg. Mikrobiol. 22, 335-347.

Mobley, H. L. T., Doyle, R. J., Streips, U. N. and Langemeier, S. O. (1982) Transport and incorporation of N-acetyl-D-glucosamine in **Bacillus subtilis**. J. Bacteriol. 150, 8-15.

Schlaeppi, J. M., Pooley, H. M. and Karamata, D. (1982) Identification of cell wall subunits in **Bacillus subtilis** and analysis of their segregation during growth. J. Bacteriol. 149, 329-337.

THE ROLE OF THE *ENVM* GENES OF *ESCHERICHIA COLI* AND *SALMONELLA TYPHIMURIUM* IN CELL MEMBRANE BIOSYNTHESIS

Friederike Turnowsky[1], Helmut Bergler[1] and Elisabeth Ingolic[2]

[1]Institut für Mikrobiologie, Universitätsplatz 2, A-8010 Graz
[2]Zentrum für Elektronenmikroskopie, Steyrergasse 17, A-8010 Graz

1. SUMMARY

Inhibition of the *env*M gene product by incubation of the conditional *env*M mutant *Escherichia coli* JP1111 at the nonpermissive temperature or by diazaborine treatment leads to pleiotropic effects on cell membrane composition. One of the most pronounced effects is the inhibition of phospholipid biosynthesis. The changes in the membrane structure lead to cell death without apparent lysis of the cells. The characteristic morphological changes are the retraction of the cytoplasmic membrane which leaves empty zones mainly at the poles of the cells. Overproduction of the wild type EnvM protein in the *env*M mutant at the nonpermissive temperature initially complements the ts phenotype but finally causes cell lysis.

2. INTRODUCTION

The *env*M gene product of both *Escherichia coli* and *Salmonella typhimurium* is the target of a group of antibacterial compounds, the diazaborines (Turnowsky *et al.*, 1989). Incubation of *E. coli* cells with diazaborine leads to pleiotropic effects, the most pronounced being the inhibition of fatty acid- and phospholipid biosyntheses, the induction of an inner membrane protein, and characteristic morphological changes. Similar effects on the envelope structure were observed in an *E. coli env*M mutant with a temperature sensitive (ts) growth phenotype at the nonpermissive temperature.

The incubation of the ts mutant *E. coli* JP1111 (Egan and Russell, 1973) at

Bacterial Growth and Lysis, Edited by M.A. de Pedro *et al.*, Plenum Press, New York, 1993

elevated temperatures resulted in gradual loss of viability without apparent lysis of the cells. Under certain growth conditions this effect could be complemented after introduction of the wild type (wt) *env*M gene on a plasmid. However, the massive overproduction of the EnvM protein also caused damage to the cells. In this communication we report about the effects of both inhibition and overproduction of the *env*M gene product on viability and cell morphology.

3. MATERIALS AND METHODS

Bacterial strains and growth conditions

Bacterial strains were *E. coli* JP1111 (*env*M ts) and *E. coli* BK80 (*env*M ts), both described previously (Turnowsky *et al.* 1989).The media used were Luria Broth (Silhavy *et al.*,1984) or 2xTY medium (Sambrook *et al.*,1989) with 0.05% NaCl. The cultures were grown at 30°C in an orbital shaker at 180 rpm to early log phase and shifted to 42°C where indicated. The optical density was measured at 600 nm. Plates for viable counts contained low salt 2xTY agar and were incubated at 30°C.

Plasmids

The recombinant plasmid pKF 402 has been described previously (Turnowsky *et al.* 1989). The *env*M gene from wt *E. coli* has been cloned as an *Eco*RV/*Sal*I fragment into the vectors pSP65 and pBR322 by standard techniques (Sambrook et al.,1989). The recombinant plasmids were designated pFT632 and pFT635, respectively (unpublished).

Inhibition of phospholipid biosynthesis

E. coli BK80 and *E. coli* BK80 [pKF402] were grown in Luria Broth at 30°C to early log phase. The cultures were shifted to 42°C or treated with 4 μg/ml of diazaborine at 30°C for 30 minutes. The labeling with radioactive orthophosphate, the extraction of phospholipids and the analysis by two-dimensional thin layer chromatography were described previously (Turnowsky *et al.* 1989).

Electron microscopy

Cells were fixed with glutaraldehyde and embedded according to Spurr (1969). Ultra-thin sections were double stained with uranyl acetate and lead citrate (Reynolds, 1963) following standard procedures.

4. RESULTS AND DISCUSSION

Inhibition of the *env*M gene product

E. coli JP1111 (Egan and Russell, 1973) with the ts mutation in the *env*M gene was cultivated both at 30°C and 42°C in low salt 2xTY medium. As shown in Fig. 1A the shift to the nonpermissive temperature slowed down the growth rate for about one hour, after which growth stopped completely without apparent lysis of the cells. At time 0 and 2, 4, and 6 hours after the temperature shift viable counts were determined. The growth inhibition observed by optical density was accompanied by a

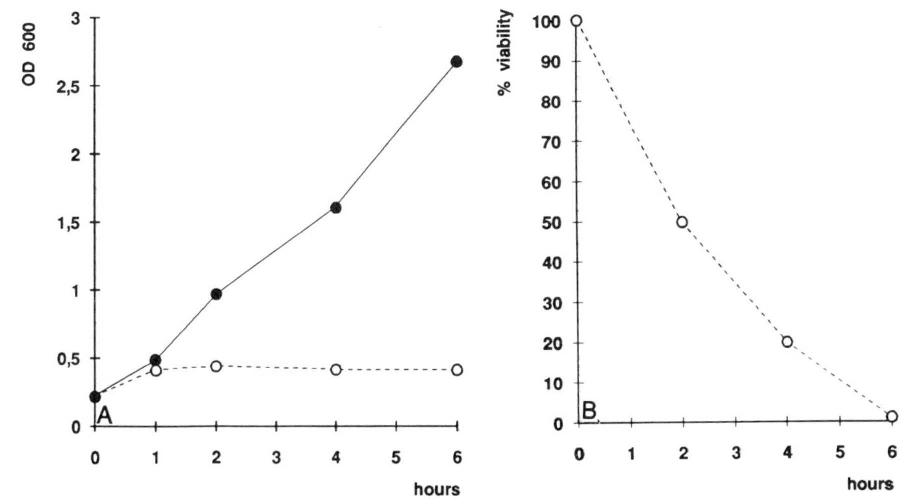

Figure 1. Growth of *E. coli* JP1111 (*env*Mts) at 30°C (•----•) and at 42°C (o- - -o) followed by optical density at 600 nm (A) and viable counts at 42°C (B).

gradual loss of viability (Fig. 1B); after 6 hours at 42°C the viable counts were 1% of the value before the temperature shift (time 0) and 4 orders of magnitude lower than in the control culture at 30°C.

We have previously shown (Turnowsky *et al.* 1989) that inhibition of the *env*M gene product either by temperature elevation in the conditional *env*M mutant or by diazaborine treatment caused changes in the envelope composition, like the strong induction of a 28 kDa protein in the cytoplasmic membrane or the inhibition of the lipopolysaccharide biosynthesis (Högenauer and Woisetschläger 1981). One of the earliest effects which we observed was that on phospholipid synthesis. The overall incorporation of radioactive phosphate into lipids was inhibited by 95% within 30 minutes; the qualitative composition of the phospholipids, however, was not altered as shown in Fig. 2. The effect of the temperature shift in the ts mutant was complemented by the *env*M gene on plasmid pKF402 (Fig. 2D). This recombinant plasmid contains the *env*M gene from a diazaborine resistant *S. typhimurium* strain (Turnowsky *et al.* 1989).

The alterations in the cell envelope composition are accompanied by characteristic morphological changes, which are shown in Fig. 3. The majority of the *E. coli* JP1111 cells after diazaborine treatment at 30°C (Fig. 3B) or after the

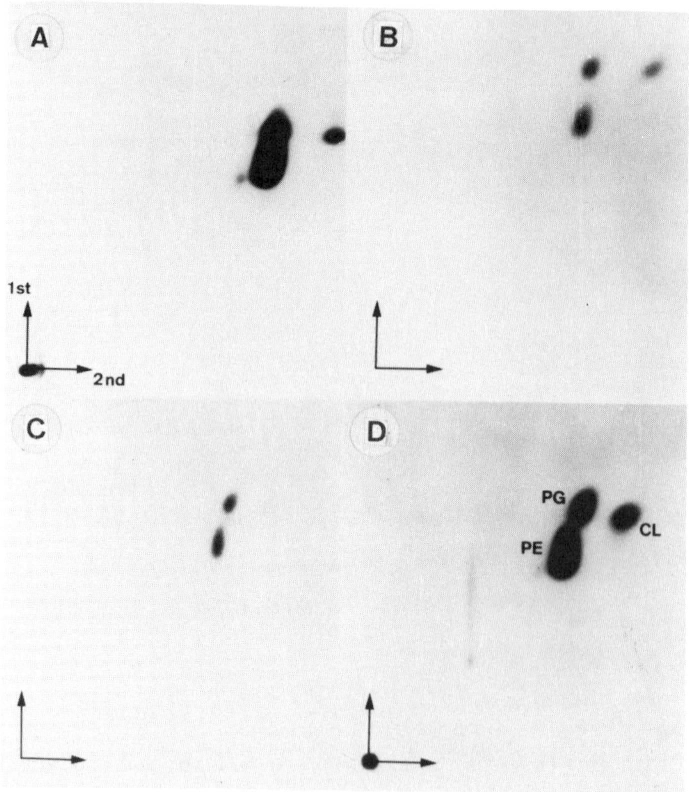

Figure 2. Effect of diazaborine or temperature in the conditional *env*M mutant *E. coli* BK 80 on phospholipid biosynthesis. Autoradiograms of two-dimensional thin layer chromatograms are shown. A: *E. coli* BK 80 at 30°C; B: *E. coli* BK 80 at 42°C; C: *E. coli* BK 80 in the presence of 4 μg/ml diazaborine at 30°C; D: *E. coli* BK 80 complemented with pKF402 at 42°C. PE: phosphatidylethanolamine; PG: phospatidylglycerol; CL: cardiolipin.

temperature shift to 42°C (Fig. 3C) showed light zones at the poles which seemed to have been formed by a retraction of the cytoplasmic membrane. Again, the effect caused by temperature treatment could be complemented by the recombinant plasmid pKF402 (Fig. 3D).

The retraction of the inner membrane could be caused by the greatly reduced level of phospholipids. Whether the 28 kDa protein, which is induced in the cytoplasmic membrane after inhibition of the EnvM protein, also contributes to the morphological alterations is presently under investigation.

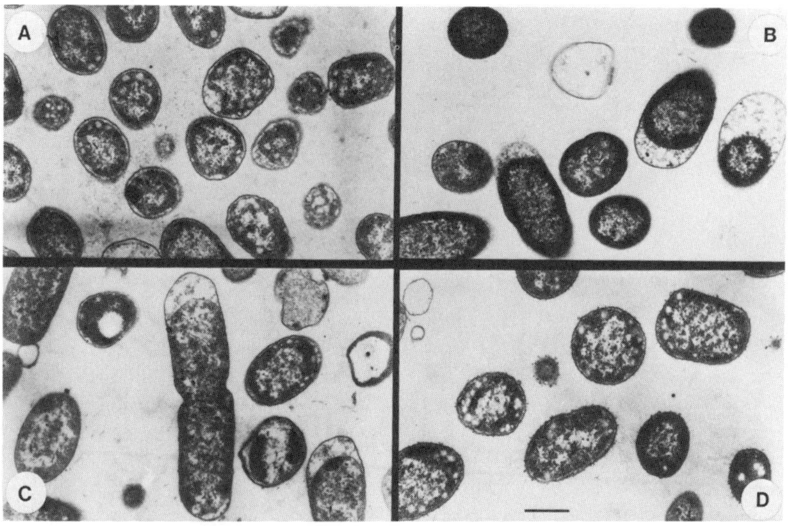

Figure 3. Effect of diazaborine or temperature on the cell morphology of the conditional *envM* mutant *E. coli* JP1111. Electron micrographs of thin sections are shown. Bar represents 500 nm.
A: *E. coli* JP1111 at 30°C; B: *E. coli* JP1111 at 30°C in the presence of 4 μg/ml of diazaborine; C: *E. coli* JP1111 at 42°C; D: *E. coli* JP1111 complemented with pKF402 at 42°C.

Overproduction of the EnvM protein

The *envM* gene of wt *E. coli* was cloned into the vectors pSP65 and pBR322; the recombinant plasmids were designated pFT632 and pFT635, respectively. *E. coli* JP1111 was transformed with both plasmids and cultivated in low salt medium at 30°C and 42°C. The plasmids differ in their copy numbers and therefore also in the amount of EnvM protein expressed (data not shown). The growth curves are shown in Fig. 4. At 30°C we did not observe any effect on either cell density or viability of *E. coli* JP1111. At 42°C, however, overproduction of wt EnvM does induce lysis of the cells after an initial period of complementation of the ts phenotype (Fig. 4A and B). The observed effect was gene dosage dependent and also dependent on the osmolarity of the medium. Previous experiments have shown that the temperature sensitive growth of *E. coli* JP1111 was complemented by the *envM* gene on a plasmid without obvious damage of the cells when they were grown in a medium of intermediate osmolarity (Turnowsky *et al.* 1989).

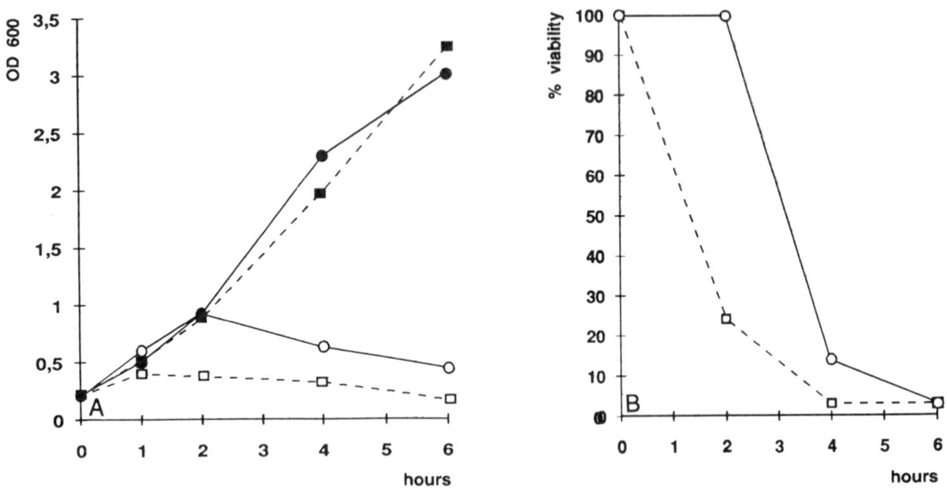

Figure 4. Effect of the overproduction of EnvM protein on growth of *E. coli* JP1111. A: optical density of cultures of *E. coli* JP1111 [pFT632] (■- -■) and *E. coli* JP1111 [pFT635] (●----●) at 30°C and 42°C (open symbols). B: viable counts of *E. coli* JP1111 [pFT632] (□ - -□) and *E. coli* JP1111 [pFT635] (o----o) at 42°C.

The electron micrographs of *E. coli* JP1111 carrying pFT632 are shown in Fig. 5. At 30°C no gross alterations on cell morphology were detected (Fig. 5A). At 42°C, however, the cells were greatly enlarged; many of them were lysed and empty (Fig. 5B).

The expression of the EnvM protein is regulated by growth rate. In late log phase cells the protein was not detected by Western blot analysis and no activity was detected in *lac*Z-fusion experiments (unpublished). The massive overproduction of EnvM from pFT632 could therefore interfere with the normal regulation of the amount of protein which is essential for growth and thus damage the cells.

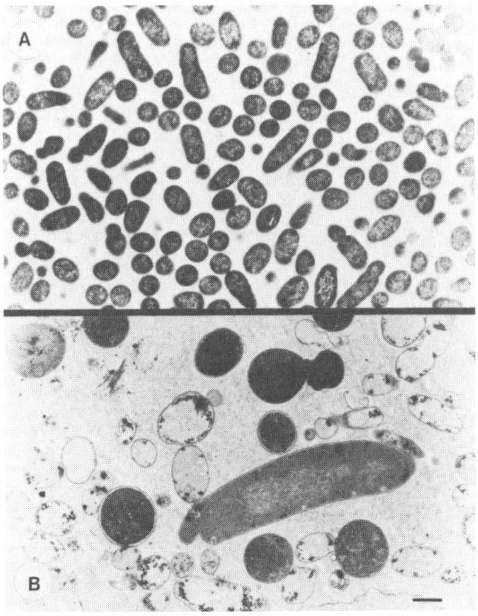

Figure 5. Effect of the overproduction of EnvM protein on the cell morphology of *E. coli* JP1111. Electron micrographs of thin sections of *E. coli* JP1111 [pFT632] at 30°C (A) and 42°C (B). Bar represents 1 μm.

ACKNOWLEDGEMENTS

We thank Petra Wallner and Karina Landl for expert help. This work was supported by the Fonds zur Förderung der wissenschaftlichen Forschung, project number P7959.

REFERENCES

Egan, A.F. and Russell, R.R.B. (1973) Conditional mutants affecting the cell envelope of *Escherichia coli* K12. Genet.Res.Camb. 21, 139-152.

Högenauer, G. and Woisetschläger, M. (1981) A diazaborine derivative inhibits lipopolysaccharide biosynthesis. Nature 293, 662-664.

Reynolds, E.S. (1963) The use of lead citrate at high pH as an electron opaque stain in electron microscopy. J. Cell Biol. 17, 208-212.

Sambrook, J., Fritsch, E.F., and Maniatis, T. (1989) "Molecular Cloning, a Laboratory Manual," 2nd edition, Cold Spring Harbor Laboratory Press, Cold Spring Harbor.

Silhavy, T.J., Berman, M.L., and Enquist, L.W. (1984) "Experiments with Gene Fusions," Cold Spring Harbor Laboratory Press, Cold Spring Harbor.

Spurr, A.R. (1969) A low-viscosity epoxy resin embedding medium for electron microscopy. J.Ultrastruct.Research 26, 31-43.

Turnowsky, F., Fuchs, K., Jeschek, C., and Högenauer, G.(1989) *envM* genes of *Salmonella typhimurium* and *Escherichia coli*, J.Bacteriol. 171, 6555-6565.

BIOSYNTHESIS OF PSEUDOMUREIN AND OTHER METHANOBACTERIAL CELL WALLS

Helmut König and Evamarie Hartmann

Angewandte Mikrobiologie
Universität Ulm
Albert Einstein-Allee 11
W-7900 Ulm

INTRODUCTION

The methanogens represent a major group of the archaea. A common feature of these otherwise diverse organisms is the production of methane. The cell wall profiles are also heterogeneous (Kandler and König, 1985). The common eubacterial cell wall polymer, the murein, is not found. Here we give a brief overview of the structure and biosynthetic pathways of three different methanobacterial cell wall polymers: a peptidoglycan (pseudomurein), a heteropolysaccharide (methanochondroitin) and a surface glycoprotein.

PSEUDOMUREIN

General Characteristics

The species of the order Methanobacteriales (*Methanobacterium, Methanobrevibacter, Methanosphaera*) and of the genus *Methanopyrus* possess an electron-dense pseudomurein layer (Kandler and König, 1985). The pseudomurein sacculi of *Methanothermus fervidus* exhibit often channels at the two cell poles, which are probably the passage holes of the flagella. (Nußer et al., 1988). *Methanothermus* and *Methanopyrus* are surrounded by an additional proteinaceous surface layer.

The isolated pseudomurein sacculi possess none of the typical murein constituents muramic acid, diaminopimelic acid, D-glutamic acid or D-alanine. They contain a set of three or four L-amino acids (Lys, Glu, Ala, Thr or Ser), the N-acetylated amino sugars glucosamine and/or galactosamine and N-acetyl-L-talosaminuronic acid (Fig. 1; Kandler and König, 1985).

The pseudomurein (Fig. 1) is made up of glycan strands consisting of alternating ß(1→3)-linked N-acetyl-D-glucosamine/N-acetyl-D-galactosamine and ß(1→3)-linked N-acetyl-L-talosaminuronic acid residues, which are crosslinked via short peptides attached to the carboxylic group of N-acetyl-L-talosaminuronic acid.

The pseudomurein represents a new, but basically different type of peptidoglycan. The main differences between pseudomurein and murein are the replacement of muramic acid by talosaminuronic acid, the partial or total substitution of glucosamine by galactosamine, the ß(1→3) linkage of the glycan components and the exclusive occurence of L-amino acids.

Bacterial Growth and Lysis, Edited by M.A. de Pedro
et al., Plenum Press, New York, 1993

As a consequence of the different composition pseudomurein is resistant to cell wall antibiotics, which interact with the cytoplasmic step of biosynthesis such as penicillin, D-cycloserine or vancomycin (Hammes et al., 1979; Hilpert et al., 1981). It is also insensitive against lysozymes and common proteases, which may be due to the different glycan composition and the occurence of ϵ- and γ-bonds in the peptide subunit.

Modifications are found in the glycan as well in the peptide moiety (Fig. 1; Kandler and König, 1985). Artificial modifications could also be induced. The amino acids glycine, threonine, ornithine, or aspartic acid are incorporated into the pseudomurein, when they occur in elevated concentrations in the culture medium (König, 1985).

Varying amounts of saccharides are present, but teichoic acid or teichuronic acid-like polymers are absent.

An autolytic enzyme, which hydrolyzes the pseudomurein sacculi, could be induced in *Methanobacterium wolfei* (König et al., 1985). The isolated lytic enzyme is a peptidase hydrolyzing the ϵ-Ala-Lys bound of the peptide subunits (Kiener et al., 1987).

Only limited information on the biological activity of pseudomurein is available. The pseudomurein is antigenic in animals (Conway de Macario et al., 1983). The immunochemical characterization with monoclonal antibodies revealed four distinctive determinants in the pseudomurein. Three determinants are the components of the glycan strand (N-acetylglucosamine, N-acetylgalactosamine, N-acetyltalosaminuronic acid) and the other includes the C-terminal sequence γ-glutamyl-alanine of the peptide moiety. Intravenous injections of rabbits with suspensions of pseudomurein from *Methanobacterium thermoautotrophicum* alter sleep and brain temperature (Johannsen et al., 1991) and intraarticular injections of high doses of pseudomurein-polysaccharide fragments caused an acute inflammation in rats (Stimpson et al., 1986).

Biosynthesis of Pseudomurein

The biosynthesis of the pseudomurein has been studied by characterizing the corresponding precursors (Fig. 1; Hartmann and König, 1990). Like the murein the pseudomurein should be synthesized in three stages occuring in the cytoplasm and the inner and outer face of the cytoplasmic membrane. As a main precursor of the cytoplasmic step a UDP-activated disaccharide composed of N-acetylglucosamine and N-acetyltalosaminuronic acid was isolated with N-acetylglucosamine at the reducing end (König et al., 1989). A monomeric talosaminuronic acid residue was missing in the cell extracts. It may therefore be assumed that N-acetyltalosaminuronic acid is formed during the synthesis of the disaccharide, probably by epimerization and oxidation of N-acetylgalactosamine.

Parallel to the synthesis of the disaccharide a UDP-activated pentapeptide should be built. The synthesis of the pentapeptide starts with a UDP-activated glutamic acid derivative followed by the formation of a UDP-activated di-, tri-, and pentapeptide. The peptide precursors lack the glycan components. UDP was directly linked to the N^{α}-amino group of a glutamic acid residue. In a final cytoplasmic step the UDP-activated pentapeptide should be linked to the disaccharide to give a UDP-activated disaccharide pentapeptide.

In the second stage of pseudomurein biosynthesis the nucleotide activated disaccharide pentapeptide is transferred to the carrier lipid and an undecaprenyl pyrophosphate activated disaccharide pentapeptide is formed. In the last stage the transpeptidation is most probably achieved by splitting off one alanine residue and connecting a glutamic acid with a lysine residue. The excretion of alanine into the medium has been observed with *Methanobacterium thermoautotrophicum* (Schönheit and Thauer, 1980).

METHANOCHONDROITIN

General Characteristics

Cells of *Methanosarcina* species form large aggregates and are surrounded by a compact or sometimes loose cell wall matrix, which is composed of methanochondroitin (Kreisl and Kandler, 1986).

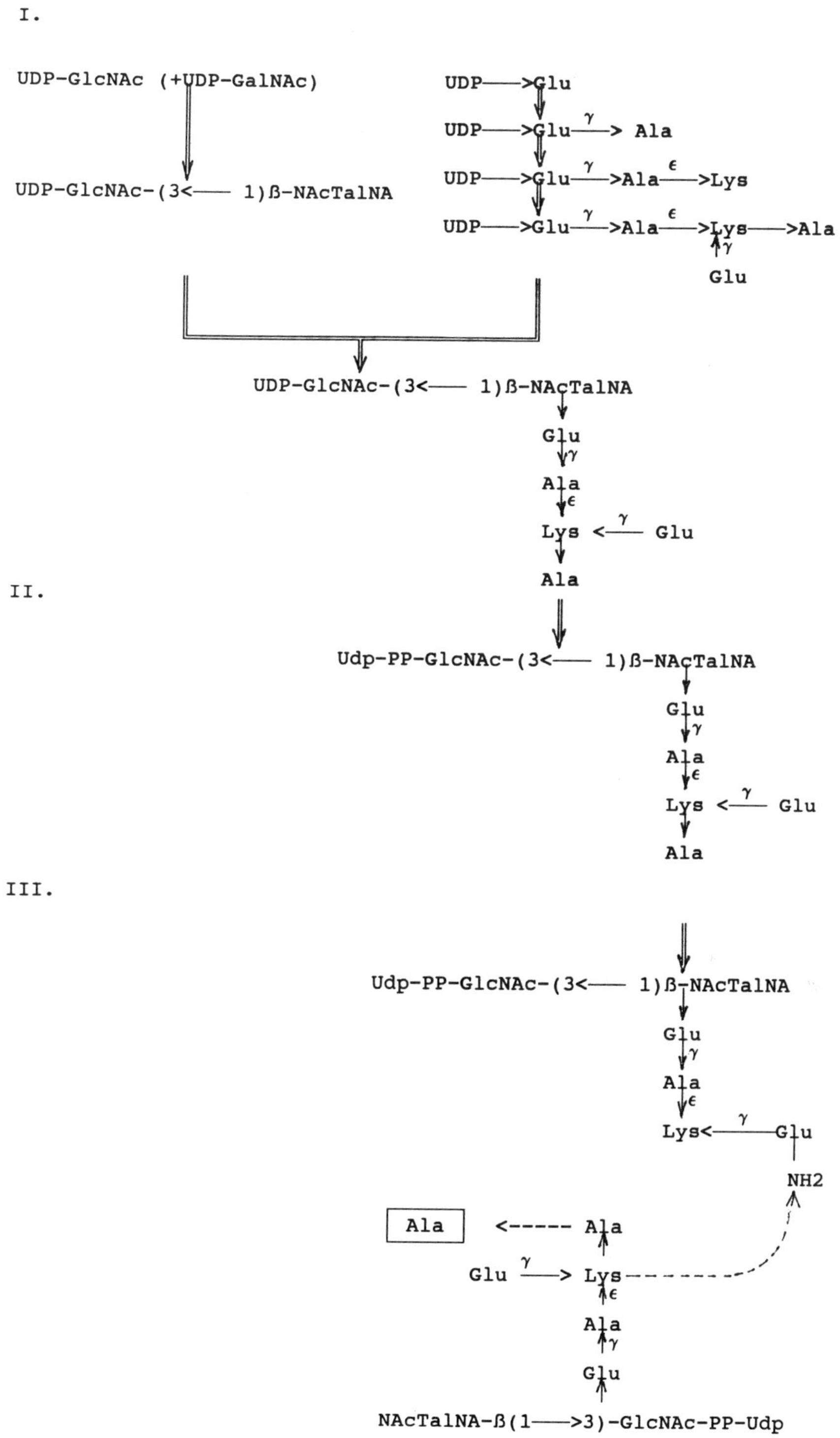
I.
UDP-GlcNAc (+UDP-GalNAc)
UDP-GlcNAc-(3<—— 1)ß-NAcTalNA
UDP——>Glu
UDP——>Glu——> Ala
UDP——>Glu——>Ala——>Lys
UDP——>Glu——>Ala——>Lys——>Ala
Glu
UDP-GlcNAc-(3<—— 1)ß-NAcTalNA
Glu
Ala
Lys <—— Glu
Ala
II.
Udp-PP-GlcNAc-(3<—— 1)ß-NAcTalNA
Glu
Ala
Lys <—— Glu
Ala
III.
Udp-PP-GlcNAc-(3<—— 1)ß-NAcTalNA
Glu
Ala
Lys<——Glu
NH2
Ala
<----- Ala
Glu ——> Lys
Ala
Glu
NAcTalNA-ß(1——>3)-GlcNAc-PP-Udp

Figure 1.

IV.

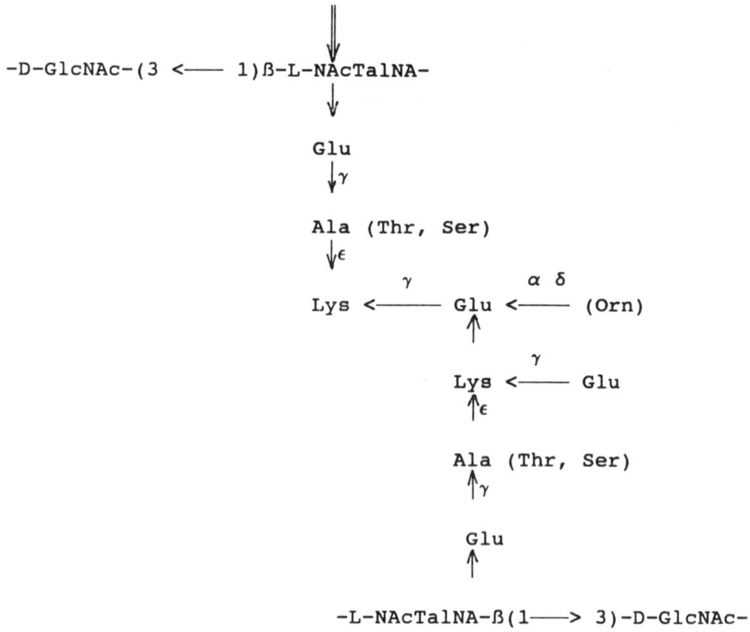

Figure 1. Proposed scheme of the three stages (I,II,III) of the biosynthesis of the pseudomurein and primary structure of the pseudomurein (IV). Modifications are also shown. Udp = undecaprenyl.

Methanochondroitin preparations contained D-galactosamine, D-glucuronic acid or D-galacturonic acid, D-glucose and acetic acid. Mannose was only present in trace amounts. The molar ratio of N-acetyl-D-galactosamine to uronic acid is generally about 2 : 1, but it is 3 : 1 in *Methanosarcina mazei*. In two strains galacturonic acid is replaced by glucuronic acid. A trimer was identified to be the building block of the glycan (Fig. 2; Kreisl and Kandler, 1986).

Thus the cell wall polymer of *Methanosarcina barkeri* is reminiscent of chondroitin sulfate, a common component of the connective tissue of human and animals. Methanochondroitin is not sulfated and contains two N-acetylgalactosamine residues in the repeating unit. In some species showing a cell cycle the methanochondroitin layer could be lost in a certain stage.

Biosynthesis of Methanochondroitin

From cell extracts of *Methanosarcina barkeri* four uridine diphosphate and one undecaprenyl pyrophosphate activated intermediates of methanochondroitin were isolated. Based on the isolated precursors a rough biosynthetic pathway of the methanochondroitin could be proposed (Fig. 2; Hartmann and König, 1991). The biosynthesis should start with the formation of UDP-N-acetylgalactosamine and UDP-glucuronic acid, which are then linked together to form UDP-N-acetylchondrosine. From UDP-N-acetylchondrosine N-acetylchondrosine should be transferred to UDP-N-acetylgalactosamine to build a UDP-activated trisaccharide. The trisaccharide intermediate (Hartmann and König, 1991) possesses the same structure as the proposed repeating unit of the intact glycan (Kreisl and Kandler, 1986). In the next step the UDP residue of the trisaccharide should be replaced by undecaprenyl pyrophosphate. The trisaccharide is then most probably polymerized on the outer face of the cytoplasmic membrane.

SURFACE LAYER GLYCOPROTEIN

General characteristics

Members of the genera *Methanothermus* and *Methanopyrus* have a double-layered cell envelope. In the case of *Methanothermus* the inner pseudomurein sacculus is surrounded by an outer layer (S-layer) of glycoprotein subunits, which are hexagonally

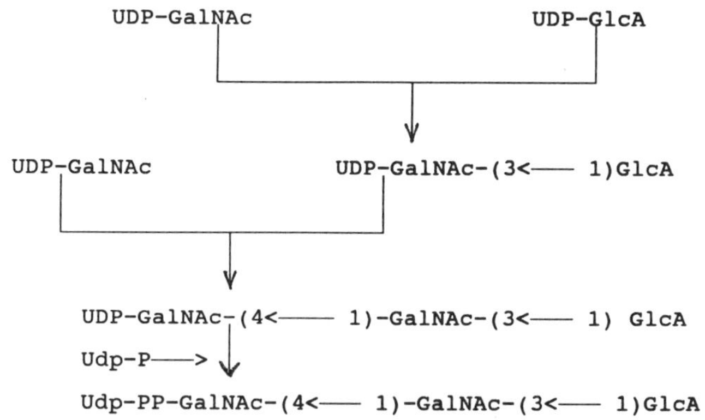

Figure 2. Proposed scheme of the biosynthesis of methanochondroitin. The trimer represents the repeating unit.

arranged. The primary sequence of the peptide (Bröckl et al., 1991) and of the a heterosaccharide chain (Kärcher and König, unpublished results) have been determined. The mature glycoprotein consists of 593 amino acids. Compared to mesophilic S-layer glycoproteins this hyperthermophilic glycoprotein possesses significant higher amounts of e. g. isoleucine, asparagine and cysteine and it has 14% more ß-sheet structures. A typical leader peptide and 20 sequon structures are present. The isolated heterosaccharide consists of mannose, 3-O-methylglucose and N-acetylgalactosamine, while in the glycoprotein also galactose and N-acetylglucosamine were found. The glycoprotein could be isolated by extracting whole cells with trichloroacetic acid (5%-10%) (Nußer et al., 1988).

Biosynthesis of the Glycan Strands

Based on isolated precursors (Fig. 3; Hartmann and König, 1989) the biosynthesis of the glycan chains of the S-layer glycoprotein of *Methanothermus fervidus* is suggested to start with the formation of the C-1-phosphate derivatives of Man, Gal, GlcNAc and GalNAc. They are then converted to the corresponding nucleotide activated derivatives, which form two types of UDP activated oligosaccharides.

One oligosaccharide contains only neutral sugars, while the other is a heterosaccharide. Glc and 3-O-methylglucose are not found as monomers, but they are constituents of the oligosaccharides. The oligosaccharides are most probably transferred to C_{55}-dolichylphosphate forming dolichyl pyrophosphate activated oligosaccharides. The lipid activated oligosaccharides contain additional Glc residues, which are removed during the further processing.

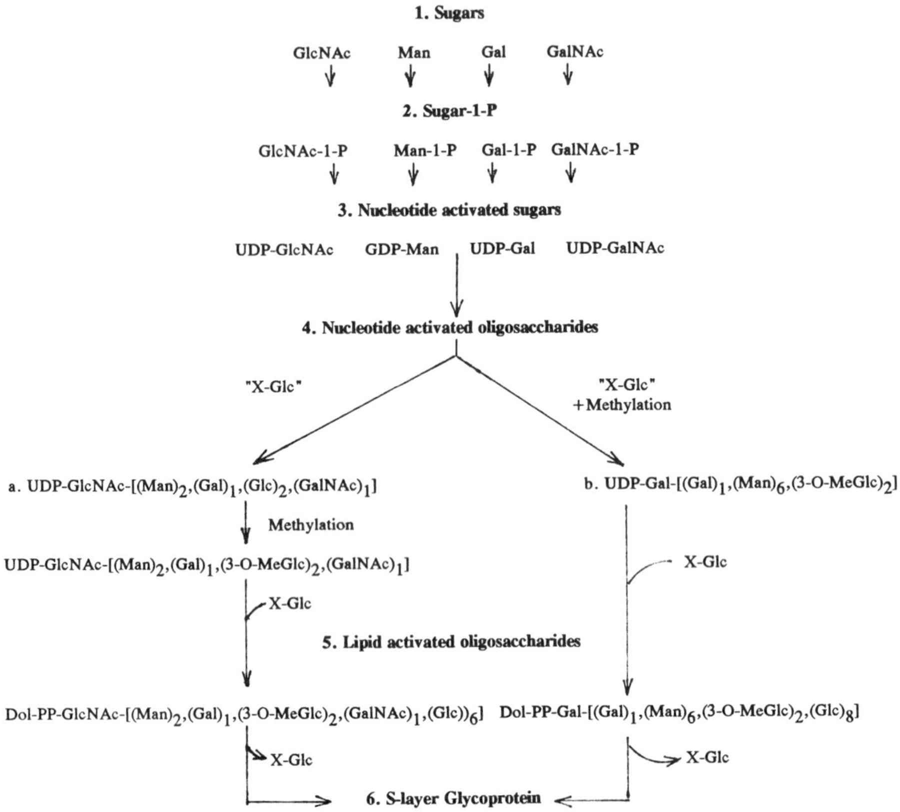

Figure 3. Proposal for the biosynthesis of the oligosaccharide chains of the S-layer glycoprotein of *Methanothermus fervidus.*

CONCLUSIONS

1. Pseudomurein: The participation of a nucleotide activated disaccharide and nucleotide activated peptide intermediates indicates that the route of biosynthesis for murein and pseudomurein follows different pathways (Hartmann and König, 1990). Usually, oligosaccharides occuring in bacterial cell wall polymers are formed at the lipid stage (Tipper and Wright, 1979; Wright and Tipper, 1979) and amino acids carrying a nucleotide residue at the N^{α}-amino group have not been found in nature so far. The different biosynthetic routes explain the insensitivity of the pseudomurein to ß-lactam antibiotics (Hammes et al., 1979; Hilpert et al., 1981) and support also the hypothesis (Kandler, 1982) that the murein of bacteria and pseudomurein of methanogenic archaea do not possess a common origin. These two peptidoglycan polymers must have been independent inventions after the formation of the two primary prokaryotic domains (Kandler 1982; Hartmann and König, 1990).

2. Methanochondroitin: In contrast to the known biosynthetic pathways of polysaccharides from prokaryotes (Tipper and Wright, 1979; Wright and Tipper, 1979) and eukaryotes, nukleotide activated oligosaccharides are involved in the biosynthesis of methanochondroitin (Hartmann and König, 1991). Also the biosynthesis of the structurally very similar eukaryotic chondroitin sulfate follows a quite different mode. The

oligosaccharides are directly formed at distinct serine or threonine residues of the growing peptide chain from their nucleotide activated monomers (Róden, 1970). The teichuronic acid of *Bacillus licheniformis* has also a very similar structure (Rogers et al., 1980), but this teichuronic acid is built like other cell wall polymers of eubacteria at the lipid stage with undecaprenyl pyrophosphate as lipid carrier.

3. S-layer glycoprotein: The occurence of UDP activated oligosaccharides (Hartmann and König, 1989) is also a difference between the biosynthesis of methanobacterial and eukaryotic glycoproteins. In eukaryotic glycoproteins the oligosaccharide precursors are synthesized at a dolichol residue (Berger et al. 1982).

4. Since nucleotide activated oligosaccharides are involved in the biosynthesis of three chemically different cell wall polymers, namely a peptidoglycan (pseudomurein; Hartmann and König, 1990) , a glycoprotein (Hartmann and König, 1989) and a polysaccharide (methanochondroitin; Hartmann and König, 1991) they seem to be a typical and common feature of the biosynthesis of methanobacterial cell envelope polymers.

5. Both, undecaprenol and dolichol are found as lipid carriers in methanogens (Hartmann and König, 1989, 1990, 1991). It seems most likely that dolichol is the common lipid carrier involved in the glycoprotein synthesis in eukaryotes and prokaryotes, while undecaprenol plays a role in the biosynthesis of prokaryotic peptidoglycans and other cell wall polymers (Ingraham et al. 1983).

REFERENCES

Berger, E.G., Buddecke, E., Kamerling, J.P., Lobata, A., Paulson, I.C. and Vliegenhart, F.G. (1982) Structure, biosynthesis and function of glycoprotein glycans. Experientia 38, 1129-1163.

Bröckl, G., Behr, M., Fabry, S., Hensel, R., Kaudewitz, H., Biendl, E. and König, H. (1991) Analysis and nucleotide sequence of the genes encoding the surface-layer glycoproteins of the hyperthermophilic methanogens *Methanothermus fervidus* and *Methanothermus sociabilis*. Eur. J. Biochem. 199, 147-152.

Conway de Macario, E., Macario, A.J.L, Magarinos, M.C., König, H. and Kandler, O. (1983) Dissecting the antigenic mosaic of the archaebacterium *Methanobacterium thermoautotrophicum* by monoclonal antibodies of defined molecular specificity. Proc. Natl. Acad. Sci. (USA) 80, 6346-6350.

Hammes, W.P., Winter, J. and Kandler, O. (1979) The sensitivity of the pseudomurein-containing genus *Methanobacterium* to inhibitors of murein synthesis. Arch. Microbiol. 123, 275-279.

Hartmann, E. and König, H. (1989) Uridine and dolichyl diphosphate activated oligosaccharide intermediates are involved in the biosynthesis of the surface layer glycoprotein of *Methanothermus fervidus*. Arch Microbiol 151, 274-281.

Hartmann, E. and König, H. (1990) Comparison of the biosynthesis of the methanobacterial pseudomurein and the eubacterial murein. Naturwissenschaften 77, 472-475.

Hartmann, E. and König, H. (1991) Nucleotide activated oligosaccharides are intermediates of the cell wall polysaccharide of *Methanosarcina barkeri*. Biol. Chem. Hoppe-Seyler 372, 971-974.

Hilpert, R., Winter, J., Hammes, W. and Kandler, O. (1981) The sensitivity of archaebacteria to antibiotics. Zentralbl. Bakteriol., Mikrobiol., Hyg., Abt. 1, Orig. C 2, 11-20.

Ingraham, J.L., Maaloe, O. and Neidhardt, F.C. (1983) Growth of the bacterial cell. Sinauer Associates, Inc. Sunderland, MA.

Johannsen, L., Labischinski, H. and Krueger, J.M. (1991) Somnogenic activity of pseudomurein in rabbits. Infect. Immun. 59, 2502-2504.

Kandler, O. (1982) Cell wall structures and their phylogenetic implications. Zentralbl. Bakteriol., Mikrobiol., Hyg., Abt. 1, Orig. C 3, 149-160.

Kandler, O. and König, H. (1985) Cell envelopes of archaebacteria, in "The Bacteria. A Treatise on Structure and Function" (Woese, C.R. and Wolfe, R.S., Eds.), Vol. VIII, Archaebacteria, pp 413-457. Academic Press, New York.

Kiener, A., König, H., Winter, J. and Leisinger, Th. (1987) Purification and use of *Methanobacterium wolfei* pseudomurein endopeptidase for lysis of *Methanobacterium thermoautotrophicum.* J. Bacteriol. 169, 1010-1016.
König, H. (1985) Influence of amino acids on growth and cell wall composition of Methanobacteriales. J. Gen. Microbiol. 131, 3271-3275.
König, H., Semmler, R., Lerp, C. and Winter, J. (1985) Evidence for the occurence of autolytic enzymes in *Methanobacterium wolfei.* Arch. Microbiol. 141, 177-180.
König, H., Kandler, O. and Hammes, W. (1989) Biosynthesis of pseudomurein: Isolation of putative precursors from *Methanobacterium thermoautotrophicum.* Can. J. Microbiol. 35, 176-181.
Kreisl, P. and Kandler, O. (1986) Chemical structure of the cell wall polymer of *Methanosarcina.* Syst. Appl. Microbiol. 7, 293-299.
Nußer, E., Hartmann, E., Allmeier, H., König, H., Paul, G. and Stetter, K.O. (1988) A glycoprotein surface layer covers the pseudomurein sacculus of the extreme thermohile *Methanothermus fervidus*, in "Crystalline bacterial cell surface layers" (Sleytr, U.B., Messner, P., Pum, D. and Sàra, M., Eds.) pp. 21-25. Springer Verlag, Berlin.
Róden. L. (1970) Biosynthesis of acidic glycosaminoglycans (mucopolysaccharides), in "Metabolic Conjugation and Metabolic Hydrolysis" (Fishman, W.H., Ed.), Vol. 2, pp. 345-442, Academic Press, New York.
Rogers, H.J., Perkins, H.R. and Ward, J.B. (1980) "Microbial Cell Walls and Membranes". Chapman & Hall, New York.
Schönheit, P. and Thauer, R.K. (1980) L-Alanine, a product of cell wall synthesis in *Methanobacterium thermoautotrophicum.* FEMS Microbiol. Lett. 9, 77-80.
Stimpson, S.A., Brown, R.R., Anderle, S.K., Klapper, D.G., Clark, R.L., Cromartie, W.J. and Schwab, J.H. (1986) Arthropathic properties of cell wall polymers from normal flora bacteria. Infect. Immun. 51, 240-249.
Tipper, K.J. and Wright, A. (1979) Structure and biosynthesis of bacterial cell walls, in "The Bacteria: A Treatise on Structure and Fuction" (Gunsalus, I.C., Sokatch, J.R. and Ornston, L.N., Eds.), Vol. 7, pp. 291-426. Academic Press, New York.
Wright, A. and Tipper, K.J.(1979) The outer membrane of gram-negative bacteria, in "The Bacteria. A Treatise on Structure and Function" (Gunsalus, J.C., Sokatch, J.R., Ornston, L.N., eds.) Vol. 7, pp. 427-485, Academic Press, New York.

PEPTIDOGLYCAN (MUREIN) HYDROLASES:

UNUSUAL ENZYMES FOR UNUSUAL SUBSTRATES

Gerald D. Shockman, Chien-Peng Chu, Reiko Kariyama, Lori K. Tepper, and Lolita Daneo-Moore

Department of Microbiology and Immunology
Temple University School of Medicine
Philadelphia, PA 19140

A broad variety of both Gram-positive and Gram-negative bacteria have been shown to have the ability to dissolve themselves (autolyze), particularly under "adverse" conditions (see Ghuysen and Shockman, 1973; Daneo-Moore and Shockman, 1977; Rogers *et al.*, 1980; Shockman and Barrett, 1983; Doyle and Koch, 1987; Höltje and Tuomanen, 1991, for reviews). This process is now known to be initiated by the action of endogenous enzymes that hydrolyze specific bonds in the insoluble, osmotically protective, shape-maintaining and essential, peptidoglycan (murein) polymer of the bacterial cell wall. Hydrolysis of a sufficient number of bonds in a restricted area of the two- or three-dimensional peptidoglycan network, or a larger number of bonds in a broader area of the wall, creates a weak spot (or a generalized weakness in the structure), so that the wall can no longer protect the protoplast from its own internal osmotic pressure, so that the protoplast then explodes out through the weakened structure. Endogenous enzymes are now known that can hydrolyze virtually every bond in the peptidoglycan including: (i) glycosidases and transglycosidases, such as N-acetylmuramoylhydrolases (muramidases) and N-acetylglucosaminidases, (ii) N-acetylmuramoyl-L-alanine amidases, and (iii) various peptidases and transpeptidases, including DD- and LD-carboxypeptidases (Rogers *et al.*, 1980).

The power of such enzymic activities to destroy the integrity of bacterial cells cannot be underestimated and is known to be the cause of death by lysis of at least some bacteria, consequent to inhibition of further cell wall peptidoglycan assembly, either by antibiotic treatment or by deprivation of an essential precursor of the cell wall peptidoglycan (see Shockman, 1965; Shockman *et al.*, 1974, 1979, 1982; Tomasz, 1979*a,b*, for reviews). The puzzling, somewhat teleological question that continues to remain incompletely answered is, why should bacteria carry around with themselves such potentially suicidal activities? Along with this question are some at least partially related questions concerning the potential roles of peptidoglycan hydrolase activities

Bacterial Growth and Lysis, Edited by M.A. de Pedro *et al.*, Plenum Press, New York, 1993

in the synthesis, assembly, and postassembly modifications of the protective wall peptidoglycan.

A number of functions, some of which are essential, have been proposed for peptidoglycan hydrolases (see Ghuysen and Shockman, 1973; Daneo-Moore and Shockman, 1977, Rogers *et al.*, 1980, for reviews). These include a role in permitting the state called competence and/or the penetration of DNA through the cell wall in genetic transformation, and various roles in modifying peptidoglycan and or dissolving the mother cell in the processes of sporulation and in spore germination. In addition, a number of potential roles for peptidoglycan hydrolases in bacterial cell surface growth and cell division have been proposed, with some indirect experimental evidence in support of many of them. These roles include (i) establishment of new wall expansion sites in the "bag-shaped" macromolecule surrounding the bacterium; (ii) hydrolysis of bonds to create suitable acceptor sites for the addition of new structural units; (iii) cell division; (iv) the final stages of cell separation; (v) morphogenesis, i.e., modifications of the shape of previously assembled wall (e.g., from linear to more rounded); (vi) biosynthesis, e.g., the measurement of the reverse of a reaction that might in fact be a biosynthetic process; and (vii) the process of loss of the outer layers of wall by the process usually referred to as "wall turnover."

For some of these potential functions, peptidoglycan hydrolases of only certain of the specificities mentioned above would be suitable. For example, for functions (i) and (ii), only muramidase, certain transglycosylases, and certain transpeptidase activities would be consistent with current information concerning wall biosynthetic reactions, whereas functions (iv) final cell separation, (v) postassembly changes in wall shape, and (vi) loss of wall via turnover could involve the action of peptidoglycan hydrolases of virtually any specificity.

For this discussion, several points seem to be of some importance. (i) Because of the nature of the peptidoglycan structure itself, i.e., its two- to three-dimensional structure resulting from the peptide cross-linking of glycosidically linked disaccharide peptide structural units to each other, the release of monomeric or oligomeric peptidoglycan fragments from the wall itself frequently involves the hydrolysis of more than one, of the same, or of different, bonds. Thus, the results of peptidoglycan hydrolase cleavage of bonds that are integral to the two- or three-dimensional peptidoglycan network might not be detected by measurements of release of peptidoglycan fragments from walls or cells or by measurements of wall dissolution. (ii) The hydrolysis of certain bonds such as the C-terminal D-alanyl-D-alanine bond, hydrolyzed by the action of a DD-carboxypeptidase, does not result in damage to the integrity of the peptidoglycan meshwork. Thus, not all peptidoglycan hydrolases are autolysins. In addition, hydrolysis of *some* bonds integrally associated with, or part of, the structural network, particularly at widely dispersed sites in the wall surface, may not result in a significant change in wall integrity, strength, or shape (nor to loss from the cell of peptidoglycan components). (iii) To our knowledge, truly autolytically *negative,* otherwise isogenic, mutants of any bacterial species or strain have not yet been obtained. However, with several bacterial species, mutants with greatly reduced abilities to autolyze have been obtained, and bacterial species and strains are known that appear to be highly resistant to cellular autolysis under conditions that result in rapid autolysis of other, sometimes closely related, strains. Thus it seems clear that the bulk of autolysin activity, as frequently measured under unbalanced conditions, is not necessary for wall surface growth and wall modification. There are a number of potential reasons for this state, including the presence of more than one enzyme of the same enzymatic specificity in some bacteria (see below). With the recent cloning, sequencing and genetic characterization of peptidoglycan hydrolases of several bacterial species (García *et al.*, 1985; Biavasco *et al.*, 1988; Potvin *et al.*, 1988; Sekiguchi *et al.*,

1988; Jayaswal *et al.*, 1990; Kuroda and Sekiguchi, 1990, 1991; Beliveau *et al.*, 1991; Croux and García, 1991; Foster, 1991; Wang *et al.*, 1991; Chu *et al.*, 1992), one can expect that expansion of these genetic approaches will clarify these anomalies in the near future.

As expected, for such potentially suicidal activities, peptidoglycan hydrolase activities must be exquisitely regulated and integrated with other biosynthetic and perhaps other degradative processes.

A variety of data suggest that peptidoglycan hydrolase activities are coupled with wall biosynthesis and with growth- and cell division-related processes. Examples of autolytic-defective (in at least some instances regulatory) mutants of *Bacillus subtilis* (Fan and Beckman, 1973; Fein and Rogers, 1976), pneumococci (García *et al.*, 1985, 1986), staphylococci (Chatterjee *et al.*, 1976; Koyama *et al.*, 1977), streptococci (Pooley *et al.*, 1972; Cornett *et al.*, 1978; Shungu *et al.*, 1979), and *Escherichia coli* that either fail to divide completely or fail to separate after division are known. Furthermore, correlations have been made between the capacity of bacterial cells to autolyze and growth rate (Hinks *et al.*, 1978), the cell division cycle in synchronized cultures (Hakenbeck and Messer, 1974; Beck and Park, 1976; Hinks *et al.*, 1978), and phase of growth (Pooley and Shockman, 1970). For example, when cultures of *Enterococcus hirae* ATCC 9790 enter the stationary phase due to exhaustion of the supply of valine or threonine (amino acids that are components of protein but not of peptidoglycan), the capacity of the bacteria to autolyze decreases rapidly, although their content of autolytic enzyme activity (probably muramidase-1, see below) remains virtually unchanged (Pooley and Shockman, 1970).

Regulation of peptidoglycan hydrolase activities may occur at several levels. First of all, such enzymes, that are, like all proteins, synthesized on cytoplasmic ribosomes, must be transported through the cytoplasmic membrane to the exterior wall of Gram positives or to the periplasm of Gram negatives to have access to the substrate. Thus, control of general access to the insoluble substrate is an important factor. For example, Hartmann *et al.* (1974) proposed a "barrier" that prevented access of peptidoglycan hydrolases to the peptidoglycan of *E. coli*. Such access could be limited to topologically located areas on the surface, such as nascent cross walls in the case of *E. hirae* (Higgins *et al.*, 1970; Joseph and Shockman, 1976). Similarly, intact, growing cells of *E. hirae* are resistant to the action of its muramidases (Joseph and Shockman, 1976; Kariyama and Shockman, 1992) even though muramidase-2 is known to be present in an active form in the culture medium (Kariyama and Shockman, 1992). In this instance an exterior molecule (e.g., a protein?) could protect the peptidoglycan from hydrolysis. Substrate specificity is also important. The presence, absence, or exact chemical structure of nonpeptidoglycan wall polymers such as teichoic acids, perhaps at specific topological sites, and O- or N-acylation of amino sugars of the peptidoglycan itself, are two examples. The presence, absence, access or specific chemical structure of ligands that interact with either enzyme or substrate, and that modify (Herbold and Glaser, 1975*a,b*) or inhibit, hydrolase activities, such as cardiolipin (Cleveland *et al.*, 1976), lipoteichoic acids (Cleveland *et al.*, 1975), and the Forssman antigen of the pneumococcus (Höltje and Tomasz, 1975), are known. Muramidase-1 of *E. hirae* is present in a latent, zymogen form, that requires proteolytic activation (Conover *et al.*, 1966; Shockman *et al.*, 1967; Pooley and Shockman, 1969). The energized state of the membrane of *B. subtilis* appears to be important in regulating its amidase activity (Jolliffe *et al.*, 1981). The ability of such enzymes to bind to and maintain contact with the insoluble substrate, particularly outside of the cellular permeability barrier, is also important. Several peptidoglycan hydrolases have been shown to bind to their substrates, with very high affinity (Shockman and Cheney, 1969; Joseph and Shockman, 1976; Walderich and Höltje, 1991; Kariyama and

Shockman, 1992). Muramidase-2 of *E. hirae* (Chu *et al.*, 1992), the autolysin of *Streptococcus faecalis* (Beliveau *et al.*, 1991), as well as the amidase and bacteriophage-coded amidases and glycosidase of pneumococci (García *et al.*, 1990), have been shown to possess long amino acid repeat sequences that could be involved with substrate binding. For example, the choline containing wall teichoic acid of the pneumococcus appears to be important in the binding of the pneumococcal amidase to the wall (García *et al.*, 1988). Finally, recently obtained data suggest that the state of protein folding (or refolding) could be an important factor in governing the activity of muramidase-2 of *E. hirae* (Kariyama and Shockman, 1992).

The Peptidoglycan Hydrolases of *E. hirae* ATCC 9790 (*Streptococcus faecium*)

A large assortment of data is completely consistent with the enzymatic hydrolysis of only one bond in the peptidoglycan of *E. hirae*. That is the bond between N-acetylmuramic acid and N-acetylglucosamine due to the action of N-acetylmuramoyl-hydrolase (muramidase) activity. (A β-lactam-sensitive DD-carboxypeptidase is also present [Coyette *et al.*, 1974], but its removal of C-terminal D-alanine residues does not affect peptidoglycan integrity.) An interesting point is that, despite the presence of only *one* specificity of peptidoglycan hydrolase activity, we have found that *two* separate and distinct enzymes are present that are very likely the products of separate genes (see below).

Muramidase-1. Muramidase-1 is synthesized as a large-molecular-weight (130-kDa) glycoprotein that appears to be first translocated to the wall (Pooley and Shockman, 1969) where it is tightly bound and then proteolytically cleaved to an 87-kDa active form (Kawamura and Shockman, 1983*b*). Because glycoproteins and especially glycoenzymes are extremely rare in Eubacteria, and the chances of contamination of purified enzyme preparations with wall polysaccharides is high, we expended considerable effort to demonstrate that polysaccharide chains were indeed covalently linked to the polypeptide (Kawamura and Shockman, 1983*b*) even though the ability of the glycoprotein to retain the ability to bind to concanavalin A (ConA)-Sepharose and to react with the periodate-Schiff (PAS) reagent survived extreme treatments such as high pH, high salt concentrations, precipitation with trichloroacetic acid, and exposure to strong detergents including SDS-gel electrophoresis. We were able to demonstrate that proteolytic hydrolysis and cyanogen bromide cleavage of the enzyme produced multiple PAS-positive glycopeptides visible after SDS-gel electrophoresis. In addition, we were able to obtain ^{3}H-labeled oligosaccharides after β-elimination and reduction in the presence of [^{3}H]$NaBH_4$. Gas-liquid chromatography of sugars released upon hydrolysis of trichloroacetic acid-precipitated enzyme showed the presence of only glucose. Thus, muramidase-1 appears to be a *gluco*enzyme.

Muramidase-1 was shown to be a muramidase (Kawamura and Shockman, 1983*b*) that processively hydrolyzes β-1,4 bonds between N-acetylmuramic acid and N-acetylglucosamine in soluble peptidoglycan chains produced and secreted into the medium by penicillin-inhibited cultures of autolytic-defective strains of *E. hirae* (Barrett and Shockman, 1984; Barrett *et al.*, 1984*a,b*). Processive hydrolysis, starting from the nonreducing ends of the glycan chains of these 42- to 47-disaccharide-peptide-long chains yielded virtually only one product–monomeric disaccharide-peptide units (Barrett *et al.*, 1984*a,b*).

Surprisingly, the ultraviolet absorption spectrum of a highly purified preparation of muramidase-1 showed a maximum at about 260 nm rather than a maximum at 280 nm expected for a pure protein (Dolinger *et al.*, 1988). Again, our first thoughts were that the enzyme preparation was contaminated with a small amount of nucleic acid.

The absorption maximum at 260 nm, as well as the presence of about 18 moles of phosphorus per mole of enzyme, remained present after a variety of drastic treatments such as SDS-gel electrophoresis, trichloroacetic acid precipitation, or exposure to 6 M guanidine-HCl. In addition to phosphorus, the enzyme was found to contain about 12 moles of ribose per mole of enzyme and about 11 moles of sulfhydryl groups per mole of enzyme in excess of those accounted for by cysteine residues in the protein. The data suggested the presence of covalently linked nucleotide, probably a thionucleotide. Thin layer and paper chromatography showed that formic acid-treated modifier, isolated after snake venom phosphodiesterase hydrolysis, had the same R_f values as 5-mercaptouracil and cochromatographed as a single peak on reversed-phase HPLC with chemically synthesized 5-mercaptouracil. Finally, direct probe mass spectroscopy was used to show that the modifier was indeed 5-mercaptouridine monophosphate (Dolinger *et al.*, 1988). Data were obtained indicating that the nucleotide was present as multiple monomeric phosphodiester substitutions linked to the polypeptide chain, probably to the hydroxyl of tyrosine residues, and that all of the approximately twelve 5-mercaptouridine monophosphates were present on the proteinase-activated, 87-kDa form of muramidase-1. Snake venom phosphodiesterase treatment of native muramidase-1 resulted in the release of about one half of the phosphorus and nucleotides along with about a 50% reduction in muramidase activity, suggesting that the nucleotide substitutions play some role in enzyme activity (Dolinger and Shockman, unpublished data).

Muramidase-2. The presence of a second enzyme activity that can hydrolyze the bond between N-acetylmuramic acid and N-acetylglucosamine in the peptidoglycan of *E. hirae* was detected relatively recently (Kawamura and Shockman, 1983*a*). This is because muramidase-2 lacks the ability to *dissolve* intact walls or cells of *E. hirae* despite the fact that it can hydrolyze some of the muramidase-sensitive bonds in the wall peptidoglycan of *E. hirae*. Muramidase-2 was first observed as an activity that rapidly dissolved walls of *Micrococcus luteus* that failed to bind to ConA-Sepharose (Kawamura and Shockman, 1983*a*). *M. luteus* walls are not dissolved by purified preparations of muramidase-1. It was then partially purified from culture supernates by its ability to tightly bind to walls of *M. luteus* and identified as a protein with a molecular weight of about 74 kDa after SDS-gel electrophoresis. Muramidase-2 apparently is not a glycoprotein, as it fails to bind to ConA-Sepharose and to stain with the PAS reagent. It differs from muramidase-1 in a number of properties (Shockman *et al.*, 1983, 1988) including its abilities to bind to and hydrolyze various wall substrates. Important to this discussion, and in a practical sense as it provides both differential and common substrates, are the observations that whereas muramidase-1 binds to and rapidly dissolves walls of *E. hirae*, muramidase-2 does not. In contrast, muramidase-2 binds to and rapidly hydrolyzes the acid-insoluble peptidoglycan fraction of walls of *E. hirae*, which are not dissolved by muramidase-1. Re-N-acetylation of the *E. hirae* peptidoglycan fraction provides a common substrate for both enzymes. Thus, muramidase-1 activity can be differentially assayed using walls of *E. hirae* as the substrate, whereas muramidase-2 activity can be differentially assayed using walls of *M. luteus* or the peptidoglycan fraction of walls of *E. hirae* as substrate.

Although it remains possible that muramidase-2 may be a nonprocessed (e.g., nonglucosylated and/or non-nucleotidylated) form of muramidase-1, several types of evidence are consistent with its being the product of a separate gene. Several years ago, using nitrosoguanidine mutagenesis, we isolated a temperature-sensitive autolysis defective mutant of *E. hirae* called Lyt-14 (Cornett *et al.*, 1978). The phenotype of this mutant is complex and it seems likely that it possesses more than one mutation. However, one property of Lyt-14 is of particular interest. Extracts of Lyt-14 were able to dissolve walls of *E. hirae* but, in contrast to the parent strain, were unable to dissolve walls of *M. luteus*. Recent studies, with carefully prepared fresh extracts, showed that the Lyt-14 activity that dissolves walls of *M. luteus* has an extremely short half-life, on the order of 2 minutes (Tepper and Shockman, unpublished data). In contrast, alkaline extracts of Lyt-14 showed the presence of levels of muramidase-1 activity (assayed on walls of *E. hirae*) indistinguishable from those present in the parent strain (Mays and Shockman, unpublished data). These and other data indicate that Lyt-14 possesses a mutation that results in a temperature-sensitive defect in muramidase-2 activity. In addition, the amino acid composition of muramidase-2 (Chu *et a..*, 1992; Kariyama, unpublished data) is not consistent with a structural relationship with a protein that has the composition of muramidase-1 (Dolinger, unpublished data). For example, as shown in Table 1, muramidase-2 contains a large number of serine (74) and threonine (73) residues, whereas muramidase-1 possesses far fewer (44 serines and 55 threonines). Similarly, muramidase-2 contains a larger number of valines (32 compared with 12 in muramidase-1) and contains 15 tryptophans and 17 prolines, neither of which were detected in muramidase-1.

Recently, we isolated and purified muramidase-2 of *E. hirae* to apparent homogeneity from broken bacterial cell preparations (Dolinger *et al.*, 1989). The enzyme activity was associated with the crude cell wall fraction from which it could be extracted, with 0.01 to 0.02 M NaOH, along with all of the penicillin-binding proteins (PBPs) normally found in membrane preparations of *E. hirae* (Coyette *et al.*, 1977). Enzyme activity was further purified from these alkaline extracts by selective affinity binding to the acid-insoluble peptidoglycan of *E. hirae* followed by elution with 0.01 N NaOH. Two SDS-gel electrophoresis-purified polypeptides, with apparent molecular weights of 125 and 75 kDa, were shown to possess muramidase activity and to have the ability to bind radioactively labeled penicillin with low affinity (Dolinger *et al.*, 1989). At that time, we carefully stated that the two enzymatically active polypeptides had "the same electrophoretic mobility as PBPs 1 and 5 present in membrane preparations of this organism, respectively." However, as discussed below, we have since obtained immunochemical (Tepper *et al.*, in preparation) and molecular (Chu *et al.*, 1992) data that clearly demonstrate that PBP 5 and the 70-kDa form of muramidase-2 are separate and distinct proteins. Immunoaffinity binding to an antimuramidase-2 monoclonal antibody column, and ion exchange chromatography of [^{14}C]penicillin-labeled, purified muramidase-2 resulted in the copurification of [^{14}C]penicillin-labeled protein and muramidase-2. Both the 125- and 75-kDa forms retained [^{14}C]penicillin. Reversed-phase chromatography was used to separate the two [^{14}C]penicillin-labeled polypeptides and to obtain evidence that the 75-kDa form was derived from the 125-kDa form, presumably by proteolysis (Dolinger *et al.*, 1989).

As mentioned above, the smaller-molecular-weight form (70 to 75 kDa) of muramidase-2 and PBP 5 were indistinguishable from each other upon fluorography of SDS gels of [^{14}C] or [^{3}H]penicillin-labeled proteins (Dolinger *et al.*, 1989). In addition, both proteins bind radioactive penicillin with low affinity, requiring high concentrations and/or long incubation periods to reveal penicillin binding. Initial studies, using a rabbit antiserum specific for PBP 5 (El Kharroubi *et al.*, 1991) and a mouse monoclonal antibody specific for muramidase-2 (Kariyama and Shockman, 1992) showed that reaction of either of the proteins with its specific antibody in Western blots did not interfere with the reaction of the other protein with its specific antibody, although the two proteins, were not truly resolved on the gels. Minor changes in the gel electrophoresis system resulted in increased resolution in the molecular weight range of the two proteins so that the two proteins could be resolved into two separate and distinct polypeptide bands. Western blots of these SDS electrophoretograms showed that the slightly larger-molecular-weight polypeptide interacted with the anti-

Table 1. Comparison of the amino acid compositions of muramidase-1 and muramidase-2

	Number of residues per mole		
	Muramidase-2		Muramidase-1
	By sequence[1]	Amino acid analysis[2]	Amino acid analysis[3]
Aspartic acid	13	} 45	} 92
Asparagine	66		
Glutamic acid	16	} 69	} 125
Glutamine	28		
Alanine	51	53	128
Arginine	4	6	66
Cysteine	1	N.D.[4]	28
Cysteic acid	-	N.D.[4]	14
Glycine	68	46	122
Serine	74	72	44
Threonine	73	76	55
Histidine	5	10	27
Isoleucine	40	39	34
Leucine	20	28	105
Lysine	42	47	120
Methionine	8	N.D.[4]	18
Phenylalanine	15	19	41
Proline	17	-[5]	-[5]
Tryptophan	15	N.D.[4]	-[6]
Tyrosine	29	21	62
Valine	32	32	12

[1]From the derived amino acid sequence (Chu *et al.*, 1992), without the putative leader sequence. Total of 617 amino acid residues.
[2]Kariyama, unpublished data.
[3]Dolinger, unpublished data.
[4]N.D. = not determined.
[5]Below level of detection.
[6]Below level of detection after hydrolysis with 3 M *p*-toluenesulfonic acid.

PBP 5 antiserum and the slightly smaller polypeptide interacted with monoclonal antibodies to muramidase-2 (Tepper, Shockman, and Daneo-Moore, unpublished data). More recently, comparison of the nucleotide and derived amino acid sequences of muramidase-2 (Fig. 1 and Chu *et al.*, 1992) with that of PBP 5 of *E. hirae* R40 (El Kharroubi *et al.*, 1991) failed to show any significant homology.

The ability of muramidase-2 to bind penicillin was unexpected and certainly unusual for a muramidase, rather than a transpeptidase, except for the possibility that, like some of the larger molecular weight PBPs of *E. coli*, we could be dealing with a "double-headed" enzyme that possesses both transpeptidase and transglycosylase activities (Nakagawa and Matsuhashi, 1982; Nakagawa *et al.*, 1984). Examination of the derived amino acid sequence of muramidase-2 (Fig. 1) showed the presence of several amino acid groupings that have been found to be characteristic motifs in serine β-lactamases and penicillin-binding proteins (Ghuysen, 1991). These include the tetrad

```
                                                         ↓
  1  MENIARKERRRLNETKRFRKVKRSAALVGTAMVGCSVAAPLIQPVQVDAD   50
 51  QTPTQFGARINTAAFIAEIATYAQPIAQANDLYASVMIAQAVVESGWGSS  100
101  ALSQAPYYNLFGIKGSYQGQTVYMDTLEYLNNKWVSKKEPFRQYPSFAES  150
151  FNDNAYVLRNTSFGNGYYYAGTWKSNTKSYTDATACLTGRYATDPGYAGK  200
201  LNNIITTYGLTKYDTPASGNAGGGVTIGNGGNTGNTSNSGSTSGNSGGSA  250
251  TTTGTTYTVKSGDSVWGISHSFGITMAQLIEWNNIKNNFIYPGQKLTIKG  300
301  GQSAGSSTTNTGNNASSGNTSGNTNTSGSTGQATGAKYTVKSGDSVWKIA  350
351  NDHGISMNQLIEWNNIKNNFVYPGQQLVVSKGSSSASGSTSNTSTGNTSS  400
401  NTANTGSTTSGSTYTVKAGESVWSVSNKFGISMNQLIQWNNIKNNFIYPG  450
451  QKLIVKGGSSSSNASTSTANNKNTASSNTSSTATGQATYTVKAGESVWGV  500
501  ANKNGISMNQLIEWNNIKNNFIYPGQKLIVKGGSSKASATATIKPTASTP  550
551  ASTTPTASSTGDTKYTVKAGESVWGVANKHHITMDQLIEWNNIKNNFIYP  600
601  GQEVIVKKGTAQSTPAKSDEKTYTVKAGESVWGVADSHGITMNQLIEWNN  650
651  IKNNFIYPGQQLIVKK  666
```

Figure 1. The derived amino acid sequence of muramidase-2 of *E. hirae* ATCC 9790. The signal sequence cleavage site is indicated by a vertical arrow. The six repeat units are underlined. The SXXK serine-active site tetrad and the four SGN and two KSG triads, considered to be characteristic of penicillin-interactive proteins (Ghuysen, 1991), are boxed. The GenBank accession number for the muramidase-2 sequence is M77639. (From Chu *et al.*, 1992)

S175NTK, the triad SGN at positions S218, S243, S317, and S321 and the triad KSG at positions K260 and K341. It should be noted that the highly homologous *S. faecalis* autolysin lacks the SXXK-active site motif, except for an S647IQK sequence very near the C-terminal end of the protein. However, the shape and distribution of hydrophobic and hydrophilic clusters along the amino acid sequences of both the *E. hirae* and *S. faecalis* enzymes differ from those found in a broad assortment of serine β-lactamases and in a number of PBPs. Thus, it is possible that the mechanism of penicillin interaction may differ from that of the typical serine-active site penicillin-interactive protein.

Recently obtained data (Kariyama and Shockman, these proceedings) are highly consistent with the specific binding of [^{3}H]penicillin to extracellular muramidase-2. Reversed-phase FPLC purification of extracellular muramidase-2, that had been partially purified by affinity binding to peptidoglycan and eluted with 6 M guanidine-HCl, showed parallel elution of protein, muramidase activity, reactivity with a monoclonal antibody to muramidase-2 in Western blots, and [^{3}H]penicillin-binding activity. Penicillin-binding activity appeared to be blocked when the muramidase-2 was bound to peptidoglycan and was enhanced by unfolding in guanidine-HCl followed by refolding. Thus, penicillin binding by muramidase-2 may require protein unfolding and refolding to expose normally "buried" penicillin-binding sites.

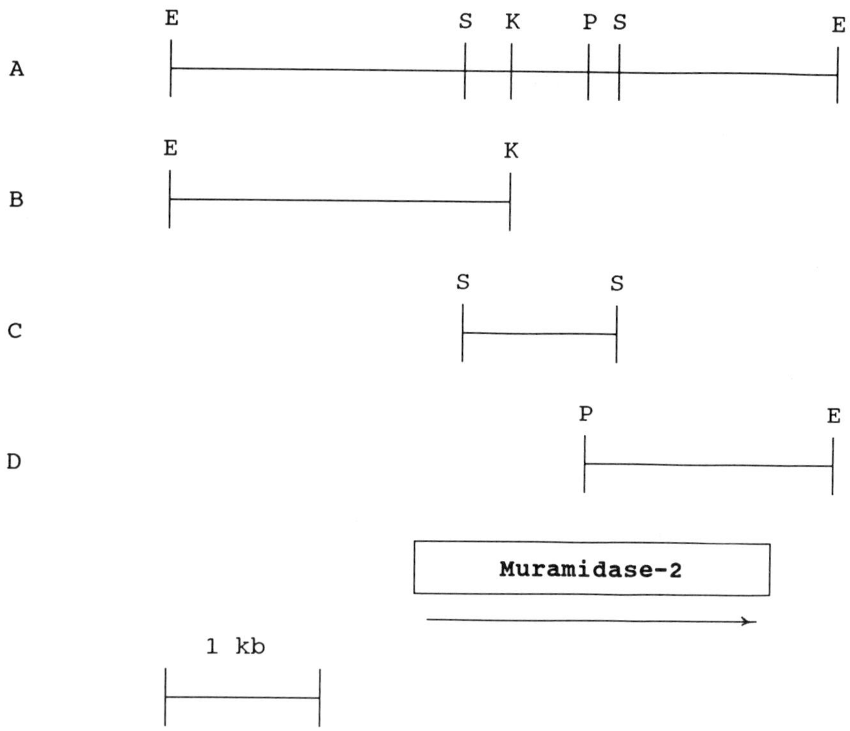

Figure 2. Restriction map of DNA in the vicinity of the muramidase-2 gene. (A) The 4.5-kb DNA insert. (B) The 2.3-kb fragment. (C) The 1.0-kb fragment. (D) The 1.5-kb fragment. The location and orientation of the muramidase-2 gene are indicated. Restriction enzyme sites are designated as follows: E, *Eco*RI; K, *Kpn*1; P, *Pst*I; S, *Sau*3AI. (From Chu *et al.*, 1992)

Cloning and Sequence Analysis of Muramidase-2

Recently (Chu *et al.*, 1992), we successfully cloned and sequenced the extracellular form of muramidase-2. A phage λgt11 clone that reacted with a monoclonal antibody (2F8) to muramidase-2 was found to contain a 4.5-kb insert that hybridized with six DNA bands at about 4.5, 5.0, 6.1, 7.1, 9.1, and 12.5 kb of EcoRI-digested chromosomal DNA of *E. hirae*. The 4.5-kb band was cut with appropriate restriction enzymes into 2.3-, 1.0- and 1.5-kb fragments which were subcloned into plasmids (Fig. 2), and the nucleotide sequences of these fragments were determined. The deduced amino acid sequence of the open reading frame (Fig. 1), containing 666 amino acid residues, showed the following features. (i) A 49-amino acid-long putative signal peptide, with a potential protease cleavage site between Ala 49 and Asp 50. The virtual identity of the determined N-terminal amino acid sequence of mature extracellular muramidase-2 with the deduced sequence, clearly identified this cleavage site. (ii) The presence of six 45-amino acid-long, highly homologous, amino acid repeat units, separated by intervening sequences that are highly enriched for serine and threonine (ca. 51% ser plus thr) were found at the C-terminal end of the protein. Although these repeats have little homology with repeat units found at the C-terminal ends of a variety of different proteins of Gram-positive bacterial species including those in the pneumococcal amidase gene (García *et al.*, 1988) and the muramidase genes of pneumococcal phages CP-1, CP-7 and CP-9 (García *et al.*, 1990), substantial homology was observed with four contiguous 68-amino acid-long repeat units at the C-terminus of an autolysin of unspecified specificity of *S. faecalis* (Beliveau *et al.*, 1991). The amino acid sequence of the *S. faecalis* protein showed other areas of substantial homology with muramidase-2 including the putative signal peptide and much of the rest of the two polypeptides except for a 104-amino acid insertion between amino acids 64 and 65 of the *E. hirae* muramidase-2; and for the serine- plus threonine-enriched sequences, between the C-terminal repeats of muramidase-2 (Joris *et al.*, 1992).

García *et al.* (1990) obtained evidence that the repeat sequences of the pneumococcal enzymes appear to be involved with binding of the enzymes to the pneumococcal wall and that the presence of choline in the wall teichoic acid is important for this binding. While it seems likely that the highly conserved C-terminal repeats of the *E. hirae* muramidase-2 (and the *S. faecalis* autolysin) are involved with binding to the wall substrate, these interactions would not involve a choline containing teichoic acids which is not present in the wall of *E. hirae*. Furthermore, muramidase-2 has been shown to bind with very high affinity to a limited number of sites on the isolated peptidoglycan fraction of walls of *E. hirae* (Kariyama and Shockman, 1992). Further studies of this possible function will be required.

In addition to the homology described above, searches of the protein sequence data bank revealed significant levels of similarity, at least at the level of defined polypeptide segments, with several other proteins (Joris *et al.*, 1992). These include homologies with *B. subtilis* phage PZA lysozyme (Paces *et al.*, 1986), the homologous *Bacillus* phage 29 gene 15 lysozyme (Garvey *et al.*, 1986; Saedi *et al.*, 1987), the *Salmonella typhimurium* FlgJ intracellular flagellar protein (Jones *et al.*, 1989), the *Listeria monocytogenes* pathogenicity associated extracellular P60 protein (Köhler *et al.*, 1990), and *Staphylococcus aureus* protein A (Uhlen *et al.*, 1984). In any event, the data obtained so far are consistent with a modular design of muramidase-2, with an N-terminal muramidase domain, perhaps a central penicillin-binding domain, and a C-terminal, putative substrate-binding domain.

Possible Functions of the Two Muramidases of *E. hirae* in Surface Growth and Division

The presence of two peptidoglycan hydrolases in *E. hirae* that hydrolyze the same bond, at the least, complicates earlier interpretations and hypotheses concerning the possible functions of autolysins in surface growth and division. On the other hand, the presence of two separate and distinct enzymes that can hydrolyze the same bond, at least contributes to an explanation of our lack of success in isolating autolytic-*negative* mutants of *E. hirae* despite considerable and varied efforts over many years. It is also somewhat surprising that we have not even obtained mutants that lack one of the activities, with the possible exception of the Lyt-14 strain (Cornett *et al.*, 1978) that appears to carry a mutation that results in the synthesis of a thermosensitive muramidase-2 activity. Growth of Lyt-14 in long chains at the nonpermissive temperature is an appealing correlation although it is highly unlikely that Lyt-14 possesses only one mutation. It is also of interest to note the dearth of autolytic-*negative* strains of other bacterial species.

While the presence of multiple enzymes that catalyze the same type of reaction suggest that the cells are carrying "insurance" for a presumably essential function, it is equally clear from the existence of species and strains, such as some Group A streptococci, that fail to autolyze, under conditions that result in the rapid autolysis of other strains, that only a low level of peptidoglycan hydrolase activity is essential to the survival and growth of the bacterial cell. Our recent success in cloning, sequencing and now expressing muramidase-2, as well as similar successes with peptidoglycan hydrolases of other bacterial species such as of pneumococci (García *et al.*, 1985), *B.*

subtilis (Potvin *et al.*, 1988; Kuroda and Sekiguchi, 1990, 1991; Foster, 1991), and Staphylococci (Biavasco *et al.*, 1988; Jayaswal *et al.*, 1990; Wang *et al.*, 1991), should lead to solving the questions of functions and essentiality of these various peptidoglycan hydrolases in the not too distant future.

Acknowledgments

This work was supported by Public Health Service grant AI05044 from the National Institute of Allergy and Infectious Diseases. We thank G. Harvey and E. Asklar for help in preparing the typescript.

REFERENCES

Barrett, J.F. and Shockman, G.D. (1984) Isolation and characterization of soluble peptidoglycan from several strains of *Streptococcus faecium*. J. Bacteriol. 159, 511-519.

Barrett, J.F., Dolinger, D.L., Schramm, V.L., and Shockman, G.D. (1984*a*) The mechanism of soluble peptidoglycan hydrolysis by an autolytic muramidase. A processive exodisaccharidase. J. Biol. Chem. 259, 11818-11827.

Barrett, J.F., Schramm, V.L. and Shockman, G.D. (1984*b*) Hydrolysis of soluble, linear, un-cross-linked peptidoglycans by endogenous bacterial N-acetylmuramoylhydrolases. J. Bacteriol. 159, 520-526.

Beck, B. and Park, J.T. (1976) Activity of three murein hydrolases during the cell division cycle of *Escherichia coli* as measured in toluene-treated cells. J. Bacteriol. 126, 1250-1260.

Beliveau, C., Potvin, C., Trudel, J., Asselin, A. and Bellemare, G. (1991) Cloning, sequencing, and expression in *Escherichia coli* of a *Streptococcus faecalis* autolysin. J. Bacteriol. 173, 5619-5623.

Biavasco, F., Pruzzo, C. and Thomas, C. (1988) Cloning and expression of the *Staphylococcus aureus* glucosaminidase in *Escherichia coli*. FEMS Microbiol. Lett. 49, 137-142.

Chatterjee, A.N., Wong, W., Young, F.E. and Gilpin, R.W. (1976) Isolation and characterization of a mutant of *Staphylococcus aureus* deficient in autolytic activity. J. Bacteriol. 125, 961-967.

Chu, C.-P., Kariyama, R., Daneo-Moore, L. and Shockman, G.D. (1992) Cloning and sequence analysis of the muramidase-2 gene from *Enterococcus hirae*. J. Bacteriol. 174, 1619-1625.

Cleveland, R. F., Höltje, J.-V., Wicken, A. J., Tomasz, A., Daneo-Moore, L. and Shockman, G.D. (1975) Inhibition of bacterial wall lysins by lipoteichoic acids and related compounds. Biochem. Biophys. Res. Commun. 67, 1128-1135.

Cleveland, R. F., Wicken, A. J., Daneo-Moore, L. and Shockman, G.D. (1976) Inhibition of wall autolysis in *Streptococcus faecalis* by lipoteichoic acid and lipids. J. Bacteriol. 126, 192-197.

Conover, M.J., Thompson, J.S. and Shockman, G.D. (1966) Autolytic enzyme of *Streptococcus faecalis*: release of soluble enzyme from cell walls. Biochem. Biophys. Res. Commun. 23, 713-719.

Cornett, J.B., Redman, B.E. and Shockman, G.D. (1978) Autolytic defective mutant of *Streptococcus faecalis*. J. Bacteriol. 133, 631-640.

Coyette, J., Perkins, H.R., Polacheck, I., Shockman, G.D. and Ghuysen, J.-M. (1974) Membrane-bound DD-carboxypeptidase and LD-transpeptidase of *Streptococcus faecalis* ATCC 9790. Eur. J. Biochem. 44, 459-468.

Coyette, J., Ghuysen, J.-M., Binot, F., Adriaens, P., Meesschaert, B. and Vanderhaeghe, H. (1977) Interactions between β-lactam antibiotics and isolated membranes of *Streptococcus faecalis* ATCC 9790. Eur. J. Biochem. 75, 231-239.

Croux, C. and García, J.-L. (1991) Sequence of the *lyc* gene encoding the autolytic lysozyme of *Clostridium acetobutylicum* ATCC 824: comparison with other lytic enzymes. Gene 104, 25-31.

Daneo-Moore, L. and Shockman, G.D. (1977) The Bacterial Cell Surface in Growth and Division, In "Cell Surface Reviews, vol. 4. The Synthesis, Assembly, and Turnover of Cell Surface Components" (Poste, G. and Nicolson, G.L., Eds.), pp. 597-715. Elsevier/North Holland, Amsterdam.

Dolinger, D.L., Schramm, V.L. and Shockman, G.D. (1988) Covalent modification of the β-1,4-N-acetylmuramoylhydrolase of *Streptococcus faecium* with 5-mercaptouridine monophosphate. Proc. Natl. Acad. Sci. USA 85, 6667-6671.

Dolinger, D.L., Daneo-Moore, L. and Shockman, G.D. (1989) The second peptidoglycan hydrolase of *Streptococcus faecium* ATCC 9790 covalently binds penicillin. J. Bacteriol. 171, 4355-4361.

Doyle, R.J. and Koch, A.L. (1987) The functions of autolysins in the growth and division of *Bacillus subtilis*. Crit. Rev. Microbiol. 15, 169-222.

El Kharroubi, A., Jacques, P., Piras, G., Van Beeumen, J., Coyette, J. and Ghuysen, J.-M. (1991) The *Enterococcus hirae* R40 penicillin-binding protein 5 and the methicillin-resistant *Staphylococcus aureus* penicillin-binding protein 2' are similar. Biochem. J. 280, 463-469.

Fan, D.P. and Beckman, M.M. (1973) Mutant of *Bacillus subtilis* with a temperature-sensitive autolytic amidase. J. Bacteriol. 14, 798-803.

Fein, J.E. and Rogers, H. J. (1976) Autolytic enzyme-deficient mutants of *Bacillus subtilis* 168. J. Bacteriol. 127, 1427-1442.

Foster, S.J. (1991) Cloning, expression, sequence analysis and biochemical characterization of an autolytic amidase of *Bacillus subtilis* 168 *trpC2*. J. Gen. Microbiol. 137, 1987-1998.

García, E., García, J.L., Ronda, C., García, P. and López, R. (1985) Cloning and expression of the pneumococcal autolysin gene of *Escherichia coli*. Mol. Gen. Genet. 201, 225-230.

García, J.L., Sanchez-Puelles, J.M., García, P., López, R., Ronda, C. and García, E. (1986) Molecular characterization of an autolysin-defective mutant of *Streptococcus pneumoniae*. Biochem. Biophys. Res. Comm. 137, 614-619.

García, E., García, J.L., García, P., Arraras, A., Sanchez-Puelles, J.M. and López, R. (1988) Molecular evolution of lytic enzymes of *Streptococcus pneumoniae* and its bacteriophages. Proc. Natl. Acad. Sci. USA 85, 914-918.

García, P., García, J.L., García, E., Sanchez-Puelles, J.M. and López, R. (1990) Modular organization of the lytic enzymes of *Streptococcus pneumoniae* and its bacteriophages. Gene 86, 81-88.

Garvey, K.J., Saedi, M.S. and Ito, J. (1986). Nucleotide sequence of *Bacillus* phage φ29 genes 14 and 15: homology of gene 15 with other phage lysozymes. Nucleic Acid Res. 14, 10001-10008.

Ghuysen, J.-M. (1991) Serine β-lactamases and penicillin-binding proteins. Ann. Rev. Microbiol. 45, 37-67.

Ghuysen, J.-M. and Shockman, G.D. (1973) Biosynthesis of Peptidoglycan, in "Bacterial Membranes and Walls" (Leive, L., Ed.), pp. 37-130. Marcel Dekker, New York.

Hakenbeck, R. and Messer, W. (1974) Activity of murein hydrolases and membrane synthesis in synchronized *Escherichia coli* B/r. Ann. Microbiol. Inst. Pasteur 125B, 163-166.

Hartmann, R., Bock-Hennig, S.B. and Schwarz, U. (1974) Murein hydrolases in the envelope of *Escherichia coli*. Eur. J. Biochem. 41, 203-208.

Herbold, D.R. and Glaser, L. (1975*a*) *Bacillus subtilis* N-acetylmuramic acid L-alanine amidase. J. Biol. Chem. 250, 1676-1682.

Herbold, D.R. and Glaser, L. (1975*b*) Interaction of *N*-acetylmuramic acid L-alanine amidase with cell wall polymers. J. Biol. Chem. 250, 7231-7238.

Higgins, M. L., Pooley, H. M. and Shockman, G.D. (1970) Site of initiation of cellular autolysis in *Streptococcus faecalis* as seen by electron microscopy. J. Bacteriol. 103, 504-512.

Hinks, R. P., Daneo-Moore, L. and Shockman, G.D. (1978) Cellular autolytic activity in synchronized populations of *Streptococcus faecium*. J. Bacteriol. 133, 822-829.

Höltje, J.-V. and Tomasz, A. (1975) Lipoteichoic acid: A specific inhibitor of autolysin activity in Pneumococcus. Proc. Natl. Acad. Sci. USA 72, 1690-1694.

Höltje, J.-V. and Tuomanen, E.I. (1991) The murein hydrolases of *Escherichia coli*: properties, functions, and impact on the course of infections *in vivo*. J. Gen. Microbiol. 137, 441-454.

Jayaswal, R.K., Lee, Y.-I. and Wilkinson, B.J. (1990) Cloning and expression of a *Staphylococcus aureus* gene encoding a peptidoglycan hydrolase activity. J. Bacteriol. 172, 5783-5788.

Jolliffe, L.K., Doyle, R.J. and Streips, U.N. (1981) Energized membrane and cellular autolysis in *Bacillus subtilis*. Cell 25, 753-763.

Jones, C.J., Homma, M. and Macnab, R.M. (1989) L-, P-, and M-ring proteins of the flagellar basal body of *Salmonella typhimurium*: Gene sequences and deduced protein sequences. J. Bacteriol. 171, 3890-3900.

Joris, B., Englebert, S., Chu, C.-P., Kariyama, R., Daneo-Moore, L., Shockman, G.D. and Ghuysen, J.-M. (1992) Modular design of the *Enterococcus hirae* muramidase-2 and *Streptococcus faecalis* autolysin. FEMS Microbiol. Lett., in press.

Joseph, R. and Shockman, G.D. (1976) Autolytic formation of protoplasts (autoplasts) of *Streptococcus faecalis*: location of active and latent autolysin. J. Bacteriol. 127, 1482-1493.

Kariyama, R. and Shockman, G.D. (1992) Extracellular and cellular distribution of muramidase-2 and muramidase-1 of *Enterococcus hirae* ATCC 9790. J. Bacteriol., in press.

Kawamura, T. and Shockman, G.D. (1983a) Evidence for the presence of a second peptidoglycan hydrolase in *Streptococcus faecium*. FEMS Microbiol. Lett. 19, 65-69.

Kawamura, T. and Shockman, G.D. (1983b) Purification and some properties of the endogenous, autolytic N-acetylmuramoylhydrolase of *Streptococcus faecium*, a bacterial glycoenzyme. J. Biol. Chem. 258, 9514-9521.

Köhler, S., Leimeister-Wachter, L., Chakraborty, T., Lottspeich, F. and Goebel, W. (1990) The gene coding for protein p60 of *Listeria monocytogenes* and its use as a specific probe for *Listeria monocytogenes*. Inf. Immun. 58, 1943-1950.

Koyama, T., Yamada, M. and Matsuhashi, M. (1977) Formation of regular packets of *Staphylococcus aureus* cells. J. Bacteriol. 129, 1518-1523.

Kuroda, A. and Sekiguchi, J. (1990) Cloning, sequencing, and genetic mapping of a *Bacillus subtilis* cell wall hydrolase gene. J. Gen. Microbiol. 136, 2209-2216.

Kuroda, A. and Sekiguchi, J. (1991) Molecular cloning and sequencing of a major *Bacillus subtilis* autolysin gene. J. Bacteriol. 173, 7304-7312.

Nakagawa, J. and Matsuhashi, M. (1982) Molecular divergence of a major peptidoglycan synthetase with transglycosylase-transpeptidase activities in *Escherichia coli*–penicillin-binding protein 1Bs. Biochem. Biophys. Res. Commun. 105, 1-11.

Nakagawa, J., Tamake, S., Tomioka, S. and Matsuhashi, M. (1984) Functional biosynthesis of cell wall peptidoglycan by polymorphic bifunctional polypeptides. J. Biol. Chem. 259, 13937-13946.

Paces, V., Uleek, C. and Urbanek, D. (1986) Nucleotide sequence of the late region of *Bacillus subtilis* phage PZA, a close relative of phi 29. Gene 44, 107-114.

Pooley, H.M. and Shockman, G.D. (1969) Relationship between the latent form and the active form of the autolytic enzyme of *Streptococcus faecalis*. J. Bacteriol. 100, 617-624.

Pooley, H.M. and Shockman, G.D. (1970) Relationship between the location of autolysin, cell wall synthesis, and the development of resistance to cellular autolysis in *Streptococcus faecalis* after inhibition of protein synthesis. J. Bacteriol. 103, 457-466.

Pooley, H.M., Shockman, G.D., Higgins, M.L. and Porres-Juan, J. (1972) Some properties of two autolytic-defective mutants of *Streptococcus faecalis* ATCC 9790. J. Bacteriol. 109, 423-431.

Potvin, C., Leclerc, D., Tremblay, G., Asselin, A. and Bellemare, G. (1988) Cloning, sequencing and expression of a *Bacillus* bacteriolytic enzyme in *Escherichia coli*. Mol. Gen Genet 214, 241-248.

Rogers, H.J., Perkins, H.R. and Ward, J.B. (1980) "Microbial Cell Walls and Membranes," pp. 437-460. Chapman and Hall, New York.

Saedi, M.S., Garvey, K.J. and Ito, J. (1987) Cloning and purification of a unique lysozyme produced by *Bacillus* phage ϕ29. Proc. Natl. Acad. Sci. USA 84, 955-958.

Sekiguchi, J., Ezaki, B., Kodama, K. and Akamatsu, T. (1988) Molecular cloning of a gene affecting the autolysin level and flagellation in *Bacillus subtilis*. J. Gen. Microbiol. 134, 1611-1621.

Shockman, G.D. (1965) Symposium on the fine structure and replication of bacteria and their parts. IV. Unbalanced cell-wall synthesis: autolysis and cell-wall thickening. Bacteriol. Rev. 29, 345-358.

Shockman, G.D. and Barrett, J.F. (1983) Structure, function, and Assembly of cell walls of Gram-positive bacteria. Ann. Rev. Microbiol. 37, 501-527.

Shockman, G.D. and Cheney, M.C. (1969) Autolytic enzyme system of *Streptococcus faecalis*. V. Nature of the autolysin-cell wall complex and its relationship to properties of the autolytic enzyme of *Streptococcus faecalis*. J. Bacteriol. 98, 1199-1207.

Shockman, G.D., Thompson, J.S. and Conover, M.J. (1967) The autolytic enzyme system of *Streptococcus faecalis*. II. Partial characterization of the autolysin and its substrate. Biochemistry 6, 1054-1065.

Shockman, G.D., Daneo-Moore, L. and Higgins, M.L. (1974) Problems of cell wall and membrane growth, enlargement and division. Ann. N.Y. Acad. Sci. 235, 161-197.

Shockman, G.D., Daneo-Moore, L., Cornett, J.B. and Mychajlonka, M. (1979) Does penicillin kill bacteria? Rev. Infect. Dis. 1, 787-796.

Shockman, G.D., Daneo-Moore, L., McDowell, T.D. and Wong, W. (1982) The Relationship Between Inhibition of Cell Wall Synthesis and Bacterial Lethality, in "The Chemistry and Biology of β-Lactam Antibiotics" (Gorman, M. and Morin, R.B., Eds.), vol. 3, pp. 303-338. Academic Press, New York.

Shockman, G.D., Kawamura, T., Barrett, J.F. and Dolinger, D. (1983) The Autolytic System of *Streptococcus faecium*, in "The Target of Penicillin - International FEMS Symposium on the Murein Sacculus of Bacterial Cell Walls - Architecture and Growth" (Hakenbeck, R., Höltje, J.-V. and Labischinski, H., Eds.), pp. 165-172. W. de Gruyter & Co., Berlin, New York.

Shockman, G.D., Dolinger, D.L. and Daneo-Moore, L. (1988) The Autolytic Peptidoglycan Hydrolases of *Streptococcus faecium*: Two Unusual Enzymes, in "Antibiotic Inhibition of Bacterial Cell Wall Assembly and Function" (Actor, P., Daneo-Moore, L., Higgins, M.L., Salton, M.R.J. and Shockman, G.D., Eds.), pp. 195-210. American Society for Microbiology, Washington, DC.

Shungu, D.L., Cornett, J.B. and Shockman, G.D. (1979) Morphological and physiological study of autolytic-defective *Streptococcus faecium* strains. J. Bacteriol. 138, 598-608.

Tomasz, A. (1979*a*) From penicillin-binding proteins to the lysis and death of bacteria: A 1979 view. Rev. Inf. Dis., 1, 434-467.

Tomasz, A. (1979*b*) The mechanism of the irreversible antimicrobial effects of penicillins: How the beta-lactam antibiotics kill and lyse bacteria. Ann. Rev. Microbiol. 33, 113-37.

Uhlen, M., Guss, B., Wilsson, B., Gatenbeck, S., Phillipson, L. and Linberg, M. (1984) Complete sequence of the staphylococcal gene encoding protein A. J. Biol. Chem. 259, 1695-1702.

Walderich, B. and Höltje, J.-V. (1991) Subcellular distribution of the soluble lytic transglycosylase in *Escherichia coli*. J. Bacteriol. 173, 5668-5676.

Wang, X., Wilkinson, B.J. and Jayaswal, R.K. (1991) Sequence analysis of a *Staphylococcus aureus* gene encoding a peptidoglycan hydrolase activity. Gene 102, 105-109.

IS MURAMIDASE-2 OF *ENTEROCOCCUS HIRAE* A PENICILLIN-BINDING PROTEIN?

Reiko Kariyama and Gerald D. Shockman

Department of Microbiology and Immunology
Temple University School of Medicine
Philadelphia, PA 19140

Highly purified muramidase-2 (M-2), obtained from broken cell preparations of *Enterococcus hirae (Streptococcus faecium)* ATCC 9790 was shown to bind penicillin G (Pen G) with low affinity (Dolinger *et al.*, 1989). Fluorography of sodium dodecyl sulfate-polyacrylamide gel electrophoresis (SDS-PAGE) gels of [^{3}H]Pen-labeled, purified M-2 showed the presence of two radioactive bands, at about 125 and 70 kDa, both of which were shown to possess peptidoglycan hydrolase activity after elution and renaturation of the protein bands. The electrophoretic mobility of the 70-kDa band was indistinguishable from that of penicillin-binding protein (PBP) 5 of membrane preparations of *E. hirae*. More recently, immunochemical and molecular studies clearly demonstrated that PBP 5 and the 70-kDa form of M-2 were separate and distinct proteins (Shockman *et al.*, these proceedings; Tepper *et al.*, in preparation). Because penicillin binding appears to be closely associated with penicillin-interactive, active-site serine proteins, such as β-lactamases, transpeptidases and carboxypeptidases (Ghuysen, 1991), binding to a muramidase seemed unusual. However, the derived amino acid sequence of cloned M-2 showed the presence of an SXXK sequence (S175NTK), as well as other amino acid motifs characteristic of penicillin-interactive proteins (Joris *et al.*, 1992; Shockman *et al.*, these proceedings). Recent biochemical studies of M-2 (Chu *et al.*, 1992; Kariyama and Shockman, 1992) made it possible to analyze penicillin binding to the extracellular form of M-2, decreasing the possibility of contamination of M-2 with membrane and cell wall proteins.

Some Unusual Properties of Penicillin Binding to Extracellular M-2

Figure 1 shows Coomassie blue staining (A) and fluorography (B) of gels after SDS-PAGE of various samples of supernatant culture medium which were incubated with [^{3}H]Pen. Lane 1 contains a sample of culture supernate concentrated by precipitation with 60% saturated ammonium sulfate. Note the strong radioactive band

Bacterial Growth and Lysis, Edited by M.A. de Pedro
et al., Plenum Press, New York, 1993

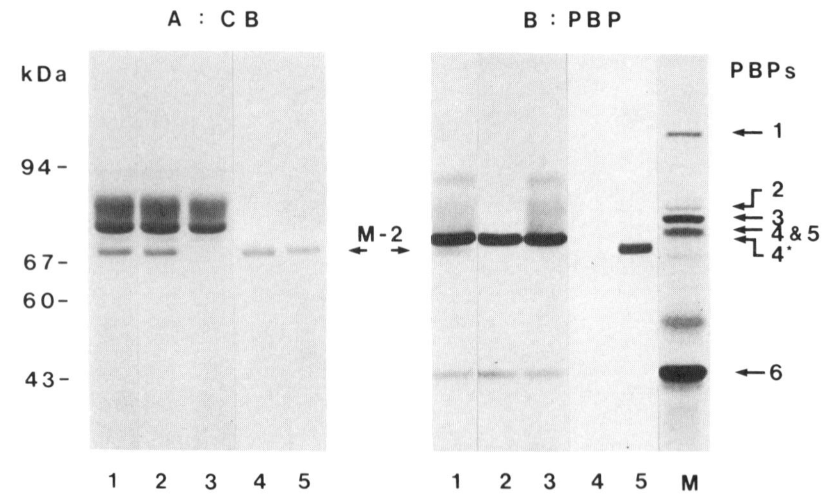

Figure 1. Some unusual properties of penicillin binding to extracellular M-2. Coomassie blue staining (A) and fluorography (B) of SDS-PAGE. Supernatant culture medium (100 ml) was separated from intact bacterial cells by centrifugation (20,000 x *g*, 10 min) during early stationary phase (1.43 mg/ml cellular dry weight). One of two 100-ml samples was treated with 20 mg *E. hirae* peptidoglycan, and kept at 0°C for 30 min. After removal of the peptidoglycan by centrifugation (20,000 x *g*, 20 min), this supernate was brought to 60% saturation with ammonium sulfate. Proteins in the sample shown in lanes 1 and 2 were immediately precipitated by the addition of ammonium sulfate to 60% saturation followed by centrifugation. The ammonium sulfate-precipitated proteins were dissolved in 2.5 ml 10 mM sodium phosphate, pH 7, and dialyzed four times in 10 mM sodium phosphate, pH 7. After dialysis, the precipitates were concentrated to 500 μl by using a Novacell (Filtron Technology Corp., Northborough, Mass.). The peptidoglycan-absorbed fraction (peptidoglycan-M-2 enzyme complex) was dissolved in 10 mM sodium phosphate, pH 7, transferred to two microcentrifuge tubes, and centrifuged (14,000 x *g*, 15 min, 4°C). The peptidoglycan-M-2 enzyme complex in one tube was suspended in 200 μl 10 mM sodium phosphate, pH 7, and a portion (50 μl) was used for the binding of [^{3}H]Pen. Protein was extracted with 250 μl 6 M guanidine-HCl from the peptidoglycan-M-2 enzyme complex in the other tube. After centrifugation, the supernate was dialyzed four times against 1 liter 10 mM sodium phosphate, pH 7, and then a portion (50 μl) was used for the binding of [^{3}H]Pen. Samples 1, 3, 4 and 5 (50 μl) were incubated at 37°C for 60 min with 2.8 μg of [^{3}H]Pen in a total volume of 70 μl. Sample 2 (50 μl) was incubated at 37°C for 60 min with one tenth the amount of [^{3}H]Pen (0.28 μg in a total volume of 70 μl). The procedure for fluorography was described by Kariyama *et al.* (1990) after SDS-PAGE on 7.5% acrylamide gel.

Lane 1, sample of the ammonium sulfate-precipitated supernatant culture medium.
Lane 2, the same as lane 1 but [^{3}H]Pen binding was performed with one tenth the concentration (4 μg/ml) of [^{3}H]Pen.
Lane 3, sample of ammonium sulfate-precipitated supernatant culture medium after exposure to peptidoglycan.
Lane 4, peptidoglycan-M-2 enzyme complex.
Lane 5, M-2 extracted with 6 M guanidine-HCl from peptidoglycan-M-2 enzyme complex.
Lane M, standard protoplast membrane preparation (Coyette *et al.*, 1978, 1980; Kariyama *et al.*, 1990).

The samples in each lane (lanes 1 to 5) were each derived from 10 ml of culture.

at about 75 kDa (Fig. 1B, lane 1) just above the less intense band at about 71 kDa (labeled M-2). This sample was labeled with a high concentration (40 μg/ml) of [^{3}H]Pen. Very little Coomassie blue-stained protein was observed at about 75 kDa (Fig. 1A, lane 1). Radiolabeling of this 75-kDa polypeptide was observed even after incubation with the relatively low concentration (4 μg/ml) of [^{3}H]Pen (Fig. 1B, lane 2). Similarly, the 43-kDa polypeptide, which corresponds to PBP 6 of *E. hirae*, was also labeled when incubated with 4 μg/ml of [^{3}H]Pen (Coyette *et al.*, 1978, 1980; Kariyama *et al.*, 1990). The 75-kDa polypeptide is considered to be PBP 4* based on its relatively high affinity for Pen G and its location on these gels, as described by Jacques *et al.* (1991). Lane 3 (Fig. 1A and 1B) contains a sample of culture supernate after it had been exposed to *E. hirae* peptidoglycan. M-2 tightly binds to the peptidoglycan and thereby is removed from culture supernate (Kariyama and Shockman, 1992). The culture supernate was concentrated by ammonium sulfate precipitation and incubated with 40 μg/ml of [^{3}H]Pen. Note the absence of both a [^{3}H]Pen-labeled band (Fig. 1B, lane 3) at 71 kDa and a corresponding Coomassie blue-stained band (Fig. 1A, lane 3), with all of the other radioactive and Coomassie blue-stained bands remaining present. The fraction that was bound to the insoluble peptidoglycan (peptidoglycan-M-2 enzyme complex) was incubated with 40 μg/ml of [^{3}H]Pen (lane 4). Note the presence of a Coomassie blue-stained band at about 71 kDa (Fig. 1A, lane 4) and the absence of [^{3}H]Pen-labeled bands (Fig. 1B, lane 4). (Prolonged exposure of the film did reveal a weak radioactive band at 71 kDa.) The sample placed in lane 5 was similar to that in lane 4, *except* that protein was first extracted from the peptidoglycan-M-2 enzyme complex with 6 M guanidine-HCl, centrifuged, the supernate dialyzed to refold extracted protein, and then the renatured protein was incubated with 40 μg/ml of [^{3}H]Pen. Note the presence of a single strongly radioactive band at 71 kDa (Fig. 1B, lane 5) corresponding to the single Coomassie blue-stained band at the same location (Fig. 1A, lanes 4 and 5).

Penicillin Binding to Reversed-Phase FPLC-Purified Extracellular M-2

Recently, we successfully purified extracellular M-2 using a new, simple and relatively efficient procedure (Chu *et al.*, 1992). As shown in Fig. 2, the peak of enzyme activity (eluted at 20% acetonitrile in 0.1% trifluoroacetic acid) corresponded to the major peak of absorbance at 280 nm. The intensity of Coomassie blue-stained protein (Fig. 2, CB), reactivity with anti-M-2 monoclonal antibody in Western blot (Fig. 2, WB), and binding of [^{3}H]Pen (Fig. 2, PBP) to the 71-kDa polypeptide present in fractions 10 to 13 all paralleled the elution of M-2 enzyme activity. Furthermore, [^{3}H]Pen bound to a sample of fraction 11 containing 2 μg of protein (about 1/10 of the amount shown in Fig. 2, PBP) was easily detected in fluorography, and [^{3}H]Pen binding was efficiently quenched by pre-exposure to excess cold Pen G (data not shown). [^{3}H]Pen binding to the protein in fraction 11 was slow (about 60 min for saturation, data not shown).

INTERPRETATION AND DISCUSSION

The 71-kDa polypeptide (previously identified as M-2 [Kariyama and Shockman, 1992]) present in ammonium sulfate concentrates of culture supernates did not bind [^{3}H]Pen very efficiently (Fig. 1A and 1B, lane 1). A large amount of protein, easily seen as a Coomassie blue-stained band (Fig. 1A, lane 1), resulted in a fairly weak radioactive band at 71 kDa (Fig. 1B, lane 1). The presence of the broad and diffuse polypeptide and [^{3}H]Pen-labeled bands at 75 to 95 kDa which, like M-2, have a low

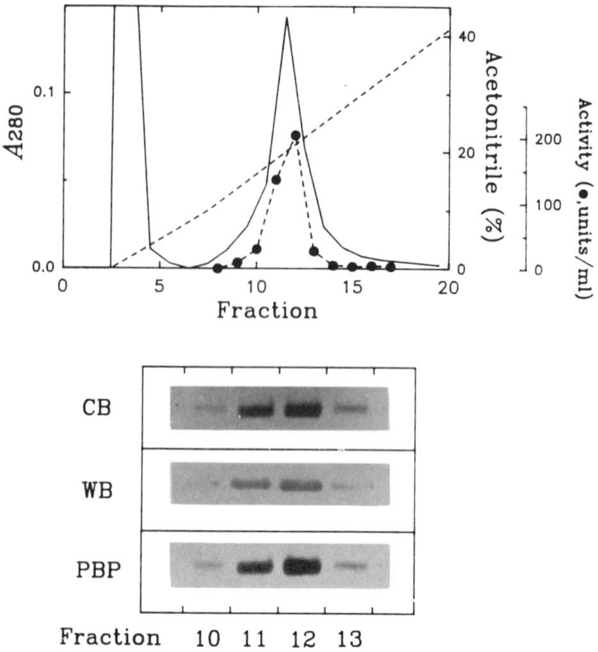

Figure 2. Penicillin binding to reversed-phase FPLC-purified extracellular M-2. Extracellular M-2 was purified as described by Chu *et al*, 1992. The guanidine-HCl extract (670 μg of protein, 580 units of M-2 activity) of the peptidoglycan-M-2 enzyme complex was injected into a ProRPC HR 5/10 column using the Pharmacia FPLC system equilibrated with 0.1% trifluoroacetic acid (TFA). The column was then eluted with a linear gradient of acetonitrile in 0.1% TFA, and elution was monitored by absorbance at 280 nm. Fractions 8 to 17 were evaporated to dryness and dissolved in 150 μl of 6 M guanidine-HCl, and 5- to 15-μl samples were assayed for M-2 activity. Portions (100 μl) of fractions 9 to 14 were dialyzed four times in 1 liter 10 mM sodium phosphate, pH 7. After dialysis, 35-μl samples of fractions 9 to 14 were incubated at 37°C for 60 min with 2.8 μg of [^{3}H]Pen in a total volume of 70 μl and then subjected to SDS-PAGE (7.5% acrylamide gel). This gel was stained with Coomassie blue and subjected to fluorography as previously described (Kariyama *et al.*, 1990). Separately, 5-μl samples of fractions 9 to 14 were subjected to SDS-PAGE. Proteins in the gels were transferred to Immobilon P membranes for Western blotting and detected with anti-M-2 monoclonal antibody 2F8 as described previously (Kariyama and Shockman, 1992).

Symbols:
—, absorbance at 280 nm of the column elute
----, acetonitrile gradient
●, M-2 activity of each 1-ml fraction
CB, Coomassie blue-stained bands of fractions 10 to 13
WB, Western blot-reactive bands of fractions 10 to 13 with anti-M-2 monoclonal antibody (2F8)
PBP, radioactive bands present in fractions 10 to 13

affinity for Pen G was observed (compare lanes 1 and 2 in Fig. 1B). Difficulty in observing [^{3}H]Pen bound to extracellular M-2 in culture supernates might be due to the presence of and penicillin binding by these abundunt proteins, and by PBP 4*, competing for [^{3}H]Pen, even at high concentrations of the antibiotic. It may also be possible that, in the concentrated culture supernates, not all of the M-2 molecules present possess fully exposed penicillin-binding sites, or that the penicillin binding sites are blocked, or partially blocked, by other extracellular substances. Furthermore, the ability of M-2 to bind penicillin was almost completely inhibited when M-2 was bound to peptidoglycan, suggesting that the penicillin-binding and peptidoglycan-binding sites are located near each other. The locations of the amino acid motifs that are considered to be characteristic of penicillin-interactive proteins (Ghuysen, 1991) are consistent with this idea (Fig. 1 in Shockman *et al.*, these proceedings) *if* the repeat sequences are indeed involved with the binding of M-2 to the peptidoglycan substrate. In contrast, penicillin-binding activity of M-2 from the same volume of culture supernate was much more pronounced when the enzyme was extracted from the peptidoglycan by unfolding followed by refolding (Fig. 1B, lane 5). The data from the reversed-phase FPLC purification experiments shown in Fig. 2 are also consistent with the specific binding of penicillin to M-2.

Although the derived amino acid sequence of cloned extracellular M-2 showed the presence of an SXXK sequence as well as other amino acid motifs characteristic of penicillin-interactive proteins, the shape and distribution of hydrophobic and hydrophilic clusters along the amino acid sequence of M-2 differ from those found in a variety of serine β-lactamases and PBPs (Joris *et al.*, 1992). Thus, extracellular M-2 may bind penicillin by a mechanism different from penicilloyl transfer involving an "active" serine residue. On the other hand, penicillin is known to bind to human or bovine serum albumin. As reported recently (Yvon *et al.*, 1989, 1990), penicillin binds to human serum albumin covalently by a mechanism which involves several lysine residues closely associated with serine residues. As discussed above, penicillin might not bind to enzymatically (muramidase) active M-2 present in the culture medium. Conformational changes may be required. The exact process and mechanism of M-2's refolding from the denatured form is not known. It remains possible that penicillin binding to M-2 requires exposure of normally "buried" sites by protein unfolding and refolding.

Acknowledgments

This work was supported by Public Health Service grant AI05044 from the National Institutes of Health. We thank P. Cassidy and A. N. Jones of Merck Sharp & Dohme Research Laboratories for the gift of [^{3}H]penicillin G. We thank G. Harvey for help in preparing this typescript.

REFERENCES

Chu, C.-P., Kariyama, R., Daneo-Moore, L. and Shockman, G.D. (1992) Cloning and sequence analysis of the muramidase-2 gene from *Enterococcus hirae*. J. Bacteriol. 174, 1619-1625.

Coyette, J., Ghuysen, J.-M. and Fontana, R. (1978) Solubilization and isolation of the membrane-bound DD-carboxypeptidase of *Streptococcus faecalis* ATCC 9790: properties of the purified enzyme. Eur. J. Biochem. 88, 297-305.

Coyette, J., Ghuysen, J.-M. and Fontana, R. (1980) The penicillin-binding proteins in *Streptococcus faecalis* ATCC 9790. Eur. J. Biochem. 110, 445-456.

Dolinger, D.L., Daneo-Moore, L. and Shockman, G.D. (1989) The second peptidoglycan hydrolase of *Streptococcus faecium* ATCC 9790 covalently binds penicillin. J. Bacteriol. 171, 4355-4361.

Ghuysen, J.-M. (1991) Serine β-lactamases and penicillin-binding proteins. Annu. Rev. Microbiol. 45, 37-67.

Jacques, P., El Kharroubi, A., Van Beeumen, J., Piras, G., Coyette J. and Ghuysen, J.-M. (1991) Mode of membrane insertion and sequence of a 32 amino acid peptide stretch of the penicillin-binding protein 4 of *Enterococcus hirae*. FEMS Microbiol. Lett. 82, 119-124.

Joris, B., Englebert, S., Chu, C.-P., Kariyama, R., Daneo-Moore, L., Shockman, G.D. and Ghuysen, J.-M. (1992) Modular design of the *Enterococcus hirae* muramidase-2 and *Streptococcus faecalis* autolysin. FEMS Microbiol. Lett., in press.

Kariyama, R. and Shockman, G.D. (1992) Extracellular and cellular distribution of muramidase-2 and muramidase-1 of *Enterococcus hirae* ATCC 9790. J. Bacteriol., in press.

Kariyama, R., Massidda, O., Daneo-Moore, L. and Shockman, G.D. (1990) Properties of cell wall-associated DD-carboxypeptidase of *Enterococcus hirae (Streptococcus faecium)* ATCC 9790 extracted with alkali. J. Bacteriol. 172, 3718-3724.

Yvon, M., Anglade, P. and Wal, J.-M. (1989) Binding of benzyl penicilloyl to human serum albumin: evidence for a highly reactive region at the junction of domains 1 and 2 of the albumin molecule. FEBS Lett. 247, 273-278.

Yvon, M., Anglade, P. and Wal, J.-M. (1990) Identification of the binding sites of benzyl penicilloyl, the allergenic metabolite of penicillin, on the serum albumin molecule. FEBS Lett. 263, 237-240.

SPECIFIC BINDING OF THE SOLUBLE LYTIC TRANSGLYCOSYLASE TO THE MUREIN SACCULUS OF *ESCHERICHIA COLI*

Tina Romeis and Joachim-Volker Höltje

Max-Planck-Institut für Entwicklungsbiologie
Abteilung Biochemie
Spemannstraße 35/II
7400 Tübingen, Germany

INTRODUCTION

Murein hydrolases, which are ubiquitous among bacteria (Ghuysen, 1968), represent a class of potentially autolytic enzymes and hence must be controlled strictly. However, little is known about the cellular control mechanism (Höltje and Tuomanen, 1991). This is in contrast to the detailed knowledge of the biochemistry of the murein synthesizing enzymes and their specific inhibition by a wide range of antibiotics. Due to this lack of information we are still far from understanding how the typical bacteriolytic response to inhibitors of murein synthesis is triggered (Tomasz, 1979).

The predominant murein hydrolytic enzyme in the soluble fraction of *E. coli* is a muramidase, called soluble lytic transglycosylase (Slt, Höltje *et al.*, 1975). It is an exo-enzyme which processively shortens the glycan strands from the N-acetylglucosamine-end by cleaving the ß-1,4-glycosidic bond between the muramic acid and the N-acetylglucosamine. Unlike lysozyme this cleavage is accompanied by an intramolecular transglycosylase reaction yielding a 1,6-anhydro-bond at the muramic acid residue of the released products (Höltje *et al.*, 1975). The gene has been cloned (Betzner and Keck, 1989) and its sequence is known (Engel *et al.*, 1991).

Interestingly, the Slt can be overproduced by a factor of about 20 without causing bacteriolysis (Betzner *et al.*, 1990). Thus even under these conditions, cellular control of the Slt can be achieved. In the following we would like to present experiments dealing with different aspects of the regulation of the Slt.

TRANSLOCATION OF THE SLT ACROSS THE CYTOPLASMIC MEMBRANE

Since the Slt is synthesized in the cytoplasm, the enzyme has to be exported into the periplasm to gain access to its substrate, the murein sacculus. A simple but efficient control mechanism would therefore be to regulate the translocation across the cytoplasmic membrane and to retain excess enzyme in the cytoplasm. However, it was known from the DNA sequence that the Slt is provided with a typical leader peptide (Engel *et al.*, 1991) and was likely to be transported via the classical sec-pathway (Wickner *et al.*, 1991).

Bacterial Growth and Lysis, Edited by M.A. de Pedro *et al.*, Plenum Press, New York, 1993

Translocation and processing of the Slt was studied by a pulse-chase experiment using the maxicell-system (Sancar *et al.*, 1979). After a 3 min pulse with radioactive methionine only the preform could be detected. During the chase a transition to the mature form took place indicating that the Slt was exported and processed. In the presence of 5mM inorganic azide maturation of the enzyme was inhibited. Since SecA is known to be sensitive towards azide (Fortin *et al.*, 1990), we concluded that the *sec*-components are indeed responsible for the translocation of the Slt.

To analyse this in some more detail we studied the export of the enzyme by pulse experiments in different *sec*-mutants. The two temperature sensitive mutants in *secA* (Oliver and Beckwith, 1981) and *secY* (Ito *et al.*, 1983) and a strain with a null-mutation

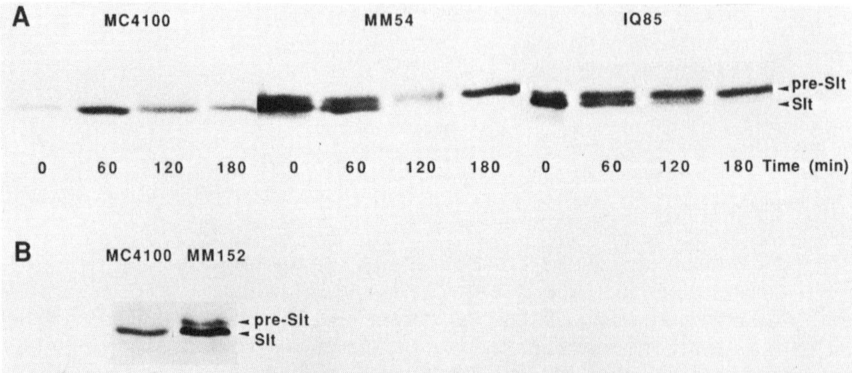

Fig.1. Accumulation of pre-Slt in various *sec*-mutants: MM54 ($SecA^{ts}$), IQ85 ($SecY^{ts}$), MM152 (SecB::Tn5) and MC4100 (control), containing pAB58, were grown in M9-glucose supplemented with 0.5% casamino acids at 30°C to early logarithmical growth phase and then shifted to 42°C (except for MM152 and the control). Aliquots of the cultures were taken at the indicated times after the temperature shift, pulse-labelled with ^{35}S-Met (25mCi) and immunmoprecipitated with Slt antiserum. Proteins were separated by SDS polyacrylamide (11%) gel electrophoresis and visualized by autoradiography.

in *secB* (Kumamoto and Beckwith, 1983) were transformed with the slt-carrying plasmid pAB58 (Betzner and Keck, 1989). The strains were grown in minimal medium at 30°C to an early logarithmical growth phase and were then shifted to 42°C, to induce the transport defect. At the times indicated in Fig. 1, samples were withdrawn, labelled with radioactive methionine, immunoprecipitated and finally separated by SDS polyacrylamide gel eletrophoresis. All mutants accumulated the Slt-preform in the cytoplasm at the restrictive temperature. Inhibition of translocation as compared with the wildtype could also be demonstrated in the *secB*-null-mutant (Fig. 1B). This indicates that SecA, SecY and SecB are involved in the export of the Slt. Hence, export of the Slt is by no means extraordinary.

SUBCELLULAR DISTRIBUTION OF THE SLT

Spheroplasts were prepared with EDTA-lysozyme in the presence of sucrose (Witholt *et al.*, 1979) to analyse the intracellular distribution of the enzyme in the different compartments of the cell. Wildtype *E. coli* MC4100, the pAB58 harboring strain and in addition cells, which had been induced for a strong overexpression of Slt using the T7-system (Tabor and Richardson, 1985) were compared. After centrifugation of the spheroplasts the supernatants containing the periplasmic fraction and the pellet containing the cell envelope and the cytoplasmic fraction of the cell were analysed by Western blotting after SDS gel electrophoresis.

The bulk of Slt was found to be located in the periplasm and only minor amounts of the enzyme remained in the pellet. This latter fraction was accessible to proteolytic degradation by pronase E and therefore could not have been present in the cytoplasm. Only with the induced T7-system was Slt-preform detected that was resistant against protease digestion indicating its location in the cytoplasm. But even in this system most of the Slt was found in the periplasm. We conclude that regulation of the Slt by a controlled export is not realized in the cell.

ENZYME STABILITY

The results so far indicate that even an excess of Slt is exported into the periplasm and becomes accessible to its substrate, the murein sacculus. Nevertheless, the bacteria do not lyse. One possible explanation would be that the enzyme activity is under control. In this respect it was of interest to find that in the Slt-overproducing strain the periplasmic enzyme fraction, which greatly surmounts the normal concentration of the wildtype, seems to be quite stable. In a pulse-chase experiment the amount of Slt-protein did not decline even after about four generations of continued growth.

However, after induction of the T7-system the enzymatic activity increased by almost a factor of ten to a maximum value and subsequently dropped, although the amount of Slt-protein in the cell remained at a high level (Fig. 2).

During induction of Slt-expression, bacterial growth slowed down but lysis did not occur. Therefore a decrease of enzyme activity due to bacteriolysis or to proteolytic degradation can be excluded. Dialysis of the sample did not restore activity (Fig. 2B). We therefore can rule out that enzyme inactivation was due to the accumulation of any unknown dialysable factor. Since the same drop in Slt activity was observed when rifampicin was added after inducing the T7-system, we also can exclude that an inhibitory protein component accumulated during the experiment.

IN VITRO SUBSTRATE BINDING STUDIES

Recently it has been shown by immunogold electron microscopy that the Slt is able to bind rather strongly to the murein sacculus (Walderich and Höltje, 1991). Although murein sacculi are prepared by boiling the bacterial cells in a 4% SDS-solution, the Slt could be detected with antibodies on isolated sacculi. However, an estimation of the amount of Slt that binds to murein reveals that this is only a very small portion, namely about one molecule out of thousand. It was tempting to speculate that the binding step of the Slt to the murein sacculus might be in context with the regulation of this enzyme. *In vitro* binding studies were performed to investigate this idea.

Isolated murein sacculi were preincubated with a protease to remove all bound protein components. These clean sacculi were then incubated with isolated Slt for a short time at 0°C. The sample was subsequently boiled in SDS-solution. The reconstituted

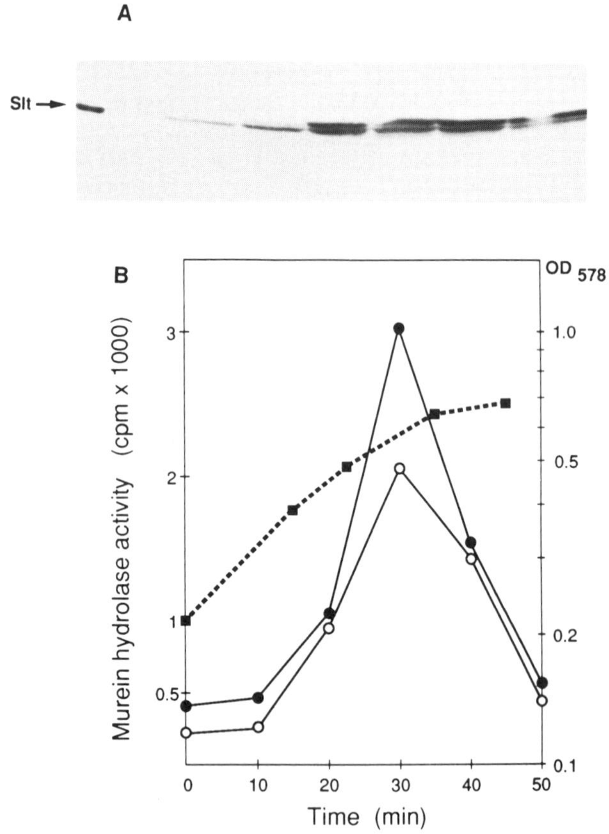

Fig.2. Overexpression of the Slt by a temperature-induced T7-system: Cells were grown exponentially at 30°C up to an OD_{578nm} of about 0.2 and then shifted to 42°C to induce T7-RNA-polymerase dependent Slt-overexpression. At the times indicated samples (25ml) were withdrawn, the cells pelleted, resuspended in 0.01 M TRIS-maleate, 0.01 M $MgSO_4$, pH 7.8 and mechanically broken with glass beads in a cell-mill.

A: Western blot analysis of the Slt protein: 60 µg of broken cell protein were separated on a 10% SDS polyacrylamide gel and transferred to nitrocellulose; the proteins were detected immunochemically using a Slt-antiserum and the peroxidase staining reaction.

B: Enzyme activity: Slt activity was determined from 30 µg protein of broken cells using $\{^3H\}$-A_2pm-labelled sacculi as a substrate (Höltje *et al.*, 1975). Enzyme activity was assayed before (o) and after (o) dialysis of the cell fractions. (o) OD_{578nm}.

murein was then pelleted and washed free from SDS. Before analyzing the material by SDS gel electrophoresis the intact murein sacculi were completely degraded by a muramidase.

Sacculi could indeed be reconstituted with Slt *in vitro*. The more Slt was added the more Slt was bound to the murein. Hovever, at high concentrations a saturation value was reached. This suggests that only a limited number of binding sites are available on the murein sacculus.

Since the Slt is an exo-muramidase, which processively degrades the sugar chains starting at the glucosamine-end (Beachey *et al.*, 1981), these ends might represent the binding-site. Consistent with this speculation, a gentle preincubation of sacculi for a short time at 0°C with an endo-muramidase, namely lysozyme, that increases the number of glucosamine ends, indeed enhanced the amount of Slt that could bind to the murein sacculi (Fig. 3).

CONCLUSION

Only a limited number of a specific recognition sites for the binding of Slt seems to be present in the murein sacculus. The binding site is likely to be the glucosamine end of the glycan strands. However, it is not known what kind of additional structural requirements have to be met for an efficient adsorption of the enzyme. Recognition of these specific sites that facilitate binding to the substrate may protect the enzyme from inactivation and degradation. It is possible that only the small fraction of Slt which binds to the murein, remains in an active state, while the bulk of Slt that cannot adsorb becomes inactivated. This could explain the decrease in enzyme activity that was observed after overexpression of the Slt in the T7-system.

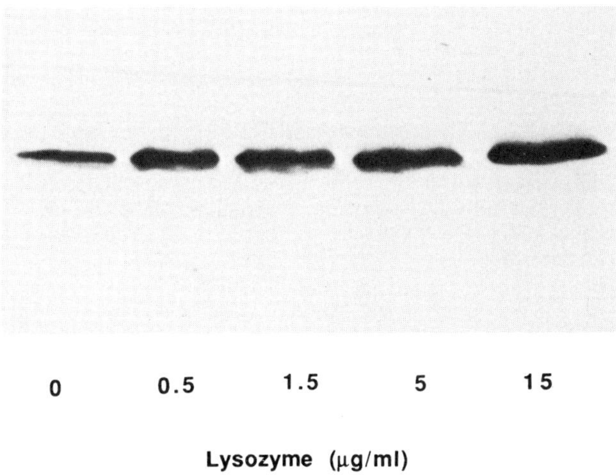

Fig.3. Western blot analysis of the *in vitro* binding of Slt to isolated murein sacculi: Protease treated murein sacculi (275µg) were pre-incubated with increasing amounts of lysozyme (as indicated) for 5min at 0°C. This was followed by the addition of isolated Slt (20µg) and incubation at 0°C was continued for another 5 min. The samples were boiled for 15 min in 1% SDS-solution, the murein was pelleted and washed free from SDS by ultracentrifugation. The whole sacculi were digested with cellosyl (20µg/ml) prior to separation by 10% SDS polyacrylamide gel electrophoresis. After transfer of the proteins onto nitrocellulose the Slt was visualized by means of a specific Slt-antiserum and a peroxidase staining-reaction.

Binding of the Slt to the murein might then be the step by which the action of this murein hydrolase can be controlled. According to this model even greatly increased amounts of Slt could be handled by the cell without risking murein dissolution. The chemistry of the murein, which reflects the physiological state of the cell, might determine the amount of murein hydrolases which can bind to the murein sacculus. Potentially autolytic enzymes might then be modulated corresponding to the need of the cell.

REFERENCES

Beachey, E.H., Keck, W., de Pedro, M.A. and Schwarz, U. (1981) Exoenzymatic activity of transglycosylase isolated from *Escherichia coli*. Eur. J. Biochem. 116, 355-358.

Betzner, A. and Keck, W. (1989) Molecular cloning, overexpression and mapping of the slt gene encoding the soluble lytic transglycosylase of *Escherichia coli*. Mol. Gen. Genet. 219, 489-491.

Betzner, A.S., Ferreira, L.C.S., Höltje, J.-V. and Keck, W. (1990) Control of the soluble lytic transglycosylase by the stringent response in *Escherichia coli*. FEMS Microbiol Lett. 67, 161-164.

Engel, H., Kazemier, B. and Keck, W. (1991) Murein-metabolizing enzymes from *Escherichia coli*: sequence analysis and controlled overexpression of the *slt* gene which encodes the soluble lytic transglycosylase. J. Bacteriol. 173, 6773-6782.

Fortin, Y., Phoenix, P. and Drapeau, G.R. (1990) Mutations conferring resistance to azide in *Escherichia coli* occur primarily in the *sec A* gene. J. Bacteriol. 172, 6670-6610.

Ghuysen, J.-M. (1968) Use of bacteriolytic enzymes in determination of wall structure and their role in cell metabolism. Bacteriol. Rev. 32, 425-464.

Höltje, J.-V. and Tuomanen, E. J. (1991) The murein hydrolases of *Escherichia coli*: properties, functions and impact on the course of infections *in vivo*. J. Gen. Microbiol. 137, 441-454.

Höltje, J.-V., Mirelman, D., Sharon, N. and Schwarz, U. (1975) Novel type of murein transglycosylase in *Escherichia coli*. J. Bacteriol. 124, 1067-1076.

Ito, K., Wittekind, M., Nomura, M., Shiba, K., Yura, T., Muira, A. and Nashimoto, H. (1983) A temperature-sensitive mutant of *E. coli* exhibiting slow processing of exported proteins. Cell 32, 789-797.

Kumamoto, C.A. and Beckwith, J. (1983) Mutations in a new gene, *sec B*, cause defective protein localization in *Escherichia coli*. J. Bacteriol. 154, 253-260.

Linnett, O.E. and Beechey, O.D. (1979) Inhibitors of the ATP synthetase system. Methods Enzymol. 55, 474-518.

Oliver, D. and Beckwith, J. (1981) *E. coli* mutant pleiotropically defective in the export of secreted proteins. Cell 25, 765-772.

Sancar, A., Hack, A.M. and Rupp, W.D. (1979) Simple method for identification of plasmid-coded proteins. J. Bacteriol. 137, 692-693.

Tabor, S. and Richardson, C.C. (1985) A bacteriophage T7 RNA polymerase/promoter system for controlled exclusive expression of specific genes. Proc. Natl. Acad. Sci. USA 82, 1074-1078.

Tomasz, A. (1979) The mechanism of the irreversible antimicrobial effect of penicillins: how the beta lactam antibiotics kill and lyse bacteria. Annual Rev. Microbiol. 33, 113-137.

Walderich, B. and Höltje, J.-V. (1991) Subcellular distribution of the soluble lytic transglycosylase. J. Bacteriol. 173, 5668-5676.

Wickner, W., Driessen, A.J.M. and Hartl, F.-U. (1991) The enzymology of protein translocation across the *Escherichia coli* plasma membrane. Annu. Rev. Biochem. 60, 101-124.

Witholt, B., Boekhout, M. Brock, M. Kingma, J. van Heerikhuizen, H. and de Leij, L. An efficient and reproducible procedure for the formation of spheroplasts from variously grown *Escherichia coli*. Anal. Biochem. 74, 160-170.

INHIBITION OF AN AUTOLYSIN TOGETHER WITH PBP3 CAUSES THE FORMATION OF BULGES: IDENTIFICATION OF THE SOLUBLE LYTIC TRANSGLYCOSYLASE IN *E. COLI* AS THE SPECIFIC TARGET OF BULGECIN

Markus F. Templin[1], David H. Edwards[2] and Joachim-Volker Höltje[1]

[1]Max-Planck-Institut für Entwicklungsbiologie, Abteilung Biochemie, Spemannstr. 35/II, 7400 Tübingen, Germany

[2]Department of Molecular Biology, University of Edinburgh, King's Building, Mayfield Road, Edinburgh EH9 3JR, Scotland

INTRODUCTION

A coordinated action of murein synthetases and murein hydrolases is assumed to be essential for proper growth of the bacterial cell. A disturbance of this balance, for example by blocking parts of the synthetic pathway turns the hydrolases into autolysins which then cause lysis of the cell (Tomasz, 1979).

Despite the fact that some of these autolytic enzymes are well characterized on a biochemical level, information about their *in-vivo* function is rather limited (for a review: Höltje and Tuomanen, 1991). The soluble lytic transglycosylase (Slt), the prominent hydrolytic activity in the soluble fraction of *E. coli*, is one of these well characterized enzymes. This muramidase is an exo-enzyme which acts on the glycan strands of the murein and cleaves off disaccharides with an unusual 1,6-anhydro group in the muramic acid (Höltje *et al.*, 1975). The gene has been cloned and sequenced (Betzner and Keck, 1989; Engel *et al.*, 1991), and efforts are under way to determine its three-dimensional structure by X-ray crystallography (Rozeboom *et al.*, 1990).

We were interested in finding out what would happen if the balance of the two enzyme systems, synthetases and hydrolases becomes disturbed by blocking the hydrolases. An obvious way to address such a question is a genetic one; in particular by trying to obtain proper mutants. For the Slt different approaches had been used, but none of them was successful.

RESULTS

Isolation of a deletion in the *slt* gene

With the introduction of a method developed for deleting essential genes, a useful system for obtaining a deletion mutant was available (Hamilton *et al.*, 1989). This approach is used

Bacterial Growth and Lysis, Edited by M.A. de Pedro *et al.*, Plenum Press, New York, 1993

to construct mutants with a thermo-null phenotype for the gene of interest. In the course of mutagenesis the chromosomal copy is destroyed while a complementing, plasmid-borne copy is generated. After a temperature-shift to 42°C plasmid replication stops, the complementing copy is lost and the mutant phenotype will appear.

To our surprise such a thermo-null mutant for the slt turned out to be without any obvious phenotype at the restrictive temperature. Southern blot analysis of the obtained strain confirmed the absence of the gene. In the mutant, the coding region for the Slt had been totally replaced by the antibiotic resistance marker for neomycin.

Phenotype of the slt deletion mutant

Assay for murein hydrolytic activity in the soluble fraction showed a massive drop in activity (Table 1). Only residual activity was found for the mutant, which is caused by other murein degrading enzymes.

Table 1. Murein hydrolytic activity in the soluble fraction.

Strain	specific hydrolytic activity (units/mg)
MUF16 (deletion mutant)	< 0.1
MC1061 (parent strain)	2.90

Murein hydrolytic activity was determined in the soluble fraction of crude cell extracts (30µg of protein) for 30 min at 37° under standard assay conditions (Höltje *et al.*, 1975).

Also it was of interest to find out whether the loss of the Sl transglycosylase shows any effect on the structure of the murein. Therefore the muropeptide composition of the slt deficient mutant and the parent strain were determined according to the method of Glauner (1988). The differences found were minor. Significant alterations in the muropeptide composition of the mutant could not be detected. However, it is interesting to note that the amount of muropeptides containing an anhydrogroup was found to be decreased for the *slt* deletion mutant (data not shown), indicating an increase in the average length of the glycan strands.

Although growth rate and morphology of the cells remained unchanged, when testing the effects of different cell wall specific antibiotics, it turned out that the mutant had changed its response to some of them. Most prominent was a change in the MIC for Mecillinam, a PBP2-specific ß-lactam, which was lowered to almost a 1/10 of the value determined for the wildtype.

However, the most striking effect was a morphological one, seen in the presence of antibiotics specific for PBP3, such as furazlocillin or azthreonam which are known to transform wild type cells into long filaments. Despite the fact that the MICs were not altered significantly towards these antibiotics a peculiar morphology of the cells could be seen. After about one generation time in presence of the antibiotic the mutant developed into elongated cells with a bulge in the middle. These cells survived at least another generation time before they started to lyse rapidly.

The very same unusual shape had been described for different situations before. Bulges are well known to be the result of a simultaneous inhibition of both, PBP2, which is involved in cell elongation and PBP3, which catalyses septum formation (Spratt, 1975). Although it is not known whether the Sl-transglycosylase takes part in cell elongation or cell division, increased sensitivity of the deletion mutant towards mecillinam may be taken as evidence for a cooperation of this hydrolytic enzyme with PBP2 in the elongation process. The morphological alterations of the *slt* deletion mutant in the presence of furazlocillin would then be the consequence of a disturbance of PBP2-specific metabolic steps in addition to inhibition of PBP3.

Interestingly, bulge formation was also observered when PBP3-specific ß-lactams were applied in combination with a class of antibiotics, called bulgecins (Imada *et al.*, 1982). Bulgecin alone does not affect growth and its target was not known up to now. An interaction with the penicillin binding proteins had been ruled out (Nakao *et al.*, 1986). It was tempting to speculate that it might inhibit the slt and thereby create an analogous situation to the Slt deletion mutant, with the result of bulge formation in the presence of PBP3-specific ß-lactams. This speculation is even more tempting because the chemical structure of bulgecin, an N-acetylglucosamine substituted with an unusual proline derivative (Shinagawa *et al.*, 1985), shows some structural analogy to muropeptides which, in turn are the subunits from which murein is made.

Inhibition of the Slt by bulgecin

In vitro Slt enzyme assays indeed showed that bulgecin inhibits the soluble lytic transglycosylase. As shown in Figure 2, already very low concentrations of this antibiotic were sufficient to block enzyme activity. Fifty percent inhibition could be obtained at a concentration of about 250 ng/ml, which is less then half a micromole of bulgecin. A test with a crude preparation of solubilized membrane bound lytic transglycosylase (Mett *et al.*, 1980) showed no inhibition, indicating that the inhibition of Slt is specific.

Detailed studies on the mode of enzyme inhibition were done by velocity measurements at various substrate levels and different inhibitor concentrations. The kinetics obtained, revealed that bulgecin inhibits the transglycosylase in a non-competetive manner. This indicates that bulgecin does not act on the active center of the enzyme. Hence, it exhibits its effect not simply by competing with the substrate or product for this site.

DISCUSSION

Since bulgecin shows some structural analogy to muropeptides, it seems at least possible, that the enzyme is capable of recognizing specific murein substructures through this bulgecin-binding site. This would then be reminiscent to other hydrolytic enzymes, which bind to macromolecular substrates via specific adsorption sites. Examples of such enzymes, consisting not only of a catalytic domain, but also of a second domain responsible for the binding to the substrate, are different chitinases and cellulases (Gilkes *et al.*, 1991). One may also think of the N-acetylmuramyl-L-alanine amidase of *Streptococcus pneumoniae* (Höltje *et al.*, 1975; Giudicelli and Tomasz, 1984).

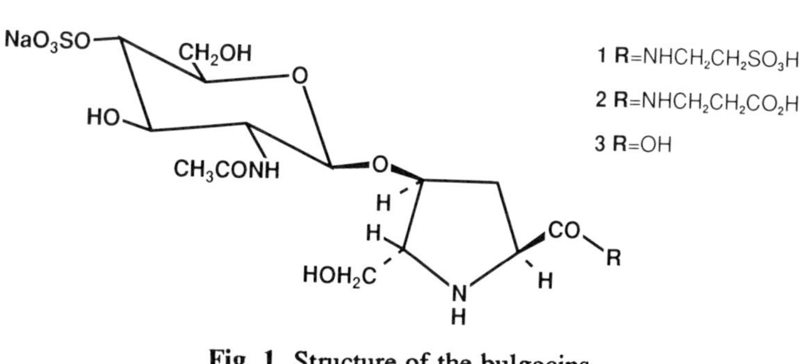

Fig. 1. Structure of the bulgecins

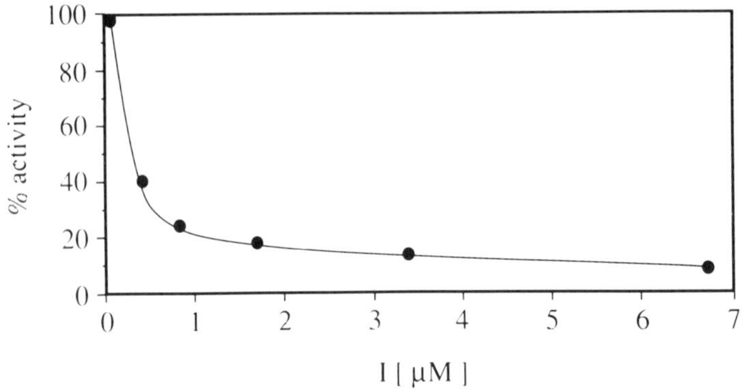

Fig. 2. Inhibition of the soluble lytic transglycosylase by bulgecin.
Bulgecin was added to the enzyme assay in the amounts indicated on the abscissa. Enzyme assays were performed as described by Höltje *et al.* (1975). The inhibition of the enzyme reaction is expressed in % of an untreated control.

The most intriguing aspect of the results presented may be the finding that specific inhibition of a murein hydrolase results in a synergistic effect when combined with certain classical murein synthesis inhibitors. Hence, murein hydrolases may be considered an interesting target for a novel class of antimicrobial agents.

REFERENCES

Betzner, A.S. and Keck, W. (1989) Molecular cloning, overexpression and mapping of the slt gene encoding the soluble lytic transglycolsylase of *Escherichia coli.* Mol. Gen. Genet. 219, 489-491.

Engel, H., Kazemier, B. and Keck, W. (1991) Murein-metabolizing enzymes from *Escherichia coli*: sequence analysis and controlled overexpression of the *slt* gene which encodes the soluble lytic transglycosylase. J. Bacteriol. 173, 6773-6782.

Gilkes, N.R., Henrissat, B., Kilburn, D.G., Miller, JR., R.C., Warren, R.A.J. (1991) Domains in microbial ß-1,4-glycanases: sequence conservation, function, and enzyme families. Microbiol. Rev. 55, 303-315.

Giudicelli, S. and Tomasz, A. (1984) Attachment of pneumococcal autolysin to wall teichoic acids, an essential step in enzymatic wall degradation. J. Bacteriol. 158, 1188-1190.

Glauner, B. (1988) Separation and quantitation of muropeptides with high performance liquid chromatography. Analyt. Biochem. 172, 451-464.

Hamilton, C.M., Aldea, M., Washburn, B.K., Babitzke, P. and Kushner, S.R. (1989) New Method for Generating Deletions and Gene Replacements in *Escherichia coli.* J. Bacteriol. 171, 4617-4622.

Höltje, J.-V., Mirelman, D., Sharon, N. and Schwarz, U. (1975) Novel type of murein transglycosylase in *Escherichia coli.* J. Bacteriol. 124, 1067-1076.

Höltje, J.-V. and Tomasz, A. (1975) Specific recognition of choline residues in the cell wall teichoic acid by the N-actylmuramyl-L-alanine amidase of pneumococcus. J. Biol. Chem. 250, 6072-6076.

Höltje, J.-V. and Tuomanen, E.I. (1991) The murein hydrolases of Escherichia coli - properties, functions and impact on the course of infections in vivo. J. Gen. Microbiol. 137, 441-454.

Imada, A., Kintaka, K., Nakao, M. and Shinagawa, S. (1982) Bulgecin, a bacterial metabolite which in concert with ß-lactam antibiotics causes bulge formation. J. Antibiot. 35, 1400-1403.

Mett, H., Keck, W., Funk, A., and Schwarz, U. (1980) Two different species of murein transglycosylase in Escherichia coli. J. Bacteriol. 144, 45-52.

Nakao, N., Yukishige, K., Kondo, M. and Imada, A. (1986) Novel morphological changes in gram-negative bacteria caused by combination of bulgecin and cefmenoxime. Antimicrob. Agents Chemother. 30, 414-417.

Rozeboom, H.J., Dijkstra, B.W., Engel, H. and Keck, W. (1990) Crystallization of the soluble lytic transglycosylase from *Escherichia coli*.K12. J. Mol. Biol. 212, 557-559.

Shinagawa, S., Maki, M., Kintaka, K., Imada, A. and Asai, M. (1985) Isolation and characterisation of bulgecins, new bacterial metabolites with bulge-inducing activity. J. Antibiot. 38, 17-23.

Spratt, B.G. (1975) Distinct penicillin binding proteins involved in the division, elongation, and shape of *Escherichia coli* K12. Proc. Natl. Acad. Sci. USA 72, 2999-3002.

Tomasz, A. (1979) The mechanism ot the irreversible antimicrobial effects of penicillins: How the beta lactam antibiotic kill and lyse bacteria. Annual Rev. Microbiol. 33, 113-137.

THE CELL WALL OF *BACILLUS SUBTILIS* IS PROTONATED DURING GROWTH

M.A. Kemper and R.J. Doyle

Department of Microbiology and Immunology
University of Louisville
Louisville, KY 40292

INTRODUCTION

The cell wall of *Bacillus subtilis* Marburg is a thick mat of peptidoglycan, teichoic acid and adventitiously bound proteins. The wall is 25-30 nm thick (Beveridge and Murray, 1979) and is composed of approximately equal amounts, by weight, of peptidoglycan and α-D-glucosylated teichoic acids. The strategy for the assembly of the wall is now beginning to be understood. The currently prevailing view is that wall precursors are exported through the cytoplasmic membrane, only to be covalently attached to pre-existing wall nearest the membrane. Upon insertion into the inner wall face, the young peptidoglycan strands are crosslinked into the wall mat. Once crosslinking has occurred, the newly-added wall components assume pressure and stretch. Further stretching occurs when even more new peptidoglycan strands have been inserted and crosslinked. The concept that the wall is a stress-bearing and stretchable covering of the bacterium was first proposed by Koch *et al.* (1982) in what is now referred to as the surface stress theory. The surface stress theory can be applied to the elongation of the cylindrical portion of the wall of *B. subtilis* (Mobley *et al.* 1984; Koch and Doyle, 1985; Kemper *et al.*, 1988). The addition of new wall at the inner wall face, the crosslinking of the newly-added wall and the subsequent addition of even more wall results in surface expansion of the organism. Very old wall (one generation is old) is the most highly stressed part of the surface fabric. The old wall must undergo mechanical scission or autolysis so as to alleviate the surface tension. In autolysin-sufficient *B. subtilis,* enzymatic cleavage of the old wall results in turnover or solubilization of wall fragments (Pooley, 1976a,b; Doyle *et al.*, 1988). In autolysin-deficient mutants or in wild-type cells grown under conditions to limit autolysin activity, the external faces of the walls are rough or serrated (Sonnenfeld *et al.*, 1985). Wall turnover seems to be a natural result of cell growth in *B. subtilis*. Wall newly assembled and crosslinked is resistant to autolysins during growth. Autolysins (the N-acetyl-muramoyl-L-alanine amidase, or more briefly the amidase) are secreted along with other

Bacterial Growth and Lysis, Edited by M.A. de Pedro
et al., Plenum Press, New York, 1993

extracellular proteins of *B. subtilis*. The amidase (approximately 35 kDa) is unable to assume an active conformation until it is fully exteriorized. It should be pointed out that each daughter of a unit cell generally possesses only one-fourth of the cell wall of the parent. In wild-type *B. subtilis* fifty percent of the wall is turned over or shed into the growth medium during each division. Cell poles tend to be more resistant to turnover than cylinders (Doyle *et al.*, 1988). The main focus of this paper is to describe the relationship between autolysin activity and the energized membrane. We take the view that energized membrane maintains a low pH in the wall matrix, thereby preventing peptidoglycan hydrolysis because the hydrogen ion concentration is too high.

ENERGIZED MEMBRANE AND AUTOLYSIN FUNCTION

In 1981, Jolliffe *et al.* reported that agents which dissipated the energized membrane resulted in the initiation of autolysis in *B. subtilis*. Metabolic poisons, such as azide, cyanide, gramicidin, ionophores and carbonyl-cyanide-m-chlorophenylhydrazone (CCCP), when added to exponential cultures of *B. subtilis,* caused cellular autolysis. In contrast, metabolic inhibitors, such as chloramphenicol, puromycin, actinomycin D and cerulenin, were unable to promote lysis. Furthermore, purging of cultures with nitrogen or removal of the carbon source, also resulted in autolysis. Jolliffe *et al.* (1981) proposed that the energized membrane influenced autolysin activity, but did not offer an explanation how the energized membrane could modify the activity of an external enzyme. Fig. 1 shows that when azide anion is added to an exponential culture of *B. subtilis* SR22, lysis ensues. When the azide is washed out, the cells then commence normal growth processes following a lag. The apparent resistance of the cells to autolysis during normal growth reflects the strong control of autolysins by energized membrane and/or other factors. Table 1 shows some of the other means reported to cause lysis in *B. subtilis*. Antibiotics, such as nafcillin, cause lysis by preventing peptidoglycan assembly. Bacteriophages and lysozyme hydrolyze pre-existing peptidoglycan. Svarachorn *et al.* (1991) have reported that cold shock activates autolysins, but the activation may have been through loss of protonmotive force. The other agents listed in Table 1 which induce autolysis in *B. subtilis* do so by interfering with proton translocation across the cytoplasmic membrane. Anaerobiosis and starvation for carbon both limit the ability of *B. subtilis* to produce protons. Furthermore, agents which reduce or destroy the ion gradients in membranes, such as ionophores, also cause lysis of *B. subtilis* because the ionophores eventually prevent proton production.

LIPOTEICHOIC ACIDS AND THE REGULATION OF AUTOLYSIN IN *BACILLUS SUBTILIS*

Several reports have described the inhibition of autolytic enzymes by lipoteichoic acids (reviewed by Doyle and Koch, 1987). Based on *in vitro* assays, LTA and phospholipids can prevent the catalytic activities of several autolysins, including those from streptococci and staphylococci (Daneo-

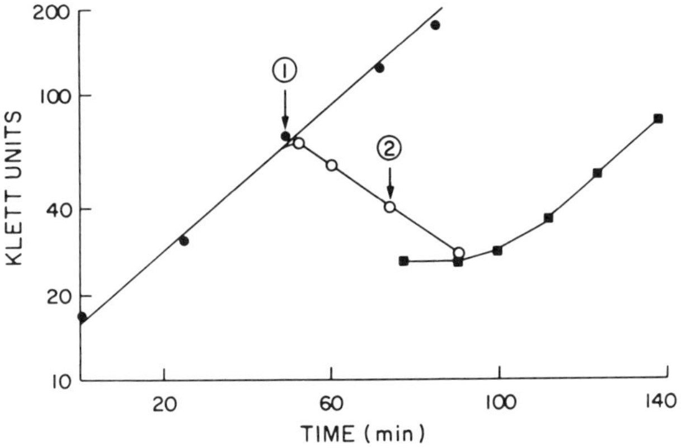

Fig. 1. Induction of autolysis in *Bacillus subtilis* SR22 by sodium azide. Sodium azide (100 mM) was added to an exponential culture in Penassay broth. Turbidity changes were monitored by use of a Klett photometer. Arrow 1, addition of azide; arrow 2, a suspension was rapidly filtered, washed with warm Penassay broth and suspended in a fresh broth.

Table 1. Induction of lysis in *Bacillus subtilis*.

Antibiotics	Prevent peptidoglycan synthesis
Bacteriophage	Lysis from within
Cold shock	Probable loss of protonmotive force
Fatty acids	Probable loss of protonmotive force
Ionophores	Ultimate loss of protonmotive force
Lipid-soluble cations	Ultimate loss of protonmotive force
Lysozyme	Degradation of peptidoglycan
Purge with nitrogen	Loss of protonmotive force
Starvation for carbon	Loss of protonmotive force
"Uncoupling" agents	Loss of protonmotive force

Moore and Shockman, 1979). When homologous LTA was added to an exponential culture of *B. subtilis*, there was no change in the rate of cell wall turnover (Fig. 2). In addition, exogenously-added autolysin was unable to promote autolysis or increased turnover. Fig. 3 shows that LTA has no effect on azide-induced autolysis of *B. subtilis*.

STRATEGY FOR SHOWING HOW ENERGIZED MEMBRANE REGULATES AUTOLYTIC ACTIVITY IN *BACILLUS SUBTILIS*

During metabolism, we view the wall as existing largely in a protonated state. The solvent in the wall would have a pH dictated by the pKa's of proton-binding sites, such as carboxylate and phosphate. As long as the cell is secreting

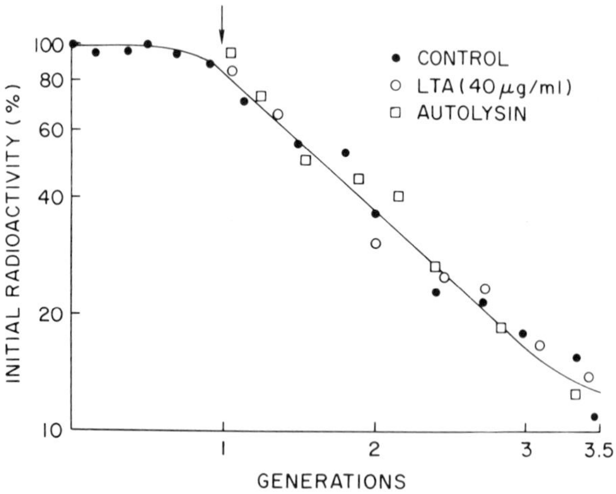

Fig. 2. Turnover of cell wall of *Bacillus subtilis* in the presence and absence of exogenously added lipoteichoic acid (LTA) and autolysin. Turnover experiments were as described by Jolliffe *et al.* (1980). Lipoteichoic acid was prepared according to Thaniyavarn *et al.* (1979). Autolysin was from a 3 M lithium chloride extract of *B. subtilis* cell walls. The dialyzed extract was used at a 100 μg/ml protein concentration and was free of proteolytic activity.

protons, the wall environment would have a pH lower than that found in most growth media. When the cell ceases to produce protons, the positive charges are neutralized and the wall environment assumes a pH equivalent to that of the medium. As one test for the premise that the wall matrix pH is relatively low, it would be predicted that positively charged ions and polymers would bind more avidly to cells in which the protonmotive force had been dissipated. In Fig. 4, it is shown that cationized ferritin (CF), a highly basic protein, readily aggregates *B. subtilis* following the addition of azide. The change in turbidity cannot be ascribed to lysis as lysis was minimal due to the conditions of the experiment.

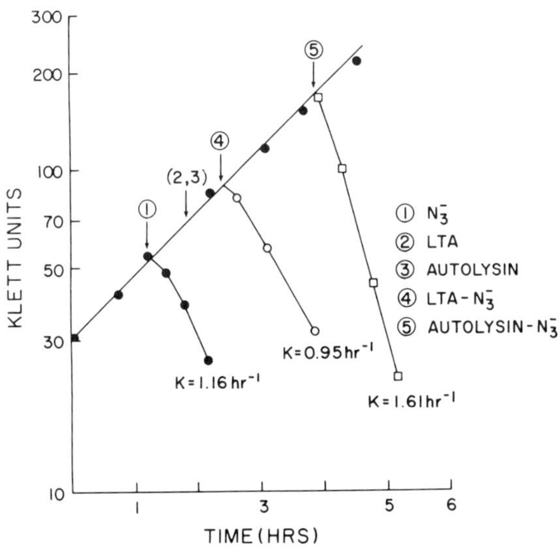

Fig. 3. Autolysis of *Bacillus subtilis* SR22 in the presence and absence of lipoteichoic acid and exogenous autolysin. To an exponential culture in Penassay broth was added various combinations of azide (100 mM), LTA (40 μg/ml) and crude autolysin (100 μg/ml). Rate constants shown on the figure are from first-order plots. Additions were made at the arrows.

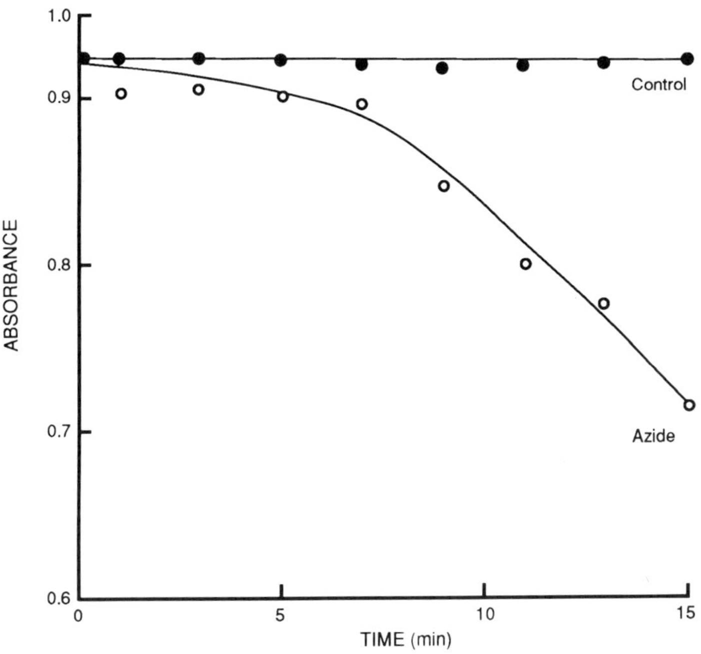

Fig. 4. Aggregation of *Bacillus subtilis* 168 by cationized ferritin. A washed suspension of *B. subtilis* in 50 mM HEPES buffer containing 1.0 mg/ml glucose was mixed with a final concentration of 1.0 mg/ml cationized ferritin. The suspension was divided and sodium azide (100 mM) added to one portion. Turbidity changes were monitored on a spectrophotometer at 20°. Closed circles, control cells without azide; open circles, azide-treated cells. Independent measurements showed that lysis of the cells by sodium azide was no more than 10 per cent under the conditions employed in the experiment.

When azide was added to autolysin-supplemented cells, the rate of autolysis exceeded that of the cells without added autolysin. We conclude that *in vivo*, LTA has no detectable effect on autolysin activity. Furthermore, exogenously added autolysin must be under some kind of control as it cannot induce spontaneous lysis in the bacteria.

In unpublished work (Kemper, Beveridge, Koch and Doyle) it has been shown that the walls of *B. subtilis* take up increased amounts of scandium ion or uranyl ion when the protonmotive force is destroyed by uncoupling agents. In further experiments, Kemper and Doyle (unpublished) have shown that carbonylcyanide-m-chlorophenylhydrazone (CCCP) induces a loss of the binding of chromate, a negatively-charged anion. Anion binding would be expected to decrease if the anions tend to complex with oppositely charged sites. Kemper *et al.* (1991) (Abstracts, Annu. Mtg, ASM) have also shown that levansucrase is inactive during its migration through the wall. Only when the levansucrase becomes fully exteriorized (approximately one generation is required) is it able to convert sucrose into levan. Finally, fluorescein diacetate binds much better to *B. subtilis* cells, as observed by fluorescence intensity, when the protonmotive force has been dissipated. The results, taken together, provide significant support for the view that the energized membrane is a component in the regulation of the major autolysin of *B. subtilis*.

A question in these studies is how universal is the capacity of the energized membrane to control cellular autolysis. It has been found that the potent uncoupling agent, carbonylcyanide-m-chlorophenylhydrazone, promotes autolysis in several bacteria (Table 2). *Branhamella catarrhalis* and *Neisseria gonorrhoeae*, closely-related Gram-negative cocci, rapidly lyse in the presence of CCCP. *Listeria monocytogenes,* a Gram-positive rod, *Pseudomonas aeruginosa*, a Gram-negative rod and *Staphylococcus epidermidis*, a Gram-positive coccus, appear to undergo partial lysis in the presence of CCCP. In contrast, *Bacillus cereus, Enterococcus hirae and Escherichia coli* seem to be refractory to CCCP. Gram-positive cells have much more cell wall with which to bind secreted protons than Gram-negative bacteria. It is surprising therefore, that several Gram-negative cells lysed upon the addition of the uncoupling agent. It is unknown if Gram-negative cells can maintain a protonated state in their periplasmic spaces during metabolism.

If our reasoning is correct, it must be considered that every enzyme secreted by *B. subtilis* (and possibly other bacteria as well) is resistant to the effects of low pH (the wall matrix is predicted to exhibit a pH of <5 >3 during active metabolism) (Kemper and Doyle, unpublished). Preliminary experiments have shown that several extracellular enzymes of *B. subtilis*, including amylase, amidase, levansucrase and subtilisin, are refractory to pHs as low as 3-4. We view the *B. subtilis* cell as a protoplast surrounded by wall bound protons.

Recent experimental results have shown that the protonmotive force has a role in bacterial adhesion. *P. aeruginosa* produces two cytoplasmic lectins, PA-I and PA-II, during stationary phase. When cultures of *P. aeruginosa* are purged with nitrogen or subjected to uncoupling agents, a portion of the cells lyse, releasing PA-I and PA-II. The released lectins can then bind any remaining intact *P.*

Table 2. Bacteria which undergo lysis or turbidity changes in the presence of carbonylcyanide-m-chloro-phenylhydrazone.[1]

Bacterium	Extent of autolysis
Bacillis cereus	< 10%
Bacillus subtilis	> 80%
Branhamella catarrhalis	> 80%
Enterococcus hirae	< 5%
Escherichia coli	< 5%
Listeria monocytogenes	20-30%
Neisseria gonorrhoeae	> 80%
Pseudomonas aeruginosa	20-30%
Staphylococcus epidermidis	10-30%

[1]Exponential cultures were subjected to 40 μM CCCP for 2 h at 37° with shaking. *E. hirae* was from a static overnight broth culture.

aeruginosa. The *P. aeruginosa* carrying PA-I and/or PA-II can then bind to glycoconjugates, causing adhesion of the bacteria to carbohydrate-containing structures (Wentworth *et al.*, 1991).

REFERENCES

Beveridge, T.J. and Murray, R.G.E. (1979) How thick is the *Bacillus subtilis* cell wall? Current Microbiol. 2, 1-4.

Daneo-Moore, L. and Shockman, G.D. (1979) The bacterial cell surface in growth and division. Cell Surface Rev. 4, 1127-1138.

Doyle, R.J. and Koch, A.L. (1987) The functions of autolysins in the growth and division of *Bacillus subtilis*. CRC Crit. Rev. Microbiol. 15, 169-222.

Doyle, R.J., Chaloupka, J. and Vinter, V. (1988) Turnover of cell wall in microorganisms. Microbiol. Rev. 52, 554-567.

Jolliffe, L.K., Doyle, R.J. and Streips, U.N. (1980) Extracellular proteases modify cell wall turnover in *Bacillus subtilis*. J. Bacteriol. 141, 1199-1208.

Jolliffe, L.K., Doyle, R.J. and Streips, U.N. (1981) Energized membrane and cellular autolysis in *Bacillus subtilis*. Cell 25, 753-763.

Kemper, M.A., Mobley, H.L.T. and Doyle, R.J. (1988) How do bacilli elongate? p 98-108, in: "Antibiotic Inhibition of Bacterial Cell Surface Assembly and Function." Actor, P., Daneo-Moore, L., Higgins, M., Salton, R.J. and Shockman, G.D. eds. American Society for Microbiology. Washington. D.C.

Koch, A.L. and Doyle, R.J. (1985) Inside-to-outside growth and the turnover of the Gram-positive rod. J. Theoret. Biol. 117, 137-157.

Koch, A.L., Higgins, M.L. and Doyle, R.J. (1982) The role of surface stress in the morphology of microbes. J. Gen. Microbiol. 128, 927-945.

Mobley, H.L.T., Koch, A.L., Doyle, R.J. and Streips, U.N. (1984) Insertion and fate of the cell wall in *Bacillus subtilis*. J. Bacteriol. 158, 169-179.

Pooley, H.M. (1976) Layered distribution, according to age, within the cell wall of *Bacillus subtilis*. J. Bacteriol. 125, 1139-1147.

Pooley, H.M. (1976) Turnover and spreading of old wall during surface growth of *Bacillus subtilis*. J. Bacteriol. 125, 1127-1138.

Sonnenfeld, E.M., Beveridge, T.J., Koch, A.L. and Doyle, R.J. (1985) Asymmetric distribution of charge on the cell wall of *Bacillus subtilis*. J. Bacteriol. 163, 1167-1171.

Svarachorn, A., Tsuchido, T., Shinmyo, A. and Takano, M. (1991) Autolysis of *Bacillus subtilis* induced by low temperature. J. Ferment. Bioeng. 4, 281-283.

Thaniyavarn, S. Doyle, R.J., Mobley, H.L.T. and Streips, U.N. (1979) Lipoteichoic acids from glucosylation-defective cell wall mutants of *Bacillus subtilis*. Carbohydr. Res. 73, 203-218.

Wentworth, J.S., Austin, F.E., Garber, N., Gilboa-Garber, N., Paterson, C.A. and Doyle, R.J. (1991) Cytoplasmic lectins contribute to the adhesion of *Pseudomonas aeruginosa*. Biofouling 4, 99-104.

SEARCHING FOR THE EVOLUTIONARY DESIGN OF THE PNEUMOCOCCAL CELL WALL LYTIC ENZYMES

Rubens López[*], José L. García, Eduardo Díaz, Jesús M. Sanz, José M. Sánchez-Puelles, Pedro García, and Ernesto García

Unidad de Génetica Bacteriana
Centro de Investigaciones Biológicas
Consejo Superior de Investigaciones Científicas
Velázquez 144, 28006 Madrid, Spain

INTRODUCTION

Peptidoglycan hydrolases represent one of the enzyme families most widely distributed in nature, since members of this family have been described from bacteriophages to man. Lysozymes, one of the best known group of enzymes at biochemical, structural and evolutionary levels, are included in this family (Jollès and Jollès, 1984). Moreover, bacterial autolysins, enzymes which have been postulated to participate in important physiological functions, are also peptidoglycan hydrolases (Rogers *et al.*, 1980). Probably, the amidase LYTA from *Streptococcus pneumoniae* is the best characterized autolysin described so far (Höltje and Tomasz, 1976; Tomasz, 1984). The gene *lytA* encoding this enzyme has been cloned and sequenced (García *et al.*, 1986), allowing us to obtain valuable information about its implication in the process of daugther cell separation (Ronda *et al.*, 1987). The cloning of *lytA* has also facilitated the isolation of the genes encoding the cell wall lytic enzymes from pneumococcal bacteriophages based on their sequence homologies (García *et al.* 1987; García *et al*, 1988; García *et al.*, 1990; Romero *et al.*, 1990 *a*). The comparison of the sequences of the host and phage genes revealed that pneumococcal cell wall lytic enzymes could be the result of the fusion of two independent functional domains, one containing the active center (N-terminal domain) and the other responsible of the recognition of the teichoic acids of the cell wall (C-terminal domain) (García *et al*, 1988; García *et al.*, 1990; Romero *et al.*, 1990*a*;). This hypothesis was based on the observation that the host LYTA and phage HBL amidases, and the phage CPL1 and CPL9 lysozymes, which depend on the presence of choline in the teichoic acids for activity, have homologous C-terminal domains. However, the phage CPL7 lysozyme that has lost this dependence, contains a completely different C-terminal domain (García *et al.*, 1990). On the other hand, the N-terminal domains of the CPL1, CPL9 and CPL7 lysozymes are homologous but different to the N-terminal domains of the LYTA and HBL amidases. These observations prompted us to test whether the fusion of these domains might create active chimeric enzymes with novel properties. This experimental approach should allow us to confirm our theory about the modular organization of the pneumococcal cell wall lytic enzymes, opening up new insights in the evolutionary design of these proteins.

CONSTRUCTION OF CHIMERIC PROTEINS BETWEEN ENZYMES THAT SHARE SEQUENCE SIMILARITY

On the basis of the known DNA sequences of the *lytA* and *cpl1* genes (García *et al.*, 1966; García *et al.*, 1988), we have constructed chimeric proteins. The new proteins were generated through *in vitro* recombination at equivalent positions of both DNAs, at sites approximating the junction zone of the N- and C-terminal domains of the enzymes, as defined by the alignment (Díaz *et al*, 1990). To facilitate the chimeric construction we created, by site-directed mutagenesis, a single *Sna* BI site in the *cpl1* gene (Figure 1). The introduction of this novel site in *cpl1* gene was neutral with respect to the reading frame of the gene and to the enzymatic activity. Plasmids pCL and pLC contain the chimeric gene coding for a pair of chimeric proteins between the phage lysozyme and the pneumococcal amidase. The biochemical characterization of the extracts obtained from *Escherichia coli* HB101(pCL) and *E. coli* HB101(pLC) demonstrated the presence of active lytic enzymes (named CLL and LCA, respectively) that degraded pneumococcal cell walls and show, as the parental enzymes, an absolute requirement for the presence of choline in the cell wall substrate for activity. The CLL enzyme, which consisted of the N-terminal domain of the CPL1 lysozyme and the C-terminal of the pneumococcal amidase (Figure 1), acts as a lysozyme and requires conversion to achieve full enzymatic activity, a peculiar property of the host pneumococcal amidase (Höltje and Tomasz, 1976). In contrast, the chimeric protein LCA, consisting of the N-terminal domain of the host amidase and the C-terminal domain of the CPL1 (Figure 1), was characterized as an amidase that did not require conversion. Other biochemical properties were also interchanged between the chimeric proteins (data not shown).

CONSTRUCTION OF THE CHIMERIC LC7 AMIDASE FROM PARENTAL ENZYMES THAT DO NOT SHARE SEQUENCE SIMILARITY

The fusion of the N-terminal domain of the choline-dependent LYTA amidase with the C-terminal domain of the choline independent CPL7 lysozyme might generate a new chimeric enzyme showing an amidase activity that should become independent of the presence of choline, i.e. able to degrade either choline or ethanolamine pneumococcal cell walls. In contrast with the previous chimeric constructions this implies the fusion of two genes that lack nucleotide homology (Díaz *et al.*, 1991). To carry out this, we created by site directed mutagenesis the appropriate restriction sites in their respective *lytA* and *cpl7* genes. We decided to generate a *PstI* restriction site within the sequence that we had

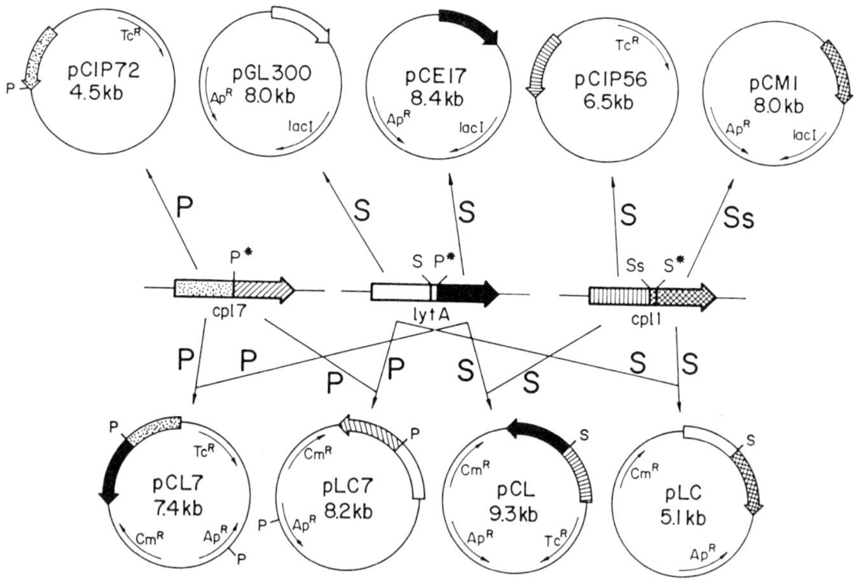

Figure 1. Diagrammatic representation of the constructions of the plasmids coding for the chimeric pneumococcal murein-hydrolases (bottom row) and for the plasmids containing the gene fragments encoding the N-terminal or the C-terminal domains of the pneumococcal murein hydrolases (top row). The headed arrows (center row) represent the phage *cpl1* and *cpl7* genes and the pneumococcal *lytA* gene. Plasmids are drawn in circles with the relevant elements and restriction sites indicated. Light line, vector plasmid; stippled and hatched boxes, 5' and 3' moieties of the *cpl7* gene, respectively; open and blackened boxes, 5' and 3' moieties of the *lytA* gene, respectively; stripped and crosshatched boxes, 5' and 3' moieties of the *cpl1* gene. The recombinant plasmids pCL7, pLC7, pCL and pLC code for the chimeric proteins CL7, LC7, CLL and LCA, respectively. Abbreviations: P, *Pst*I; S, *Sna*B1; Ss, *Ssp*I; Ap, ampicillin; Cm, chloramphenicol; *lacI*, *lac* represor gene; Tc, tetracycline. Asterisks indicate a new restriction sequence introduced by site-directed mutagenesis.

postulated to be the junction region of the N- and C-terminal modules (García et *al.*, 1988; García et *al.*, 1990) (Figure 1B). The introduction of the restriction site did not produce any significant changes in the mutant enzymes as judged by the fact that they maintained unaltered their catalytic activities (data not shown). Using the *PstI* restriction site, the chimeric *lc7* and *cl7* genes were constructed following the procedure depicted in Figure 1. One of the resulting plasmids, pLC7, was used to transform competent cells of *Escherichia coli* HB101 and the cell extract prepared from *E. coli* [pLC7] was tested for the presence of a pneumococcal cell wall lytic activity. As indicated in Table 1, this extract showed a lytic activity that hydrolyzed pneumococcal cell walls containing either choline or

Table 1. Degradation of pneumococcal cell walls by extracts prepared from different *E. coli* strains[a]

Strain	Substrate[b]	Activity[c]
HB101[pBR325]	C-cw	0
HB101[pLC7]	C-cw	410
HB101[pBR325]	EA-cw	0
HB101[pLC7]	EA-cw	6,173

[a]Lysates were obtained by sonication.
[b]C-cw, [*methyl*-^{3}H]choline-labeled cell walls; EA-cw, L-4,5-[^{3}H]lysine-labeled ethanolamine-containing cell walls.
[c]U/mg protein.

ethanolamine in the teichoic acids. In addition, this extract did not require the process of "conversion" to achieve full enzymatic activity, an absolute requirement for the host amidase (Höltje and Tomasz, 1976) (data not shown). To determine whether the chimeric enzyme was a lysozyme or an amidase, we analysed the fragments resulting from the digestion of the pneumococcal cell wall with the chimeric enzyme by chromatography on Sephadex G-75 according to the procedure described before (García et *al.* 1987; Tomasz, 1981). The resulting pattern indicated that the chimeric enzyme acted as an amidase (data not shown). Taken together these results, we could conclude that we have created a new active enzyme, designed as LC7 amidase hereafter, that presents the hybrid properties expected to find according to our previously postulated modular theory (García et *al.*, 1990). On the other hand, plasmid pCL7 (Figure 1) coded for the CL7 lysozyme that required "conversion" by choline and shows biochemical properties similar to that of the chimeric CLL lysozyme described above.

To further characterize the chimeric LC7 amidase and to evaluate its specific activity, this enzyme was overproduced in *E. coli* under the control of strong promoters and purified to electrophoretic homogeneity by conventional chromatographic procedures (data not shown). The specific activity of the LC7 amidase (1.8 U/ g) was lower than those of the parental enzymes LYTA amidase (870 U/ g) and CPL7 lysozyme (37 U/ g). This result can be explained on the basis that the new enzyme did not have the opportunity to increase its efficiency as did the parental ones which have been subjected to selective pressure during evolution.

ACTIVITY OF LC7 AMIDASE ON PNEUMOCOCCAL CELLS

The purified LC7 amidase was also used to determine whether this enzyme was also active in vivo. To carry out this experiment we grew the pneumococcal strain M31 (Sánchez-Puelles *et al.*, 1986) in the presence of LC7 amidase. *S. pneumoniae* M31 has a complete deletion of the *lytA* gene and it cannot autolyse at the end of the stationary phase of growth but it has been well established that this strain can be "cured" by addition of LYTA amidase to the culture medium (Sánchez-Puelles *et al.*, 1986). *S. pneumoniae* M31 lysed in the stationary phase in the presence of the LC7 amidase which indicates that this enzyme is also able to degrade pneumococcal cell wall in vivo (data not shown).

CLONING AND EXPRESSION OF GENE FRAGMENTS ENCODING THE N-TERMINAL AND THE C-TERMINAL DOMAINS OF THE PNEUMOCOCCAL MUREIN HYDROLASES.

We have cloned in *E. coli* the 3' moieties of the *lytA* and *cpl1* genes (plasmids pCE17 and pCM1, respectively, in Figure 1) and expressed the C-terminal domains of the pneumococcal amidase and of the CPL1 lysozyme without the N-terminal domain (Sánchez-Puelles *et al.*, 1990). Biochemical and circular dichroism analyses indicated that these are the domains responsible for the specific recognition of the choline-containing pneumococcal cell walls by the lytic enzymes and suggested that these choline-binding domains can function independently of their catalytic domains. Taking advantage of the strong affinity of these polypeptides for choline or some choline analogues like DEAE we have developed a new single-step system to immobilize and /or purify fusion proteins containing the choline-binding domain of the pneumococcal murein hydrolases (Sánchez-Puelles et *al.*, 1992).

On the other hand, we have constructed recombinant plasmids pGL300, pCIP56 and pCIP72 (Figure 1) that express in *E. coli* the DNA regions encoding the N-terminal domains of the choline-dependent LYTA amidase and CPL1 lysozyme and of the choline-independent CPL7 lysozyme, respectively, showing that these domains also adopt an active conformation even in the absence of their C-terminal domains (Sanz *et al.* , 1992).

DISCUSSION

The recent finding that the N-terminal domains of the pneumococcal murein hydrolases were active and degraded choline- or ethanolamine-containing pneumococcal cell walls with similar efficiency, a result which should be expected of murein hydrolases that do not contain the choline-binding domain, provides an experimental argument in favour of the hypothesis that the N-terminal domains of these enzymes might have constituted independent proteins during the evolution of these murein hydrolases. The lower activity shown by the catalytic domains when compared with that of the complete enzymes suggested that the acquisition of a substrate-binding domain should represent a remarkable advantage for enzymes that interact with polymeric substrates, allowing to improve their catalytic efficiency (Sanz et *al.*, 1992).

We have recently shown that the existence of regions of high nucleotide similarity between phage and bacterial DNAs promotes site-specific recombination which should allow restructuring and evolution of the bacterial and phage genomes (Romero et *al*, 1990 *b*) as already suggested for other systems (Stroynowsky, 1981). In addition, it has been pointed out that phage genomes came probably from the host gene pool and known mechanisms of gene rearrangement are adequated to ensure that phage genomes can acquire new genes from that pool as occasion demands (Campbell, 1988). Illegitimate recombination should be the dominant force that converts the phage genomes in a mosaic of genes from non-phage sources (Hunkapiller et *al.*, 1982). In this sense, the preparation of chimeric enzymes from the genes coding for the lytic enzymes of *S. pneumoniae* and its bacteriophages provides an example of the type of new genes that the bacteria or the phage can create to allow them to become adapted to new environmental situations. This conclusion reinforces the generally accepted view that viral evolution is based on the potentialities of illegitimate recombination mechanisms for creating DNA fusions.

Acknowledgements

We thank M. Carrasco and E. Cano for technical assistance. The art work of A. Hurtado is gratefully acknowledged. This work was supported by a grant from Dirección General de Investigación Científica y Técnica (PB90-0069).

REFERENCES

Campbell, A. (1988) Phage evolution and speciation. in The bacteriophages (Calendar, R., Ed.), pp. 1-14. Plenum Publishing Corp.. New York.

Díaz, E., López, R. and García, J.L. (1990) Chimeric phage-bacterial enzymes, a clue in the molecular evolution of genes. Proc. Natl. Acad. Sci. USA 87, 8125-8129.

Díaz, E., López, R. and García, J.L. (1991) Chimeric pneumococcal cell wall lytic enzymes reveal important physiologycal and evolutionary traits. J. Biol. Chem. 266, 5464-5471.

García, E., García, J.L., García, P., Arrarás, A., Sánchez-Puelles, J. M. and López, R. (1988) Molecular evolution of lytic enzymes of *Streptococcus pneumoniae* and its bacteriophages. Proc. Natl. Acad. Sci. USA. 85,914-918.

García, J.L., García, E., Arrarás A., García, P., Ronda, C. and R. López. (1987) cloning, purification and biochemical characterization of the pneumococcal bacteriophage Cp-1 lysin. J. Virol. 61, 2573-2580.

García, P., García, J.L., García, E. and López, R. (1986) Nucleotide sequence and expression of the pneumococcal autolysin gene from its own promoter in *Escherichia coli*. Gene 43, 265-272.

García, P., García, J.L., García, E., Sanchez-Puelles, J.M. and López, R. (1990) Modular organization of the lytic enzymes of *Streptococcus pneumoniae* and its bacteriophages. Gene 86, 81-88.

Höltje, J.V. and Tomasz, A. (1976) Purification of the pneumococcal *N*-acetylmuramyl-L-alanine amidase to biochemical homogeneity. J. Biol. Chem. 251, 4199-4207.

Hunkapiller, T. H., Huang, H., Hoong, L. and Campbell, J. H. (1982) The impact of modern genetics on evolutionary theory. In: R. Millman (ed.) Sinauer, Sunderland, MA.

Jollès, P. and Jollès, J. (1984) What's new in lysozyme research?. Mol. Cell. Biochem. 63, 165-189.

Rogers, H.J., Perkins, H.R. and Ward, J.B. (1980) Microbial cell walls and membranes. Chapman and Hall, London.

Romero, A., López, R. and García, P. (1990 *a*) Characterization of the pneumococcal bacteriophage HB-3 amidase: Cloning and expression in *Escherichia coli*. J. Virol. 64, 137-142.

Romero, A., López, R. and García, P. (1990 *b*) Sequence of the *Streptococcus pneumoniae* bacteriophage reveals high homology with the mayor host autolysin. J. Bacteriol. 172, 5064-5070.

Ronda, C., García, J.L., García, E., Sanchez-Puelles, J.M. and López, R. (1987) Biological role of the pneumococcal amidase. Cloning of the *lytA* gene in *Streptococcus pneumoniae*. Eur. J. Biochem. 164, 621-624.

Sánchez-Puelles, J. M., Sanz, J, García, J.L. and García, E. (1990) Cloning and expression of gene fragments encoding the choline-binding domain of pneumococcal murein hydrolases. Gene 89, 69-75.

Sánchez-Puelles, J. M., Ronda, C., García, J. L., García, P., López, R. and García, E. (1986) Searching for autolysin functions. Characterization of a pneumococcal mutant deleted in the *lytA* gene. Eur. J. Biochem. 158, 289-293.

Sánchez-Puelles, J. M., Sanz, J., García, J.L. and García, E. (1992) Immibilization and single-step purification of fusion proteins using DEAE-cellulose. Eur. J. Biochem. 203, 153-159.

Sanz, J. M., Díaz, E. and García, J.L. (1992) Studies on the structure and function of the N-terminal domain of the pneumococcal murein hydrolases. Mol. Microbiol. 6 (in press)

Stroynowsky, I. T. (1981) Distribution of bacteriophage 3T homologous deoxyribonucleic acid sequences in *Bacillus subtilis* 168, related bacteriophages, and other *Bacillus* species. J. Bacteriol. 148, 91-100.

Tomasz, A. (1981) Surface components of *Streptococcus pneumoniae*. Rev. Infect. Dis. 3:190-211.

Tomasz, A. (1984) Building and breaking of bonds in the cell wall of bacteria- the role of autolysins, p, 3-12. In C. Nombela (ed.), Microbial cell wall synthesis and autolysis. Elsevier Science Publishers, Amsterdam.

MOLECULAR CHARACTERISTICS OF THE CELL WALL LYTIC ENZYMES CODED BY PNEUMOCOCCAL PHAGES

Pedro García*, Ernesto García, Alicia Romero, Christian Croux, Concepción Ronda, Rubens López and José L. García

Unidad de Genética Bacteriana
Centro de Investigaciones Biológicas
Consejo Superior de Investigaciones Científicas
Velázquez 144, 28006 Madrid, Spain

INTRODUCTION

The biochemical characteristics of the enzymes postulated to participate in the lysis of *Streptococcus pneumoniae* infected by phages have been studied in our laboratory in some detail (García *et al.*, 1984, 1987; Romero *et al.*, 1990). Independently of their enzymatic specificity, either amidases or lysozymes, most of the lytic enzymes found in pneumococcus and its bacteriophages share important properties, especially the absolute requirement for choline-containing teichoic acid in cell wall substrate (a peculiar characteristic of the pneumococcal peptidoglycan) for activity (Höltje and Tomasz, 1975). It has been postulated that this choline dependence of the pneumococcal lysins may have served as an element of selective pressure during evolution that should have preserved the structures of these enzymes (García *et al.*, 1988 *a*).

To investigate these assumptions we have carried out an experimental approach based in the determination and comparison of the nucleotide sequences of a series of genes coding for pneumococcal phage lysins. On the other hand, studies on the localization of the insertion site of two pneumococcal temperate phages have revealed that the *att* sites are located near the lytic genes of both phages. The overall analyses of these findings provide interesting information on the organization of the lytic enzymes of the pneumococcal bacteriophages and also highlight some evolutionary relationships between phages and bacteria.

MUREIN HYDROLASES OF PNEUMOCOCCAL LYTIC PHAGES

The use of the *lytA* gene (García *et al.*, 1986), coding for the mayor murein hydrolase present in pneumococcus (LYTA amidase), as genetic probe has allowed to localize, clone and

Bacterial Growth and Lysis, Edited by M.A. de Pedro *et al.*, Plenum Press, New York, 1993

sequence a fragment of phage Cp-1 DNA, containing the gene *cpl1*, coding for a lysozyme, designed as CPL1 hereafter (García *et al.*, 1988 *a*). Comparison of the nucleotide sequences encoding these murein hydrolases revealed that their 3' regions shows a considerable sequence similarity and the deduced amino acid sequences of these lysins are practically identical. Attempts to extend these findings by studying the molecular characteristics of the lytic genes present in other pneumococcal phages, demonstrated that the pneumococcal *lytA* gene hybridized with fragment of the DNA of phage Cp-9 whereas non-homology was detectable with fragments of Cp-7 DNA in spite of the fact that several biological and biochemical tests indicated that Cp-7 also codes for a lytic enzyme active against *S. pneumoniae* cell walls. To define the location of the gene coding for the lytic activity of Cp-7 phage we successfully used as probe an *AccI-SphI* fragment of Cp-1 DNA that embraces the 5'-flanking region of the *cpl1* gene (García et *al.*, 1990). An appropriate fragment from Cp-7 DNA was cloned and sequenced to determine the nucleotide sequence of the *cpl7* gene that codes for the murein hydrolase of phage Cp7 (CPL7). In the comparison of the *cpl1, cpl7* and *cpl9* genes we found that altogether the *cpl9* gene has much greater similarity to the *cpl1* gene than to the *cpl7* gene. In *cpl9,* 52 of the first 600 nucleotides (8.7%) differed from the *cpl1* gene, whereas only 20 changes were found in the remaining 417 nucleotides (4.8%). The most noticeable differences between the *cpl7* gene and the *cpl1* and *cpl9* genes were found within the 400 bp preceding the stop codon of the *cpl7*, where the sequences diverge completely, whereas only 17.3% of the first 600 nucleotides were different.

Based on the analysis of the nucleotide sequences of *cpl1*, *cpl7,* and *cpl9* we deduced that these genes encodes proteins of 39,146 Da, 38,461 Da and 39,107 Da corresponding to 339, 342 and 339 amino acid residues, respectively (Figure 1). Only 10 amino acids were different between CPL-1 and CPL-9. Nine of these changes were localized in the N-terminal moiety of the proteins, whereas only one amino acid differed in the part of the molecule that has previously found to be homologous to the host amidase (García et *al.*, 1988 *a*). On the other hand, 34 out of the the first 204 amino acids were different between CPL-1 and CPL-7. The amino acids Asp-10 anf Glu-37 that have been proposed to be involved in the catalytic process (García et *al.*, 1988 *a*) are conserved in the three phage lytic proteins. Most remarkably, the C-terminal of CPL-7 was completely different. Interestingly, the set of six modular repetitions (P1 to P6) found in the C-terminal ends of the host LYTA amidase, and in the phage CPL1 and CPL9 lysozymes (García et *al.*, 1990), each about 20 amino acids long, was replaced in CPL7 by a 2.8 perfect tandem repetitions, each 48 amino acids long (M1 to M3). Based on amino acid sequence comparisions, we had previously suggested that the C-terminal domains of the CPL1 and of the host pneumococcal amidase might be responsible for the recognition of the choline residues present in the pneumococcal cell wall, whereas the active center of these enzymes should be located in their N-terminal domains (García et *al.*, 1988 *a*). If this hypothesis were correct, the biological and biochemical properties of the enzyme CPL7 should completely differ from those reported for the other lytic enzymes of *S. pneumoniae* and its bacteriophages. In fact, CPL7, that was biochemically characterized as a lysozyme, was able to degrade *in vitro* pneumococcal cell walls containing either choline of ethanolamine in the teichoic acids. Furthermore, Cp-7 phage, containing the gene *cpl7,* turned out to be the only pneumococcal phage capable of successfully infecting and lysing EA-grown pneumococcal cells, liberating productive progeny of phages.

MUREIN HYDROLASES OF TEMPERATE PNEUMOCOCCAL PHAGES

Bernheimer (1977) first reported the isolation of pneumococcal phages from

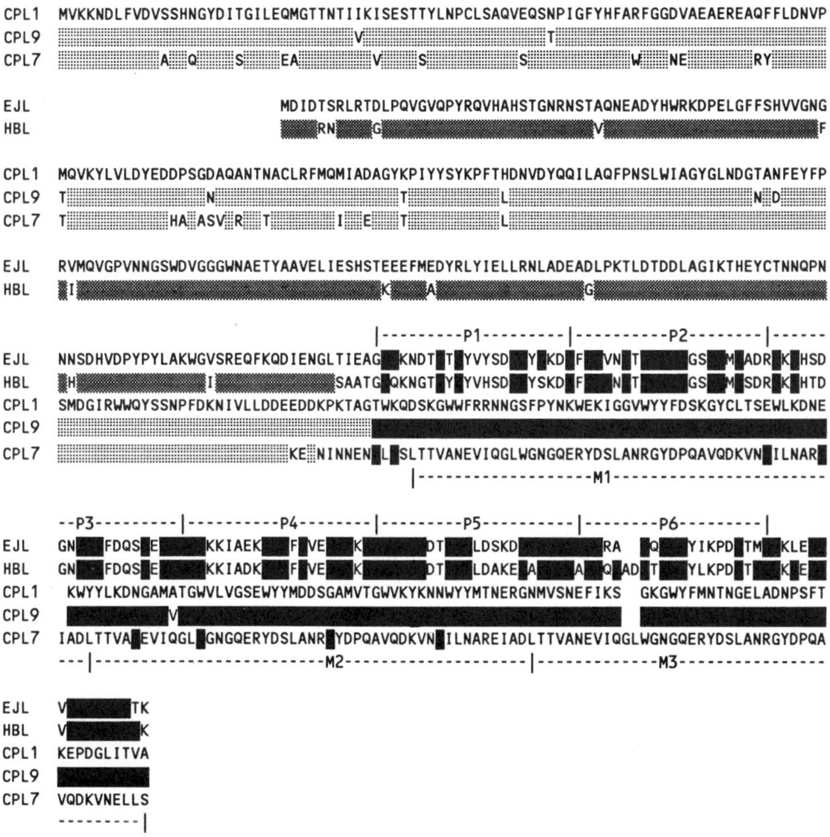

Figure 1. Comparison of the deduced amino acid sequences of the bacteriophage HB1 and EJL amidases and the bacteriophage CPL1, CPL7 and CPL9 lysozymes. ▒ and ░ represent identical amino acids when the N-terminal of the EJL or the N-terminal of the CPL1 were used as references, respectively. █ represents identical amino acids compared to the C-terminal of the CPL1 lysozyme. P1-P6 and M1-M3 represent the sets of repeated sequences.

encapsulated *S. pneumoniae,* and suggested that lysogeny was associated with strain of capsular types most frequently causing pediatric infections. One of these phages, HB-3, codes for an *N*-acetylmuramoyl-L-alanine amidase (HBL) which has been cloned in *E. coli* using the *lytA* as a probe. The phage HBL amidase is a choline-dependent enzyme that, as in the case of the host LYTA amidase, requires the process of "conversion" (Höltje and Tomasz, 1975) to be fully active. The nucleotide sequence of the *hbl* gene revealed a remarkable similarity (87%) when compared with the host amidase which represents the highest similarity found between phage and bacterial genes (Romero *et al.*, 1990). This sequence similarity dissapears at the coding boundaries. As expected, the two amidase are 90% similar in amino acid composition (Figure 1) being the N-terminal the more different region. Twenty-one of the 31 amino acid changes were observed in the first 141 amino acids. Most important five of these changes were observed in the first six amino acids.

The genes coding for the pneumococcal host and phage amidases have provided us an appropriate system for investigating the requirements for obtaining successful recombination between bacterial and phage DNAs. Combined genetic transformation and Southern blotting experiments have demonstrated the integration of *hbl* gene into the bacterial *lytA* gene and we have proposed that this integration seems to depend on the extent of similarity between the donor and recipient DNAs (Romero *et al.*, 1990).

More recently, we have isolated from a clinical atypical pneumococcal strain another temperate phage, EJ-1, that presents morphological characteristics that has not been described before for phages of streptococcal species. Following an experimental approach similar to that described above, we have also localized in the genome of this phage a lytic gene (*ejl*), that has been cloned and sequenced. The *ejl* gene codes for an choline-dependent amidase (EJL) that shows a modular organization (Figure 1) similar to those of the HBL and LYTA amidases. It is interesting to note that EJL presents in the P6 motif a deletion of two amino acids with respect to the sequence of the LYTA or HBL amidases. This deletion is equivalent to that found in the CPL1 and CPL9 lysozymes. The comparison of the structural regions of the genes *ejl* and *lytA* revealed an 80% similarity whereas the sequences of their flanking regions diverge completely. This result reinforces the idea that the coding regions defines the evolutionary concept of interchangeable functional modules, whereas the flanking, noncoding regions, should be involved in regulatory functions and would not be interchanged since they are specific for bacterial and phage systems.

LOCALIZATION OF THE INSERTION SITES OF THE PNEUMOCOCCAL TEMPERATE PHAGES NEAR THE LYTIC GENES

To better understand the process of integration of temperate pneumococcal bacteriophages we have recently mapped the *attP* sites of HB-746, that belong to the HB family, and EJ-1 phages (Romero *et al.*, 1992; Díaz, *et al.*, 1992). It is noteworthy that in both models the *attP* sites lie downstream of the 3' end of the structural region of the corresponding lytic gene (Figure 2). Furthermore, we have identified hairpin structures located at almost the same distance from the lytic genes. The regions flanking the hairpin structures share noticeable sequence similarity; e.g. the sequence GGGGCA that has been postulated to participate in the binding of the integrase (Lee and Buranen, 1989) is repeated four times.

It is interesting to point out that in the case of the pneumococcal strain 8R1, the bacteria lysogenized by HB-746, the bacterial attachment site (*attB*) has also been mapped downstream

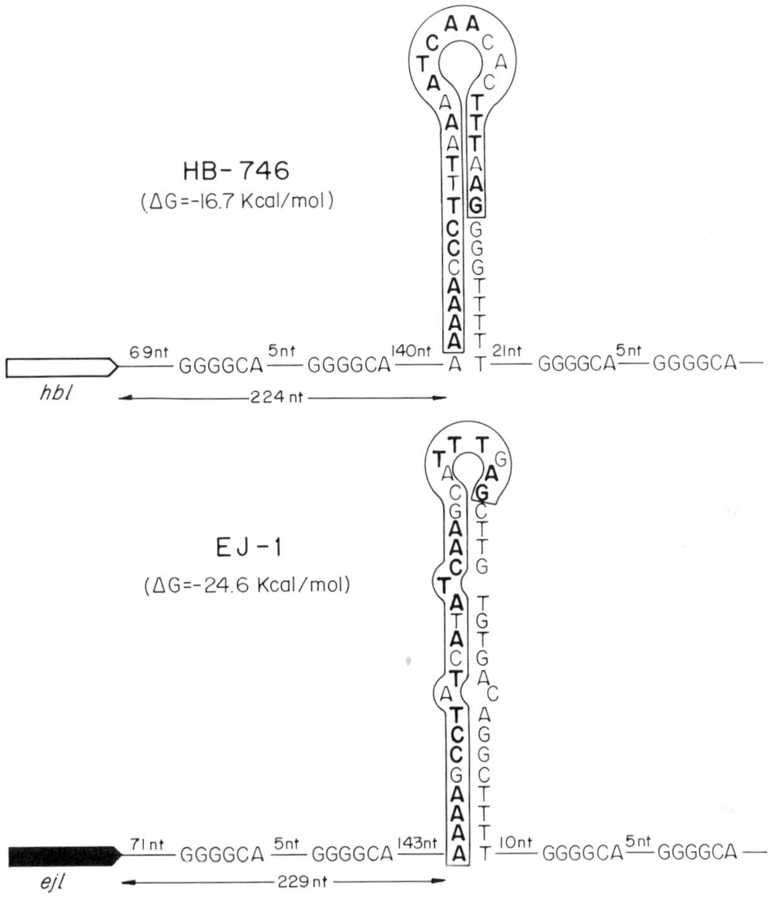

Figure 2. Possible stem-and loop structures located in the *attP* sites. *hbl* and *ejl* indicate the structural lytic gene of HB-746 and EJ-1 bacteriophages, respectively. The lines enclosing stem-and-loop structures denote homologous sequences where identical nucleotides between these structures are represented in boldface. The distances between the lytic genes, the repeated sequence (GGGGCA) and the stem-and-loop structure are indicated in nucleotides (nt). (ΔG) represents the free energy of the hairpin structures.

of the *hbl*R, a lytic gene, or part of it, that appears to represent a phage remnant in the pneumococcal host chromosome (Romero *et al.*, 1992).

DISCUSSION

An overall analysis of the results presented here reveals that the specificity for binding to the cell wall substrate resides in the C-terminal domain of the lytic enzymes and that this domain always contains a repeating motif whereas the N-terminal domain determine the type of enzyme activity. It has been proposed that some essential genes, like the lytic genes, have arisen in phages from host genes (Campbell, 1988), and, as pointed out above, recombination between genes coding for cell wall lytic enzymes of phages and bacteria in *S. pneumoniae* and *Bacillus subtilis* has been demonstrated in the laboratory (Romero *et al.*, 1990; Stroynowsky, 1981). This type of genetic event appears to be facilitated in the pneumococcal system by the close relationship between genes that exhibit the highest similarity described so far between phages and bacteria (Figure 1) and might be the instrument to introduce diversity in the genes coding for lytic enzymes in pneumococcus. Some differences in the primary structure of these enzymes, as the two amino acid deletion in the P6 motif found in the EJL amidase and the CPL1 and CPL9 with respect to the host LYTA amidase, might account for some biochemical peculiarities of those phage lytic enzymes although a precise relationship between primary structures and biochemical properties requires further investigation. Taking all these generalities together we propose a modular organization in the structure of the lytic enzymes of pneumococcus and its bacteriophages. This modular design suggests a number of combinatorial possibilities and could be a widespread model for many types of organisms as it has been recently documented for an amidase of *Staphylococcus aureus* (Wang *et al.*, 1991). Also, repeat units have been found recently in the C-terminus of the muramidase-2 of *Enterococcus hirae* (Chu *et al.*, 1992) and of the autolytic lysozyme of *Clostridium acetobutylicum* ATCC824 (Croux and García, 1991) suggesting that this region could play a role in the recognition of the polymeric substrate. In addition, preliminary analysis of the deduced amino acid sequence of a gene coding for a choline-binding protein of *C. acetobutylicum* CECT 806, a species that contains choline in the teichoic acids of its cell wall (García *et al.*, 1988 *b*), also shows the presence in the C-terminal domain of four homologous repeat units very similar to those of the CPL1 lysozyme (unpublished data). Interestingly, analysis of the sequence for the gene encoding PspA, the pneumococcal surface protein A, revealed the existence of a C-terminal domain that has a remarkable similarity with the C-terminal repeat regions of the LYTA and CPL1 proteins (Yother and Davis, 1992). It appears that PspA is attached to the pneumococcal surface by a mechanism unique among known surface proteins of gram-positive bacteria. We postulate that this protein should be probably bound to the cell surface through the choline residues present in the pneumococcal teichoic acids.

The mechanisms of integration of bacteriophages in pneumococcus have not been studied until now and there are few reports on the mechanisms of phage integration in Gram-positive bacteria. The results discussed here demonstrate that the *att*P site of the phages HB-746 and EJ-1, that belong to two different families of pneumococcal phages, lies in the vicinity of the lytic genes *hbl* and *ejl*, respectively (Figure 2). It also appears that the regions involved in the integration of a phage in pneumococcus contains elements like stem-and-loop structures and repeats which have been postulated to be needed to provide fidelity to the recombination process (Landy, 1989; Rausch and Lehmann, 1991). Whether the pneumococcal lytic enzymes play any role in the integration mechanisms of temperate phages remains to be investigated. The fact that phage remnant characterized on the basis of its nucleotide sequence as lytic gene (unpublished data), has been localized close to the *attB*, invites to speculate with the possibility that these phage debris may be vestiges of early interactions between the phage and its host.

Acknowledgements

We thank M. Carrasco and E. Cano for technical assistance. The art work of A. Hurtado is gratefully acknowledged. This work was supported by a grant from Comisión Interministerial de Ciencia y Tecnología (SAL91-0898-C02-01).

REFERENCES

Bernheimer, H.P. (1977) Lysogeny in pneumococci freshly isolated from man. Science 195, 66-68.

Campbell, A. (1988) Phage evolution and speciation. in The bacteriophages (Calendar, R., Ed.), pp. 1-14. Plenum Publishing Corp.. New York.

Chu, C-P. Kariyama, R., Daneo-Moore, L. and Shockman, G. (1992) Cloning and sequence analysis of the muramidase-2 gene from *Enterococcus hirae*. J. Bacteriol. 174, 1619-1625.

Croux, C. and García, J.L. (1991) Sequence of the *lyc* gene encoding the autolytic lysozyme of *Clostridium acetobutylicum* ATCC824: comparison with other lytic enzymes. Gene 104, 25-31.

Díaz, E., López, R. and García, J.L. (1992) EJ-1: a *myoviridae*, temperate bacteriophage of *Streptococcus pneumoniae*. Submitted for publication.

García, E., García, J.L., García, P., Arrarás, A., Sánchez-Puelles, J. M. and López, R. (1988 *a*) Molecular evolution of lytic enzymes of *Streptococcus pneumoniae* and its bacteriophages. Proc. Natl. Acad. Sci. USA. 85, 914-918.

García, J.L., García, E., Sánchez-Puelles, J.M. and López, R. (1988 *b*) Identification of a lytic enzyme of *Clostridium acetobutylicum* that degrades choline-containing pneumococcal cell walls. FEMS Microbiol. Lett. 52, 133-138.

García, J.L., García, E., Arrarás A., García, P., Ronda, C. and R. López. (1987) Cloning, purification and biochemical characterization of the pneumococcal bacteriophage Cp-1 lysin. J. Virol. 61, 2573-2580.

García, P., Méndez, E., García, E., Ronda, C. and López, R. (1984) Biochemical characterization of a murein hydrolase induced by bacteriophage Dp-1 in *Streptococcus pneumoniae*: comparative study between bacteriophage-associated lysin and the host amidase. J. Bacteriol. 159, 793-796.

García, P., García, J.L., García, E. and López, R. (1986) Nucleotide sequence and expression of the pneumococcal autolysin gene from its own promoter in *Escherichia coli*. Gene 43, 265-272.

García, P., García, J.L., García, E., Sánchez-Puelles, J.M. and López, R. (1990) Modular organization of the lytic enzymes of *Streptococcus pneumoniae* and its bacteriophages. Gene 86, 81-88.

Höltje, J.V. and Tomasz, A. (1976) Purification of the pneumococcal *N*-acetylmuramyl-L-alanine amidase to biochemical homogeneity. J. Biol. Chem. 251, 4199-4207.

Höltje, J.V. and Tomasz, A. (1976) Purification of the pneumococcal *N*-acetylmuramyl-L-alanine amidase to biochemical homogeneity. J. Biol. Chem. 251, 4199-4207.

Landy, A. (1989) Dynamic, structural, and regulatory aspects of site-specific recombination. Ann. Rev. Biochem. 58, 913-949.

Lee, C.Y. and Buranen, S.L. (1989) Extent of the DNA sequence required in integration if staphylococcal bacteriophage L54a. J. Bacteriol. 171, 1652-1657.

Rausch, H. and Lehmann, M. (1991) Structural analysis of the actinophage C31 attachment site. Nucl. Acids Res. 19, 5187-5189.

Romero, A., López, R. and García, P. (1990 *a*) Characterization of the pneumococcal bacteriophage HB-3 amidase: Cloning and expression in *Escherichia coli*. J. Virol. 64, 137-142.

Romero, A., López, R. and García, P. (1990 *b*) Sequence of the *Streptococcus pneumoniae* bacteriophage reveals high homology with the mayor host autolysin. J. Bacteriol. 172, 5064-5070.

Romero, A., López, R. and García, P. (1992) The insertion site of the temperate phage HB-746 is located near the phage remnant in the pneumococcal host chromosome. J. Virol. 66 (in press).

Stroynowsky, I.T. (1981) Distribution of bacteriophage 3T homologous deoxyribonucleic acid sequences in *Bacillus subtilis* 168, related bacteriophages, and other *Bacillus* species. J. Bacteriol. 148, 91-100.

Wang, X., Wilkinson, B.J. and Jayaswal, R.K. (1991) Sequence analysis of a *Staphylococcus aureus* gene encoding a peptidoglycan hydrolase activity. Gene 102, 105-109.

Yother, J. and Briles, D.E. (1992) Structural properties and evolutionary relationships of PspA, a surface protein of *Streptococcus pneumoniae*, as revealed by sequence analysis. J. Bacteriol. 174, 601-609.

MOLECULAR CLONING, OVEREXPRESSION OF RZ LYSIS GENE OF PHAGE λ AND SUBCELLULAR LOCALIZATION OF ITS PROTEIN PRODUCT

Barbara Hanych[1], Sabina Kędzierska[1], Brigitte Walderich[2] and Alina Taylor[1]

[1]Department of Biochemistry, University of Gdańsk
Kładki 24, 80-822 Gdańsk, Poland

[2]Abteilung Biochemie, Max-Planck-Institut für
Entwicklungsbiologie, D-7400 Tübingen, Germany

INTRODUCTION

Bacteriophage λ encodes three genes involved in bacterial lysis: S R and Rz localized in the right arm of λ DNA. These genes belong to the late operon, which is transcribed rightward from the late p'R promoter. Its transcription requires the Q protein for the antitermination (Friedman and Gottesman, 1983).

The λR gene encodes a soluble 17.5 kD transglycosylase degrading the bacterial murein layer (Bieńkowska-Szewczyk et al., 1981). The R protein has no signal sequence, so its transport across IM depends on the functions of other λ genes, most probably the S gene. It was proposed that the S gene product oligomerizes in IM thus forming pores which serve as transportation channels (Zagotta and Wilson, 1990) for the transglycosylase (gpR) to facilitate its access to the substrate, the periplasmic murein layer.

The Rz gene was discovered by Young et al. (1979) by the Tn903 transposition mutagenesis. Later, the Rz gene has been localized in the position between bp No 45966 and 46425 in the λ genome, it codes for a 17 kDa protein (Taylor et al., 1983) which, however, has never been detected on SDS-PAGE.

The Rz gene activity has not yet been elucidated. It is required for bacterial membranes destabilization prerequisite for lysis if the concen-

Bacterial Growth and Lysis, Edited by M.A. de Pedro
et al., Plenum Press, New York, 1993

tration of Mg^{2+} ions exceedes 10 mM. The λ Rz mutants convert the host cells into stable spheroplasts in the presence of Mg^{2+} ions, but they do lyse the cells in the absence of Mg^{2+} (Young et al., 1979). These observations seem to point to a role of the Rz protein in counteracting the stabilization of bacterial membranes by Mg^{2+}.

MATERIAL AND METHODS

Bacterial Strains and Plasmids. *E.coli* BL21(DE3) harboring a defective lambda prophage that contains the phage T7 RNA polymerase gene under the control of the *lac* represor (Studier et al., 1990) was used for the expression of cloned Rz gene. Plasmid pT7-7 containing the T7 RNA polymerase promoter φ10 and the translation start site for the T7 gene 10 protein in front of a polylinker sequence, was obtained from S. Tabor (Dept. Biol. Chem., Harvard Medical School, Boston) and used for cloning the Rz gene. This resulted in pBH21 recombinant plasmid (manuscript in preparation). *E.coli* W3350(λcI857*Sam*7) was obtained from R. Young.

Growth Conditions. BL21(DE3)pBH21 was grown at 37°C in 1 ml of M9 medium (Maniatis et al., 1982) containing ampicilin 0.1 mg to OD_{578}= 0.75, then 500 nmol IPTG was added, 15 min later the culture was treated with 0.3 mg rifampicin and after another 10 min the proteins were labelled by the addition of 10μCi/ml of [^{35}S]-methionine (1000 Ci/mmol). Incorporation was terminated after 15 min by chilling the culture. The cells were sedimented and either resuspended and lysed in the sample loading buffer (Laemmli, 1970) or subjected to freeze-thawing as described by Beremand and Blumenthal (1979).

Envelope Separation. The envelope fraction was prepared from a 120 ml culture of BL21(DE3)[pBH21] grown in the conditions described above. IPTG was added but the culture was incubated for 2 h longer. Cells were converted into spheroplasts (Witholt et al. 1976) and sonicated. Unbroken cells were removed, and the supernatant was fractionated according to Ishidate et al. (1986) on a two-step sucrose gradient (0.8 ml of 60% sucrose and 4.5 ml of 25% sucrose, both in 50 mM sodium phosphate buffer, pH 7.1) by centrifugation for 1.5 h at 37,000 rpm in a Beckman SW41 rotor. The supernatant contained a soluble fraction of cytoplasmic and periplasmic proteins (CP). The opaque band of the membranes fractions was collected, diluted with 50 mM sodium phosphate buffer, pH 7.1 to diminish the sucrose concentration to about 25% and layered on a discontinuous sucrose gradient of 0.5 ml of 60%, 1 ml of 55%, 2.2 ml of 50%, 2.2 ml of 45%, 2.2 ml of 40%, 1.4 ml of 35% and 1 ml of 30% sucrose in 50 mM sodium phosphate buffer (pH 7.1) and centrifuged in a Beckman SW41 rotor at

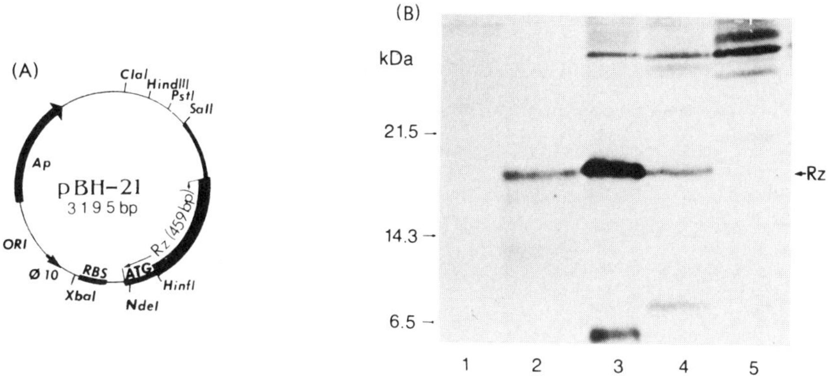

Figure 1. (A) Plasmid pBH21 carrying the λ Rz lysis gene. A fragment of λ DNA bp No 45966-46721 containing the Rz lysis gene (459 bp) was cloned into the expression vector pT7-7 resulting in pBH21. Distal to the 3'end of the Rz gene there are 297 bp of λ DNA. (B) Expression of the λ Rz gene in vivo and in vitro. Conditions for BL21(DE3)[pBH21] growth, electrophoresis in 0.1%SDS-20%PAGE and transfer on the nitrocellulose are described in Methods. The nitrocellulose was exposed to Kodak X-OMAT AR film. Lane 1) a control of the in vitro system; 2) expression of the Rz gene in vitro from the pBH21 CCC DNA as a template. 3) and 4) induced and uninduced BL21(DE3)[pBH21] respectively; 5) induced BL21(DE3)[pT7-7], a control.

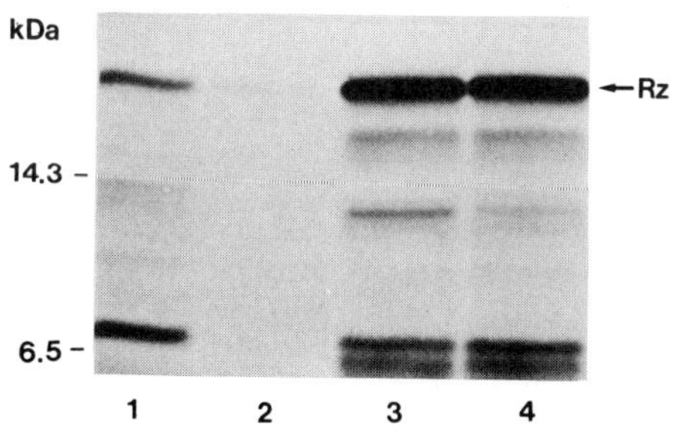

Figure 2. Preliminary subcellular fractionation of the Rz protein. 1 ml of E.coli BL21(DE3)[pBH21] culture was labeled with $[^{35}S]$-methionine and fractionated into soluble and insoluble fractions (Beremand and Blumenthal, 1979). Proteins were separated in a 0.1%SDS-20%PAGE (Laemmli, 1970), transferred to nitrocellulose (Szewczyk and Kozloff, 1985) and exposed to Kodak X-OMAT AR film. Lane 1) uninduced culture, (negative control); Lane 2) soluble fraction (CP);Lane 3) insoluble envelope fraction; Lane 4) induced culture (positive control). On the left side proteins MW markers are indicated ("Rainbow", Amersham, International, England).

39,000 rpm for 16 h. The gradient was fractionated into 30 subfractions (350 μl each). Aliquots were removed for the determination of NADH-oxidase activity (IM marker) (Owen et al., 1982), protein concentration (Bradford, 1976) and refractive index.

SDS-PAGE and Immunoblotting. Proteins were separated by 0.1%SDS-20%PAGE according to Laemmli (1970) and transferred to nitrocellulose (Szewczyk and Kozloff, 1985). Filters were blocked overnight with 0.1% Tween 20 and incubated with rabbit antiserum against a synthetic 16 aas antigenic determinant of the Rz protein (manuscript in preparation). Peroxidase coupled to goat anti-rabbit antibodies (Promega, Madison, USA) was used for the detection of the antigen-antibody complex, with 1-chloro-4-naphtol and H_2O_2 as substrates.

RESULTS

Expression of the Cloned Rz Gene. pBH21 was costructed by cloning the Rz gene directly under the strong promoter and translation signals of the pT7-7 (Fig. 1A). The recombinant plasmid was sequenced (Sanger et al. (1977) to confirm the correct cloning. The Rz gene expressed from pBH21 complemented the λRz^- lysis defect up to the level of the λ wt phage.

Expression of pBH21 *in vivo*, in BL21(DE3) or *in vitro*, in the cell free coupled transcription-translation system performed according to the manufacturers instruction (Prokaryotic DNA-directed KIT, Amersham, England) resulted in the Rz protein of M_r 17 kDa as predicted from the gene nts sequence (Fig. 1B).

Preliminary Localization of the Rz Protein in the Envelope Fraction. The Rz gene sequence inspected for the possible presence of a signal sequence in the coded protein by "Mac Molly" program, (Max-Planck-Institut, Tübingen, Germany) showed the presence of typical features of such sequences (Gierasch, 1989; von Heijne, 1984). The expression of the Rz gene *in vivo* and *in vitro* (Fig. 1B) showed that the Rz protein of M_r about 17 kDa had not been processed by the signal endopeptidase of *E. coli*. Based on the suggestion of von Heijne (private communication) that the Rz protein might have an N-terminal anchor segment serving both as an uncleavable signal sequence and a membrane anchor, was our interest to check if the protein fractionates with the membrane fractions.

In a preliminary experiment, cells were gently lysed according to Beremand and Blumenthal (1979), and fractionated by centrifugation to separate the insoluble envelope fraction from soluble cellular proteins (Fig.2)

Localization of Rz protein in the membranes. Since the labeled Rz protein was localized in the envelope it was of interest to determine the specific distribution of this protein among the cell envelope fractions, the inner (IM) and outer (OM) membranes. The membranes were separated by centrifugation in sucrose density gradient and analysed by Western blotting for the presence of the Rz protein. It was found to be distributed in IM and OM, but most of it was present in the fraction A of an intermediate density between those of IM and OM that represents adhesion zones of the membranes (Fig.3, lane 4).

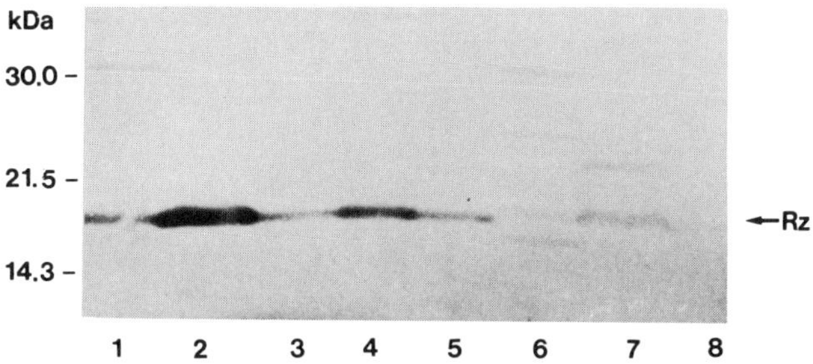

Figure 3. Distribution of the λ Rz protein in the membranes of E.coli. Membranes of E.coli BL21(DE3)[pBH21] were fractionated by sucrose density gradient centrifugation (Ishidate et al., 1986). Proteins of each of the three fractions: OM, IM and A (adhesion zones) were separated by 0.1%SDS-20%PAGE electrophoresis and the Rz protein was visualized by Western blotting using antiserum raised against a synthetic 16aas antigenic determinant of the Rz protein. Lane 1) induced culture of BL21(DE3)[pBH21]; Lane 2) total envelope fraction; Lane 3) IM; Lane 4) A, (adhesion zones);. Lane 5) OM; Lane 6) soluble, (CP) proteins; Lanes 7) and 8) W3350 (λcI857Sam7) cells, induced and uninduced respectively.

DISCUSSION

Cloning of the Rz gene in pT7-7 plasmid directly under strong transcriptional and translational signals (submitted elsewhere) allowed for the visualization of the gene's product by autoradiography or by immunodetection with antibodies raised against a synthetic antigenic determinant of the protein. Thus, it was shown that the Rz gene's product is a 17 kDa protein, as predicted from the nts sequence of the gene (Taylor et al. 1983). The N-terminal sequence of the protein bears all of the features of signal sequences yet it is uncleavable as was shown by the comparison of products synthesized *in vivo* and *in vitro*. Apparently the sequence acts as a membrane anchor since the protein proved to be localized in envelope fraction of the cell. Kucharczyk et al., (1991) have shown recently that after the induction of λ prophage dynamical changes occur in bacterial membranes which involve a considerable enlargement of the fraction of intermediate density between those of IM and OM, containing adhesion zones of the membranes (Bayer, 1979; Ishidate et al., 1986), denoted by us (A). The enlargement was found to be dependent on the λ Q gene product necessary for the antitermination of the late transcription but the gene directly responsible for this has not spotted yet. Most of the Rz protein was found in the A fraction but it was distributed also in OM and IM. It is not known if the protein is involved in the A fraction enlargement, like it was shown for L protein of MS2 phage of *E. coli* (Walderich et al., 1988). Moreover the conditions of overproduction of the Rz protein do not allow for a clear statement. The role of the Rz protein in counteracting bacterial membrane stabilization by Mg^{2} ions has not been elucidated yet. One can speculate that it might resemble that described for *pnd* or *srnB* genes of F or R483 plasmids respectively. The proteins coded by these genes insert into bacterial membranes causing intracellular Mg^{2+} release resulting in stable RNA degradation and cells death (Sakikawa et al., 1989). It is remarkable that external Mg^{2+} allows for cells survival.

ACNOWLEDGEMENTS

The financial support from Prof. U. Schwarz and Prof. J.-V. Höltje, from Max-Planck-Institut für Entwicklungsbiologie, Tübingen, Germany, is greatfully acknowledged.

REFERENCES

Bayer, M.E. (1979) The fusion sites between outer membrane and cytoplasmic membrane of bacteria: their role in membrane assembly and virus infection, in "Bacterial Outer Membranes: Biogenesis and Functions" (Inouye, M., Ed.), pp. 167-202. John Wiley & Sons Inc., New York.

Beremand, M.N. and Blumenthal, T. (1979) Overlapping genes in RNA phage: a new protein implicated in lysis. Cell. 18, 257-266.

Bieńkowska-Szewczyk, K., Lipińska, B. and Taylor, A. (1981) The R gene product of bacteriophage λ is the murein transglycosylase. Mol. Gen. Genet. 184, 111-114.

Bradford, M.M. (1976) A rapid and sensitive method for the quantitation of microgram quantities of protein utilizing the principle of protein-dye binding. Anal. Biochem. 72, 248-254.

Friedman, D.J. and Gottesman, M. (1983) Lysis mode of lambda development, in "Lambda II" (Hendrix, R.W., Roberts, J.W., Stahl, F.W. and Weisberg, R.A., Eds.), pp. 21-51. CSH, Cold Spring Harbor Laboratory, New York.

Gierasch, L.M. (1989) Signal sequences. Biochemistry. 28, 923-930.

Ishidate, K., Creeger, E.S., Zrike, J., Deb, S., Glauner, B., Mac Alister, T.J. and Rothfield, L.I. (1986) Isolation of differentiated membrane domains from *Escherichia coli* and *Salmonella typhimurium*, including a fraction containing attachment sites between the inner membrane and outer membranes and the murein skeleton of the cell envelope. J. Biol. Chem. 261, 428-443.

Kucharczyk, K., Laskowska, E. and Taylor, A. (1991) Response of *Escherichia coli* cell membranes to induction of λcI857 prophage by heat shock. Mol. Microbiol. 5, 2935-2945.

Laemmli, U.K. (1970) Cleavage of structural proteins during the assembly of the head of bacteriophage T4, Nature. 227, 680-685.

Maniatis, T., Fritsch, E.F. and Sambrook, J. (1982) "Molecular Cloning: A Laboratory Manual", Cold Spring Harbor, Cold Spring Harbor Laboratory Press, New York.

Owen, P., Graeme, K.A., Crowe, B.A. and Condon, C. (1982) Bacterial membranes: preparative techniques and criteria of purity, in "Techniques in lipid and membrane biochemistry" (Hasketh, T.R., Kornberg, H.L., Metcalfe, J.C., Northcote, D.H., Pogson, C.I. and Tipton, K.F., Eds.), pp. 1-69. Elsevier, North Holland Scientific Publishers Ltd.

Sakikawa, T., Akimoto, S. and Ohnishi, Y. (1989) The pnd gene in *E. coli* plasmid R16: nucleotide sequence and gene expression leading to cell Mg^{2+} release and stable RNA degradation. Biochim. Biophys. Acta. 1007, 158-166.

Sanger, F., Nicklen, S. and Coulson, A.R. (1977) DNA sequencing with chain terminating inhibitors. Proc.Natl.Acad.Sci. 74, 5463-5467.

Studier, F.W., Rosenberg, A.L., Dunn, J.J. and Dubendorf, J.W. (1990) Use of T7 RNA polymerase to direct expression of cloned genes. Methods in Enzymol. 185, 60-89.

Szewczyk, B., and Kozloff, L.M. (1985) A method for the efficient blotting of strongly basic proteins from sodium dodecyl sulfate-polyacrylamide gels to nitrocellulose. Anal. Biochem. 150, 403-407.

Taylor, A., Benedik, M. and Campbell, A. (1983) Location of the Rz gene in bacteriophage lambda. Gene. 26, 159-163.

Von Heijne, G. (1984) How signal sequences maintain cleavage specificity. J. Mol. Biol. 173, 243-251.

Walderich, B., Ursinus-Wössner, A., van Duin, J. and Höltje, J.-V. (1988) Induction of the autolytic system of *Escherichia coli* by specific insertion of bacteriophage MS2 lysis protein into the bacterial cell envelope. J. Bacteriol. 170, 5027-5033.

Witholt, B., Boekhout, M., Brock, M., Kingma, J., van Heerikhuizen, H. and de Leij, L. (1976) An efficient and reproducible procedure in formation of spheroplasts from variously grown *Escherichia coli*. Anal. Biochem. 74, 160-170.

Zagotta, M.T. and Wilson, D.B. (1990) Oligomerization of the bacteriophage lambda S protein in the inner membrane of *Escherichia coli*. J. Bacteriol. 172, 912-921.

PATHWAY OF PHIX174 PROTEIN E MEDIATED LYSIS OF ESCHERICHIA COLI

Angela Witte, Edith Brand, Gerald Schrot, Werner Lubitz

Institute of Microbiology and Genetics
University of Vienna
Dr. Bohr-Gasse 9
A-1030 Vienna, Austria

INTRODUCTION

Genetic analysis of phage PhiX174 mutants (Hutchison and Sinsheimer 1966) identified a single lysis gene of the phage, E, and it was shown that its expression from a plasmid is sufficient to cause bacterial lysis (Henrich et al., 1982; Young and Young, 1982). A summary of specific features of protein E-mediated lysis is given in table 1. Gene E codes for a membrane protein of 91 amino acids (Barrell et al., 1976) detectable in the inner and outer membrane fractions of E. coli (Bläsi et al., 1983; Altman et al., 1985; Bläsi et al., 1989). It exerts its lytic effect by a process which is dependent on the proton-motive-force of the cells (Witte et al., 1987) and on cellular factors which are coupled with the growth phase of the cells as well as with the regulation of the cells autolytic system (Lubitz et al., 1984a,b). E-lysed cells show discrete holes in their cell envelopes which are preferentialy located in potential division zones of the cells (Witte et al., 1990a). Questions which remain to be answered are concerned with the molecular architecture of the E-specific lysis tunnel, its dependence on cellular proteins and on indirect effects of local membrane disturbance.

RESULTS AND DISCUSSION

From inspection of numerous protein E-lysed E. coli cells by scanning and transmission electron microscopy we feel confident in concluding that the location of the E-specific transmembrane lysis tunnel within the envelope is localized at potential division zones of the cells (Witte et al., 1992). The predominant location of the E-lysis tunnel is in the middle of E. coli cells, and in about one third of E-lysed cells in the region of the periseptal annuli (MacAlister et al., 1983). The E-lysis tunnel is characterized by a small opening in the envelope structure

Bacterial Growth and Lysis, Edited by M.A. de Pedro *et al.*, Plenum Press, New York, 1993

Tab. 1. Molecular characterization of gene/protein E and E-mediated lysis

- gene E overlaps gene D with +1 reading frame, has a poor ribosome binding site and rare codon usage (Barrell et al., 1976; Bläsi et al., 1990)
- E is a hydrophobic polypeptide, 91 aa, mw ca. 10 kD (Barrell et al., 1976; Bläsi et al., 1989)
- growth phase of cells and pH of the culture medium influence E-mediated lysis (Lubitz et al., 1984a)
- no enzymatic activity associated with protein E (Eigner et al., 1963; Markert and Zillig, 1965)
- E protein detected in inner and outer membranes but not in cytoplasm, E oligomerizes to multimers (Bläsi et al., 1983,1989; Altman et al., 1985)
- pmf, regulation of autolytic system and/or cell division activities necessary for E-mediated lysis (Lubitz et al., 1984a,b; Witte et al., 1987)
- E-specific limited degradation of peptidoglycan (Witte, 1990)
- E-specific induction of phospholipase activities (Lubitz and Pugsley, 1985)
- breakdown of membrane potential, efflux of K-ions, influx of Na-ions, release of cytoplasmic solutes, enzymes and nucleic acids, limited liberation of periplasmic enzymes at onset of lysis (Witte and Lubitz, 1989)
- transmembrane tunnel formation through envelope complex with fused inner and outer membranes at its borders (Witte et al., 1990a,b)
- lysed cells retain the structural integrity of the envelope complex (Witte et al., 1990a)
- lysis can be inhibited by high concentrations of $MgSO_4$ (Witte et al., 1991)

with borders of fused inner and outer membranes sealing the periplasmic space (Witte et al., 1990a). This feature explains why only 10% of periplasmic enzymes are found to be released during E-mediated lysis (Witte and Lubitz, 1989). The diameter and shape of the lysis tunnel fluctuates between border values of about 40 nm and 200 nm, and between round openings and elongated clefts (Witte et al., 1992). These variations seen in the structure of different E-lysis tunnels indicate that protein E does not assemble into rigid structures like a tube or pilus. It is more likely that the diameter of the lysis tunnel is influenced by the osmotic pressure at the time of lysis as well as the extent of local autolysis and as a consequence of the mesh size within the peptidoglycan net. In PhiX174 infected E. coli cells it has been shown that in the middle of the infection cycle a process is initiated which leads to rapid degradation of newly synthesized peptidoglycan (Lubitz and Plapp, 1980). More recent investigations showed that the peptidoglycan hydrolysis which preceeds E-mediated lysis is more accurately described as stimulation of the peptidoglycan turnover rate (Witte, 1990). As the structure of the sacculus is not impaired such that it could be detected by electron microscopy and since the size distribution of meshes within the sacculus excludes the existence of pores larger than 20 nm (Kellenberger, 1990), we interpret the observed stimulation of the peptidoglycan turnover rate and the varying size distribution of E-lysis tunnels as an E-mediated local enlargement of meshes within the sacculus at potential areas of lysis.

It could well be that the initial effect of tunnel formation is the E-mediated fusion of inner and outer membranes at membrane adhesion sites, and that secondary effects like local activation of phospholipase activities due to membrane perturbations caused by integration of protein E in the envelope complex (Lubitz and Pugsley, 1985), increase the membrane fluidity such that localized areas can no longer withstand the inner osmotic pressure. The consequence of these perturbations is the disruption of the envelope barrier which leaves a hole with fused membranes at its border.
Dynamic reconstructions of bacterial ghosts from serial ultrathin sections of individual E-lysed E. coli cells showed that expulsion of cytoplasm through the E-specific lysis tunnel is very vigorous often followed by inward bending of the inner membrane during the course of cell lysis. Areas of the inner membrane which are not tightly attached to the envelope are extended inward to compensate for the negative pressure produced by the outstreaming cytoplasm. Uptake studies of ^{3}H-water during the time course of E-mediated lysis indicate that only part of the water filling the interior of lysed cells is aquired from diffusion through the inner membrane during lysis and that most of the water originates from influx through the tunnel structure after cell lysis. This indicates that E-mediated lysis is a very fast and vigorous process. However, depending on the host strains used, variations in the time point of lysis onset between 2.5 min and 45 min after gene E expression have been observed (fig. 1).

If our current model of the pathway of E-mediated lysis (fig. 2), is correct the question arises as to what enzymatic activities are responsible for the enlargement of the peptidoglycan net. As protein E has no known enzymatic activity, autolysis is the simplest answer (Lubitz et al., 1984a,b). Many E. coli mutant strains with defects in peptidoglycan metabolism or autolysis have been investigated

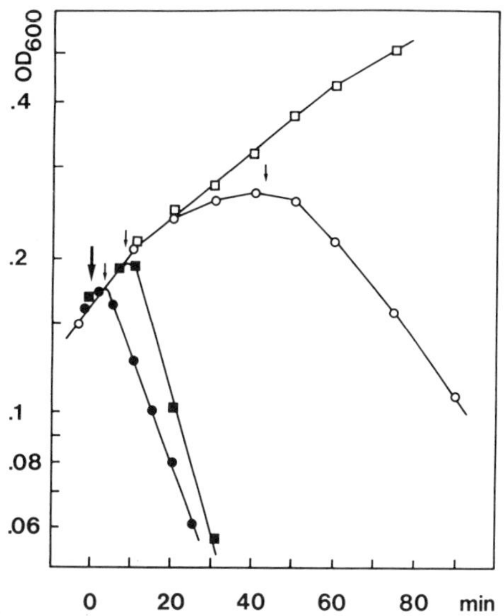

Fig.1. Expression of gene E in different E. coli strains. At time 0 min (large arrow) samples of E. coli strains KN126 (pSB12,pcI857) (■ - ■), VC30 (pSB12,pcI857) (● - ●), and JE841930 (pSB12,pcI857) (○ - ○) grown in rich medium with supplements at 28°C were shifted to 42°C to inactivate cI857 repressor molecules and allow expression of gene E cloned under lambda promoter P_L control carried on plasmid pSB12 (Bläsi et al., 1985). (□ - □): uninduced control of strain JE841930 (pSB12,pcI857). The relevant genotypes are: strain KN126 (wt; Nagata and Horiguchi, 1973), VC30 (lytA; Harkness and Ishiguro, 1983), and JE841930 (mepA,B; Iida et al., 1983). Onset of E-mediated lysis (small arrows) occurs at 2.5 min for strain VC30, 10 min for strain KN126, and 45 min for strain JE841930.

Tab. 2. Host cell factors which affect protein E-mediated lysis

inhibition	retardation	no effect
stationary growth[1]	growth rate[1]	lytB[2]
pmf[3]	fabB, fadE[4]	mrcA(pbp1a)[5]
envC + rle[4]	envC[4]	mrcB(pbp1b)[5]
lytA[6]	mepA,B[5]	pbpA(pbp2)[5]
slyD[7]	wee[8]	pbpB(pbp3)[5]
pxlA[8]		dacB(pbp4)[5]
ftsZ[9]		ftsA3, ftsQ[9]
ftsA12[9]		relA[10]
		pldA1[11]

[1]Lubitz et al., 1984a; [2]Halfmann and Lubitz, 1986; [3]Witte et al., 1987; [4]Wadle et al., 1986; [5]Witte et al., 1990b; [6]Lubitz et al., 1984b; [7]Maratea et al., 1985; [8]Halfmann, 1989; [9]Brand, Witte and Lubitz, unpublished; [10]Witte, 1990; [11]Lubitz and Pugsley, 1985.

to elucidate the enzymatic activities necessary for E-mediated lysis. None of the mutants investigated defective in single peptidoglycan metabolic enzymes exhibits an inhibitory effect on E-mediated lysis (Tab. 2). Only E. coli mutant strains selected for E tolerance e.g., slyD (Maratea et al., 1985), rle in combination with envZ (Wadle et al., 1986), or pxlA (Halfmann, 1989), or mutant strains with defects in regulatory functions of autolysis like lytA (Lubitz et al., 1984b), or functions involved in initiation of cell division like ftsZ and ftsA12 (Brand, Witte and Lubitz unpublished) exhibit an E-lysis negative phenotype (Tab. 2).
Modulatory functions which affect onset of E-mediated lysis

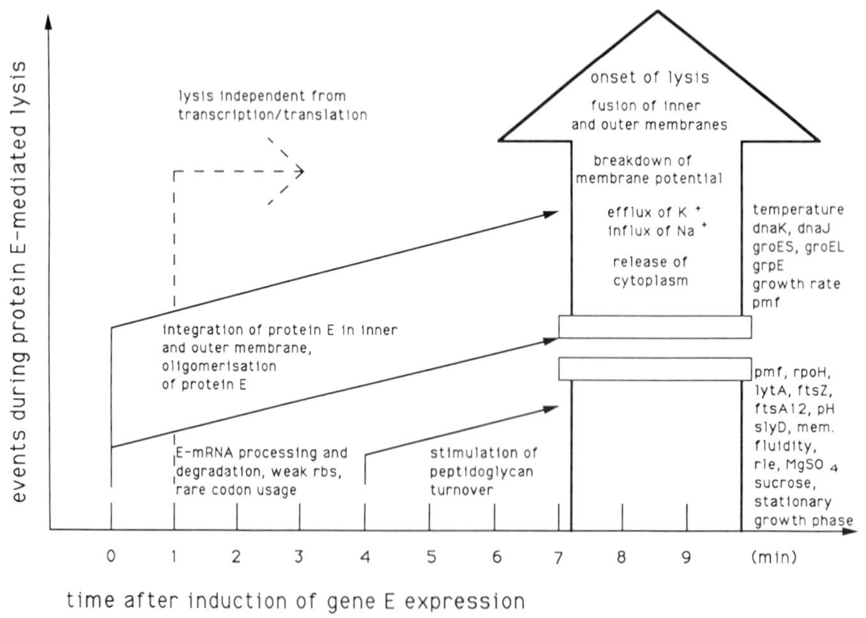

Fig.2. Events during course of protein E-mediated lysis after induction of gene E expression. *Most events represented on the flow chart are described in the text. Effects of dnaK, dnaJ, groES, groEL, grpE, and rpoH were described by Young et al., (1989); E-mRNA processing and degradation is described by Halfmann (1989).*

or efficiency of lysis have also been described and are summarized in figure 2.

The mutant strains investigated and cellular prerequisites determined for E-mediated lysis do not resolve a clear picture of the molecular mechanism of E-mediated lysis. Although the puzzle has still many pieces missing, our current model of E-mediated tunnel formation, which is based on immunelectron microscopy and protease K protection studies, is presented in figure 3. According to the model, protein E integrates co-translationally into the inner membrane (stage 1), and by lateral diffusion, oligomerizes and targets the cell division initiation complex (stage 2). In these areas of membrane attachment protein E somehow achieves fusion of the inner and outer membranes resulting in tunnel formation (stage 3). Whether this process is dependent on a conformational change of protein E (C-terminus of protein E is accessible to protease K from the outside) or whether protein E molecules are also distributed in the outer membrane fusion is still a question to be resolved.

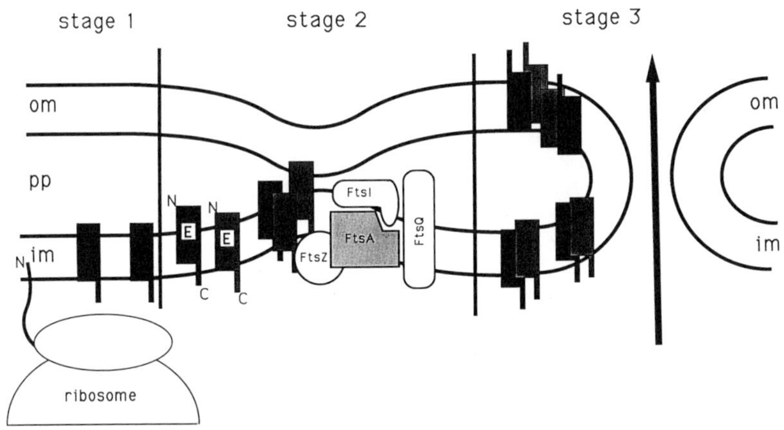

Fig.3. Model of protein E-mediated lysis. Events during the different stages of E-mediated lysis are descibed in the text. im, inner membrane; om, outer membrane; pp, periplasmic space. Dark and shadowed flags represent protein E molecules with the orientation of the N- and C-termini indicated (N and C, respectively); FtsZ, FtsA, FtsI and FtsQ represent the cell division complex. The large arrow gives the direction of outstreaming cytoplasm from E-lysed cells.

Acknowledgements

We are grateful to Dr. K. Tedin for critical reading of the manuscript. This work was supported by a grant from the Austrian Fonds zur Förderung der wissenschaftlichen Forschung (P8465-MOB).

References

Altman E, Young KD, Garret J, Altman R, Young R (1985) Subcellular location of lethal lysis proteins of bacteriophages lambda and PhiX174. J Virol 53:1008-1011.

Barrel BG, Air GM, Hutchison III CA (1976) Overlapping genes in bacteriophage PhiX174. Nature 264:34-41.
Bläsi U, Geisen R, Lubitz W, Henrich B, Plapp R (1983) Localization of the bacteriophage PhiX174 lysis gene product in the cell envelope of Escherichia coli. In: Hakenbeck R, Höltje JV, Labischinski H (eds) Target of Penicillin. de Gruyter, Berlin, New York, pp 205-210.
Bläsi U, Henrich B, Lubitz W (1985) Lysis of Escherichia coli by cloned PhiX174 gene E depends on its expression. J Gen Microbiol 131:1107-1114.
Bläsi U, Linke RP, Lubitz W (1989) Evidence for membrane-bound oligomerization of bacteriophage PhiX174 lysis protein E. J Biol Chem 264:4552-4558.
Bläsi U, Nam K, Lubitz W, Young R (1990) Translational efficiency of PhiX174 lysis gene E is unaffected by upstream translation of the overlapping gene D reading frame. J Bacteriol 172:5617-5623.
Eigner J, Stouthamer AH, Van der Sluys J, Cohen JA (1963) A study of the 70S component of bacteriophage PhiX174. J Mol Biol 6:61-84.
Halfmann G, Leduc M, Lubitz W (1984) Different sensitivity of autolytic deficient Escherichia coli mutants to the mode of induction. FEMS Microbiol Lett 24:205-208.
Halfmann G, Lubitz W (1986) Different induction of Escherichia coli autolysis by penicillin and the bacteriophage PhiX174 gene E product. J Bacteriol 66:683-685.
Halfmann G (1989) Einfluß zellulärer Mutationen auf die PhiX174 Protein E-induzierte Lyse von Escherichia coli. PhD-Thesis, University of Munich.
Harkness RE, Ishiguro EE (1983) Temperature-sensitive autolysis-defective mutants of Escherichia coli. J Bacteriol 155:15-21.
Henrich B, Lubitz W, Plapp R (1982) Lysis of Escherichia coli by induction of cloned PhiX174 genes. Mol Gen Genet 185:493-497.
Hutchison III CA, Sinsheimer RL (1966) The process of infection with bacteriophage PhiX174. Mutations in a PhiX174 lysis gene. J Mol Biol 18:429-447.
Iida K, Hirota Y, Schwarz U (1983) Mutants of Escherichia coli defective in penicillin-insensitive murein DD-endopeptidase. Mol Gen Genet 189:215-221.
Kellenberger E, (1990) The 'Bayer bridges' confrontated with results from improved electron microscopy methods. Mol Microbiol 4:697-705.
Lubitz W, Plapp R (1980) Murein degradation in Escherichia coli infected with bacteriophage PhiX174. Curr Microbiol 4:301-304.
Lubitz W, Halfmann G, Plapp R (1984a) Lysis of Escherichia coli after infection with PhiX174 depends on the regulation of the cellular autolytic system J Gen Microbiol 130:1079-1087.
Lubitz W, Harkness RE, Ishiguro EE (1984b) Requirement for a functional host cell autolytic system for lysis of Escherichia coli by bacteriophage PhiX174. J Bacteriol 159:385-387.
Lubitz W, Pugsley AP (1985) Changes in cell phospholipid composition of PhiX174 gene E product. FEMS Microbiol Lett 30:171-175.

MacAlister TJ, MacDonald B, Rothfield LI (1983) The periseptal annulus: an organelle associated with cell division in Gram-negative bacteria. Proc Natl Acad Sci USA 80:1372-1376.
Maratea D, Young Y, Young R (1985) Deletion and fusion analysis of the phage PhiX174 lysis gene E. Gene 40:39-46.
Markert A, Zillig W (1965) Studies on the lysis of Escherichia coli by bacteriophage PhiX174. Virology 25:88-97.
Nagata T, Horiuchi T (1973) Isolation and characterization of a temperature-sensitive amber suppressor mutant of Escherichia coli K12. Mol Gen Genet 123:77-88.
Wadle D, Henrich B, Plapp R (1986) Effects of mutations in genes fadR, fabB, fadE, and envC of Escherichia coli on the action of the lysis gene of bacteriophage PhiX174. Curr. Microbiol. 14:65-69.
Witte A (1990) Untersuchungen zur Wirkung von PhiX174 Protein E sowie davon abgeleiteter chimärer Proteine auf den Zellwandkomplex von Escherichia coli. PhD-Thesis, University of Munich.
Witte A, Lubitz W, Bakker EP (1987) Proton-motive-force -dependent step in the pathway to lysis of Escherichia coli induced by bacteriophage PhiX174 gene E product. J Bacteriol 169:1750-1752.
Witte A, Lubitz W (1989) Biochemical characterization of PhiX174-protein-E-mediated lysis of Escherichia coli. Eur J Biochem 180:393-398.
Witte A, Wanner G, Bläsi U, Halfmann G, Szostak M, Lubitz W (1990a) Endogenous transmembrane tunnel formation mediated by PhiX174 protein E. J Bacteriol 172:4109-4114.
Witte A, Bläsi U, Halfmann G, Szostak M, Wanner G. Lubitz W (1990b) PhiX174 protein E-mediated lysis of Escherichia coli. Biochimie 72:191-200.
Witte A, Narendja F, Säckl W, Eckerstorfer M, Szostak M, Bläsi U, Halfmann G, Lubitz W (1991) Multifunctional safety vector systems for DNA cloning, controlled expression of fusion genes, and simplified preparation of vector DNA and recombinant gene products. BTF5, Advances Mol Gen 3:219-239.
Witte A, Wanner G, Sulzner M, Lubitz W (1992) Dynamics of PhiX174 protein E-mediated lysis of Escherichia coli. Arch. Microbiol. 157, 381-388.
Young KD, Young R (1982) Lytic action of cloned PhiX174 gene E. J Virol 44:993-1002.
Young KD, Anderson RJ, Hafner RJ (1989) Lysis of Escherichia coli by the bacteriophage PhiX174 E protein: inhibition of lysis by heat shock proteins. J Bacteriol 171:4334-4341.

GENETIC CONTROL OF FUNGAL CELL WALL AUTOLYSIS

César Nombela, Gloria Molero, Humberto Martín, Rosa Cenamor, María Molina and Miguel Sánchez

Departamento de Microbiología II
Facultad de Farmacia. Universidad Complutense
28040-Madrid. Spain

INTRODUCTION

The major structural components of fungal cell walls are usually polysaccharide homopolymers, such as glucans and chitin, with chemical structures simpler than those of bacterial peptidoglycan. However, fungal cell walls are also thick and complex envelopes playing the same role as bacterial cell walls. For a number of years we have been addressing the question of the biological role of fungal autolysins, enzymes that can be recognized biochemically as being capable of degrading components of the producer's cell wall. The autolytic action of these enzymes on fungal walls has always been envisaged as the controlled hydrolytic modification of the wall polymers in many situations that might require some kind of modification of the wall structure. Therefore fungal autolysins have been proposed as agents that may be critical for fungal morphogenesis, and they must represent basic elements contributing to the dynamics of the wall structure, that involves a series of changes throughout the yeast mitotic cell cycle and apical extension of hyphae, as well as in other processes such as sexual conjugation, meiosis, dimorphism, etc.

A set of different enzyme activities has been characterized as potential wall hydrolases (Fleet, 1984), but the two major and probably most relevant ones are glucanases and chitinases. This would imply that the fungal wall autolytic systems are simple and accessible. However, the multiplicity of enzymes with some of these activities, such as glucanase, produced by a single organism, and the diversity in the properties of specific hydrolases found in different species, as judged from biochemical data, has suggested that the phenomenom of fungal cell wall autolysis is not a simple one to characterize from the point of view of its biological role (Nombela *et al.*, 1988).

In any case, it is worth mentioning that the clarification of the relevance of fungal cell wall autolysins is interesting not only from the point of view of basic knowledge. Several other aspects, such as the modification of cell wall dynamics to promote secretion and/or release of heterologous proteins from yeast of biotechnological interest, should benefit substantially from the clarification of the role of enzymes hydrolyzing fungal cell wall

Bacterial Growth and Lysis, Edited by M.A. de Pedro *et al.*, Plenum Press, New York, 1993

components. These studies could also contribute to the identification of potential targets for the new antifungal agents that are needed in view of the increasing incidence of mycotic infections, which might be capable of inducing cell wall autolysis, thus leading to a fungicidal effect (Nombela *et al.*, 1992). This new knowledge is beginning to emerge mainly through studies at the molecular level and will be the subject of this discussion.

ARE WALL HYDROLASES NEEDED FOR GROWTH OF THE FUNGAL CELL?

The finding that bacterial and fungal cells produce hydrolases of their cell wall components has led to different hypotheses about the role of these enzymes. Classical views regarding the function of fungal autolysins were based on the ideas previously developed for bacterial systems and postulated that any aspect of growth and morphogenesis of the cell wall would require not only the biosynthesis and transport of new materials to the structure but the assembly of them at the proper location, which must involve specific mechanisms. According to these models, the assembly of new components in the wall structure demanded the opening of areas for deposit of new materials, due to the hydrolytic action of the wall degrading enzymes, that would act at the proper positions and in a very well controlled manner (Bartnicki-García, 1973).

The consideration of the data regarding wall hydrolases that had been characterized in most detail, such as glucanases of *Saccharomyces cerevisiae*, made it easy to apply these ideas in general terms. For example, a substantial number of glucan hydrolases are produced by this species (Nombela *et al.*, 1988). Endo- and exo-glucanases, acting at intermediate or end points of the polymer chains respectively, are secreted by the cell and were postulated to act on the modification of glucan, either at growth areas or at any other location. The pattern of regulation of glucanase production in this yeast also fits in the current models since enzyme production was essentially growth-associated (del Rey *et al.*, 1979), which might mean that yeast glucanases were essential for morphogenetic processes taking place during growth. In addition to the aforementioned evidence, it was also documented that significant processes in yeast biological cycle, such as meiosis and sporulation, and yeast-to-mycelium transition (in *C.albicans*), led to significant modification of the pattern of glucanase production, such as induction of new enzymes or repression of others produced during vegetative growth.

However, experiments designed to address the question of the role of specific hydrolases in a more direct manner have sometimes led to unexpected results suggesting that the problem is more complex than anticipated. The most significant finding in this regard was that *S.cerevisiae* mutants, deficient in the most abundant glucanase produced by this species, displayed an absolutely normal phenotype. The dispensability of the major exo-glucanase, EXG1, of *S.cerevisiae* has been fully confirmed by disruption of the corresponding gene (Nebreda *et al.*, 1986). Contrarily, the fundamental role as wall hydrolase of another *S.cerevisiae* enzyme, namely chitinase CTS1, has been fully confirmed by molecular cloning and disruption of the coding gene (Kuranda and Robbins, 1991). The cells deficient in this activity did not separate at the end of cell division, and the enzyme was unequivocally shown to act by hydrolyzing chitin at the septum (Cabib *et al.*, 1992).

Therefore, the answer to the question of the role of fungal autolysins is not a simple one. Specific enzymes with capacities to degrade fungal components may or may not be critical for growth and the roles played by different enzymes may also be different. In any case, the variety of enzymes under consideration, and the fact that many of these enzymes identified as potential autolysins are produced in very small amounts, requires multiple experimental approaches to this question. One is, of course, the cloning of genes coding for enzymes hydrolyzing cell wall components, based on the isolation of the enzyme itself.

The other is the characterization of mutants isolated on the basis of alterations in the functions related to cell wall metabolism.

MUTANTS ALTERED IN CELL WALL AUTOLYSIS

The work of Cabib and Duran (1975) describing a procedure for isolating yeast mutants with alterations in the formation of a stable cell wall opened the way for genetic approaches to the general question of the role of yeast cell wall autolysis, as well as its regulation and significance for growth of the cells. Conditional autolytic mutants that express a lethal phenotype leading to the release of the intracellular content, usually at a non-permissive temperature, can be isolated by various strategies. The feasibility of this isolation and the corresponding characterization of the mutants isolated, has led to the identification of yeast genes that can be considered relevant for the formation of an osmotically stable cell wall. Thus the relevant characteristic of autolytic mutants is the alteration in the cell wall structure that can be recognized by the release of the intracellular content. Expression of the autolytic mutations generally requires active growth of the cells that lyse after reaching a terminal stage, characteristic of each mutant. The fact that the cells from some of the mutants lyse at the same stage suggests that the gene affected could be involved in the control of the mitotic cycle.

The existence of many different autolytic mutants could mean that the formation of a stable cell wall is a complicated process, that requires proper controls, and is very sensitive to alterations in the corresponding mechanisms that are operative during cell growth. The fact that the terminal phenotype is cell lysis, suggests that some kind of controlled autolysis of cell wall components is essential for growth and morphogenesis, and it must be precisely regulated so that the corresponding mutation eventually alters the equilibrium and triggers an umbalanced hydrolytic action on the cell wall. Of course, the nature of the genetic defect in some autolytic mutants could be directly related to the regulation of autolytic mechanisms or, alternatively, it might be a deficiency in some other aspect of cellular mechanisms such as synthesis of cell wall components, that eventually leads to autolysis. Characterization of these mutants is beginning to unravel some of the regulatory mechanisms controlling cell wall autolysis and, in general, the requirements for the formation of a stable cell wall as an essential process in fungal morphogenesis. At this point it would be premature to try to devise any definitive model as emerging from the studies on autolytic mutants. But the possibilities of this type of analysis are revealed by the increasing number of genes related to these functions that are beginning to be characterized by different means.

Two categories can be established among yeast autolytic mutants depending on whether the lethal phenotype can be rescued or not by osmotic stabilization with sorbitol. Obviously the former can affected in functions more related to the cell wall, whereas it is conceivable that the cell wall alterations in the latter could, at least hypothetically, be attributed to a pleiotropic effect. However, the existence of direct cell wall alterations in non-osmotic remedial mutants cannot be ruled out.

The frequency of autolytic mutations is high enough so that systematic surveys of yeast mutagenized populations to search for autolytic mutants allows the isolation of a significant number of interesting mutant strains. In addition to the original mutants described by Cabib and Durán (1975), a series of autolytic strains have been isolated in *Candida albicans* (Payton and de Tiani, 1990), *Schizosaccharomyces pombe* (Ribas *et al.*, 1991) and *Aspergillus nidulans* (Borgia and Dodge, 1992). San Segundo and del Rey (personal communication) have isolated a series of *S.cerevisiae* sorbitol-rescued autolytic mutants defining at least five complementation groups. Recent efforts in our laboratory have led successfully to the identification of autolytic mutations that are dominant in *S.cerevisiae* diploid cells (V. Jimenez, unpublished). The high frequency and feasibility of inducing autolytic mutations represents a clear indication that the number of genes involved in functions, more or less directly, related to cell wall autolysis and its regulation is high. The characterization of such a high number of genes is going to represents obvious difficulties, but the combination of the available methodologies available, especially those based on Flow Cytometric analysis of the lytic phenotype (de la Fuente *et al*, 1992) and on techniques of molecular cloning should be an efficient strategy.

It is remarkable that many of the mutants reported (see below) that clearly fit in the category of osmotic-remedial autolytic mutants have not been isolated in the course of a screening designed to search for these mutants, but by disruption of genes that had been cloned by other strategies.

GENES RELATED TO FUNGAL CELL WALL AUTOLYSIS

Recent advances in the characterization of autolytic mutants are leading to a remarkable amount of data regarding several genes that may be related to functions of cell wall autolysis. The phenotypes of many of these mutants suggest that the genes have an implication in morphogenesis and/or in the control of cell division during the mitotic cycle as well as other processes that include cell wall modifications, such as mating. The nature of the genes themselves suggests that they are regulatory. Therefore, as it will be shown, it is difficult to present a complete picture regarding the role of fungal cell wall autolysis, that must be postponed for the outcome of many of the studies in progress in different laboratories. However, a revision of these data should be helpful to understand the present trend in this field. For simplicity we shall consider four groups of genes in this regard. The most signifcant ones are shown in Table 1.

Structural genes for fungal cell wall hydrolases

Cloning of genes coding for exo-1,3-ß-glucanases has advanced substantially and led to the conclusion that the major exo-glucanase (EXG1) of *S.cerevisiae*, that amounts to more than half of the total secreted activity in this yeast, is dispensable, no alterations being observed in null mutants of gene *EXG1* (Nebreda *et al.*, 1986). Despite the dispensability of gene *EXG1* at least two other exo-glucanase genes, coding for another exo-glucanase of low abundance, EXG2, and for the sporulation-specific enzyme SSG1 (del Rey *et al.*, 1980) have been isolated and are being characterized (F. del Rey, personal communication). A gene coding for a cell wall protein of the same species has also been cloned and shown to code for a protein with exo-glucanase activity (Klebl and Tanner, 1989).

It is difficult to explain why a single species has developed a whole set of enzymes with the capacity to hydrolyze glucan at the ends of the polymer chain, some of them dispensable, especially if we consider that several endo-glucanases are also produced (Nombela *et al.*, 1988). But, in any case, it should be a reflection of the complexity of cell wall autolytic systems that may require the isolation and characterization of many genes to obtain definite data about their role. Studies are on the way in our laboratory to clone several genes for endo-glucanases that have been purified. Until definite results are obtained regarding the functions of fungal glucanases the indication that glucanases may not only be hydrolases but they could act as transferases (Sullivan *et al.*, 1991) represents an attractive hypothesis.

The situation seems to be much simpler with regard to chitin hydrolases, at least in *S.cerevisiae* that carries a limited amount of chitin in the cell wall. Kuranda and Robbins (1991) have recently carried out a very complete characterization of gene *CTS1* that codes for a chitinase. Several domains, including the active center and substrate binding regions, have been characterized in the protein, which is critically required for cell separation at the end of cell division. It is also interesting that this chitinase has been implicated in cell wall autolysis, under altered conditions such as deficiency in chitin synthesis due the disruption of the chitin synthetase gene *CHS1* (Cabib *et al.*, 1992). These observations have also contributed data suggesting that in addition to chitinase other gene products are required

Table 1. Some *S.cerevisiae* genes related to phenomena of cell wall autolysis.

Gene	Product	Characteristics	Reference(s)
EXG1	Most abundant form of vegetative exo-1,3-ß-glucanase	Dispensable for growth, mating and sporulation	Nebreda *et al.*, 1986
CTS1	Chitinase	Septum chitin hydrolysis for cell separation	Kuranda and Robbins, 1991
LYT1	Unknown	Osmotically unstable *lyt1* conditional autolytic mutants not complemented by sorbitol	Nombela and Santamaría, 1984
LYT2	Unknown	Osmotically unstable *lyt2* conditional autolytic mutants complemented by sorbitol	Torres *et al.*, 1991
SLT2	Protein kinase 56-KD	*slt2* mutants osmotically unstable. Gene suppresses *lyt2* mutations	Torres *et al.*, 1991
BCK1/ SLK1	Protein kinase 163-KD	*bck1/slk1* mutants osmotically unstable. Mutations in this gene suppress other autolytic phenotypes	Lee and Levin, 1992; Costigan *et al.*, 1992
PKC1	Protein Kinase C1	*pkc1* mutants osmotically unstable and suppressed by mutations in *BCK1/SLK1*	Lee and Levin, 1992
SRB1	Unknown	*srb1* mutants dependent on an osmotic stabilizer	Stateva *et al.*, 1991
SMP3	Unknown	*smp3* mutants osmotically unstable	Irie *et al.*, 1991

for autolysis and represent a clear confirmation of the usefulness of autolytic mutants as systems for characterization of proteins involved in cell wall autolysis.

Genes affected in non-osmotic remedial autolytic mutants

Our group is currently characterizing one of these mutants that has enabled us to identify gene *LYT1*. Other mutants of this type have also been isolated (V. Jimenez, unpublished). Thermosensitive *lyt1* mutants grow to the stage that the bud reaches the size of almost a daughter cell, and lyse at the non-permissive temperature of 37°C (de la Fuente *et al.*, 1992). Gene *LYT1* has not been cloned so far, but a suppressor of the mutant phenotype determined by *lyt1* mutations has been isolated and identified as being the *S.cerevisiae* sporulation gene *SPO12* (Molero, Nombela and Sánchez, unpublished). Multiple copies of *SPO12* are required for suppression. The function of *SPO12* gene product is not known and this makes it more difficult to interpret the significance of these findings. In any case, it is remarkable that a meiotic gene, such as *SPO12*, can be expressed during the mitotic cycle and complement mitotic functions, when carried in a multicopy plasmid. Cloning and characterization of *LYT1* gene, which is also in progress, should enable us to clarify the functions involved in these phenomena.

Genes characterized in studies on osmotic-remedial autolytic mutants

Studies on autolytic mutants rescued under osmotic stabilization with sorbitol have provided significant information on many genes even at the molecular level. These conditional mutants lyse when actively growing, usually at non-permissive temperature, but viability and growth are restored by the addtion of 1M sorbitol to the medium. It follows that the corresponding genetic traits determine a failure in the capacity to assemble a stable cell wall.

Ribas *et al.* (1991) have identified, by complementation analysis, up to fourteen loci of *Sch.pombe.* The study of the different mutant strains showed mainly alterations in cell wall composition, specially a reduction in some components such as ß-glucan or galactomannan that correlate with changes in the shape of cells that grow, for example, as rounded forms. Borgia and Dodge (1992) identified mutations in two genes that lead to autolytic phenotypes in *A.nidulans*. Again various alterations in cell wall composition of the mutants were observed, leading to reduction in chitin or α-1,3-glucan; and remediation of the lytic phenotype was not only achieved by restoring osmotically favourable conditions with sorbitol, but by addittion of N-acetylglucosamine. In the absence of any further genetic studies it is not possible to conclude anything regarding the functions of the genes affected in these mutants. But, it can be emphasized that physiological studies on the activity of wall synthetases in these mutants do not show the disappearance of particular synthetases but alterations in the levels of these enzymes. It follows that the mutations must determine changes mainly in the regulation patterns of a complex set of mechanisms.

S.cerevisiae continues to be the fungal species most accessible to many types of genetic analysis and therefore the characterization of several of these genes have advanced more than in any other systems. Gene *LYT2* is involved in growth functions so that *lyt2* mutations lead to a thermosensitive autolytic phenotype, that can be followed by the release of intramembrane enzymes such as alkaline phosphatase as well as by the observation of lysed cells, most of them with a small bud, suggesting that growth is arrested at initial stages of mitotic cycle (Torres *et al.*, 1991). Sorbitol protection was clearly demonstrated by precise analysis of large populations carried out by means of Flow Cytometry (de la Fuente *et al.*, 1992). Attempts to clone the structural gene *LYT2* so far have not met success, but led to the isolation of a suppressor of *lyt2* mutation, namely *SLT2* which has opened the way for the identification of protein kinase genes with an involvement in morphogenesis and cell wall functions.

Protein kinase genes related to yeast autolytic mutants

Gene *SLT2* is a homolog of protein kinase genes that was isolated on the basis of complementation of the lytic phenotype of *lyt2* mutants of *S.cerevisiae* (Torres *et al.*, 1991). The complementation occurred in *lyt2* strains transformed with either multicopy or low copy plasmids with gene *SLT2*. Deletion of *SLT2* in wild type strains determined a thermosensitive autolytic phenotype. However, the segregation pattern of *lyt2* and *slt2* mutations, in diploids heterozygous for both alleles, clearly indicated that *LYT2* and *SLT2* are different genes. The characteristics of *slt2* mutants clearly showed that protein kinase SLT2 is needed for growth at 37°C, the lack of the enzyme causing the formation of osmotically sensitive cells that can be protected from lysis by 1 M sorbitol. The requirement of SLT2 protein for stable growth is further emphasized by observations with transformants carrying plasmids that allow regulation of the expression of *SLT2*. Both overexpression and inhibition of expression of this gene affected growth and viability of the cells (H. Martín, unpublished) suggesting that the level of the protein must be precisely regulated during the mitotic cycle. *SLT2* was the first protein kinase gene which has been shown to be critically needed for the formation of a osmotically stable cell wall. However, a previous report showed that a protein kinase, *Kin1*$^{+}$ of *Schz.pombe*, is required for morphogenesis so that cells lacking the gene product grow as rods and display alterations in cell wall composition (Levin and Bishop, 1990). Protein kinase genes must obviously be involved in regulatory functions related to this phenomenon. Therefore these findings should open the way to the clarification of some crucial regulatory events for the assembly of a stable cell wall, based on the identification of whatever substrate(s) are phosphorilated by the corresponding kinases.

SLT2 is not the only protein kinase required for stable cell wall formation. In very recent reports two groups have independently characterized another gene encoding a 163-KD protein that carries all the characteristic domains of protein kinases. Mutations in this gene, named *BCK1* (Lee and Levin, 1992) and *SLK1* (Costigan *et al.*, 1992), bypass the requirement for growth of another well characterized protein kinase, PKC. Or, on the contrary, they lead to a strict requirement of the product of gene SPA2, dispensable in wild type, for vegetative growth. PKC participates in the yeast cell response to extracellular signals and protein SPA2 localizes at sites involved in polarize cell growth in budding and mating cells, and seems to be important mainly for mating. Therefore the protein kinase BCK1/SLK1 must be part of functional pathways, so far unknown, relevant to morphogenesis.But for this discussion is worth mentioning that deletion of this gene resulted in a temperature-sensitive cell lysis defect which is suppressed by osmotic stabilizing agents. Therefore the phenotype of these mutants was very similar to that of *slt2* mutants. Since disruption of PKC can also generate autolytic mutants we can conclude that there are several protein kinases critically required for growth of *S.cerevisiae* at high temperature, so that the lethal phenotype is produced by the lack of formation of an osmotically stable cell wall.

The relationship between the functions controlled by all these protein kinases remains to be established. BCK1 and PKC1 may normally function in the same pathway, as shown by the results obtained with the corresponding mutants (Lee and Levin, 1992). But we have observed that gene *PKC1* (kindly provided by Dr. M. Payton) did not complement either *lyt2* or *slt2* mutations. We conclude that at least two pathways controlled by protein kinases, involving both PKC1 and BCK1/SLK1, and SLT2 must be operative in the mechanisms that are eventaully required for osmotically stable cell wall formation. Further characterization of these genes and the cloning of *LYT2*, should eventually lead to further understanding of their relationship.

Other genes related to cell wall autolysis

The isolation of other mutants with different phenotypes have led to the identification of some other genes that can be included in the groups that are being discussed here. Venkov *at al.* (1974) described fragile mutants of *S.cerevisiae* that, in contrast to the conditional ones that lyse at non-permissive temperature, they are unstable under all conditions and need osmotic stabilization because they lyse spontaneously when transferred to hypotonic medium. Cell wall functions must be altered in these mutant strains. The cloning of a DNA fragment complementing some of these phenotypic alterations in one of the fragile mutants led to the isolation of gene *SRB1* coding for a product unknown so far (Stateva *et al.*, 1991). Other studies not directly related to cell wall functions may also led to autolytic mutants. This is the case of *smp3* mutant, selected on the basis stable maintenance of heterologous plasmid in *S.cerevisiae* (Irie *et al.*, 1991). This mutant lyses at the restrictive temperature and the temperature-sensitive growth is suppressed by the addition of 1M sorbitol (Irie, personal communication).

FINAL COMMENTS AND OUTLOOK

The role of fungal enzymes, involved in autodegradation of cell wall components, and in the mechanisms of cell wall autolysis in fungal cells, continues to be a general question. Only in the case of chitinase of *S.cerevisiae* some unequivocal evidence has been obtained that implies this hydrolytic enzyme in specific functions required cell separation (Kuranda and Robbins, 1991). The rest of the putative fungal autolysins, that have been biochemically characterized, still remain to be enzymes in search of a function in the controlled modification of the cell wall structural components. Outstanding among these are ß-glucanases, in view of the abundance and variety of these enzymes. It is expected the cloning of glucanase genes, which is rapidly advancing, will produce the data required to establish their function.

The possibilities of the new experimental approaches, based on molecular genetic techniques, are beginning to yield information that suggests the involvement of many complex mechanisms of cell wall modification by autolysis in morphogenesis and growth of the fungal cells. The most accessible fungal systems to this type of approaches, namely the yeasts *S.cerevisiae* and *Schz.pombe*, are now receiving significant attention in this regard. Autolytic mutants, isolated from these species and properly characterized, are providing new evidence regarding the genes controlling cell wall autolysis. The identification of protein kinases as some of these genes seeems to be particularly relevant, since these regulatory enzymes might be controlling crucial events in the development of a stable cell wall. The wealth of new data appearing in very recent times indicates that the field is advancing rapidly and we can expect to develop new ideas in models in the next future.

Acknowledgements

Work in the authors laboratory was supported by a grant from Comisión Interministerial de Ciencia y Tecnología. The support from Glaxo Spain is gratefully acknowledged.

REFERENCES

Bartnicki-García, S. 1973. Fundamental aspects of hyphal morphogenesis. *Symposia of the Society for General Microbiology* 23:245.

Borgia, P.B. and Dodge, C.L. 1992. Characterization of *Aspergillus nidulans* mutants deficient in cell wall chitin or glucan. *J. Bacteriol.* 174:377.

Cabib, E. and Duran, A. 1975. Simple and sensitive procedure for screening yeast mutants that lyse at non-permissive temperatures. *J. Bacteriol.* 124:1604.
Cabib, E., Silverman, S.J. and Shaw, J.A. 1992. Chitinase and chitin synthase 1: counterbalancing activities in cell separation of *Saccharomyces cerevisiae. J. Gen. Microbiol.* 138:97.
Costigan, C., Gehrung, S. and Snyder, M. 1992. A synthetic lethal screen identifies SLK1, a novel protein kinase homolog implicated in yeast cell morphogenesis and cell growth. *Mol. Cell. Biol.* 12:1162.
de la Fuente, J., Alvarez, A., Nombela, C. and Sánchez, M. 1992. Flow cytometric analysis of *Saccharomyces cerevisiae* autolytic mutants and protoplasts. *Yeast* 8:39.
del Rey, F., García-Acha, I. and Nombela, C. 1979. The regulation of ß-glucanase synthesis in fungi and yeast. *J. Gen. Microbiol.* 110:83.
del Rey, F., Santos, T., García-Acha, I. and Nombela, C. 1980. Synthesis of ß-glucanases during sporulation in *Saccharomyces cerevisiae*: formation of a new sporulation-specific 1,3-ß-glucanase. *J. Bacteriol.* 146:621.
Fleet, G.H. 1984. The occurence and function of endogenous wall degrading enzymes in yeasts, *in*: "Microbial cell wall synthesis and autolysis", C.Nombela (ed.). Elsevier Science Publishers.
Irie, K., Araki, H. and Oshima, Y. 1991. Mutations in a *Saccharomyces cerevisiae* host showing increased holding stability of the heterologous plasmid pSR1. *Mol. Gen. Genet.* 225:257.
Klebl, F. and Tanner, W. 1989. Molecular cloning of a cell wall exo-1,3-ß-glucanase from *Saccharomyces cerevisiae. J. Bacteriol.* 171:6259.
Kuranda, M.J. and Robbins, P.W. 1991. Chitinase is required for cell separation during growth of *Saccharomyces cerevisiae. J. Biol. Chem.* 266:19758.
Lee, K.S. and Levin, D.E. 1992. Dominant mutations in a gene encoding a putative protein kinase (*BCK1*) bypass the requirement for a *Saccharomyces cerevisiae* protein kinase C homolog. *Mol. Cell. Biol.* 12:172.
Levin, D.E. and Bishop, J.M. 1990. A putative protein kinase ($Kin1^+$) is important for growth polarity in *Schizosaccharomyces pombe. Proc. Nat. Acad. Sci.* 87:8272.
Nebreda, M., Villa, T.G., Villanueva, J.R. and del Rey, F. 1986. Cloning of genes related to exo-ß-production in *Saccharomyces cerevisiae*: characterization of an exo-ß-glucanase structural gene. *Gene* 47:245.
Nombela, C. and Santamaría, C. 1984. Genetics of yeast cell wall autolysis, *in*: "Microbial cell wall synthesis and autolysis", C.Nombela (ed.). Elsevier Science Publishers. Amsterdam.
Nombela, C., Molina, M., Cenamor, R. and Sánchez, M. 1988. Yeast ß-glucanases: a complex system of secreted enzymes. *Microbiol. Sci.* 5:328.
Nombela, C., Pla, J., Herreros, E., Gil, C., Molina, M. and Sánchez, M. 1992. Novel targets for antifungal drugs, *in* "New strategies in fungal disease," J.E.Bennet, R.J.Hay and P.K.Peterson, eds., Churchill Livinstone, London.
Payton, M. and de Tiani, M. 1990. The isolation of osmotic-remedial conditional lethal mutants of *Candida albicans. Curr. Genet.* 17:293.
Ribas, J.C., Diaz, M., Duran, A. and Pérez, P. 1991. Isolation and characterization of *Schizosaccharomyces pombe* mutants defective in cell wall (1-3)ß-D-Glucan. *J. Bacteriol.* 173:3456.
Stateva, L.I., Oliver, S.G., Trueman, L.J. and Venkov. P.V. 1991. Cloning and characterization of a gene which determines osmotic stability in *Saccharomyces cerevisiae*. Mol. Cell. Biol. 11:4235.
Sullivan, P.A., Emerson, M.J., Broughton, M.J. and Stubs, H.J. 1991. Transglucosylation

catalysed by the exo-ß-glucanase of *Candida albicans*, *in*: "Proc. FEMS Symposium on Candida and Candidamycosis, E.Tumbay, H.P.R.Seeliger and O.Ang eds., Plenum Press. New York.

Torres, L., Martín, H., García-Sáez, M.I., Arroyo, J., Molina, M., Sánchez, M. and Nombela, C. 1991. A protein kinase gene complements the lytic phenotype of *Saccharomyces cerevisiae lyt2* mutants. *Mol. Microbiol.* 5:2845.

Venkov, P.V., Hadjiolov, A.A., Battaner, E. and Schlessinger, D. 1974. *Saccharomyces cerevisiae* sorbitol dependent fragile mutants. *Biochem. Biophys. Res. Commun.* 56:559.

THE USE OF A MERCURY-PENICILLIN V DERIVATIVE TO LOCALIZE PENICILLIN-BINDING PROTEINS IN *ESCHERICHIA COLI*

Terry R. Paul[1], Terry J. Beveridge[1], Noreen G. Halligan[2], Larry C. Blaszczak[2], and Tom R. Parr, Jr[2]

[1]Department of Microbiology, College of Biological Sciences, University of Guelph, Guelph, Ontario, Canada N1G 2W1
[2]Lilly Research Laboratories, Eli Lilly and Company, Lilly Corporate Center, Indianapolis, Indiana, USA 46285

Precise ultrastructural localization of penicillin-binding protein (PBP)-antibiotic complexes in a wild-type strain and two PBP overproducing mutants of *Escherichia coli* JM101 was investigated by high resolution electron microscopy (EM). A novel mercury-penicillin V (Hg-pen V) derivative was used as a heavy-metal labeled, electron dense probe for accurately localizing PBPs *in situ* in single bacterial cells grown to exponential growth-phase. Results obtained from susceptibility tests, lysis experiments and PBP affinity assays revealed few differences between strains and between the antimicrobial activities of Hg-pen V and the parent compound, penicillin V. When cells were grown in the presence of Hg-pen V, Hg-pen V-PBP complexes could readily be visualized by EM of unstained whole mounts as distinct randomly situated electron dense particles. Thin sections of Hg-pen V-treated bacteria revealed similar electron dense particles located predominantly on the plasma membrane and seldom in the cytoplasm. Particles positioned on the plasma membranes were occasionally shown to protrude into the periplasmic space thereby demonstrating the high resolution of the Hg-pen V probe. In addition, some particles were observed free in the periplasm, suggesting, for the first time, that a proportion of PBPs may not be restricted to the plasma membrane but may be tightly associated with the peptidoglycan for higher efficiency of peptidoglycan assembly. All controls were devoid of the electron dense particles. The presence of electron dense particles in cells of the wild-type strain, demonstrated that our probe could identify PBPs in a naturally-occuring strain without inducing PBP overproduction.

INTRODUCTION

Localization of penicillin-binding proteins (PBPs) in bacterial cells has typically relied upon disruption of large populations of cells followed by membrane fractionation and electrophoretic separation techniques in order to isolate various envelope fractions and separate the PBPs; PBPs can then be assigned back to the different cell fractions (Spratt, 1977; Rodriguez-Tébar *et al.*, 1985). However such studies overlook the fact that the structural and functional characteristics of proteins removed from their cellular domains may not truely reflect *in situ* activity or location. To accurately localize PBPs within a single bacterial cell, high resolution electron microscopy (EM) techniques are necessary.

Bacterial Growth and Lysis, Edited by M.A. de Pedro *et al.*, Plenum Press, New York, 1993

Conventional EM techniques so far have had limited success in localizing membrane proteins *in situ*. For example, immunogold labeling has been shown to be a valuable technique for revealing the cellular distribution of murein hydrolases (Hobot and Rogers, 1991), thioredoxin (Bayer *et al.*, 1988) and PBP 1b (Bayer *et al.*, 1990; den Blaauwen and Nanninga, 1990) in individual *E. coli* cells. Several problems associated with this technique still persist that may profoundly influence accurate interpretation of data; including loss of antigenicity of target proteins and their epitopes during conventional embedding, and steric hindrance effects by the gold particle size which in turn may affect the frequency of successful labeling (Craig and Goodchild, 1982; Bendayan *et al.*, 1987)

To circumvent many of the problems associated with the above techniques we introduce a novel approach for the identification and accurate localization of PBPs in bacteria. We have studied the potential of EM to trace the progress of a heavy metal-labeled antibiotic as it enters a single bacterial cell and interacts with its target sites. Penicillin V was chosen because of its chemical stability and broad PBP activity. Since antibiotic molecules tend to be indistinguishable from biological tissue when examined by EM, it was necessary to chemically synthesize mercury-penicillin V (Hg-pen V) so that the mercury is covalently bonded to the ß-lactam side chain to increase its electron density and promote visualization (Figure 1). Using Hg-pen V as our PBP-specific probe and high resolution EM, we show that PBPs can be accurately distinguished *in situ* within single bacterial cells.

MATERIALS AND METHODS

Bacterial Strains and Culture Conditions

Three *E. coli* strains were used: the wild-type strain JM101 ATCC 33876 (Messing *et al.*, 1977), strain JM101 (pBS96), which constitutively overproduces PBP1b (Broome-Smith *et al.*, 1985*a*) and strain JM101 (pPH116), which overproduces PBP 3 upon induction with isopropyl-ß-D-thio-galactopyranoside (IPTG) (Broome-Smith *et al.*, 1985*b*). All strains were grown under agitation (180 rpm) in Luria broth (LB), pH 7.0 at 37°C. Strain JM101 (pBS96) was grown in LB containing kanamycin (25 μg/ml), strain JM101 (pPH116) was grown in

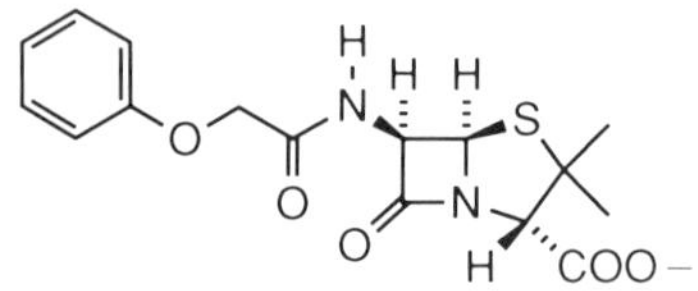

phenoxyacetamido penicillin
(penicillin V)

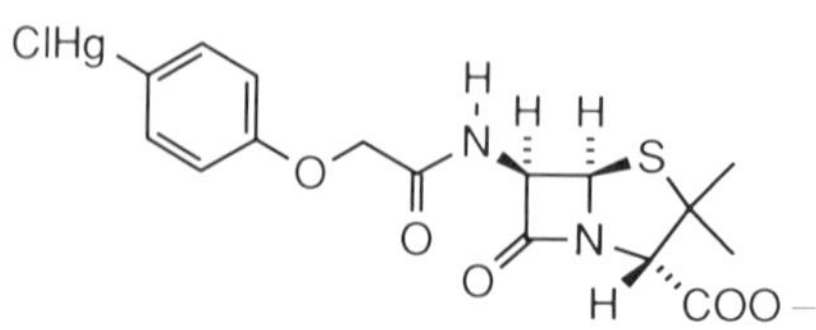

p-(chloromercuric)-phenoxyacetamido penicillin
(Hg-pen V)

Figure 1. Chemical structures of ß-lactams.

LB containing kanamycin and 1 mM IPTG, and the wild-type strain was grown in unmodified LB.

Penicillin-Binding Protein (PBP) Assay

Binding of ß-lactams to membrane fractions and competitive binding experiments were carried out using the method of Spratt (1977) with modifications as described by Preston *et al.* (1990). p-[^{125}I] phenoxyacetamido penicillin (IPV) (40 μg/ml stock concentration, specific activity 37.3 Ci/mmol) was used as the radiolabeled ß-lactam for visualizing PBP profiles as described previously (Blaszczak *et al.*, 1989; Preston *et al.*, 1990) because of its high specific activity which allowed shorter exposure times. Phenoxyacetamido penicillin (penicillin V) and p-(chloromercuric)-phenoxyacetamido penicillin (Hg-pen V) (Figure 1) were synthesized at Eli Lilly & Co., Indianapolis. Hg-pen V was prepared by electrophilic transmetallation of sodium p-(trimethylstannyl)phenoxyacetamido penicillin using mercuric chloride and sodium chloride.

Detection of Hg-pen V in Bacterial Cells

Mid-exponential growth-phase cultures of Hg-pen V-treated bacteria were harvested and washed four times in 0.1 M hydroxyethylpiperazine-N'-2-ethanesulfonic acid (HEPES) buffer (Research Organics Inc.), pH 7.2 at 4°C. Lyophilized bacterial pellets were then digested with 1 ml concentrated nitric acid for 18h at 22°C and diluted with high resistance (~5 MΩ) deionized distilled water. Solubilized cells, culture supernatants and HEPES washes were analyzed for the mercury substituent in Hg-pen V using an inductively coupled plasma-atomic emission spectrometer (ICP-AES) (LECO Instruments Ltd., Ontario, Canada) (Boumans, 1987) to determine the Hg-pen V concentration. All plasticware was acid-leached overnight prior to use.

Electron Microscopy

Intact cells were examined as whole mounts with no staining, or processed by a conventional fixation and embedding protocol. Thin-sections of Epon 812-embedded bacteria were viewed unstained or post-stained with uranyl acetate and lead citrate before examination. EM was performed on a Philips EM 300 electron microscope at an operating voltage of 60 kV equipped with a liquid nitrogen cold trap.

RESULTS AND DISCUSSION

Biochemical Comparison of Hg-pen V and Penicillin V

All strains possessed a minimum inhibitory concentration (MIC) of 32 μg/ml for both ß-lactams. Both antibiotics revealed differences in binding affinities for PBPs in each strain. For example, Hg-pen V bound PBPs 1b, 2, 4 and 6 whereas penicillin V bound PBPs 1b and 3 in strain JM101 (pPH116). Hg-pen V possessed affinity for PBPs 1b and 3, and penicillin V bound PBPs 1b and 4 in the wild-type strain, whereas both antibiotics had affinities for PBPs 1b, 2 and 3 in strain JM101 (pBS96). These differences did not influence the magnitude of Hg-pen V binding to cells because differences in PBP binding affinities could not be distinguished at the EM level using the Hg-pen V probe.

Even though the PBPs of our *E. coli* strains possessed relatively low binding affinities for both ß-lactams, the binding to a broad range of PBPs, particularly the essential proteins 1a, 1b, 2 and 3, was sufficient to inhibit growth and induce lysis in all strains studied. Experiments designed to study the effects of protein synthesis inhibition by chloramphenicol demonstrated that *de novo* protein synthesis was essential for Hg-pen V- or penicillin V-induced lysis in *E. coli.*

To detect Hg-pen V by its mercury constituent and to determine the concentration of antibiotic bound to cells, we used a highly sensitive analytical technique, ICP-AES, which allows the detection of small concentrations of sample elements through atomization of its constituent free atoms or ions. Greater than 94% of Hg-pen V added to cultures was bound to cells of all strains following extensive washing, whereas only trace amounts were found in culture supernatants and no antibiotic was found in successive washes.

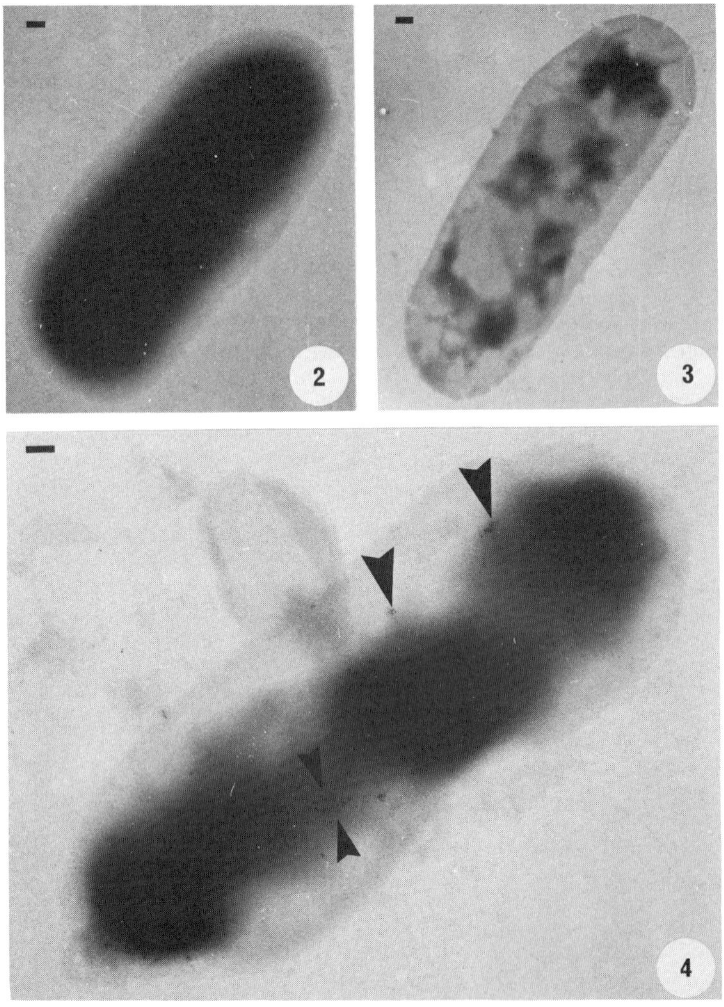

Figures 2-4. Unstained whole mounts of *E. coli* JM101 (pPH116) prepared from cells grown in the absence of antibiotics (Figure 2), in the presence of 5 x MIC penicillin V (Figure 3) or Hg-pen V (Figure 4) for 3h. Note the presence of single electron dense particles (large arrowheads) and clusters of single particles (small arrowheads) in Figure 4. Magnification bar = 100nm.

Ultrastructural Examination of Antibiotic-treated Cells

To visualize penicillin in single bacterial cells at the ultrastructural level we relied on the relative atomic mass of the mercury atom in Hg-pen V to impart sufficient electron density in order to distinguish it from the cellular substance. Since β-lactams are covalently attached to PBPs forming a stable inactive acylated complex, we believe this stable complex would be identified as electron dense particles which in turn represents the actual location of PBPs. EM examination of unstained Hg-pen V-treated exponential growth-phase whole cells revealed electron-dense particles (7.1 to 28.0 nm in diameter) in 20% to 40% of the cell population of each strain. Figures 2-4 depict cells of strain JM101 (pPH116) which was chosen as representative of all the strains examined. The fact that particles were observed in cells of the wild-type strain demonstrates that our Hg-pen V probe can resolve PBPs in a naturally-occuring strain without inducing PBP overproduction. Particles were apparent in cells exposed to antibiotic concentrations corresponding to 1 x MIC, 5 x MIC and 10 x MIC, and were not observed in the external milieu but instead were always cell associated. No particles were found in penicillin V nor untreated control cells (Figures 2 and 3).

The majority of individual particles were not confined to sites of cell constriction, elongation or pole formation, but instead were randomly distributed throughout a single cell. Furthermore, clusters comprising 5 to 10 individual particles were frequently observed in a single site on a cell (Figure 4). These tantalizing results may be the first indication of the complex spatial organization of PBPs in a single cell. We believe that clusters of single particles could consist of a single PBP species or multiple PBP species in close interaction to allow better efficiency for incorporation of newly synthesized peptidoglycan at distinct sites of wall growth. Complexes between some PBP 1b molecules (den Blaauwen and Nanninga, 1990) and between PBPs 1, 3 and 5 (Said and Höltje, 1983) on plasma membranes of *E. coli* have been reported previously but their physiological significance requires further investigation.

To determine the precise intracellular location of PBPs in a single cell we examined thin-sections of conventional embedded Hg-pen V-treated cells. To ensure retention of the Hg-pen V-PBP complex during processing, we included glutaraldehyde in our embedding protocol to provide additional covalent stability to the antibiotic-protein complex. We also monitored the amount of Hg-pen V retained by cells during different stages of conventional embedding using ICP-AES and found that 49 to 54% of antibiotic added to cell cultures remained cell-associated following fixation, dehydration and embedding in Epon 812 plastic resin. It appears that the mercury extracted during processing represented Hg-pen V-PBP complexes

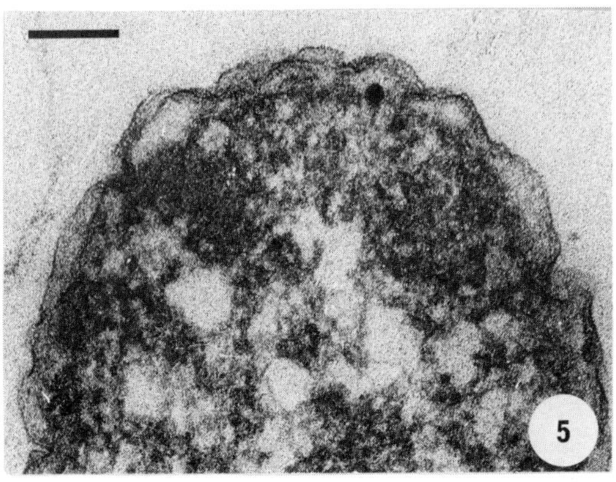

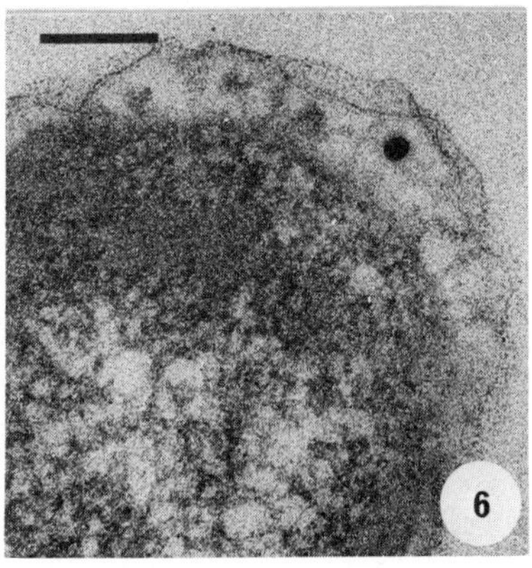

Figures 5 and 6. Thin-sections of cell envelopes of *E. coli* JM101 (pPH116) grown in 5 x MIC Hg-pen V for 1h. Large electron dense particles representing Hg-pen V-PBP complexes were located on the plasma membrane or free within the periplasm. The large particle size probably comprises several smaller individual particles that have congealed together to form a single mass. Magnification bar = 100nm.

detached from the cell envelope, however sufficient numbers of complexes were retained to allow visualization by EM.

Thin-sections of Hg-pen V-treated cells of all strains revealed distinct electron dense particles located predominantly on the plasma membranes but rarely in the cytoplasm (Figures 5 and 6). The protrusion of some plasma membrane-bound particles into the periplasmic space demonstrated the greater precision accomplished with the Hg-pen V probe than with other techniques (Figure 5). Since a thin-section possesses only a small fraction of the total cell mass, the number of particles per section was low (1 to 4) and was dependent on cell alignment within the section. Penicillin V-treated and -untreated bacteria were devoid of electron dense particles.

Our images provide ultrastructural evidence that PBPs of *E. coli* are anchored to the periplasmic face of the plasma membrane so that most of the protein is exposed to the periplasm (Pratt *et al.*, 1981, 1986; den Blaauwen and Nanninga, 1990). Interestingly, particles were also observed free within the periplasm (Figure. 6) which suggests that PBP distribution is not confined to the plasma membrane but may also be tightly associated with peptidoglycan to confer proficient cross-linking and insertion of new wall polymers into the developing murein sacculus.

We have shown that PBPs can be identified *in situ* at the ultrastructural level by EM. Our probe endowed a high resolution image of the precise region of Hg-pen V-PBP complexes within a single bacterial cell. We believe that this novel approach could be expanded to study the behaviour of other heavy-metal labeled ß-lactams with good binding affinity for a single PBP to further explore the topology of PBPs during specific events in the cell growth cycle.

ACKNOWLEDGMENTS

We thank Graham Spiers for performing the ICP-AES analysis. This research was supported by a Medical Research Council of Canada University - Industry grant to T. J. B. We thank our industrial partners, Eli Lilly & Co. and Eli Lilly Canada Inc., for their support throughout this research and B. Spratt for his kind donation of bacterial strains. The microscopy was performed in the NSERC Guelph Regional STEM Facility located in the Department of Microbiology, the College of Biological Sciences, University of Guelph. This facility is partially supported by an infrastructure grant from the Natural Sciences and Engineering Research Council of Canada.

REFERENCES

Bayer, M. H., Haberer, S. and Bayer, M. E. (1988) Distribution of membrane-associated proteins: photo-cross-linking, postembedding-immunolabeling, in "Antibiotic inhibition of bacterial cell surface assembly and function" (Actor, P., Daneo-Moore, L., Higgins, M. L., Salton, M. R. J. and Shockman, G. D., Eds.), pp. 320-331. American Society for Microbiology, Washington D.C.

Bayer, M. H., Keck, W. and Bayer, M. E. (1990) Localization of penicillin-binding protein 1b in *Escherichia coli* : immunoelectron microscopy and immunotransfer studies. J. Bacteriol. 172, 125-135.

Bendayan, M., Nanci, A. and Kan, F. W. K. (1987) Effect of tissue processing on colloidal gold cytochemistry. J. Histochem. Cytochem. 35, 983-996.

Blaszczak, L. C., Halligan, N. G. and Seitz, D. E. (1989) Radioiododestannylation. Convenient synthesis of a stable penicillin derivative for rapid penicillin binding protein (PBP) assay. J. Labelled Compd. Radiopharmacol. 27, 401-406.

Boumans, P. J. W. M. (1987) "Inductively coupled plasma emission spectroscopy" part 1, pp. 1-492. John Wiley & Sons, New York.

Broome-Smith, J. K., Edelman, A., Yousif, S. and Spratt. B. G. (1985a) The nucleotide sequences of the pon A and pon B genes encoding penicillin-binding proteins 1A and 1B of *Escherichia coli* K12. Eur. J. Biochem. 147, 437-446.

Broome-Smith, J. K., Hedge, P. J. and Spratt, B. G. (1985b) Production of thiol-penicillin-binding protein 3 of *Escherichia coli* using a two primer method of site-directed mutagenesis. EMBO J. 4, 231-235.

Craig, S. and Goodchild, D. J. (1982) Post-embedding immunolabelling. Some effects of

tissue preparation on the antigenicity of plant proteins. Eur. J. Cell Biol. 28, 251-256.
den Blaauwen, T. and Nanninga, N. (1990) Topology of penicillin-binding protein 1b of *Escherichia coli* and topography of four antigenic determinants studied by immunocolabeling electron microscopy. J. Bacteriol. 171, 71-79.
Hobot, J. A. and Rogers, H. J. (1991) Intracellular location of the autolytic N-acetylmuramyl-L-alanine amidase in *Bacillus subtilis* 168 and in an autolysis-deficient mutant by immunoelectron microscopy. J. Bacteriol. 173, 961-967.
Messing, J., Gronenborn, B., Muller-Hill, B. and Hofschneider, H. (1977) Filamentous coliphage M13 as a cloning vehicle: insertion of a Hind II fragment of the lac regulatory region in M13 replicative form *in vitro*. Proc. Natl. Acad. Sci. USA 74, 3642-3646.
Pratt, J. M., Holland, I. B. and Spratt, B. G. (1981) Precursor forms of penicillin-binding proteins 5 and 6 of *Escherichia coli* cytoplasmic membrane. Nature (London) 293, 307-309.
Pratt, J. M., Jackson, M. E. and Holland, I. B. (1986) The C-terminus of penicillin-binding protein 5 is essential for localisation to the *Escherichia coli* inner membrane. EMBO J. 5, 2399-2405.
Preston, D. A., Wu, C. Y. E., Blaszczak, L. C., Seitz, D. E. and Halligan, N. G. (1990) Biological characterization of a new radioactive labeling reagent for bacterial penicillin-binding proteins. Antimicrob. Agents Chemother. 34, 718-721.
Rodriguez-Tébar, A., Barbas, J. A. and Váquez, D. (1985) Location of some proteins involved in peptidoglycan synthesis and cell division in the inner membranes and outer membranes of *Escherichia coli*. J. Bacteriol. 161, 243-248.
Said, I. M. and Höltje, J-V. (1983) Topological interrelation of penicillin-binding proteins in *Escherichia coli*- a study using cleavable cross-linker, in "The target of penicillin. The murein sacculus of bacterial cell walls architecture and growth" (Hakenbeck, R., Höltje, J-V. and Labischinski, H., Eds.), pp. 439-444. Walter de Gruyter, Berlin.
Spratt, B. G. (1977) Properties of the penicillin-binding proteins of *Escherichia coli* K12. Eur. J. Biochem. 72, 341-352.

A FAR UPSTREAM REGION IS REQUIRED FOR EXPRESSION OF THE *ftsI* GENE CODING FOR PENICILLIN-BINDING PROTEIN 3 OF *ESCHERICHIA COLI*

Hiroshi Hara and James T. Park

Department of Molecular Biology and Microbiology
Tufts University School of Medicine
Boston, Massachusetts 02111, U.S.A.

INTRODUCTION

The *ftsI* (*pbpB*) gene of *Escherichia coli* codes for penicillin-binding protein 3 (PBP 3), a cytoplasmic membrane protein that functions as DD-transpeptidase for the formation of a septum of the murein sacculus and is the principal lethal target of β-lactam antibiotics. PBP 3 was shown to be essential for cell division by analyzing thermosensitive *ftsI* mutants and their revertants (Spratt, 1977). The *ftsI* gene is located at 2 min on the *E. coli* genetic map, a little distal to *leu*, and is among a large cluster of murein biosynthetic genes (*mur* genes and *ddl*) and cell division genes (*fts* genes and *envA*). The DNA sequence of 28 kilobases (kb) of this region including the whole cluster has been determined (EMBL database, entry name EC2MIN).

In an attempt to generate and characterize various *ftsI* mutant alleles on a plasmid, we realized that it would be useful to have a null allele of this gene on the chromosome. In this report we constructed a deletion/insertion mutant and unexpectedly found that, when *ftsI* is cloned on a single-copy plasmid, a chromosomal fragment more than 1.9-kb upstream is required to complement the null mutation.

RESULTS AND DISCUSSION

Construction of a Null Allele of *ftsI*

Since *ftsI* was shown essential, disruption of the chromosomal allele of this gene should be carried out in the presence of a second copy of the wild-type allele. We chose to use *polA*(Ts) strains, in which a null mutation constructed on a ColE1-type plasmid can be exchanged with a chromosomal copy by integration and resolution of the plasmid and the wild-type copy can be kept on the plasmid after the exchange. The procedure for the gene disruption is summarized in Figure 1. First, a deletion/insertion mutation of the *ftsI* gene was constructed on plasmid pHR4 (Ap^r; Figure 2), which carries a 9.7-kb *Bam*HI-*Eco*RI fragment containing *ftsI* cloned into pBR322: this fragment has a long flanking region on either side of *ftsI* that would give a higher frequency of recombination with the chromosome. The 1.5-kb *Mlu*I-*Nar*I fragment within the *ftsI* gene (nucleotides 124-1628 out of its 1764-nucleotide open reading frame [ORF]) was replaced with a 1.3-kb *Hae*II fragment of

Bacterial Growth and Lysis, Edited by M.A. de Pedro *et al.*, Plenum Press, New York, 1993

pACYC184 containing the chloramphenicol acetyltransferase (*cat*) gene. The *cat* gene was placed in the same orientation as the *ftsI* gene. We will refer to this deletion/insertion as Δ*ftsI::cat-1*. A *polA*(Ts) HfrP4X strain was transformed with this plasmid at 30°C and then grown at 42°C in the presence of ampicillin and chloramphenicol. Since *polA* function is required for replication of pBR322 replicon, survivors should carry the plasmid integrated into the chromosome by homologous recombination around *ftsI*. This was confirmed by conjugational transfer of the integrated plasmid to a *polA*(Ts) *str* F^- strain at 42°C with selection for Ap^r Cm^r Sm^r. The exconjugant was subsequently grown at 30°C, a permissive temperature for replication of the plasmid. Since excessive replication from the origin of the

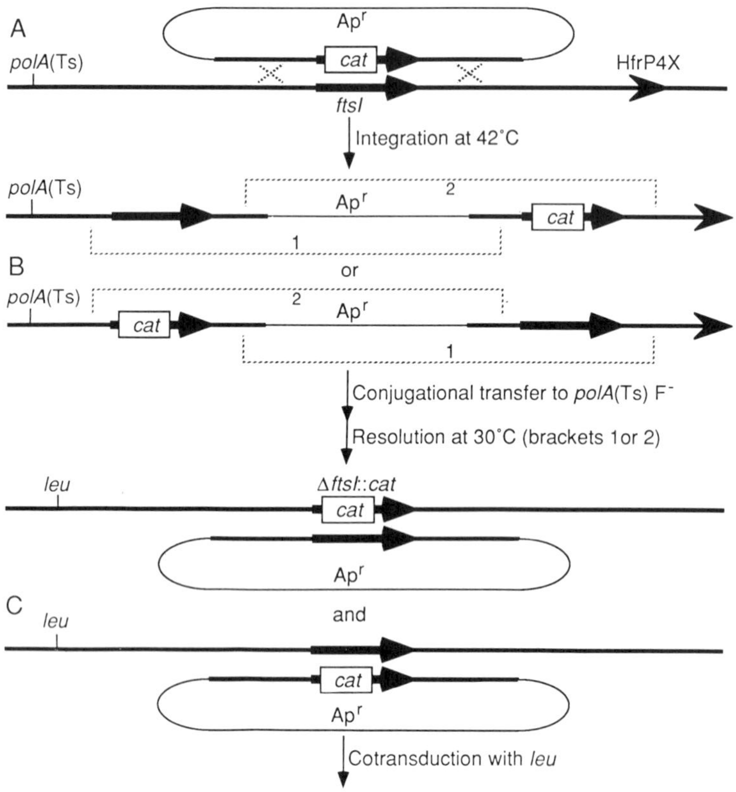

Figure 1. Construction of a strain containing a deletion/insertion mutation of the *ftsI* gene on the chromosome. Bold and thin lines denote the chromosomal and vector DNA, respectively. (Not drawn to scale.)

integrated multicopy plasmid is deleterious to the cell (Yamaguchi and Tomizawa, 1980), cells that had the plasmid excised from the chromosome by a second recombination event grow better and dominate the culture. Depending on the second recombination site, Δ*ftsI::cat-1* would either be retained on the chromosome or reappear on the plasmid. The disrupted allele on the chromosome was transferred to a third strain by cotransduction with a *leu* marker: a phage P1 lysate prepared on such a culture was used to transduce W3110*leu*::Tn*10* harboring pHR223 (Figure 2), an *ftsI*-carrying pACYC177 derivative (Km^r), to Leu^+ Cm^r. The transductants were tested for drug resistance and one of Tc^s Cm^r Ap^s Km^r transductants, which should have received the disrupted *ftsI* gene, was used for further experiments.

ftsI is an Essential Cell Division Gene

A P1 transduction experiment using W3110Δ*ftsI*::*cat-1* harboring pHR223 as a donor and W3110*leu*::Tn*10* carrying either no plasmid or pHR4 as a recipient with selection for Leu+ (Table 1, Experiment 1) showed that the *ftsI* null allele could be cotransduced with *leu* only if a recipient strain carried additional copies of wild-type *ftsI* on a plasmid. Next, to obtain conditional expression, the *ftsI* gene cut at 15 base-pairs upstream of the putative ribosome-binding site (Nakamura *et al.*, 1983) was fused to the *lac* operator-promoter sequence of pUC19 and then cloned into pHR277, a mini-F vector carrying *lacI^q*. When W3110*leu*::Tn*10* harboring the resultant plasmid pHR295 (Figure 2) was transduced to Leu+, Δ*ftsI*::*cat-1* could be cotransduced with *leu* only in the presence of a *lac* inducer IPTG (Table 1, Experiment 2). The transductants thus obtained grew normally in the presence of IPTG, but failed to divide and formed long filaments when incubated without the inducer. These results provide final proof that *ftsI* is an essential cell division gene.

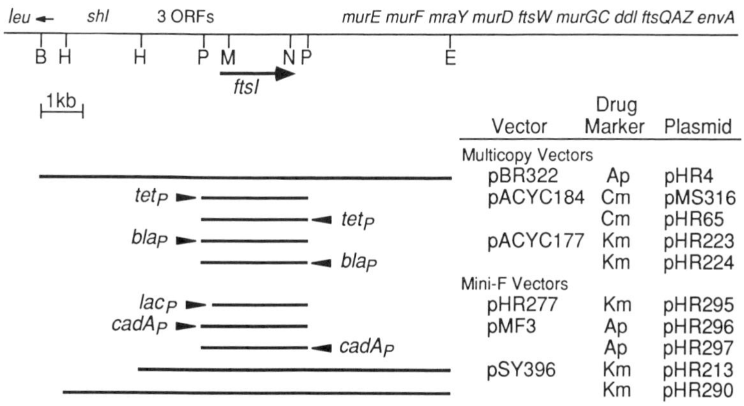

Figure 2. Plasmids carrying the *ftsI* gene. Only cloned chromosomal regions are shown. Arrowheads denote the direction of the nearby promoters. The relevant restriction sites (B, *Bam*HI; E, *Eco*RI; H, *Hin*dIII; M, *Mlu*I; N, *Nar*I; and P, *Pvu*II), the location and orientation of *ftsI*, and the approximate locations of other genes are shown at the top.

A Far Upstream Region is Required for a Single Copy of *ftsI* to Complement the Null Allele

A 2.6-kb *Pvu*II fragment containing the entire *ftsI* ORF could complement an *ftsI*(Ts) mutation even when it was inserted in the opposite orientation into a drug resistance gene of multicopy vectors (pHR65 and pHR224; Figure 2). There are at least two promoter-like sequences identified in this fragment (Nakamura *et al.*, 1983; Ishino *et al.*, 1989). This fragment was cloned on a mini-F vector pMF3 in either orientation to construct pHR296 and pHR297 (Figure 2). When introduced to W3110*leu*::Tn*10* by transformation, only a transformant with pHR296, in which *ftsI* was placed under the *cadA* promoter in the vector (Hara *et al.*, 1991), could receive Δ*ftsI*::*cat-1* by cotransduction with *leu*. A transformant with pHR297, which has the insert in the opposite orientation, could not receive Δ*ftsI*::*cat-1* (Table 1, Experiment 3). A complementation test of *ftsI*-carrying mini-F plasmids using an *ftsI*(Ts) mutant did not give a clear result because of thermosensitive maintenance of F replicon (Stadler and Adelberg, 1972). Next, we constructed pHR213 (Figure 2) by cloning the 7.3-kb *Hin*dIII-*Eco*RI fragment containing *ftsI* into a mini-F vector pSY396 (a gift from Seiichi Yasuda). In this vector there seems to be no promoter activity that flows into the cloning site like the *cadA* promoter in pMF3. The cloned fragment contained 1.9 kb of DNA upstream of *ftsI*, yet pHR213 was unable to complement the null allele of *ftsI* as shown by a transduction experiment (Table 1, Experiment 4). When recloned into pBR322, the insert

Table 1. Cotransduction of Δ*ftsI*::*cat* with *leu* by phage P1.[1]

Experiment	Donor	Recipient	Cm^r transductants
	W3110Δ*ftsI*::*cat-1* harboring:	W3110*leu*::Tn*10* harboring:	
1	pHR223 (Km^r)	no plasmid	0/12
		pHR4 (Ap^r)	7/11
2	pHR4 (Ap^r)	pHR295 (Km^r)	0/12 in the absence of IPTG
			11/12 in the presence of IPTG
3	pHR223 (Km^r)	pHR296 (Ap^r)	7/10
		pHR297 (Ap^r)	1/12[2]
4	pHR4 (Ap^r)	pHR223 (Km^r)	9/12
		pHR213 (Km^r)	0/12
		pHR290 (Km^r)	9/11
	polA(Ts) after integration and resolution of pHR4 (Ap^r) containing:	W3110*leu*::Tn*10* harboring:	
5	Δ*ftsI*::*cat-1*	pHR290 (Km^r)	4/15
	Δ*ftsI*::*cat-2*	pHR290 (Km^r)	0/16

[1]Transductants were selected for Leu$^+$. The numbers of Cm^r transductants out of the transductants that retained the plasmid of the recipient and did not receive the plasmid from the donor (*e.g.* Cm^r Km^r Ap^s Tc^s/[Cm^r Km^r Ap^s Tc^s + Cm^s Km^r Ap^s Tc^s] in Experiment 2) are shown.

[2]We tried to use this Cm^r transductant as a donor to transduce W3110*leu*::Tn*10* harboring pHR290, but Cm^r was not cotransduced with *leu*.

of pHR213 could complement an *ftsI*(Ts) mutation, showing this unexpected result was not due to an accidental loss of *ftsI* function during the cloning procedure. Complementation by an *ftsI*-carrying mini-F plasmid was observed when a fragment containing an additional 1.8-kb *Hin*dIII fragment of a further upstream region was cloned into pSY396 (pHR290; Figure 2), as shown by cotransduction with *leu* (Table 1, Experiment 4). Complementation ability of pHR213 and pHR290 (both Km^r) was also tested by trying to displace pHR296 (Ap^r) in a *recA*$^-$ derivative of W3110Δ*ftsI*::*cat-1* by transformation: since mini-F plasmids are incompatible with each other, Km^r transformants should lose an Ap^r mini-F plasmid. While pHR290 could displace pHR296 in a Δ*ftsI*::*cat-1* background, pHR213 could not give a Km^r transformant, showing that it cannot complement the disrupted *ftsI*. Either plasmid gave many Km^r transformants in a control experiment in which an isogenic *ftsI*$^+$ strain harboring pHR296 was transformed.

The strains harboring pHR213 and pHR290 were assayed for their PBP 3 content. An *ftsI730*(Ts) mutant, whose PBP 3 is defective in penicillin-binding activity *in vitro* even at temperatures permissive for growth, was used as a host. As shown in Figure 3, pHR213 did not lead to the production of a detectable level of PBP 3, whereas a pHR290-harboring strain contained a normal level of PBP 3.

These results indicate that the region more than 1.9-kb upstream is necessary for a single-copy *ftsI* to be functional. Since the upstream region is intact on the chromosome in the strain containing Δ*ftsI*::*cat-1*, there seems to be a sequence required in *cis*, most likely a promoter(s) essential for the *ftsI* expression. In multicopy plasmids that carry a smaller fragment but still complement an *ftsI*(Ts) mutation, such as pHR65 and pHR224, weak expression might be compensated by a high copy number, or readthrough transcription from somewhere in the vectors might contribute to the expression. In the 1.8-kb *Hin*dIII fragment that was cloned in pHR290 and not in pHR213, the *shl* gene and its potential promoter were

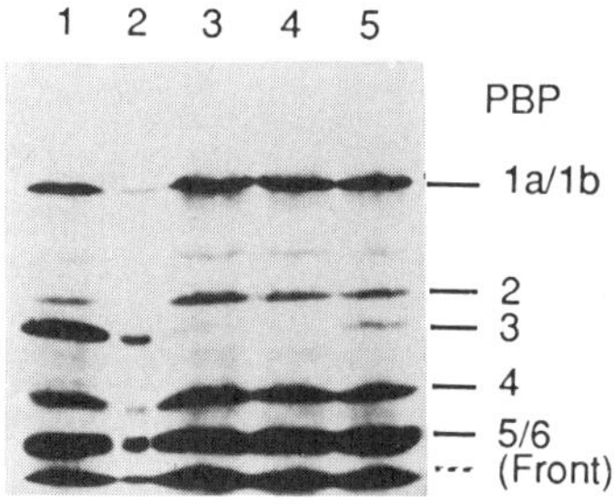

Figure 3. Penicillin-binding assay of an *ftsI730*(Ts) strain that harbors mini-F plasmids containing the *ftsI* ORF. PBPs in envelope fractions were labeled with [^{3}H]benzylpenicillin and analyzed by SDS-polyacrylamide gel electrophoresis and fluorography. Lanes 1 and 2, pMS316, a multicopy plasmid carrying *ftsI* (a tenth amount of the labeled sample was applied to lane 2); lane 3, pSY396, a mini-F vector; lane 4, pHR213; lane 5, pHR290.

present (Leclerc *et al.*, 1990). Also in this fragment, between *shl* and three possible ORFs found just upstream of *ftsI*, there is a 0.6-kb region with no ORF of significant length, which may contain a promoter yet to be identified.

Is There an 18 kb Transcript?

It was reported that a null allele of the *ftsZ* gene could not be complemented by a λ transducing phage containing 6 kb of DNA upstream of *ftsZ* (Dai and Lutkenhaus, 1991). Since all the genes from *ftsI* through *ftsZ* and *envA* are tightly clustered in the same orientation without a recognizable transcription terminator, the authors proposed a possibility of a long transcript required for full expression of *ftsZ*. Transcription originating more than 1.9 kb upstream of *ftsI,* as suggested in this report, might proceed for 18 kb (*i.e.* beyond *ftsZ*) to the only known terminator just downstream of *envA* and be responsible for the *ftsZ* expression.

Another deletion/insertion mutation of *ftsI* (Δ*ftsI*::*cat-2*), in which *cat* was substituted for the internal segment of *ftsI* as in Δ*ftsI*::*cat-1* but in the opposite orientation, was constructed on pHR4. We attempted to introduce it into the chromosome. The *polA*(Ts) F^{-} strain that had received the integrated plasmid at 42°C and undergone its resolution at 30°C was used as a donor in P1 transduction of W3110*leu*::Tn*10* harboring pHR290 to Leu^{+}. Although Δ*ftsI*::*cat-1* having *cat* in the same orientation as *ftsI* was cotransduced with *leu* at the frequency expected if a little less than half of the donor cells had retained the disrupted gene on the chromosome (stage C in Figure 1), Δ*ftsI*::*cat-2* could not be cotransduced (Table 1, Experiment 5). This result can be explained if a hypothetical long transcript of the clustered genes is essential and if its synthesis is hindered by Δ*ftsI*::*cat-2*. The insertion of the *cat* gene in the opposite orientation might give a polar effect eliminating expression of the downstream genes, which cannot be complemented by pHR290 that covers only up to *murF* (Figure 2). It will be interesting to determinine whether a mini-F plasmid that carries a larger fragment extending from a hypothetical promoter more than 1.9-kb upstram of *ftsI* through *ftsZ* can complement Δ*ftsI*::*cat-2* and a null allele of *ftsZ*.

ACKNOWLEDGMENTS

We thank Drs. Kensuke Horiuchi, Carol Kumamoto, Yuthana Limpa-Amara, Michael Malamy, Debabrata RayChaudhuri, and Seiichi Yasuda for helpful advice and discussion. This work was supported, in part, by grant AI05090 from the National Institutes of Health.

REFERENCES

Dai, K. and Lutkenhaus, J. (1991) *ftsZ* is an essential cell division gene in *Escherichia coli*. J. Bacteriol. 173, 3500-3506.

Hara, H., Yamamoto, Y., Higashitani, A., Suzuki, H. and Nishimura, Y. (1991) Cloning, mapping, and characterization of the *Escherichia coli prc* gene, which is involved in C-terminal processing of penicillin-binding protein 3. J. Bacteriol. 173, 4799-4813.

Ishino, F., Jung, H. K., Ikeda, M., Doi, M., Wachi, M. and Matsuhashi, M. (1989) New mutations *fts-36, lts-33*, and *ftsW* clustered in the *mra* region of the *Escherichia coli* chromosome induce thermosensitive cell growth and division. J. Bacteriol. 171, 5523-5530.

Leclerc, G., Noël, G. and Drapeau, G.R. (1990) Molecular cloning, nucleotide sequence, and expression of *shl*, a new gene in the 2-minute region of the genetic map of *Escherichia coli*. J. Bacteriol. 172, 4696-4700.

Nakamura, M., Maruyama, I. N., Soma, M., Kato, J., Suzuki, H. and Hirota, Y. (1983) On the process of cellular division in *Escherichia coli*: nucleotide sequence of the gene for penicillin-binding protein 3. Mol. Gen. Genet. 191, 1-9.

Spratt, B. G. (1977) Temperature-sensitive cell division mutants of *Escherichia coli* with thermolabile penicillin-binding proteins. J. Bacteriol. 131, 293-305.

Stadler, J. and Adelberg, E. A. (1972) Temperature dependence of sex-factor maintenance in *Escherichia coli* K-12. J. Bacteriol. 109, 447-449.

Yamaguchi, K. and Tomizawa, J. (1980) Establishment of *Escherichia coli* cells with an integrated high copy number plasmid. Mol. Gen. Genet. 178, 525-533.

INVOLVEMENT OF THE NH_2- AND COOH-TERMINAL ENDS OF PBP3 OF *ESCHERICHIA COLI* ON β-LACTAM BINDING, MEMBRANE LOCALIZATION AND FUNCTION OF THE PROTEIN

Manuel J. Gómez, Lourdes R. Desviat, Rafael Merchante, and Juan A. Ayala

Centro de Biología Molecular. CSIC - UAM
Canto Blanco, 28049 Madrid

INTRODUCTION

PBP3 of *Escherichia coli* is a membrane protein that has an essential function at the time of septum formation. This protein is involved in the last steps of incorporation of precursors into the septal peptidoglycan layer. Both transglycosylase and transpeptidase activities have been predicted for the protein, although only the transpeptidase activity has been repeatedly proved by "in vitro" assays. The protein binds the β-lactam antibiotics due to the structural homology of these compounds with the natural substrate for the enzyme, and they remain covalently linked due to the cyclic nature of the lactam ring and the mechanism of reaction.

PBP3 is synthesized as a precursor and processed to the mature form by proteolytic cleavage of the eleven COOH-terminal amino acids. The NH_2-terminal end have the consensus structure of a signal peptide, consisting of a sequence of basic residues (aa 1-23) followed by an hydrophobic sequence (aa 24-43). Also, sequences that approximate to the consensus for signal peptidases I ($L_{32}LALA_{36}$) and II ($L_{27}CGC_{30}$) are found. It has been described by Hayashi *et al.* (1988) that about 15% of the total PBP3 molecules are modified with glycerol and fatty acids, in an overproducing strain, but this modification is not sensitive to globomycin and consequently is not accompanied by proteolytic processing.

Analysis of PBP3-β-lactamase fusions suggested that the amino terminal end of the protein (aa 1-40) is the only segment spanning the cytoplasmic membrane, being the bulk of the protein (aa 41-577) almost free in the periplasm (Bowler and Spratt, 1989).

Several proteins have been predicted to interact with PBP3 in the septum formation complex (septosome): FtsA, due to the inhibition of the β-lactam binding capacity of PBP3

Bacterial Growth and Lysis, Edited by M.A. de Pedro *et al.*, Plenum Press, New York, 1993

in a *ftsA*(ts) background (Tormo *et al.*, 1986); FtsZ, due to the wrong location of a β-galactosidase-PBP3 fusion protein in a *ftsZ*(ts) mutant strain at the restrictive temperature (Ayala *et al.*, 1988); and FtsW, due to the increase of lipid II precursor and transglycosylation products in a strain overproducing this protein (Matsuhashi *et al.*, 1990).

In this work we have addressed the possible role of the NH_2- and COOH-terminal ends of PBP3 in the maturation and function "in vivo" of the protein. To do so, we have studied the behaviour of mutant proteins, obtained by site-directed mutagenesis, on β-lactam binding, protein topology, COOH-terminus processing and complementation of the *pbpB*(ts) phenotype.

RESULTS

Comparison of *pbpB-Bla* and *pbpB-PhoA* gene fusions

After analysis of 61 different PBP3-β-lactamase protein fusions, Bowler and Spratt (1989) suggested that the bulk of the protein (aa 41-577) is located in the periplasm in an essentially free, globular form. This topology was also deduced from proteolysis of spheroplasts and inverted membrane vesicles of a wild type strain. However, we have found that, after "in situ" proteolytic analysis of PBP3 in EDTA-permeabilized cells, a protected fragment of about 40 KD, retaining the β-lactam binding capacity, appears. This fragment was systematically found associated to the membrane fraction in all PBP3 overproducing strains tested. Nagasawa *et al.* (1989) showed it to start at L_{211} of PBP3.

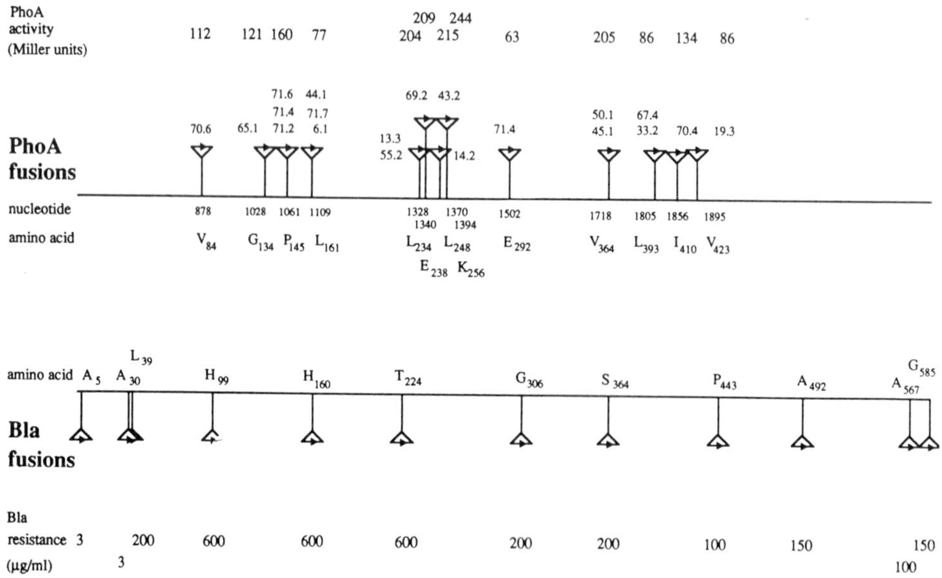

Figure 1. Comparison of the enzymatic activity and position for PBP3-PhoA and PBP3-Bla protein fusions. Data for β-lactamase fusions are from Bowler and Spratt (1989).

Our analyses on 80 *pbpB-phoA* gene fusions (Figure 1) showed that only 20 were in-frame insertions. The insertion sites mapped between residues V_{84} and V_{423}, and showed a high alkaline phosphatase activity.

Comparison of the position and activity of PBP3-PhoA and PBP3-Bla fusion proteins reveals that no in-frame PBP3-PhoA fusions were found in the COOH-terminal sequence and that β-lactamase activity of the PBP3-Bla fusions between amino acids 400 and 588 was severely reduced. Moreover, between residues V_{364} and P_{443} only two PBP3-Bla fusions were found at residues P_{402} and S_{408} (Bowler and Spratt, 1989). The domain from A_{413} to P_{443} corresponds to a highly hydrophobic region that is highly conserved in all the HMW-PBPs (high-Mr penicillin-binding proteins) of class B (Ghuysen, 1991) and would be located within the 3D-structure model for DD-peptidases into the region α7 - α8 consensus helices between the SSN and KTG motifs.

Construction of point mutations and deletions in the gene *pbpB*

From the comparison of the position and enzymatic activity of PBP3-Bla and PBP3-PhoA protein fusions, and from "in situ" proteolytic analysis of wild type PBP3, we may predict that domains of the COOH-terminal end could have some implication in the recognition of the substrate emerging from the membrane and/or in the interaction with some component of the membrane.

On the other hand, there is a controversy about the domain implicated in the insertion of the protein within the membrane and its lipoprotein nature (Bowler and Spratt, 1989; Hayashi *et al.*, 1988).

We have addressed the analysis of both questions by site directed mutagenesis of specific residues at the NH_2- and COOH-terminal ends of the protein. Figure 2 shows the point mutations and deletions designed to reach the following objectives: i) To produce structural and polar modifications in the hydrophilic region of the putative signal peptide [A1 (P_{10}>L), and A2 (K_9PK_{11}>ILI)], ii) To introduce polar modifications in the hydrophobic core of the putative signal peptide [A12 (Δ $A_{17}NFISWR_{23}$), A14 (Δ $R_{41}VAWLQVISP_{50}$), and A4 (R_{41}>P)], iii) To alter the putative recognition site for signal peptidase II (SPase II) [A5 (C_{30}>R), A6 ($L_{27}CG_{29}$>FCS), and A19 (C_{30}>A)], and iv) To remove putative sites of recognition, modification, or interaction at the carboxy terminus [A15 (Δ $Y_{540}YGGAVSAPV_{549}$), A16 (Δ $F_{550}GAIMGGVLR_{559}$), A7 (Δ $R_{559}TMNIEP_{565}$), A8 (Δ $P_{565}DALTTG_{571}$), and A9 (Δ $G_{571}DKNEFV_{577}$)].

Expression analysis of the mutant proteins

Expression of the different mutant forms of PBP3 was carried out in the strain TG1 transformed with the appropriate plasmid or infected with the corresponding M13 phage (Figure 3). Cells were broken by sonication and membrane and cytoplasmic fractions were prepared by differential centrifugation. Total proteins were fractionated in SDS-PAGE acrylamide 8% gels and detected by immunoblotting using a polyclonal anti-PBP3 serum or by labelling with ^{125}I-Ampicillin. Mutant forms were named by the number of the mutation introduced, i.e. PBP3A5 is the mutant protein containing the mutation A5.

Modifications at the amino terminal end (mutations A1, A2, A4, A5, A6, A12, A14, and A19). All these mutations at the putative signal peptide gave no accumulation of mutant protein in the cytoplasm, and both precursor and mature mutant forms were found in the particulate fraction. The electrophoretic mobilities of the mutant forms were those expected from their amino acid compositions.

PBP3A6 and PBP3A19 mutant proteins, that contain a modified recognition site for SPase II, had the same behaviour (pattern of precursor and mature bands, electrophoretic mobility, and β-lactam binding capacity) as wt PBP3. This indicates that a modification at the putative signal-peptidase consensus site at the amino terminus, and consequently SPase II processing, are not taking part in the export of PBP3.

In strain TG1 infected with the corresponding M13 phage, PBP3A2 and PBP3A5 were expressed at lower level than the wt PBP3. Normal levels of PBP3A2 were restored when the protein was expressed in a *secA* or *secY* mutant strain containing the mutant alele cloned in plasmid pSPA2 (unpublished observations). However, normal level of PBP3A5 was also observed when the protein was expressed in a wild type strain containing plasmid pSPA5.

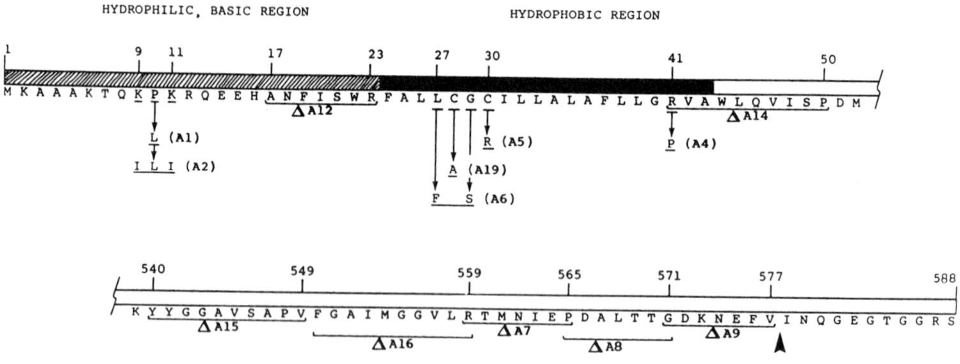

Figure 2. Point amino acid replacements and deletions obtained at amino and carboxy terminal ends of PBP3. Site directed mutagenesis was used to modified the single stranded DNA of phage M13mp9PS3, that contains the wild type *pbpB* gene.

Modifications at the carboxy terminal end (mutations A15, A16, A7, A8, A9). Mutant proteins were found in the particulate fraction after centrifugation of broken cells at 150,000 x g for 60 min (Figure 3). However they were found in the soluble fraction obtained by centrifugation at 35,000 x g for 30 min (Figure 5). This results indicate that a certain percentage of the protein, even for the wild type, become localized in very small membrane vesicles once the cells are broken by sonication.

Mutant mature forms PBP3A7, A8, A9, A15 and A16 may have a higher electrophoretic mobility in an SDS-PAGE gel than the mature wild type PBP3. With the analytical (immunoblotting or ^{125}I-Ampicillin labelling) and protein fractionation (SDS-PAGE acrylamide 8% gel) systems used, we systematically find a small amount of precursor form, seen as a shadow above mature PBP3. However a single form was found for PBP3A7, A8, A9, and A15 mutant proteins that migrates with an electrophoretic

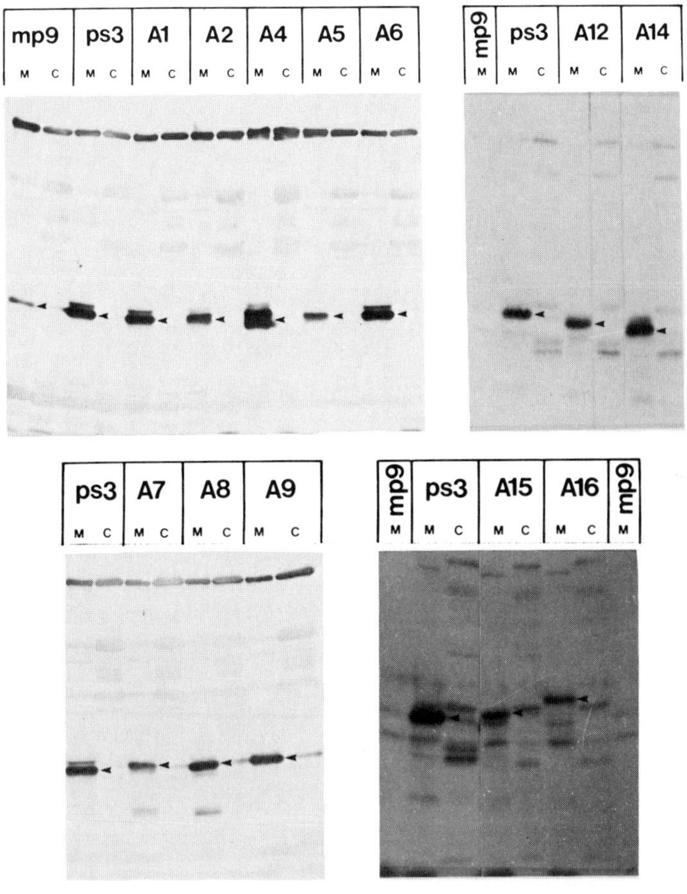

Figure 3. Immunodetection of mutant forms of PBP3 obtained by site directed mutagenesis. Cultures of TG1 strain were grown in LB medium at 37°C. M13 phage suspensions (m.o.i.=10) and IPTG (1 mM) were added when the cultures reached exponential phase. Cells were collected 60 minutes later and broken by sonication. Membrane (M) and soluble (C) fractions were obtained by ultracentrifugation (200.000 g, 30 min). Proteins were separated by SDS-PAGE (acrylamide concentration of 8%), transferred to a PVDF membrane and immunodetected. The amount of sample loaded in each track is that corresponding to 2.5 x·10^8 cells.

mobility congruent with the size of the precursor of the mutant protein. This result indicates that these deletions block processing of the protein. Processing at the carboxy terminus is apparently not required for export of the protein, because the precursor was found in the membrane fraction. Blokade of processing at the COOH terminus was also observed when mutations A16 and A7 were placed in an OmpA-PBP3 fusion protein (unpublished results).

Anomalous electrophoretic mobility of mutant form PBP3A16 could not be explained by simple differences in charge, because its experimental isoelectric point is quite similar to wt PBP3. We assume that some modification that normally must occur in wt PBP3 was suppressed in PBP3A16, although binding of lower amounts of SDS to the mutant protein could not be completely excluded.

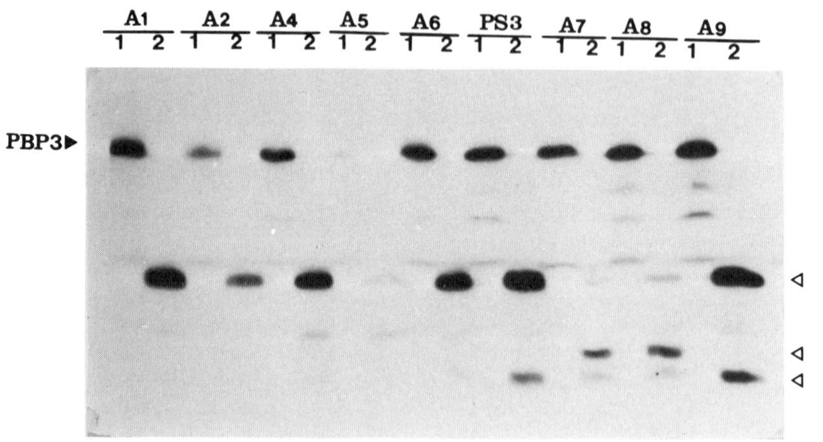

Figure 4. Trypsin accessibility of several mutant forms of PBP3 encoded by M13mp9 derivatives. Cultures of TG1 strain were grown and infected as in Figure 3. Cells were collected 60 minutes after infection, permeabilized with EDTA, and incubated (30 min, 4°C) without (1) or with (2) trypsin (100 μg/ml). Proteins were separated by SDS-PAGE (acrylamide concentration of 10%), transferred onto a PVDF membrane and immunodetected. M13mp9PS3 encodes the wild type PBP3 (PS·3). Protected fragments are indicated with an open arrowhead.

Proteolytic analysis of mutant proteins in EDTA-permeabilized cells

To characterize the topology of the mutant proteins in the cytoplasmic membrane, we performed tryptic digestion in cells permeabilized with EDTA (Figure 4). After digestion of wt PBP3 with 100 μg/ml of trypsin for 30 min at 4 °C, two protected fragments of 40 KD and 25 KD were detected by immunoblotting analysis. The protected fragments were completely degraded when EDTA-permeabilized cells were treated with 1% Triton-X100 and 10 μg/ml lysozyme before addition of the protease. Mutant forms corresponding to modifications in the amino terminus (A1, A2, A4, A5 and A6) gave the same pattern as wt PBP3, with a slight increase of resistance to proteolysis for the 40 KD fragment. Pattern for mutant protein PBP3A9 was indistinguible from the wt PBP3. Then, topology and interactions of these mutant forms in the membrane must be analogous to the wild type PBP3. However, mutant PBP3A7 and PBP3A8 presented a completely different proteolytic pattern when compared with the wt PBP3; 40 KD and 25 KD fragments were absent, and a new 26 KD protected fragment appeared. This change in accessibility to proteolysis indicates a different topological organization or different interactions in the membrane for these mutant forms.

β-lactam binding capacity of the mutant proteins

As an indicator of the potential functionality of mutant proteins in the transpeptidase reaction we analyzed the capacity to bind covalently a labelled β-lactam antibiotic (^{125}I-ampicillin). Expression of the mutant protein was carried out in cultures of strain TG1 induced for 60 min in the presence of 1 mM IPTG. Figure 5 shows the labelling pattern of cytoplasm and particulate fractions after broken cells were fractionated by centrifugation at 35,000 x g for 30 min. The labelled form found in the cytoplasmic fraction disappeared when the particulate fraction was obtained at 150,000 x g for 60 min. All the mutant forms, except PBP3A15 and PBP3A16, were able to bind the antibiotic, and the amount of labelled protein correlated with the amount of protein found by immunoblotting.

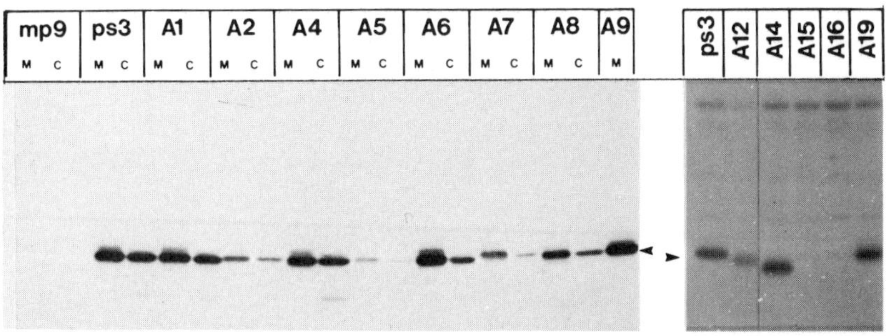

Figure 5. Binding of ^{125}I-ampicillin to membrane (M) and soluble (C) fractions prepared from strain TG1 infected with M13mp9 derivatives encoding different mutant forms of PBP3. Cultures of strain TG1 were grown and infected as in Figure 3. The subcellular fractionation of the samples of the left panel was done by centrifugation (35.000 g, 30 min). In the right panel only membrane fractions are shown, and were obtained by ultracentrifugation, as in Figure 3. Binding conditions are those described by Spratt (1975). Proteins were separated by SDS-PAGE and detected by autoradiography.

Complementation analysis

To study the "in vivo" functionality of mutant proteins modified in different domains, the mutant genes were transferred from the replicative form of the M13 phage to a plasmid. The 2.6 Kbp BamHI-EcoRI fragment that contains the mutant gene was subcloned into pEMBL19 (pSPAX plasmids) or pPR328 (pPRAX plasmids) vectors. Plasmids containing the mutant gene are named according the mutation and the plasmid vector used, i.e. plasmid pSPA7 contains the mutant gene *pbpB*A7 cloned in the pEMBL19 vector. Thermosensitive mutant strain AT51 (*ftsI*$_{2158}$) was transformed with each plasmid, and the ability to complement the thermosensitive phenotype was analyzed in plates and liquid cultures grown at 30 or 42 °C. The results obtained are summarized in Table 1. Mutant forms PBP3A7, A12, A15, A16 and A19 were not able to restore normal growth at 42 °C of strain AT51.

Table 1. Growth of the thermosensitive mutant strain AT51 (*ftsI*$_{2158}$) transformed with plasmids expressing mutant forms of PBP3.

Plasmid	Agar plate[1]		Liquid culture[2]	
	30 °C	42 °C	30 °C	42°C
pEMBL19	+	-	N	F
pSPME1	+	+	N	N
pSPA1	+	+	N.D[3]	N.D.
pSPA2	+	+	N.D.	N.D.
pSPA4	+	+	N.D.	N.D.
pSPA5	+	+	N.D.	N.D.
pSPA6	+	+	N.D.	N.D.
pSPA7	+	-	N	F
pSPA8	+	+	N	N
pSPA9	+	+	N	N
pSPA12	+	-	N	F
pSPA19	+	-	N	F
pPR328	+	-	N	F
pMM7	+	+/-	N	N
pPRA15	+	-	N	F
pPRA16	+	-	N	F

[1]No colonies (-) or visible colonies after 12 hours (+) or 24 hours (+/-) appear in LB agar plates supplemented with 100 µg/ml of Ampicillin.
[2]Normal cells (N) or filaments (F) are found in an exponentially growing culture in LB medium.
[3]N.D. not determined.

DISCUSSION

The analysis of mutant forms of PBP3 shown in this work (results summarized in Table 2) lead us to assume the following conclusions about the export of the wild type protein and the functional involvement of small domains of the amino and carboxy termini.

i) The amino terminal end of PBP3 behaves as a non-processed, non-modified signal peptide, driving the protein to the membrane and anchoring PBP3 to the cytoplasmic membrane. PBP3A5 mutant protein that can not be modified by the glycosylation-acylation system and SPase II, due to the change of C_{30} to R, was able to complement a thermosensitive *ftsI* strain. This mutant protein, although expressed at lower level in the infected cells, was located in a topological conformation equivalent to that of the wild type. Moreover, the lack of complementation for the mutant PBP3A19, that contains the consensus LACG sequence, may indicate that glycosylation-acylation at C_{30} may lead the inactivation of the protein.

Table 2.

Mutation	Expression[1]	Processing[2]	Topology[3]	Binding[4]	Functionality[5]
A1	++	+	+	++	+
A2	+	+	+	+	+
A4	++	+	+	++	+
A12	++	+	N.D.	++	-
A14	++	+	N.D.	++	N.D.
A5	+/++	+	+	+	+
A6	++	+	+	++	+
A19	N.D.[6]	+	N.D.	++	-
A15	++	-	N.D.	-	-
A16	++	-	N.D.	-	-
A7	++	-	-	++	-
A8	++	-	-	++	+
A9	++	-	+	++	+

[1]Level of expression determined by Western blotting; (++) level equivalent to wt protein; (+), decreased level.
[2]Processing at the COOH-end of PBP3; (+), processing; (-), no processing.
[3]Topology of the protein based on the comparison of the trypsin accessibility in EDTA-permeabilized cells; (+), same accessibility that wt protein; (-), different accessibility.
[4]Penicillin binding activity determined by ^{125}I-ampicillin labelling; (++), equivalent to wt protein; (+), decreased binding activity; (-), no binding.
[5]Functionality of the protein assigned by complementation analysis.
[6]N.D. not determined.

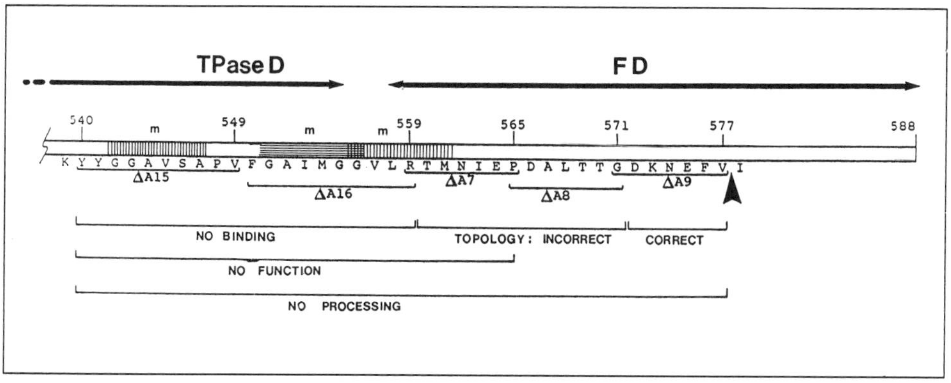

Figure 6. Schematic drawing of PBP3 COOH-terminal end showing the results obtained for the corresponding deletion mutant. The transpeptidase domain (**TPase D**) is located between the residues 240 to 557. The last 20-25 amino acids could be considered as a different domain involved in functionality (**F D**). The position of the deletions obtained is indicated in the figure, as well as the detected effects. The processing site is indicated with an arrow. The shadowed areas localize the miristylation sites (m) predicted with MOTIFS program of the GCG package.

ii) The results for the carboxy-terminus deletions, summarized in Figure 6, indicate that: a) processing at the COOH terminus is not essential for the function of PBP3, b) residues up to 556 form part of the transpeptidase domain, and c) the domain defined by amino acids 550 to 565 is required for the "in vivo" function of the protein. Some modification and/or interaction may take place in this region.

iii) These facts indicate that the integrity of the topological domains, at amino and carboxy termini, in addition to the active site for the enzymatic activity, are required for the "in vivo" function of the protein.

REFERENCES

Ayala, J., J. Plá, L. Desviat, and M. A. de Pedro. (1988) A *lacZ-pbpB* gene fusion coding for an inducible hybrid protein that recognizes localized sites in the inner membrane of *Escherichia coli.* J. Bacteriol. 170,3333-3341.

Bowler, L. D., and B. G. Spratt. (1989) Membrane topology of penicillin-binding protein 3 of *Escherichia coli.* Mol. Microbiol. 3(9),1277-1286.

Ghuysen, J.- M. (1991) Serine β-lactamases and penicillin-binding proteins. Annu. Rev. Microbiol. 45,37-67.

Hayashi, S., H. Hara, H. Suzuki and Y. Hirota. (1988) Lipid modification of *Escherichia coli* penicillin-binding protein 3. J. Bacteriol. 170,5392-5395.

Matsuhashi, M., M. Wachi and F. Ishino. (1990) Machinery for cell growth and division: penicillin-binding proteins and other proteins. Res. Microbiol. 141,89-103.

Nagasawa, H., Y. Sakagami, A. Suzuki, H. Suzuki, H. Hara, and Y. Hirota. (1989) Determination of the cleavage site involved in C-terminal processing of penicillin-binding protein 3 of *Escherichia coli.* J. Bacteriol. 171(11),5890-5893.

Spratt, B. G. (1975) Distinct penicillin binding proteins involved in the division, elongation, and shape of *Escherichia coli* K12. Proceedings of the National Academy of Sciences USA. 72(8),2999-3003.

Tormo, A., J. A. Ayala, M. A. de Pedro, M. Aldea, and M. Vicente. (1986) Interaction of FtsA and PBP3 proteins in the *Escherichia coli* septum. J. Bacteriol. 166(3),985-992.

MODULAR DESIGN OF THE BI(MULTI ?)FUNCTIONAL PENICILLIN-BINDING PROTEINS

Serge Englebert, Aboubaker El Kharroubi, Graziella Piras, Bernard Joris, Jacques Coyette, Martine Nguyen-Distèche and Jean-Marie Ghuysen

Centre d'Ingénierie des Protéines, Université de Liège, Institut de Chimie, B6, B-4000 Sart Tilman (Liège 1), Belgium

INTRODUCTION

The penicilloyl serine transferases are proteins or protein domains that catalyse rupture of the ß-lactam amide bond of penicillin and transfer of the penicilloyl moiety to an essential serine. The serine-ester-linked penicilloyl derivatives formed by reaction with the ß-lactamases are hydrolytically labile. Those formed by reaction with the penicillin-binding proteins (PBPs) are hydrolytically inert. Penicilloylation of the essential serine of some PBPs produces a signal that is transmitted to the cytosol. Penicilloylation of other PBPs causes loss of a catalytic function related to wall peptidoglycan synthesis. For a recent review and list of references, see Ghuysen (1991).

The defensive ß-lactamases are secretory proteins. Some of them are produced in maximal amounts only in the presence of ß-lactam antibiotics. The high-Mr PBP BlaR is the penicillin sensory-transducer responsible for the induction of ß-lactamase synthesis in Bacillus licheniformis. The penicillin-binding C-terminal domain, i.e. the sensor, is extracellular and is fused at its amino side to a three-transmembrane segment transducer. The extracellular domain of the transducer is responsible for the reception of the signal generated by penicilloylation of the active-site serine of the sensor and the cytosolic domain is responsible for the generation of an intracellular signal that launches ß-lactamase synthesis.

As with the reactions they catalyse in vitro on D-alanyl-D-alanine-terminated peptides, the monofunctional low-Mr PBPs are peptidoglycan-remodelling DD-peptidases and help control the extent of peptidoglycan crosslinking. Some monofunctional PBPs are secretory proteins. Most of them are membrane-bound. The extracellular penicillin-sensitive DD-peptidase domain has a C-terminal extension the end of which serves as membrane anchor.

The high-Mr PBPs involved in wall peptidoglycan synthesis contain a penicillin-binding C-terminal domain and a several hundred amino acid residue N-terminal domain. Membrane anchoring is at the amino terminus so that the bulk of the protein is extracellular. As with the reactions they catalyse in vitro on lipid-linked disaccharide-(D-alanyl-D-alanine-terminated) pentapeptide precursors, the isolated PBPs 1A and 1B of Escherichia coli (Broome-Smith et al., 1985) are peptidoglycan-synthe-

Bacterial Growth and Lysis, Edited by M.A. de Pedro *et al.*, Plenum Press, New York, 1993

sizing enzymes. The N-terminal domain functions as a penicillin-insensitive transglycosylase (performing glycan chain elongation) while the C-terminal domain functions as a penicillin-sensitive DD-transpeptidase (performing peptide crosslinking). The isolated PBP 2 (Asoh et al., 1986) and PBP3 (Nakamura et al., 1983) of E. coli are inert in vitro on lipid-linked peptidoglycan precursors. They play important roles in cell morphogenesis in conjunction with intrinsic membrane proteins. The pair PBP2-RodA is responsible for the rod shape of the cell and the pair PBP3-FtsW is responsible for septum formation and cell division. PBP 2' of Staphylococcus aureus (Song et al., 1987), PBP5 of Enterococcus hirae R40 (El Kharroubi et al., 1991) and PBP3r of E. hirae S185r (Piras et al., 1990, and unpublished data) confer penicillin resistance to the relevant strains. They are thought to be able to take over the functions needed for wall peptidoglycan assembly under conditions in which the other PBPs are inactivated. PBPs 1a (Martin et al., 1992), 2X (Laible et al., 1989) and 2b (Dowson et al., 1989) of Streptococcus pneumoniae, PBP2 of Neisseria meningitidis (Zhang and Spratt, 1989) and PBP2 of Neisseria gonorrhoeae (Spratt, 1988) are also of known primary structure.

Four ß-lactamases and, to a lesser resolution, one monofunctional secretory PBP are of known 3-D structure. They consist of two structural domains : an all-α domain and an α/ß domain whose five-stranded ß-sheet is protected on both faces by additional helices. As a result of the folding of the polypeptide chain, three conserved amino acid groupings are brought close to each other and form part of the active site. Using the numbering of the ß-lactamases of class A, the essential serine of the S70XXK motif is at the N-terminus of helix α2 of the all-α domain, the S130DN motif, on a loop connecting helices α4 and α5 of the all-α domain, occurs on one side of the cavity, and the K234T(S)G motif, on the innermost ß3 strand of the ß-sheet, is on the other side of the cavity.

Based on homology searches and hydrophobic cluster analysis, the ß-lactamases of class C and D, the monofunctional DD-peptidases/PBPs and the BlaR penicillin sensor for which no X-ray data are available, have, in all likelihood, the same folding and active-site defining motifs (or analogues) as the ß-lactamases of class A used as reference for structural modelling (Joris et al., 1991; Palomeque-Messia et al., 1991; Granier et al., 1992). One peculiarity of the ß-lactamases of class A is the presence of a fourth marker, the E166XELN pentapeptide. This marker occurs on a loop connecting helices α7 and α8 of the all-α domain at the entrance of the active site. It is an important component of the catalytic machinery (Lamotte-Brasseur et al., 1991).

In order to shed light on their molecular organization, the twelve high-Mr bifunctional PBPs of known primary structure, mentioned above, have been examined by hydrophobic-cluster analysis.

MATERIAL AND METHODS

Hydrophobic cluster analysis (HCA) is a powerful method for comparing proteins that are weakely related in the primary structure (Gaboriaud et al, 1987). It rests upon a duplicated representation of the amino acid sequences on an α-helical two-dimensional pattern (in which the hydrophobic residues tend to form clusters) and compares the distribution of the hydrophobic clusters along the sequences. In this representation, the hydrophobic residues F, I, L, M, V, W, Y are encircled, the hydrophobic clusters are also delineated and different symbols are used for P (✱), G (◆), S (⊡), T (□) and C (©). The shapes of the clusters are usually associated with definite secondary structures. Clusters of similar shapes, sizes and relative positions express similarity in the polypeptide folding. When compared to methods based only on single amino acid property/identity, HCA allows distant information to become visible more readily (since the six residues adjacent to the amino acid i are i-4, i-3, i-1, i+1, i+3 and i+4) and allows deletions/insertions to be introduced more easily between the secondary structures.

By analogy with the ß-lactamases and secretory monofunctional DD-peptidases/PBPs, the penicillin-binding C-terminal domain of the high-Mr PBPs is assumed to start ≈ 60 residues upstream from the SXXK motif. On this basis, the HCA plots have been divided into two parts which extend upstream and downstream from the amino end of the C-terminal domain, respectively. In these representations, the hydrophobic residues and clusters occurring at equivalent places in pairs or groups of proteins are in bold; the amino acid residues A, C, D, E, G, H, K, N, P, Q, R, S, T occurring as strict identities are marked by scattered points; and the SXXK, SDN and KT(S)G motifs (or analogues) are in dashed spots.

There is a high extent of similarity in the amino acid sequence between the E. hirae PBPs 5 and 3r (73 % and 85 % identities for the N-terminal and C-terminal domains, respectively) and between the N. meningitidis PBP2 and N. gonorrhoeae PBP2 (97 % and 99 % identities). Consequently, the HCA plots shown for the E. hirae PBP5 and N. meningitidis PBP2 apply to the E. hirae PBP3r and N. gonorrhoeae PBP2, respectively.

RESULTS

The twelve high-Mr PBPs under study are proteins of varying molecular masses. The number of constituent amino acid residues range from about 580 (the N. meningitidis PBP2 and E. coli PBP3) to about 850 (the E. coli PBPs 1A and 1B). Hydrophobic cluster analysis leads to the following observations.

Based on the identification of large polypeptide segments that have significant similarity in the distribution of the hydrophobic and hydrophilic clusters, the high-Mr PBPs fall into groups. Members of a given pair or triad possess several homologous segments, i.e. modules, whose conserved clusters align easily and cover a large proportion of the amino acid sequences.

The modules thus defined help solve the molecular organization of the PBPs. The simplest pattern is that of a three-module protein. The membrane anchor, usually 40 to 60 amino acid residues long, is linked to the N-terminal domain which is linked to the C-terminal domain. Moreover, the carboxy end of the C-terminal domain, about 60 to 80 amino acid residues downstream from the KT(S)G motif, is the carboxy end of the protein. This pattern is found in the E. coli PBP3 (Fig. 1a,b), N. meningitidis PBP2 (Fig. 1a,b), S. pneumoniae PBP2b (Fig. 1a,b) and E. coli PBP2 (Fig. 2a,b).

The other PBPs are of a more complex design due to the presence of large inserts. The E. coli PBP1B has the hydrophobic membrane anchor (the G64-L87 transmembrane α-helix) in the middle of a 185 amino acid residue N-terminal extension (Fig. 3a). The S. aureus PBP2' and E. hirae PBP5 have a highly homologous ≈ 100 amino acid residue polypeptide inserted between the carboxy side of the membrane anchor and the amino side of the N-terminal domain (Fig. 2a). Note that the N-terminal domains of the S. aureus PBP2' and E. hirae PBP5 are similar to the N-terminal domain of the E. coli PBP2 (Fig. 2a). The E. coli PBP1A has a 90 amino acid residue intervening sequence that links the carboxy side of the N-terminal domain to the amino side of the C-terminal domain (Fig. 3a). The C-terminal domains of the S. pneumoniae PBP2X (Fig. 1b), E. coli PBP1A, E. coli PBP1B and S. pneumoniae PBP1a (Fig. 3b) bear a tail, 80 to 100 amino acid residues long. Inserts - 40 to 50 amino acid residues long-also occur in the N-terminal domain of the S. pneumoniae PBP2b (Fig. 1a) and in the C-terminal domains of the S. pneumoniae PBP2X (Fig. 1b) and E. coli PBP1A (Fig. 3b).

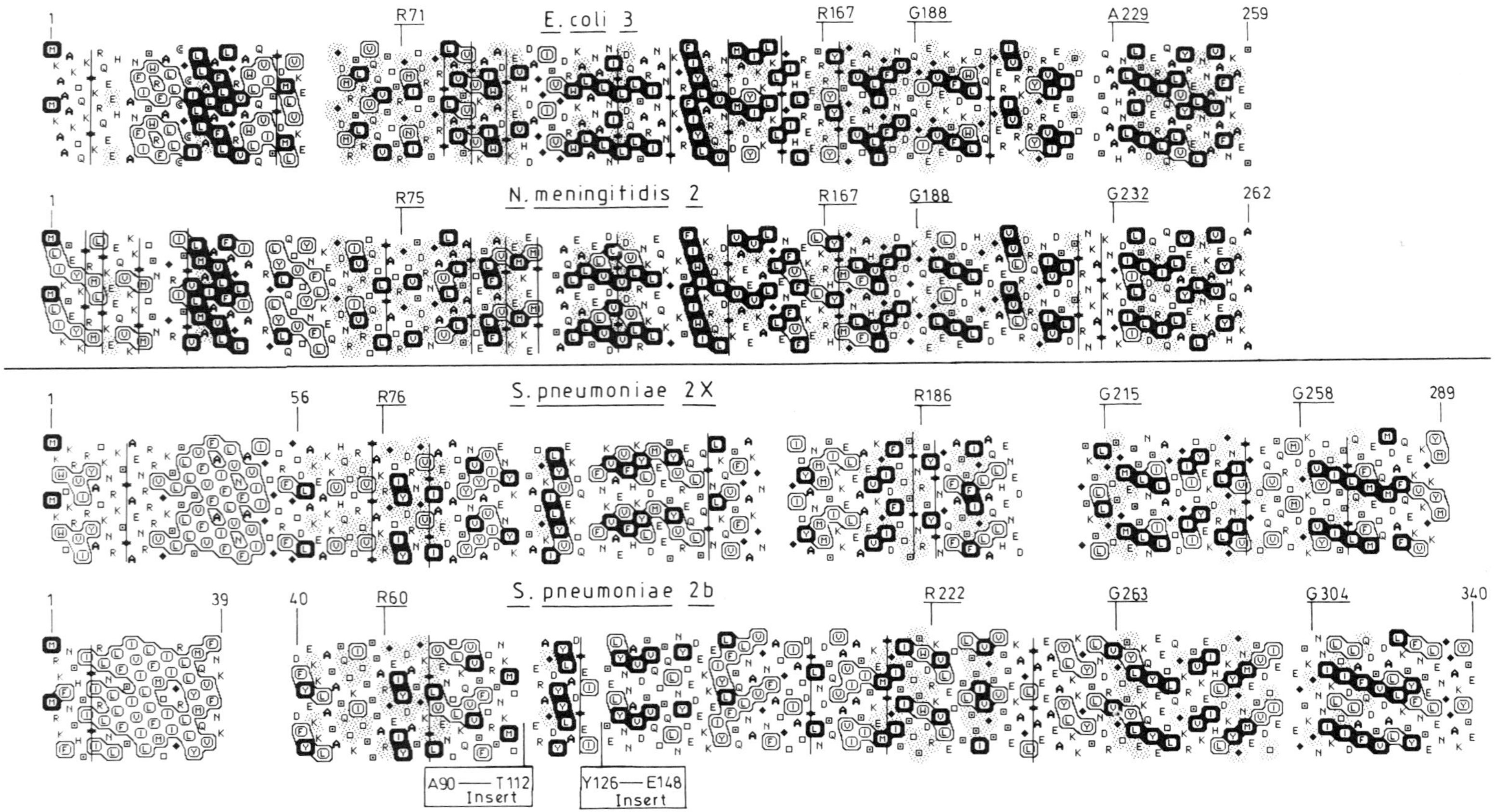

Fig. 1a. HCA plots of the pair E.coli PBP3-N.meningitidis PBP2 and the pair S. pneumoniae PBPs 2X and 2b. Analysed sequences : from the amino end of the PBPs to the carboxy end of the N-terminal domains.

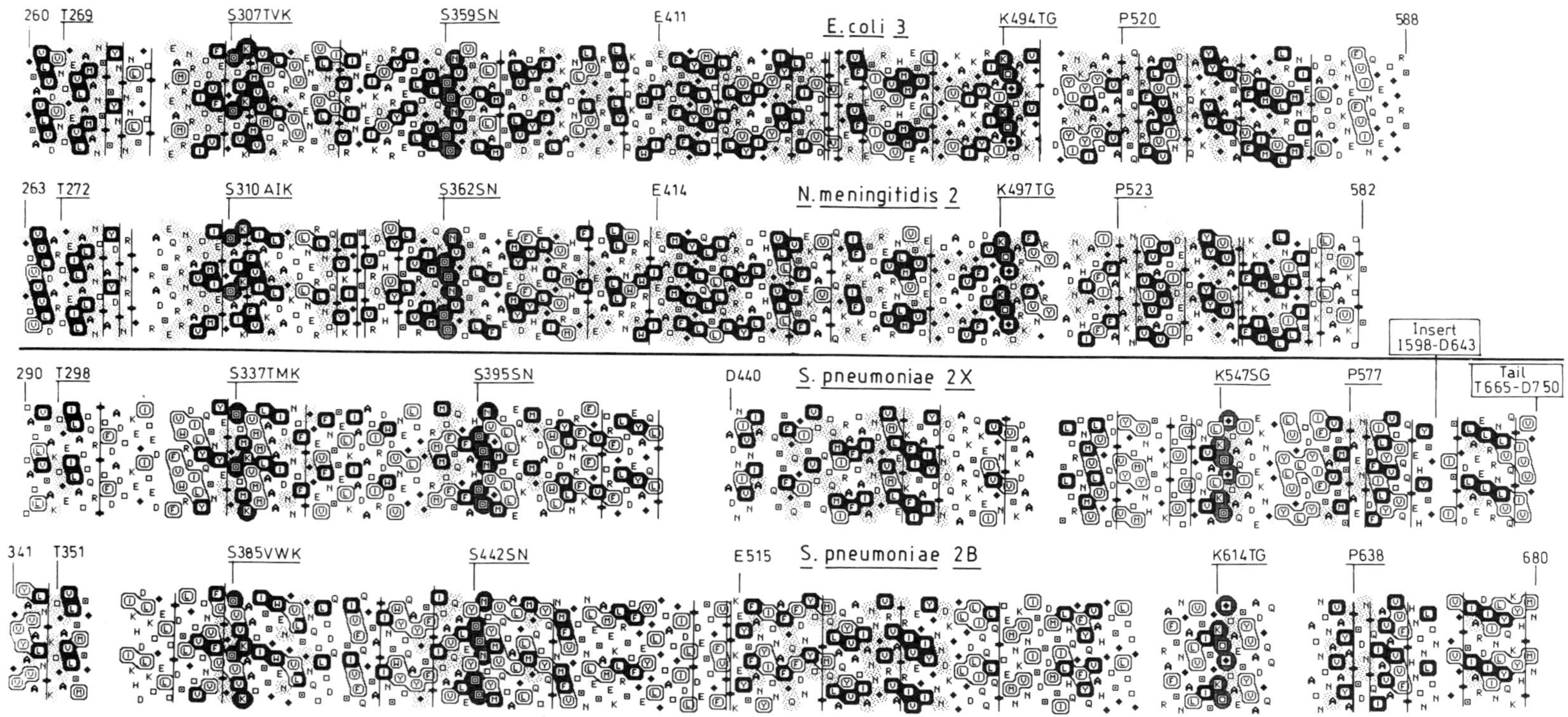

Fig. 1b. HCA plots of the pair E.coli PBP3-N.meningitidis PBP2 and the pair S. pneumoniae PBPs 2X and 2b. Analysed sequences : from the amino end of the C-terminal domains to the carboxy end of the PBPs.

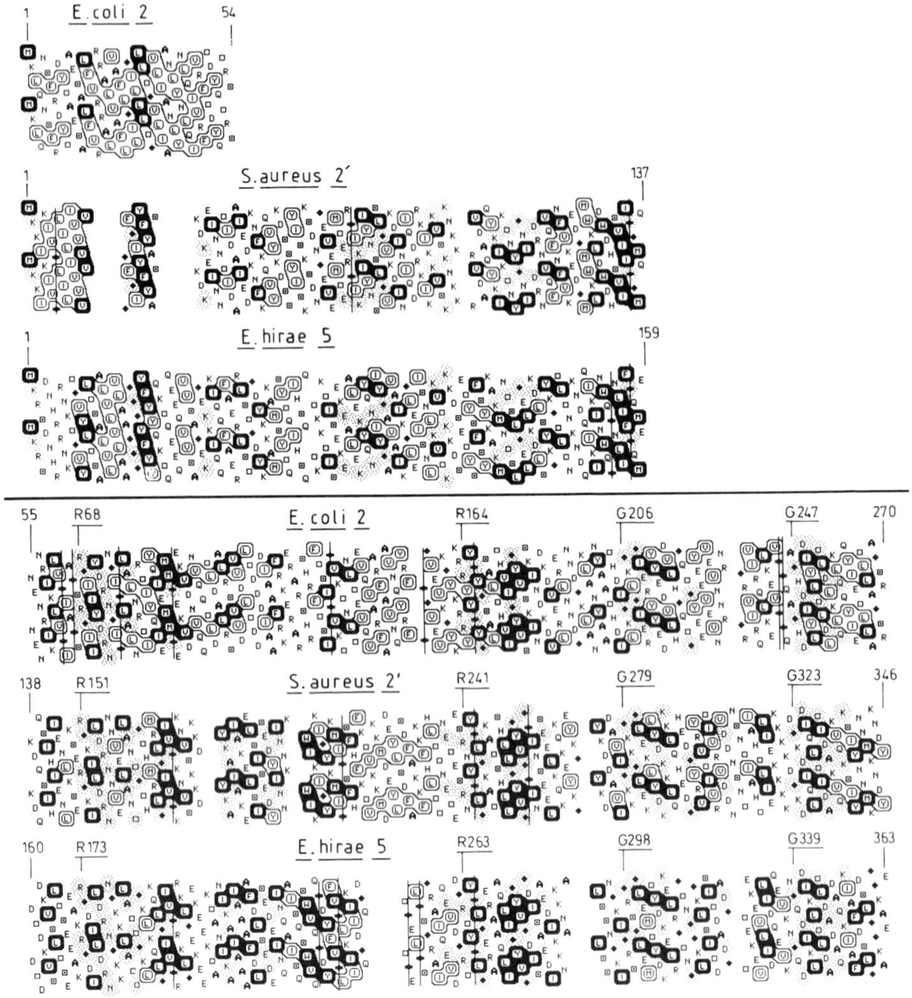

Fig. 2a. HCA plots of the pair S.aureus PBP2'-E.hirae PBP5 and the E. coli PBP2 used as reference. Analysed sequences : from the amino end of the PBPs to the carboxy end of the N-terminal domains.

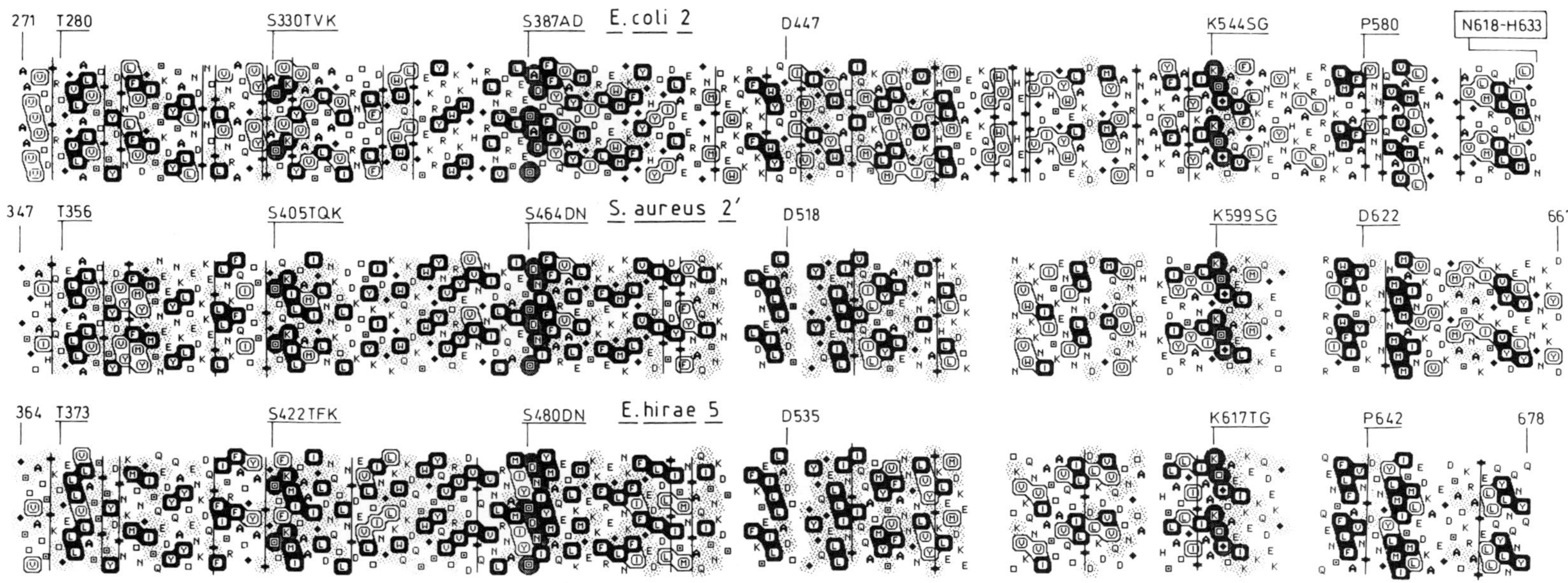

Fig. 2b. HCA plots of the pair S.aureus PBP2'-E.hirae PBP5 and the E. coli PBP2 used as reference. Analysed sequences : from the amino end of the C-terminal domains to the carboxy end of the PBPs.

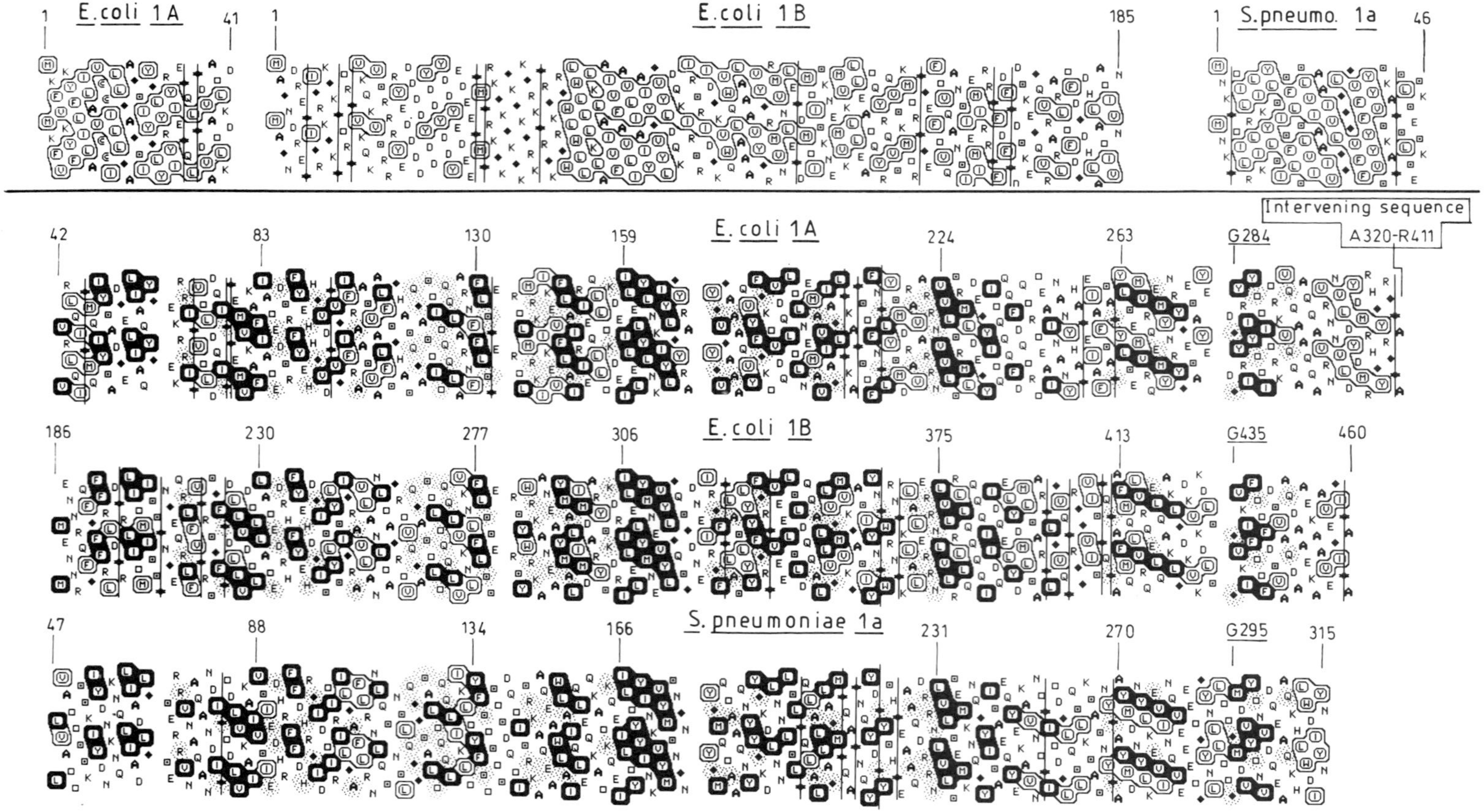

Fig. 3a. HCA plots of the triad E.coli PBP1A-E.coli PBP1B-S.pneumoniae PBP1a. Analysed sequences : from the amino end of the PBPs to the carboxy end of the N-terminal domains.

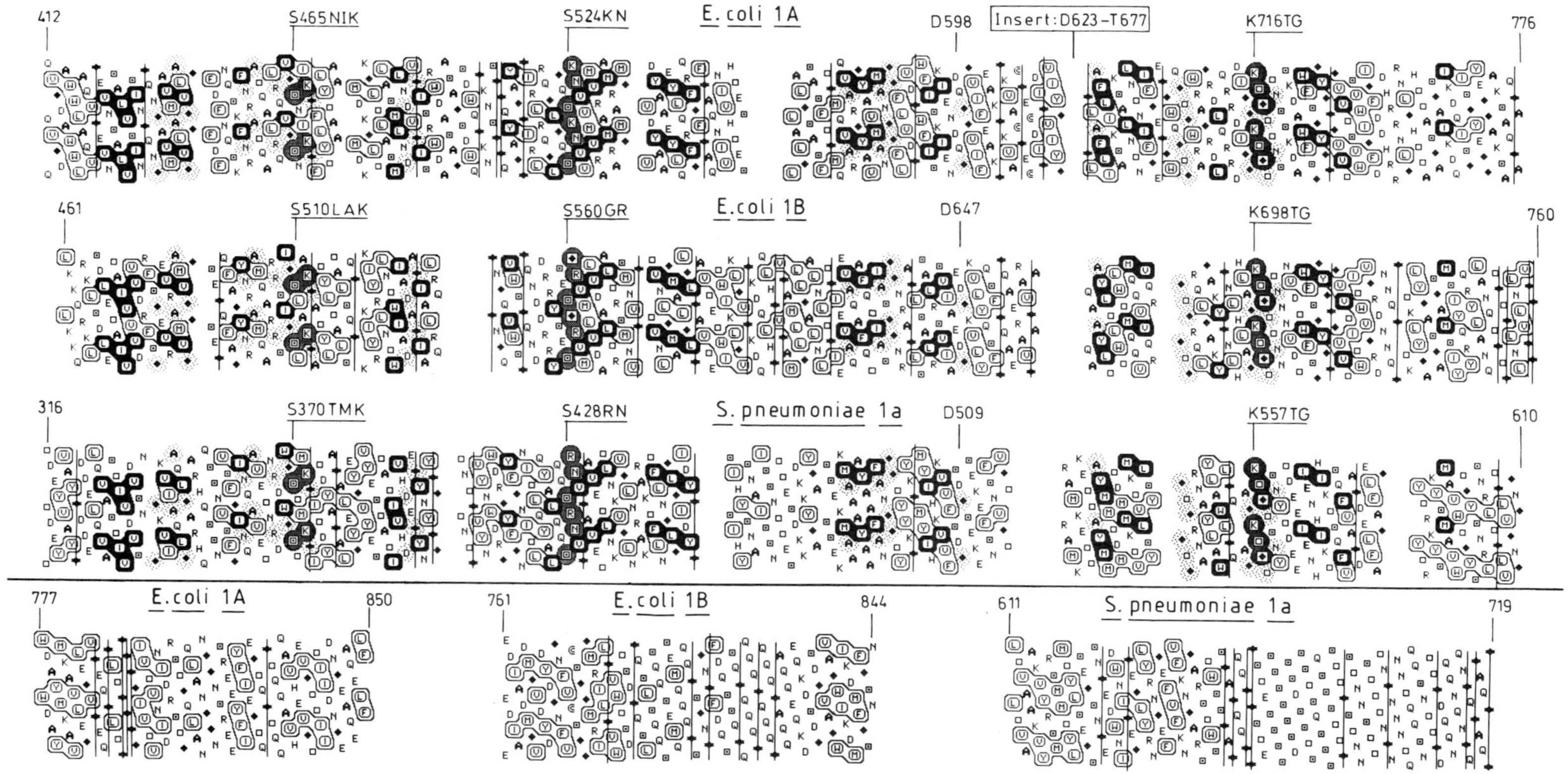

Fig. 3b. HCA plots of the triad E.coli PBP1A-E.coli PBP1B-S.pneumoniae PBP1a. Analysed sequences : from the amino end of the C-terminal domains to the carboxy end of the PBPs.

Depending on the markers borne by the N-terminal domains, the high-Mr PBPs fall into two classes. The E. coli PBPs 1A and 1B and S. pneumoniae PBP1a belong to class A. The E. coli PBPs 2 and 3, N. meningitidis PBP2 (and N. gonorrhoeae PBP2), S. aureus PBP2', E. hirae PBP5 (and PBP3r) and S. pneumoniae PBPs 2X and 2b belong to class B.

The markers of the N-terminal domains of the PBPs of class B are four amino acid groupings or boxes (Figs 4 and 5). Altogether, they provide several conserved amino acids with potentially important functional side-chains : one arginine residue and one aspartic acid residue in box 1, another arginine residue in box 2, one glutamic acid residue in box 3, one glutamine residue in box 4, and one serine or threonine residue also in box 4. Using the numbering of the E. coli PBP2, R68 and D73 of box 1, R164 of box 2, E211 of box 3 and Q259 of box 4 may well be within the active site of the domain.

The N-terminal domains of the PBPs of class A have some amino acid groupings which show remote similarity to boxes 1, 3 and 4 (Fig. 5). These groupings may provide a diamino acid residue, R or H, and a dicarboxylic amino acid, E or D, equivalent to the conserved arginine and aspartic acid residues of box 1.

The C-terminal domains of all the high-Mr PBPs possess the three SXXK, SXN and KT(S)G markers (or analogues) characteristic of the penicilloyl serine transferases (boxes 6, 7 and 8 in Fig. 5). E. coli PBP1B, however, has the unusual triad S560GR (Fig. 3b). This triad aligns perfectly with S524KN of the E. coli PBP1A and S428RN of the S. pneumoniae PBP1a, and the three markers have comparable hydrophobic environments. Note, however, that the E. coli PBP1B possesses a S572MN triad 12 residues downstream from the S560GR motif.

The C-terminal domains of all the high-Mr PBPs also possess several dicarboxylic acid residues in the polypeptide stretches that extend between the SXN (or equivalent) and KT(S)G motifs. Among these, E411 and E414 in the pair E.coli PBP3-N.meningitidis PBP2 (Fig. 1b), D447, D518 and D535 in the triad E.coli PBP2-S.aureus PBP2'-E.hirae PBP5 (Fig. 2b), and D598, D647 and D509 in the triad E.coli PBP1A-E.coli PBP1B-S.pneumoniae PBP1a (Fig. 3b) occur as strict identities in the aligned amino acid sequences. D440 in the S. pneumoniae PBP2X is two residues upstream from E515 in the S. pneumoniae PBP2b (Fig. 1b). Each of these dicarboxylic amino acid residues may be equivalent to E166 of the E166XPELN pentapeptide that forms part of the immediate boundary of the active site of the class A ß-lactamases. As shown by site-directed mutagenesis, D447 is important for the functioning of the E. coli PBP2 (Adachi et al., 1992).

The C-terminal domains of the PBPs of class B possess two additional boxes, 5 and 9 (Fig. 5). These boxes are not found in the PBPs of class A. Box 5 is located upstream from the SXXK motif close to amino end of the C-terminal domain. It contains a vertical hydrophobic cluster that is reminiscent of strand ß1 of the ß-lactamases (Fig. 4). Box 9 is located downstream from the KT(S)G motif, close to the carboxy end of the domain. Box 9, probably a turn, is sandwiched between two vertical hydrophobic clusters that are reminiscent of strands ß4 and ß5 of the class A ß-lactamases (Fig. 5).

DISCUSSION

The high-Mr PBPs illustrate the principle that protein molecules frequently are constructed from modules that are linked together in a single polypeptide chain. Each module is folded into one domain that is responsible for a particular function.

High-Mr PBPs exist which consist of a hydrophobic membrane anchor and two domains, the N- and C-terminal domains, suggesting bifunction-

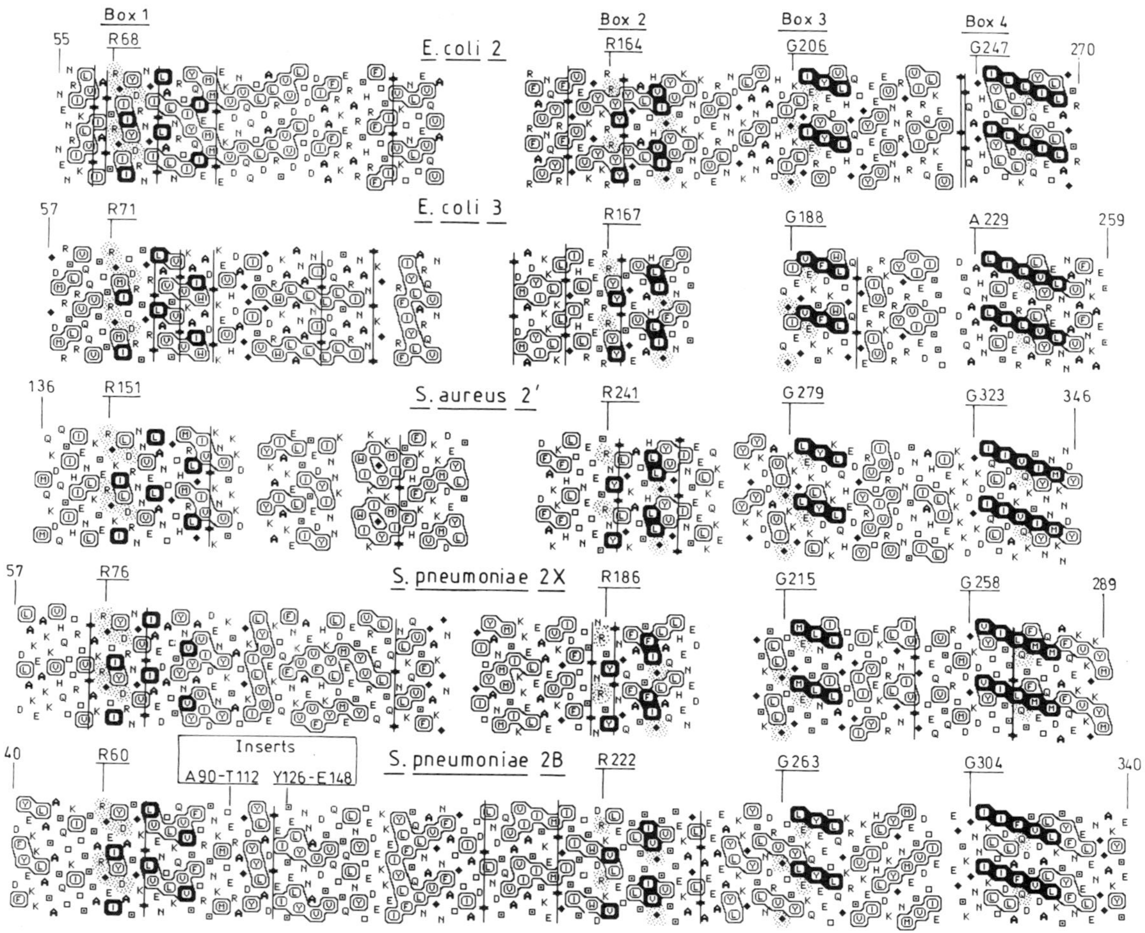

Fig. 4. HCA plots of the N-terminal domains of high-Mr PBPs of class B. Identification of the domain-specific boxes 1 to 4.

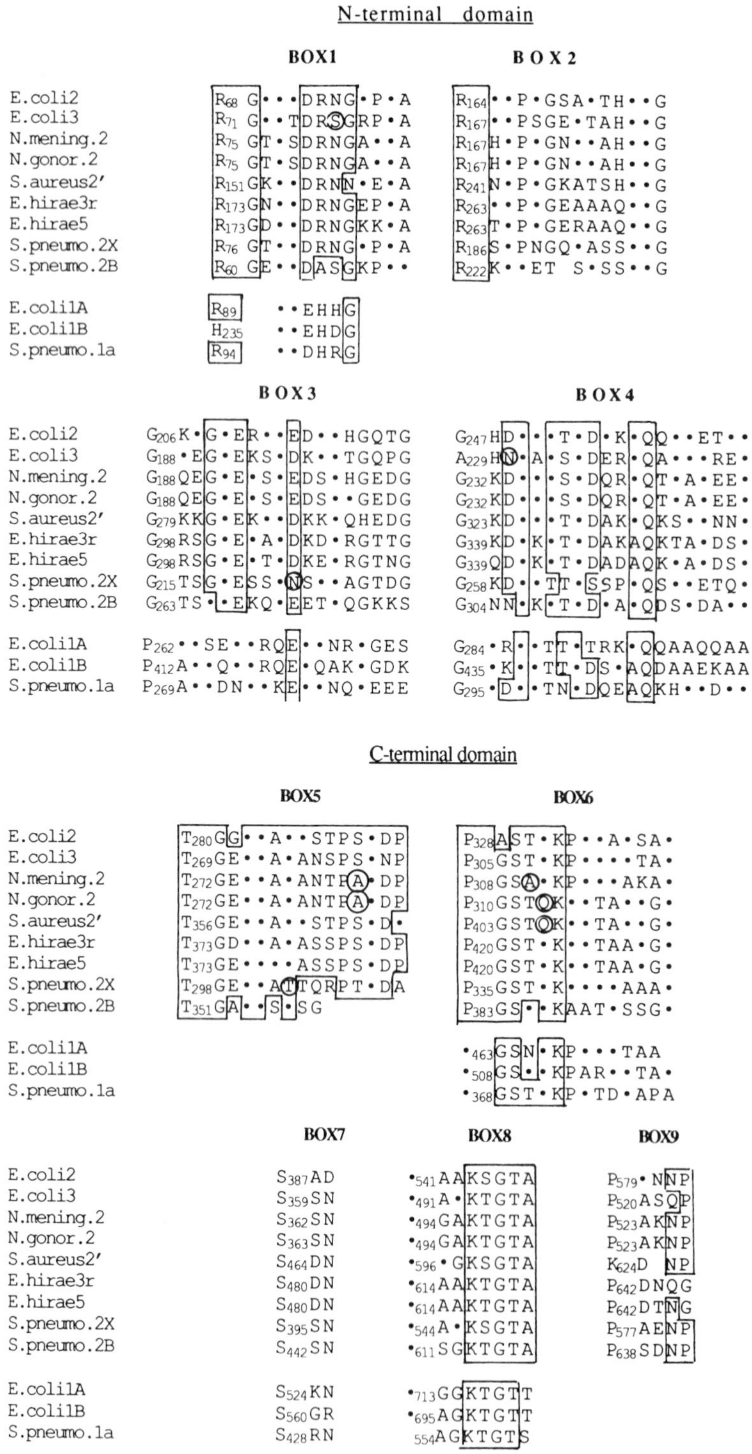

Fig. 5 Domain-specific boxes of the hight-Mr PBPs. Hydrophobic residues are shown in black dots.

ality. Other high-Mr PBPs have additional inserts, 80 to 100 amino acid residues long. These inserts may occur upstream from the N-terminal domain, between the N- and C-terminal domains or downstream from the C-terminal domain. They are large enough to give rise to additional domains each of which possessing a particular folding and performing a separate function. These high-Mr PBPs are probably multifunctional.

The penicillin-binding C-terminal domains of all the high-Mr PBPs studied possess, at the expected relative positions along the amino acid sequences, the active-site defining markers of the penicilloyl serine transferases. However, by reference to the ß-lactamases (and the mono-functional PBPs and BlaR penicillin sensor), the C-terminal domains of the high-Mr PBPs have diverged so far that other traces of homology in the primary structure have almost completely disappeared.

The N-terminal domains of the high-Mr PBPs of class B have markers in the form of four conserved amino acid groupings occurring in distinct hydrophobic environments and possessing amino acid residues that are potentially important for catalytic function. The N-terminal domains of the high-Mr PBPs of class A do not possess this type of fingerprint, except, perhaps, for the presence of a conserved diamino acid, R or H, and a dicarboxylic acid, D or E, close to the amino end of the domain. Whether or not the N-terminal domains of the high-Mr PBPs of class A and B have different foldings and functionalities remains an open question.

The question of which functions the high-Mr PBPs perform remains a major challenge and a major stumbling-block in the field. The E. coli PBPs 1A and 1B are the only known PBPs which catalyse in vitro transglycosylation and transpeptidation reactions and are able, as isolated proteins, to carry out peptidoglycan assembly (see the Introduction). Water-soluble derivatives of the E. coli PBP3 (and other high-Mr PBPs) catalyse acyl transfer reactions on acyclic thioester compounds such as C6H5-CO-D-Ala-S-CH2-COOH in a way that mimics transpeptidation (Adam et al., 1991) but they are devoid of activity in terms of peptidoglycan assembly. High-Mr PBPs may be inactive unless they associate with other host proteins such as RodA for the E. coli PBP2 and FtsW for the E. coli PBP3. By analogy with the high-Mr PBP BlaR which provides an example of how domains with different functions can combine and give rise to a new function, i.e. control of gene expression, high-Mr PBPs may be components of multi-molecular systems in which they function as signalling devices of one kind or another.

Acknowledgements

This work was supported in part by the Belgian programme on Interuniversity Poles of attraction initiated by the Belgian State, Prime Minister's Office, Science Policy Programming (PAI n°19), the "Fonds de la Recherche Scientifique Médicale" (contract n°3.4531.92), and a "Convention tripartite" between the "Région wallonne", SmithKline Beecham, U.K., and the University of Liège. B.J. is "chercheur qualifié" of the "Fonds National de la Recherche Scientifique", Belgium.

REFERENCES

Adam, M., Damblon, C., Jamin, M., Zorzi, W., Dusart, V., Galleni, M., El Kharroubi, A., Piras, G., Spratt, B.G., Keck, W., Coyette, J., Ghuysen, J.M., Nguyen-Distèche, M. and Frère, J.M. (1991) Acyltransferase activities of the high-molecular-mass essential penicillin-binding proteins. Biochem. J. 279, 601-604.

Adachi, H., Ishiguro, M., Imajoh, S., Ohta, T. and Matsuzawa, H. (1992) Active-site residues of the transpeptidase domain of penicillin-binding protein 2 from Escherichia coli : similarity in the catalytic mechanism to class A ß-lactamases. Biochemistry 31, 430-437.

Asoh, S., Matsuzawa, H., Ishino, F., Strominger, J.L., Matsuhashi, M. and Ohta, T. (1986) Nucleotide sequence of the pbpA gene and characteristics of the deduced amino acid sequence of penicillin-binding protein 2 of Escherichia coli K12. Eur. J. Biochem. 160, 231-238.

Broome-Smith, J.K., Edelman, A., Yousif, S. and Spratt, B.G. (1985) The nucleotide sequence of the ponA and ponB genes encoding penicillin-binding proteins 1A and 1B of Escherichia coli K12. Eur. J. Biochem. 147, 437-446.

Dowson, C.G., Hutchinson, A. and Spratt, B.G. (1989) Nucleotide sequence of the penicillin-binding protein 2B gene of Streptococcus pneumoniae strain R6. Nucl. Ac. Res. 17, 7518.

El Kharroubi, A., Jacques, P., Piras, G., Coyette, J., Van Beeumen, J. and Ghuysen, J.M. (1991) The Enterococcus hirae R40 penicillin-binding protein 5 and the methicillin-resistant Staphylococcus aureus penicillin-binding protein 2' are similar. Biochem. J. 280, 463-469.

Gaboriaud, C., Bissery, V., Benchetrit, T. and Mornon, J.P. (1987) Hydrophobic cluster analyses : an efficient new way to compare and analyse amino acid sequences. FEBS Lett. 1, 149-155.

Ghuysen, J.M. (1991) Serine ß-lactamases and penicillin-binding proteins. Annu. Rev. Microbiol. 45, 37-67.

Granier, B., Duez, C., Lepage, S., Englebert, S., Dusart, J., Dideberg, O., Van Beeumen, J., Frère, J.M. and Ghuysen, J.M. (1992) Primary and predicted secondary structures of the Actinomadura R39 extracellular DD-peptidase, a penicillin-binding protein (PBP) related to the Escherichia coli PBP4. Biochem. J. 228, 781-788.

Joris, B., Ledent, P., Dideberg, O., Fonzé, E., Lamotte-Brasseur, J., Kelly, J.A., Ghuysen, J.M. and Frère, J.M. (1991) Comparison of the sequences of class A ß-lactamases and of the secondary structure elements of penicillin-recognizing proteins. Antimicrob. Ag. Chemother. 35, 2294-2301.

Laible, G., Hakenbeck, R., Sicard, M.A., Joris, B. and Ghuysen, J.M. (1989) Nucleotide sequence of the pbpX genes encoding the penicillin-binding proteins 2X from Streptococcus pneumoniae R6 and a cefotaxime-resistant mutant, C506. Mol. Microb. 3, 1337-1348.

Lamotte-Brasseur, J., Dive, G., Dideberg, O., Charlier, P., Frère, J.M. and Ghuysen, J.M. (1991) Mechanism of acyl transfer by the class A serine ß-lactamase of Streptomyces albus G. Biochem. J. 279, 213-221.

Martin, C., Briese, T. and Hakenbeck, R. (1992) Nucleotide sequence of genes encoding penicillin-binding proteins from Streptococcus pneumoniae and Streptococcus oralis with high homology to Escherichia coli penicillin-binding proteins 1A and 1B. J. Bacteriol., in press.

Nakamura, M., Maruyama, I.N., Soma, M., Kato, J.I., Suzuki, H. and Horota, Y. (1983) On the process of cellular division in Escherichia coli : nucleotide sequence of the gene for penicillin-binding protein 3. Mol. Gen. Genet. 191, 1-9.

Palomeque-Messia, P., Englebert, S., Leyh-Bouille, M., Nguyen-Distèche, M., Duez, C., Houba, S., Dideberg, O., Van Beeumen, J. and Ghuysen, J.M. (1991) Biochem. J. 279, 223-230.

Piras, G., El Kharroubi, A., Van Beeumen, J., Coeme, E., Coyette, J. and Ghuysen, J.M. (1990) Characterization of an Enterococcus hirae penicillin-binding protein 3 with low penicillin affinity. J. Bacteriol. 172, 6856-6862.

Song, M.D., Wachi, M., Doi, M., Ishino, F. and Matsuhashi, M. (1987) Evolution of an inducible penicillin-target protein in methicillin-resistant Staphylococcus aureus by gene fusion. FEBS Lett. 221, 167-171.

Spratt, B.G. (1988) Hybrid penicillin-binding proteins in penicillin-resistant strains of Neisseria gonorrhoeae. Nature 332, 173-176.

Zhang, Q.Y. and Spratt, B.G. (1989) Nucleotide sequence of the penicillin-binding protein 2 gene of Neisseria meningitidis. Nucl. Ac. Res. 17, 5383.

PENICILLIN-BINDING PROTEINS 1A AND 3 IN *STREPTOCOCCUS PNEUMONIAE*: WHAT ARE ESSENTIAL PBP'S?

Regine Hakenbeck[1], Heinz Ellerbrok[2], Christiane Martin[1], Gioavanna Morelli[1], Cordelia Schuster[3], Anatol Severin[4], and Alex Tomasz[4]

[1]Max-Planck Institut für molekulare Genetik
Ihnestr. 73, D-1000 Berlin 33
[2]present address: INSERM U152, Institut de Génétique Moléculaire
27, Rue du Faubourg St. Jacques, 75674 Paris Cedex
[3]present address: University of Wales
Bangor, Gwynedd LL57 2DG
[4]The Rockefeller University, Microbiology Department
1230 York Avenue, New York 10021

INTRODUCTION

Penicillin-binding proteins (PBPs) exist in every penicillin-susceptible bacterial species. Onc of the primary steps in the mechanism of action of ß-lactams that finally lead to growth inhibition, lysis and death of the cell, is the interaction of penicillin with PBPs, the target enzymes for this group of antibiotics. Penicillin, believed to structurally resemble the actual substrate of PBPs - the C-terminal D-Ala-D-Ala moiety of muropentapeptides - becomes covalently bound to a serine residue located in the active site of all PBPs. This penicilloyl-PBP complex is enzymatically inactive, and therefore correct assembly of the murein layer which is vital for cell growth can no longer be ensured (for review, see Ghuysen, 1991).

THE INTRINSIC MECHANISM OF PENICILLIN-RESISTANCE

Several pathogenic bacterial species have evolved intrinsic penicillin-resistance, a mechanism in which PBPs are altered in such a way that interaction with penicillin requires much higher concentrations compared to sensitive bacteria, and the minimal

Bacterial Growth and Lysis, Edited by M.A. de Pedro *et al.*, Plenum Press, New York, 1993

inhibitory concentrations (MIC values) increases accordingly (spratt, 1989). Since the binding constants are different for each PBP and for each ß-lactam, a complicated process is necessary to guarantee high level resistance. Alteration of the PBP which is the most sensitive for a given ß-lactam to a low affinity PBP form shifts the MIC value only to that antibiotic concentration at which another PBP becomes inhibited. Thus, the PBP representing the killing target for a ß-lactam varies during the course of resistance development. This becomes especially evident in pneumococci where four out of five high M_r PBP are successively altered with increasing resistance (Laible *et al.*, 1991). Furthermore, a given PBP may well be the primary target for one ß-lactam but not for another one, and depending on the ß-lactam used for selecting resistant mutants, different PBPs are primarily affected as has been documented in piperacillin- versus cefotaxime-resistant laboratory mutants of *S. pneumoniae* (Laible and Hakenbeck, 1987).

The fact that some PBPs bind ß-lactams far below the respective MIC values and that they are not altered into low affinity forms even in highly penicillin resistant strains has been used as an argument for differentiating between essential and non essential PBPs: non essential ones need not to be altered, since their inhibition by ß-lactams apparently has no adverse effects on cell physiology.

S. PNEUMONIAE PBP'S 3 AND 1a

The *S. pneumoniae* PBP 3 is an example for such an apparently "nonessential" PBP whereas PBP 1a is supposed to play a vital role. Several ß-lactams saturate PBP 3 *in vivo* far below MIC values (Williamson *et al.*, 1980), and PBP 3 in resistant strains has the same penicillin-affinity as in the wildtype (with the exception of one cefotaxime resistant laboratory mutants). In contrast, PBP 1a is modified into a low affinity form in resistant clinical isolates.

MUTANTS WITH INSERTIONALLY INACTIVATED PBP 3 OR PBP 1a

The *E. coli* vector pJDC9 can not replicate in *S. pneumoniae* (Chen and Morrison, 1988). However, it carries an erythromycin resistance marker that can be selected in *S. pneumoniae*, provided that pneumococcal DNA has been cloned into pJDC9 allowing for integration of the vector fragment into the chromosome by homologous recombination via the inserted DNA fragment. Since for both, the PBP 3 (*dacA*) (Schuster *et al.*, 1990) and the PBP 1a (*ponA*) gene (Martin *et al.*, 1992), internal gene fragments had been cloned into pJDC9, these constructs could be used for isolating mutants interrupted in the respective PBP genes. Those mutants should be able to synthesize peptides that terminate at the 3' end of the cloned DNA fragment (Fig. 1).

The *dacA* Mutant

In case of PBP 3, the cloned internal gene fragment represented the C-terminal end of the penicillin-binding domain, including the KTG box and lacking the entire C-terminal end of the protein. A schematic drawing of the truncated PBP 3 is shown in Fig. 2. Since D,D-carboxypeptidases (CPases) are associated with the membrane via their C-terminal end, the PBP 3 mutant derivative was found merely in the medium although still active in terms of ß-lactam binding. No cell associated PBP 3 was

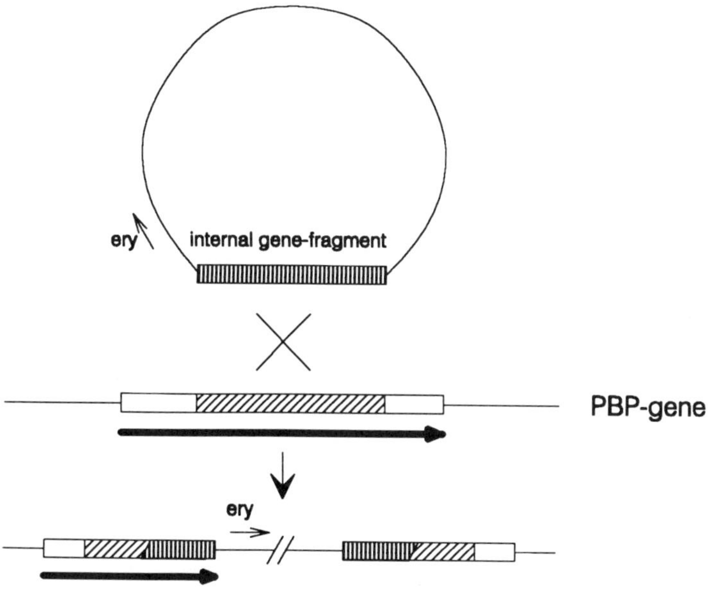

Figure 1. Schematic representation of the disruption of the chromosomal PBP gene (boxed area) by a cloned internal gene fragment (hatched box). The crosshatched area indicates the corresponding region in the chromosomal gene. The erythromycin marker of the vector is indicated.

detectable. The unexpected finding was not that these mutants could be isolated, but that the *dacA*$^-$ phenotype was associated with unusual cell morphology. The most striking effect was a misplacement of septa, leading to irregular division sites and completely misshaped cells. In addition, the generation time of the mutants was more than 50 min compared to 36 min of the parental strain R6 (Schuster *et al.*, 1990).

Analysis of the murein chemistry agreed with a lack of cell-associated D,D-CPase, in that the major monomers of the mutants were linear pentapeptides paralleled by a disappearance of linear tripeptides, in contrast to wildtype cells where trimers constitute the predominant monomer (A. Severin, C. Schuster, R. Hakenbeck, and A. Tomasz, submitted). It is more difficult to explain the inability of PBP 3$^-$ cells to form septa at

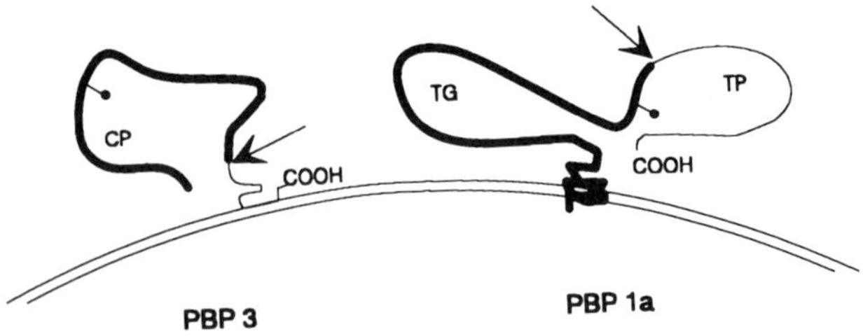

Figure 2. Schematic representation of the N-terminal peptides synthesized by the mutants with insertionally inactivated PBP 3 and PBP 1a genes. The arrow indicates the C-terminal end of the PBP peptide (thick line). Membrane association is mediated by the C-terminal extension in the wildtype PBP 3, and via an N-terminal hydrophobic domain in PBP 1a.

the correct sites. PBP 3 molecules were found to be distributed over the entire cell membrane and no preference for septal sites was apparent (Fig. 3), thus the morphological effects cannot be related to the localization of PBP 3 itself. However, in the central growth zone of wildtype pneumococci tripeptides are preferentially found in nascent cell walls (Laitinen and Tomasz, 1990). These tripeptides, functioning as potential acceptor molecules, could serve as a marker for the localized action of transpeptidases. Since they are not generated in the mutant this mechanism can not be maintained.

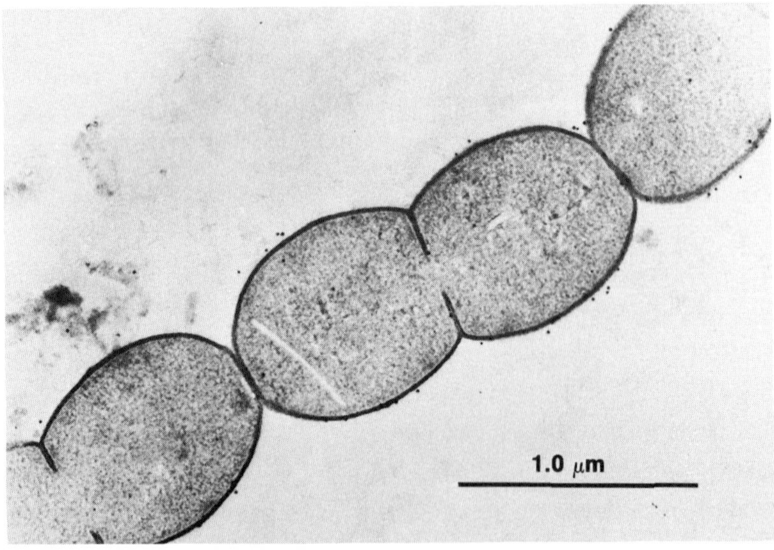

Figure 3. Localization of PBP 3 in the membrane. Electron microscopy of ultrathin sections of *S. pneumoniae* R6. Cells were grown in the presence of ethanolamine, and sections were incubated with anti-PBP 3 IgG followed by protein A-gold labeling. The gold particles mark the position of PBP 3.

The *ponA* Mutant

The cloned PBP 1a gene fragment used for construction of insertionally inactivated *ponA* mutants included the N-terminal domain and extended 60 bp downstream the TCA codon of the active site serine. The putative truncated PBP 1a derivative in the mutants is expected to be bound to the membrane via the N-terminal hydrophobic peptide (see Fig. 2), and due to the lack of the main portion of its penicillin-binding domain should be devoid of penicillin-binding activity and other enzymatic functions related to this domain. After transformation of *S. pneumoniae* R6 with pCMC2, erythromycin resistant colonies were obtained, and no PBP 1a was detectable by the penicillin-binding assay after fluorography (Fig. 4). The mutant cell grew slower than the parental strain. No major defect in morphology was apparent in the light microscope. Penicillin-induced lysis was somewhat slower than in the parental strain, reminiscent to observations of an *E. coli* PBP 1a mutant (Spratt and Jobanputra, 1977).

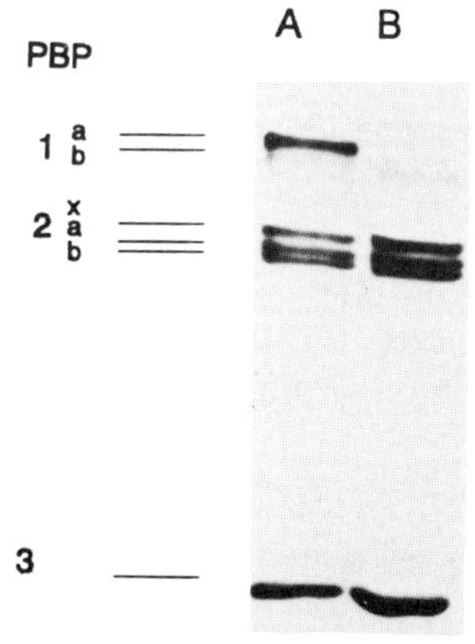

Figure 4. Penicillin-binding proteins in the parental strain R6 (A) and a mutant (B) with insertionally inactivated PBP 1a. PBPs were labeled in cell lysates with 3H-propionylampicillin, prepared for SDS-polyacrylamide gelelectrophoresis, and a fluorograph was prepared. The positions of the R6 PBPs are indicated on the left.

Although we do not know whether the N-terminal peptide is in fact synthesized (so far we were unable to trace it with anti-PBP 1a antibodies after immunoblotting), the results clearly show that a least the penicillin-binding domain can be deleted from the cells, and that such cells are viable under laboratory conditions. Analysis of the murein chemistry of the PBP 1a mutant may help to clarify the actual in vivo function of this protein.

CONCLUDING REMARKS

The description of PBPs as essential or non essential has been used on the basis of their role during development of penicillin resistance. In this context it is assumed that the cell tries to change those PBPs into low affinity forms which perform an important function. However, this apparently does not imply that the respective PBP cannot be deleted in susceptible wildtype strains, at least under laboratory conditions, or that deletions in a PBP gene that is not involved in penicillin resistance has no severe physiological consequences.

REFERENCES

Chen, J.-D. and Morrison, D.A. (1988) Construction and properties of a new insertion vector, pJDC9, that is protected by transcriptional terminators and useful for cloning of DNA from *Streptococcus pneumoniae*. Gene 64, 155-164.

Ghuysen, J.-M. (1991) Serine ß-lactamases and penicillin-binding proteins. Annu. Rev. Microbiol. 45, 37-67.

Laible, G.,, Spratt, B.G. and Hakenbeck, R. (1991) Interspecies recombinational events during the evolution of altered PBP 2x genes in penicillin-resistant clinical isolates of *Streptococcus pneumoniae*. Mol. Microbiol. 5, 1993-2002.

Laible, G. and Hakenbeck, R. (1987) Penicillin-binding proteins in ß-lactam resistant laboratory mutants of *Streptococcus pneumoniae*. Mol. Microbiol. 1, 355-363.

Laitinen, H. and Tomasz, A. (1990) Changes in composition of peptidoglycan during maturation of the cell wall in pneumococci. J. Bacteriol. 172, 5961-5967.

Martin, C., Briese, T. and Hakenbeck, R. (1992) Nucleotide sequences of genes encoding penicillin binding proteins from *Streptococcus pneumoniae* and *Streptococcus oralis* with high homology to *Escherichia coli* penicillin-binding proteins 1A and 1B. J. Bacteriol., in press.

Schuster, C., Dobrinski, B. and Hakenbeck, R. (1990) Unusual septum formation in *Streptococcus pneumoniae* mutants with an alteration in the D,D-carboxypeptidase penicillin-binding protein 3. J. Bacteriol. 172, 6499-6505.

Spratt, B.G. and Jobanputra, V. (1977) Mutants of *Escherichia coli* which lack a component of penicillin binding protein 1 are viable. FEMS Lett. 79, 374-378.

Spratt, B.G. (1989) Resistance to ß-lactam antibiotics mediated by alterations of penicillin binding proteins, in "Handbook of Experimental Pharmacology" (Bryan, L.E., Ed.) pp 77-100. Springer, Berlin.

Williamson, R., Hakenbeck, R. and Tomasz, A. (1980) In vivo interaction of ß-lactam antibiotics wiht the penicillin-binding proteins of *Streptococcus pneumoniae. Antimicrob. Agents Chemother.* 18, 629-637.

DISTURBANCE OF PEPTIDOGLYCAN SYNTHESIS BY GLYCINE AND D-METHIONINE CREATES A SIGNAL FOR THE *ampG*-MEDIATED INDUCTION OF AmpC-ß-LACTAMASE IN *ESCHERICHIA COLI*

Hans H. Martin* and Herbert Schmidt

Institut für Mikrobiologie, Technische Hochschule
D-6100 Darmstadt, Germany

INTRODUCTION

It has been reasoned that bacterial ß-lactamases must have a "normal" physiological function, apart from inactivating ß-lactam antibiotics, probably connected to the metabolism of cell wall peptidoglycan (Abraham and Chain, 1940; Saz and Lowery, 1979). In the same context, evidence has been brought forward which points to an essential role of peptidoglycan metabolism in the induced synthesis of ß-lactamase in Gram-positive and Gram-negative bacteria (Collins, 1979; Tuomanen et al., 1991).

ß-Lactam antibiotics have long been known as specific inducers of ß-lactamase (Citri and Pollock, 1966). In addition, "unspecific" induction of the enzyme in certain complex growth media and by a variety of low molecular weight compounds including glycine, D-methionine and other D-amino acids has been described (Cullman et al., 1984; Gatus et al., 1986, Ottolenghi et al., 1991, Ottolenghi and Ayala, 1991; Tuomanen et al. 1991). Like ß-lactams, glycine and D-amino acids interfere with peptidoglycan synthesis, although by different mechanisms (Hammes et al., 1973; Trippen et al., 1976) Thus, possibly, the action of both types of inhibitors leads to the formation of similar if not identical peptidoglycan metabolites which function as signals in a common pathway of ß-lactamase induction.

In different enterobacteria it has been shown that a typical chromosomal lactamase is constitutively expressed at low level from the structural gene *ampC*, and that the enzyme is further inducible by ß-lactam antibiotics in the presence of gene *ampR*, encoding a transscriptional regulator (Lindberg et al., 1985; Honoré et al., 1986). As reported recently, the induction also depends on the function of a novel gene *ampG* (Korfmann and Sanders, 1989). This gene presumably establishes the function of a transmitter which relays the still unknown induction signal to the *ampC-ampR* operon (Lindquist et al., 1992).

In the present report we will show that induction of ß-lactamase from cloned *ampC-ampR* of *Citrobacter freundii* (Lindberg et al., 1985) also takes place during growth of *Escherichia coli* with glycine and D-methionine at concentrations causing cell wall damage, and that this process also requires the function of *ampG*.

Bacterial Growth and Lysis, Edited by M.A. de Pedro
et al., Plenum Press, New York, 1993

METHODS

Bacterial Strains and Plasmids

E.coli SN0301 (ampD1) was described by Lindberg et al. (1987) and Lindquist et al. (1989). *E.coli* SN0301-1 (*amp*D1, *amp*G1), defective in *amp*G, was described by Schmidt (1992).

Plasmid pNU305 (tet^r), containing cloned *amp*C-*amp*R of *C. freundii*, was described by Lindberg et al. (1985).

Growth Medium and Conditions for ß-Lactamase Induction

Transformants *E.coli* SN0301/pNU305 and SN0301-1/pNU305 were grown in M9CA medium (Lindberg et al. 1987)) containing the following supplements per liter, glucose (2 g), casamino acids (Difco Laboratories, 2 g), uracil (50 mg), L-tryptophan (50 mg), thiamine (1 mg), and tetracycline (10 mg).

Shake cultures with 50 ml medium were inoculated with cells from overnight cultures in the same medium to give an initial OD_{578} of 0.1, and were incubated at 37°C with rotation at 120 r.p.m.. For induction, glycine and D-methionine were added to the shake cultures from concentrated stock solutions prior to inoculation. Cells for ß-lactamase assays were sedimented from 10 ml samples taken from the shake cultures at different times.

Induction by 6-aminopenicillanic acid, preparation of ß-lactamase extracts and assay of ß-lactamase were carried out as described by Lindberg et al. (1985). However, cephalothin at 200 μmol l^{-1} was used as substrate, and hydrolysis of the substrate was measured spectrophotometrically at wavelength 262 nm. One unit (U) of ß-lactamase hydrolyses 1 μmol of substrate min^{-1} in phosphate buffer (0.05 mol l^{-1}, pH 7) at 30°C.

RESULTS AND DICUSSION

E.coli Strains with Intact and Defective *ampG*

For assays of ß-lactamase induction in the presence of intact *amp*G we used *E.coli* SN0301(*amp*D1)/pNU305(*amp*C^{+},*amp*R^{+}) described by Lindberg et al. (1987). Gene *amp*G, essential for ß-lactamase induction, was identified in this strain at 9.85 min on the *E.coli* genetic map (Schmidt, 1992; Lindquist et al., 1992). As described by Lindberg et al. (1987) and Lindquist et al. (1989), *E.coli* SN0301 is partially defective in the chromosomal gene *amp*D, coding for a negative modulator which down-regulates the expression of AmpC-ß-lactamase in the absence of inducer. Therefore, in this organism, an elevated level of constitutive ß-lactamase is expressed from cloned *amp*C-*amp*R of *C. freundii* on plasmid pNU305. In addition, the *amp*D-defect causes the enzyme to be "hyperinducible" by ß-lactam to a much higher content. Thus, a strong response to the unspecific inducers could also be expected in this strain.

The effect of non-funtional ampG was tested in the isogenic, *amp*G-defective *E.coli* SN0301-1(*amp*G1)/pNU305. This organism was obtained from *E.coli* SN0301/pNU305 by chemical mutagenesis and selection for non-inducibility of ß-lactamase from intact *amp*C-*amp*R by ß-lactam. The strain carried a missense point mutation in the open reading frame of the *amp*G mutant allele (Schmidt, 1992).

Cell Damage and Growth Inhibition by Glycine and D-Methionine in *E.coli* SN0301

In previous studies on different Gram-positive and Gram-negative bacteria, high concentrations of glycine and D-methionine in the range of 1 - 10 mg ml^{-1} and above were found to be necessary for the inhibition of growth, damage to the shape and stability of cell walls, formation of spheroplasts

(Lark and Lark, 1959, Dienes et al., 1950; Dienes and Zamecnik, 1952), and for replacing the normal constitutent D-alanine in the stem-peptides of peptidoglycan (Hammes et al., 1973; Trippen et al., 1976). An equally high, growth inhibitory concentration of glycine was also needed for ß-lactamase induction in *Enterobacter cloacae* (Gatus et al.,1988), suggesting a relation between cell wall damage and enzyme induction.

Growth of *E.coli* SN0301/pNU305 and SN0301-1/pNU305 was not inhibited by glycine at 5 and 10 mg ml^{-1}, but elongation of cells and partial formation of filaments took place (data not shown).

D-Methionine at concentrations of 1, 2 and 4 mg ml^{-1} caused increasing formation of spheroplasts, but growth was inhibited only at the highest concentration.

Fig.1 illustrates inhibition of growth of *E.coli* SNO301/pNU305 in M9CA-minimal medium by D-methionine at 4 mg ml^{-1}, in contrast to non-inhibited growth with the same concentration of L-methionine.

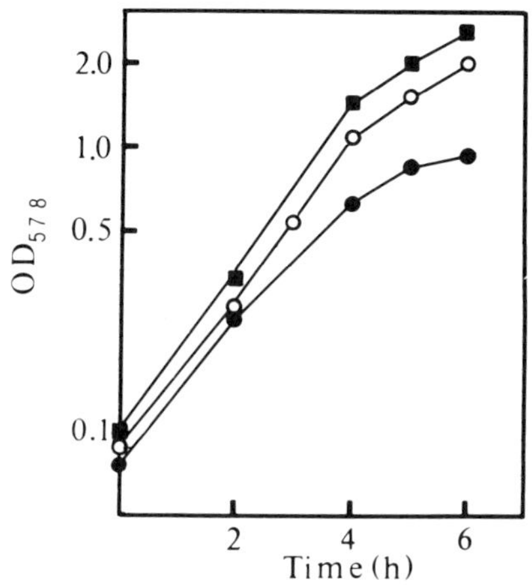

Fig.1. Growth Inhibition of E. coli SN0301/pNU305 by D-Methionine. Growth curves of shake cultures in M9CA-medium, control (○), and in M9CA with 4 mg ml^{-1} of D- (●) and L- (■) methionine, respectively

Induction of ß-Lactamase by Glycine and D-Methionine

In a typical experiment of induction with ß-lactam (Lindberg et. al.,1987), *E. coli* SN0301/pNU305 was exposed to 6-aminopenicillanic acid at a concentration of 0.25 mg ml^{-1} for 90 min. In agreement with data of Lindquist et al.(1989), the specific activity of ß-lactamase was increased more than tenfold to 16.4 U (mg $protein)^{-1}$, in comparison to 1.4 U (mg $protein)^{-1}$ in the non-induced control.

Similar short time experiments with glycine and D-methionine revealed no increase of ß-lactamase activity when these compounds acted on exponentially growing bacteria for 1 or 2 hours. However, during growth in shake cultures containing glycine or D-methionine, time-dependent induction of ß-lactamase, quantitatively comparable to that by ß-lactam, occurred from mid-log phase to early stationary phase of growth. As shown in Tab.1, enzyme activities rose significantly above the constitutive level after 3 hours and reached a maximum after 5 hours. Only with D-methionine at 4 mg ml^{-1}, the specific activity of ß-lactamase did not increase appreciably after this longest time of exposure. This coincided with the marked growth inhibition and spheroplast-formation under these conditions and, thus, presumably resulted from extensive lysis of cells and partial release of the periplasmic ß-lactamase into the medium.

TABLE 1. Induction of ß-Lactamase by Glycine and D-Methionine in E.coli SN0301/pNU305, containing intact ampG

INDUCER mg ml^{-1}	ß-LACTAMASE* Specific Activity [U (mg Protein)$^{-1}$] after Different Times of Growth with Inducer: 3 Hours	4 Hours	5 Hours
Control noninduced	0.9	1.4	2.0
Glycine			
5	12.4	14.9	22.6
D-Methionine			
1	8.3	11.2	20.8
2	12.5	18.9	30.3
4	11.8	17.5	21.1

* Values are representative results of duplicate assays

A moderate increase of ß-lactamase activity to values 1.5 - 2 times those of the non-induced control also took place during growth with 4 mg ml^{-1} of L-methionine for 3 - 5 hours.

In the *amp*G-defective *E.coli* SN0301-1/pNU305, it was shown previously that ß-lactamase is expressed constitutively at an average specific activity of 0.5 U (mg protein)$^{-1}$, and that this low activity remains unchanged under the same treatment with ß-lactam which yields high induction of ß-lactamase in the *amp*G^{+} parent strain (Schmidt, 1992).

We also found now that no induction of ß-lactamase was detectable in *E. coli* SN0301-1/pNU305 under any condition of growth with glycine or D-methionine. Virtually identicial specific activities of 0.2 - 0.4 U (mg protein)$^{-1}$ were measured after 3, 4 and 5 hours of growth with 4 mg ml^{-1} of D-methionine and in the non-induced control culture.

The results showed that glycine and D-methionine were as efficient as ß-lactams in the initiation of the *amp*R-regulated induction of enterobacterial AmpC ß-lactamase. Also, both types of inducers depended on the function of *amp*G. Therefore , presumably, in spite of the drastically different inhibitory mechanisms of glycine/D-methionine and of ß-lactams on peptidoglycan synthesis, similar or identical metabolites are eventually formed, which *amp*G or its gene product can recognize as induction signals.

From the nucleotide sequences of *E.coli* wild-type *amp*G (Lindquist et al.,1992) and mutant *amp*G1 (Schmidt, 1992), a hypothetical gene product AmpG can be deduced as a strongly hydrophobic protein which contains several membrane spanning sequences, and is devoid of the specific ß-lactam-binding -ser-X-X-lys- motif present in the active-site-serine penicillin-recognizing proteins (Joris et al., 1988). Thus, AmpG is probably membrane-bound and is likely to perform an essential signal-transmitting function in ß-lactamase induction, but not as a direct and specific sensor of an inducing ß-lactam. The eventual confirmation of an indirect induction pathway, with a key interaction

of AmpG with a peptidoglycan signal-metabolite, would, in fact, make it unnecessary to assume the function of a specific receptor, possibly a penicillin-binding protein, as induction sensor for ß-lactams.
On the other hand, it may still be possible to explain the previous observation of an exceptionally high potency of certain ß-lactams, such as imipenem and different cephamycins, as inducers of chromosomal ß-lactamases in Gram-negative bacteria, as the result of target specificity (Martin et al., 1988). The preferred binding of these antibiotics to certain penicillin-binding proteins may initiate a particularly efficient production of the AmpG-interacting peptidoglycan metabolites.

ACKNOWLEDGEMENTS: This study was supported by the Deutsche Forschungsgemeinschaft and the Fonds der Chemischen Industrie.

REFERENCES

Abraham, E.P. and Chain, E.B. (1940) An enzyme from bacteria able to destroy penicillin. Nature (Lond.) 146, 837-959

Citri, N. and Pollock, M.R. (1966) The biochemistry and function of ß-lactamases (penicillinase). Advanc. Enzymol. 28, 237-323

Collins, J.F. (1979) The *Bacillus licheniformis* ß-lactamase system, in "Beta-Lactamases", (Hamilton-Miller, J.M.T. and Smith, J.T., Eds.) pp. 351-368, Academic Press, London

Cullmann, W., Dalhoff, A. and Dick, W. (1984) Nonspecific induction of ß-lactamase in *Enterobacter cloacae*. J. Gen. Microbiol. 130, 1781-1786

Dienes, L., Weinberger, J. H. and Madoff, S. (1950) The transformation of typhoid bacilli into L-forms under various conditions. J. Bacteriol. 59, 755-764

Dienes, L. and Zamecnik, P.C. (1952) Transformation of bacteria into L-forms by amino acids. J. Bacteriol. 64, 770-771

Gatus, B.J., Bell, S.M. and Jimenez, A.S. (1986) Comparison of glycine enhancement with cefoxitin induction of ß-lactamase production in *Enterobacter cloacae* ATCC 13047. J. Antimicrob. Chemother. 21, 163-170

Hammes, W., Schleifer, K.H. and Kandler, O. (1973) Mode of action of glycine on the biosynthesis of peptidoglycan. J. Bacteriol. 116, 1029-1053

Honoré, N.,Nicolas, M.H. and Cole, S.T. (1986) Inducible cephalosporinase production in clinical isolates of *Enterobacter cloacae* is controlled by a regulatory gene that has been deleted from *Escherichia coli*. EMBO J. 5, 3709-3714

Joris, B., Ghuysen, J.M., Dive, G., Renard, A., Dideberg, O., Charlier, P., Frère, J.M., Kelly, J.A., Boyington,J.C., Moews, P.C. and Knox, J.R. (1988) The active-site-serine penicillin-recognizing enzymes as members of the *Streptomyces* R61 DD-peptidase family. Biochem. J. 250, 313-324

Korfmann, G. and Sanders, C.C. (1989) *amp*G is essential for high-level expression of AmpC ß-lactamase in *Enterobacter cloacae*. Antimicrob. Agents Chemother. 33, 1946-1951

Lark, C. and Lark, K.G. (1959) The effects of D-amino acids on *Alcaligenes fecalis*. Can. J. Microbiol. 5, 369-379

Lindberg, F., Westman, L. and Normark, S. (1985) Regulatory components in *Citrobacter freundii* ampC ß-lactamase induction. Proc. Natl. Acad. Sci. USA 82, 4620-4624

Lindberg, F., Lindquist, S. and Normark, S. (1987) Inactivation of the *amp*D gene causes semi-constitutive overproduction of the inducible *Citrobacter freundii* ß-lactamase. J. Bacteriol. 169, 4620-4624

Lindquist, S., Galleni, M., Lindberg, F. and Normark, S. (1989) Signalling protein in enterobacterial AmpC ß-lactamase regulation. Mol. Microbiol. 3, 1091-1102

Lindquist, S., Korfmann, G., Sanders, C.C., Schmidt, H., Martin, H.H., Weston-Hafer, K., and Normark, S. (1992) *amp*G, a signal transducer in chromosomal ß-lactamase induction, submitted for publication

Martin, H.H., Schmidt, B., Bräutigam, S., Noguchi, H. and Matsuhashi, M. (1988) Initiation of induction of chromosomal ß-lactamase by binding of inducing ß-lactam antibiotics to low molecular-weight penicillin-binding proteins, in "Antibiotic Inhibition of Bacterial Cell Surface Assembly and Function" (Actor, P., Daneo-Moore, L., Higgins, M.L., Salton, M.R.J. and Shockman, G.D., Eds.), pp. 494-501. American Society for Microbiology, Washington, D.C.

Ottolenghi, A.C.and Ayala, J.A. (1991) Induction of a class I ß-Lactamase from *Citrobacter freundii* in *Escherichia coli* requires active *fts*Z but not *fts*A or *fts*Q products. Antimicrob. Agents Chemother. 35, 2359-2365

Ottolenghi, A.C., Caparros, M. and De Pedro, M.A. (1991) Class I ß-Lactamase induction in *Enterobacter cloacae* ATCC 13047 by D-methionine, glycine and D-tryptophan. Program Abstr. 31st Intersci. Conf. Antimicrob. Agents Chemother., abstr. 21, p.102

Saz, A.K. and Lowery, D.L. (1979) Do ß-lactamases have a biological function, in "Beta-Lactamases", (Hamilton-Miller, J.M.T. and Smith, J.T., Eds.) pp. 465-479, Academic Press, London

Schmidt, H. (1992) Doctoral Dissertation, Technische Hochschule Darmstadt

Trippen, B., Hammes, W., Schleifer, K.H. and Kandler, O. (1976) Die Wirkung von D-Aminosäuren auf die Struktur und Biosynthese des Peptidoglycans. Arch. Microbiol. 109, 247-261

Tuomanen, E., Lindquist, S., Sande, S., Galleni, M., Light, K., Gage, D. and Normark, S. (1991) Coordinate regulation of ß-lactamase induction and peptidoglycan composition by the *amp*D operon. Science 251, 201-203

IN *ENTEROBACTER CLOACAE* ALTERATIONS INDUCED BY GLYCINE AND D-AMINO ACIDS IN THE COMPOSITION AND STRUCTURE OF PEPTIDOGLYCAN ARE ACCOMPANIED BY INDUCTION OF CHROMOSOMAL β-LACTAMASE: A MODEL INVOLVING *ftsZ* AND SEPTATION

Abramo C. Ottolenghi

Department of Medical Microbiology and Immunology, The Ohio State University, Columbus, OH 43210 (USA)

INTRODUCTION

The induction of Class I β-lactamase (Bla) in the *Enterobacteriaceae* and other Gram negative organisms has been studied extensively but as yet no information is available as to the nature of the event(s) which triggers it nor of the specific molecule that might act as the inducer. It is the purpose of this paper to present evidence and to suggest a model which makes septation/division the controlling factor for the induction of Bla.

A number of genes which regulate Bla induction have been described. First there is *ampR* which is essential for transcription of, and closely linked with the structural gene *ampC*. *ampR* is described as the sensor which leads to the transcriptional activation of *ampC* (Lindberg *et al.*, 1986). Korfmann and Sanders (1989) have described *ampG* another gene, essential for induction. AmpG is a membrane bound protein.

Two genes with regulatory properties but not essential for Bla production have also been described, *ampD* a 25 kDa cytoplasmically located product down-regulates the production of Bla and, the closely linked *ampE* (Lindberg *et al.*, 1987) whose 32.1 kDa product is located in the cytoplasmic membrane. The level of AmpE in the presence of the AmpR transcriptional regulator affects the basal level of Bla expression in the absence of AmpD.

In addition to being initiated by some, but not all β-lactam antibiotics, Bla induction has been observed in *Enterobacter*, *Pseudomonas* and *Proteus* with glycine, hemin, tryptophane, folic acid, and in various complex media without necessitating the addition of β-lactam antibiotics (Gatus *et al.*, 1986, 1988; Cullmann *et al.*, 1983; Dalhoff and Cullmann 1984). Bla induction in *Pseudomonas* is affected by the presence of body fluids such as serum and urine (Letendre and Turgeon 1989).

Bacterial Growth and Lysis, Edited by M.A. de Pedro *et al.*, Plenum Press, New York, 1993

Such a variety of inducing conditions points to the existence of a global type of response triggered by a number of different stimulii. The involvement of the very specific β-lactam antibiotics points to the murein sacculus, as being important but also to morphogenesis and septation/division as possible "trigger" sites for this putative global response.

RESULTS AND DISCUSSION

When *Enterobacter cloacae* was exposed to different β-lactam antibiotics the alteration in muropeptide composition and structure showed little correlation with Bla induction (Table 1). For example, induction by cefoxitin or cefmetazol was accompanied by an increase in pentapeptides, as might be expected fron the results seen in *Pseudomonas* with cefoxitin by Martin *et al.* (1988). No increase was seen when induction was by ampicillin or imipenem.

Table 1. β-Lactamase induction by β-lactam antibiotics. Effect on peptide composition and structure of the muropeptide[a].

Agent	Bla[b]	Penta peptides	Tetra peptides	Tri peptides	Lipo protein	Cross-linked	
						ala-dap	dap-dap
none	1	1.37	81	15.99	4.98	32.48	4.36
cefoxitin	30	16.14	71	15.30	4.06	36.2	3.55
ampicillin	20	1.44	77	19.44	7.78	35.2	4.49
cefmetazole	8	8.72	73	17.85	5.53	35.7	2.84
imipenem	39	2.02	75	20.54	9.47	36.5	3.74
ceftazimidine.	1	1.3	79	17.54	5.63	33.1	4.27
mecillinam	1	1.09	72	24.61	9.02	33.2	3.76
piperacillin	1	1.08	78	19.16	8.82	39.7	4.02
M-4**	1800	1.2	77	19.39	7.9	30.5	2.94

E. cloacae 12 (Findell and Sherris 1976) was grown in a phosphate limited peptone medium to $O.D._{550}$ ca. 0.5 at which time the culture was split and 10 μg/ml antibiotic added (0.5 μg/ml for imipenem). After 20 minutes (ca. 1/2 generation) the cells were cooled, and muropeptide prepared. No antibiotic was added to the M4 mutant.

a) Moles % .

b) Activity relative to uninduced wild type

**Bla hyper-producing mutant of *E. cloacae* 12

In solid medium, excess glycine alters the structure of peptidoglycan (Hammes *et al.*, 1973) and induces Bla in *E.cloacae* (Gatus *et al.*, 1986). Glycine is known to take the place of D-alanine in the peptidoglycan tetra and pentapeptides (Hammes *et al.*, 1973). In a preliminary report Ottolenghi, *et al.*, (1991) showed that D-methionine also induced Bla.

E. cloacae grown in the presence of this D amino acid exhibited marked morphologic changes in addition to the production of Bla and alterations in the composition of the peptidoglycan. (Ottolenghi, *et al.*, manuscript in preparation). Incomplete septation is the most obvious of these morphologic alterations. Significant morphological alterations as a result of culturing *Alcaligenes faecalis* in the presence of D-amino acids have been reported by Lark and Lark (1959).

When *Escherichia coli* is grown in the presence of D-methionine, this amino acid is incorporated into the peptidoglycan. It is added to form part of tetrapeptide moieties (Caparrós and de Pedro this volume). The changes in the peptide content and extent of cross-linking of the muropeptide are different in nature when obtained from cells grown in glycine, D-methionine or D-tryptophane, nevertheless, Bla is induced with all three agents (Table 2). Martin and Schmidt (this volume) report Bla induction with glycine and D-methionine in *E. coli* carrying the *ampR-ampC* plamid pNU305 (Lindberg *et al.*, (1985).

Table 2. β-Lactamase induction by glycine, D-methionine and D-tryptophane. Effect on peptide composition and structure of the muropeptides[a].

Agent	Bla[b]	Penta peptide	Tetra peptide	Tri peptides	Lipo protein	Cross-linked	
						ala-dap	dap-dap
None	1	1.49	73	19	11.4	25.2	5.5
Glycine	18	5.8	69	21	4.7	16.6	3.3
D-meth.	17	0.18	77	20	9.6	31.0	1.7
D-trypt.	20	6.0	77	16	6.7	27.8	1.2

a) Moles %. (May not add to 100 because of small unidentified peaks).
b)Activity relative to uninduced cells
*E. cloacae*ATCC 13047 was grown in a phosphate limited peptone medium to $O.D._{550}$ ca 0.3 at which time the amino acid was added to the culture. The culture was maintained in logarithmic phase by dilution. Data represents values obtained 3 hours after the addition of the amino acids. Glycine , 50 mM. D-methionine and D-tryptophane 4mg/ml.

We observed that the induction by the antibiotics was much more rapid (minutes) than that seen with the amino acids (hours). Similar induction kinetics were reported by Gatus *et al.* (1988) for glycine and cefoxitin induction in solid media.

The reports and findings we have described above speak against peptidoglycan synthesis and composition being *per se* the controlling parameters for Bla induction. The chemostat experiments described in Fig. 1 and the morphological changes seen when D-methionine is included in the medium point to alterations in septation division as important elements in this induction.

When Bla was induced in *E. cloacae* by cefoxitin in rate varying chemostats, the rate of induction was dependent on the growth rate (Fig. 1a). Wu and Livermore (1990) have reported the same growth rate dependence in *Pseudomonas.* When the data in Fig 1a was re-examined on the basis of generation number it was seen that there was no difference in the level of Bla induced at the two growth rates tested (Fig. 1b). This outcome is consistent with dependence on a gearbox control as described by Aldea *et al.*, (1990).

Using *E. coli* mutants with temperature sensitive mutations at the gearbox controlled (Aldea *et al.*, 1990) division genes *ftsA, ftsQ* and *ftsZ,* Ottolenghi and Ayala (1991) showed that FtsZ , but not FtsA nor FtsQ, is required for induction of Bla. This finding implicates the initiation of septation as a required step for the induction of Bla. The observation that nalidixic acid inhibits the induction of Bla in *E. cloacae* (Table 3) is a further indication of the possible connection between Bla induction, *ftsZ* and thus septation/division.

Nalidixic acid initiates the SOS response which has as one of its hallmarks filamentation and decreased FtsZ activity. The probable involvement of FtsZ in the induction of Bla is strengthened further by the observation (Ottolenghi and Ayala, 1991) that induction in *ftsZ*(Ts) mutants was dependent on the time that the culture was held at the non-permissive temperature before the addition of the inducer. Dai and Lutkenhaus (1991) showed that FtsZ is a stable molecule.

It is thus possible to view the various alterations in peptidoglycan composition brought about by D-amino acids, glycine or the β-lactam antibiotics as inhibiting the correct initiation of the septation/division process. In the case of glycine and the D-

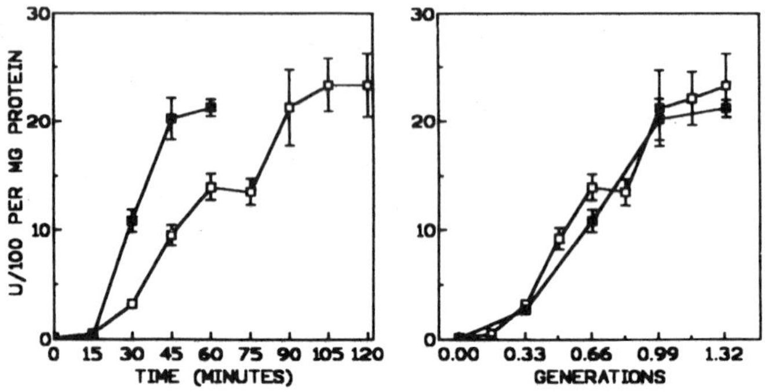

Figure. 1. Induction of β -lactamase in chemostats. *E. cloacae* 12 (Findell and Sherris, 1976) was grown at 37° C in a chemostat using a modification of the peptone medium of Torriani (1960) with limiting Pi (0.32 mM rather than 3-glycerol phosphate). Cultures were allowed to grow for 8 hrs before addition of the medium was started. Fast growth (closed squares) was D=60 minutes. Slow growth (open squares) was D=120 minutes. Cefoxitin addition (10μg/ml) was started 7 hrs. and 10 hrs. later for the fast and slow growing cultures respectively. Fig. 1b = data in Fig. 1a replotted to show relationships to generation times .(Gt.). Gt.= D x 0.693. Points shown are mean and S.E. of three replicate experiments.

Table 3. Effect of nalidixic acid on β-lactamase induction by cefoxitin in *E. cloacae*

O.D.550*	Nalidixic	Cefoxitin	β-lactamase**
0.45	0	-	0.15
0.5	0	+	16.6
0.52	10 μg	-	0.23
0.52	10 μg	+	6.1
0.51	15 μg	-	0.32
0.48	15 μg	+	3.15
0.540	20 μg	-	0.6
0.48	20 μg	+	3.7

An overnight culture of *E. cloacae* 12 (Findell and Sherris 1976) in a low peptone, low Pi medium was diluted 1/10 ($O.D._{550}$ 0.155) in pre-warmed fresh medium. Portions of the suspension were transferred to flasks containing the appropriate amount of nalidixic acid. Following 30 min incubation at 37°C with agitation, cefoxitin (10 μg/ml) was added to the appropriate flasks. The incubation was continued for 20 minutes at which time 50 ml samples were collected on super-cooled (-70°C) 0.85% NaCl solution. *O.D. at end of incubation. ** Specific activity (μmol/100 cefaloridine hydrolyzed/min/mg protein.

amino acids possibly through an accumulation of a sufficient number of altered residues which may be required for induction,while in the case of the inducing antibiotics the inhibition would take place through inhibition of the action of some penicillin binding protein(s) (PBP), which in turn interfere with "proper" septation/division.

Our Proposed Model

Whatever model is presented for the induction of Bla it must account minimally for the following observations:

1) the requirement for *ampG* and its association with the membrane (Korfmann and Sanders, 1990);
2) the dose effect of *ampD* (Peter *et al.*, 1989);
3) the requirement for an active *ftsZ* (Ottolenghi and Ayala, 1991);
4) the possibility of inducing Bla in minicells (Linberg *et al.*, 1986)
5) the appearance of hyperproducing partially inducible mutants which are not *ampD* (Lindberg *et al.*,1986);
6) the ability to induce Bla with glycine and D-amino acids (Cullman *et al.*,1985; Ottolenghi *et al.*,(1991);

Korfmann and Sanders (1989) proposed a model for Bla induction. They suggested that AmpG acts on a substrate and modifies it in such a way that it becomes an activator for *ampR* permitting the transcription of *ampC*. AmpD would modulate the activation of AmpR by destroying the product of the AmpG function. Our model proposes a possible source for the substrate for AmpG.

FtsZ appears to be transcribed continuously with transient increases at the time of division (Robin *et al.*, 1990). At this division time, and through the action of the product of *minB* FtsZ is directed to specific division sites (Bi and Lutkenhaus, 1991). Combination of FtsZ with a membrane component or a product arising from a reaction initiated by this combination.

It is our proposal (Fig. 2) that FtsZ binds to an appropriate site and a complex that we (together with all other model makers) shall call X, is formed. Complex X can be either the combination of FtsZ with a membrane component or a product arising from a reaction initiated by this combination. It may even be FtsZ itself.

During the cell cycle period between the initiation of septation cycles, FtsZ is mainly cytoplasmic and no substrate is formed for AmpG. At the time of septation FtsZ is directed to specific sites around the septal ring, as shown elegantly by Bi and Lutkenhaus (1991). This is when we propose that Complex X is formed.

Under normal circumstances, Complex X is consumed during septation, AmpG would be sufficient to process any "normal" excess and AmpD would be present in sufficient amounts to prevent activation of AmpR. Complex X is not "utilized" appropriately under conditions of induction, (or hyperproduction for that matter) thus creating an "excess" of substrate for AmpG. The amount of AmpG in the cell is sufficient to process this "excess" and produce AmpR activator. Unless the amount of AmpD is sufficient to destroy all (or most) of this AmpR activator *ampC* is transcribed. This would account for the gene dose effect seen with *ampD* (Peter *et al.*, 1989).

Since Complex X, which depends on the level of FtsZ, would be the rate limiting element in the action of AmpG, the lack of dose effect shown by *ampG* (Korfmann and Sanders 1989) is accounted for. In our model, the amount of AmpR activator depends on the level of FtsZ and not on the level of AmpG. Since both peptidoglycan assembly and septation occur with the participation of the membrane the membrane association of AmpG is accounted for.

The induction of Bla in minicells reported by Lindberg *et al.* (1986) indicates that the presence of the chromosome is not necessary for induction to occur as long as the *ampR-ampC* complex is present. The strain used by those authors was a *recA*strain. *recA* strains that form minicells are postulated to have excess FtsZ which plays an important role in their formation. Thus, there is reason to believe that "excess" FtsZ was present in the cytoplasm of the minicells. Reeve (1977) showed peptidoglycan synthesis in minicells thus, there is reason to believe that Complex X could have been formed.

Our model accounts for the appearance of hyperproducing, partially inducible mutants which are not *ampD* (as is the case for our M4 mutant mentioned earlier in Table 1), since any mutation that affects the utilization of Compound X would result in this phenotype. This factor in itself would account for the very high rate of apperance of hyperpoducing mutants that has been observed. Bla induction by glycine and D-amino acids is accounted for because the changes in peptidoglycan composition and structure can be envisioned as interfering with the proper utilization of Coumpound X.

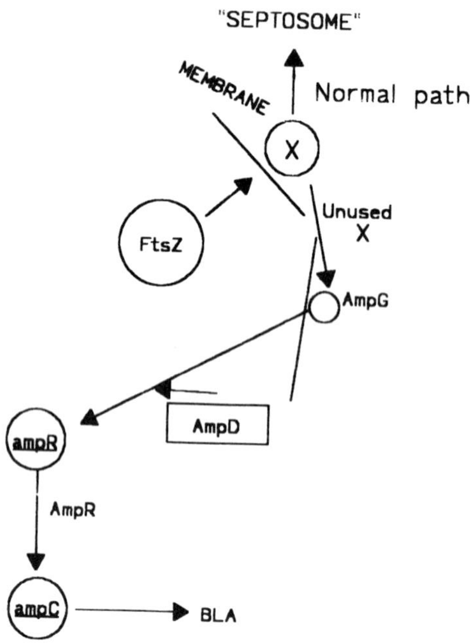

Figure. 2 Model for Bla induction

In our model, induction is triggered by some event in septation between the action of FtsZ and those of FtsA, FtsQ and PBP3. Ottolenghi and Ayala (1991) have shown that neither *ftsA* nor *ftsQ* are needed for induction. Oliva *et al.* (1989) have shown that PBP3 is also is not needed. In addition, imipenem and furazlocillin, antibiotics which have a predilection for PBP3, are inducers of Bla. Although this model is based on the model of Korfmann and Sanders (1990), it also accomodates the concepts advanced by Donachie *et al.*(1990) and the model for septation proposed by Nanninga (1991).

Teleologically, (always a dangerous exercise) the model fits the need for an alarm and repair mechanism for agents affecting the cell wall. As proposed, this mechanism would be triggered by the presence of agents affecting cell wall synthesis, and through it septation and/or division. It would form part of the diverse mechanisms for adaptation available to the cell

Acknowledgements

Thanks are due to M. Caparrós for help with the HPLC analysis and to M.A. de Pedro for hospitality and helpful comments. This work was supported in part by a grant from the Bremmer Foundation, The Ohio State University College of Medicine.

REFERENCES

Aldea, M, Garrido, T., Pia, J. and Vicente, M. (1990) Division genes in *Escherichia coli* are expressed coordinately to cell septum requirements by gearbox promoters. Embo J. 9 3787-3794.

Begg, J.K., Takasuga, A., Edwards, D.H.,Dewar, S.J., Spratt, B.G., Adachi, H., Ohta, T., Matsuzawa, H. and Donachie, W.D. (1990) The balance between different peptidoglycan precursors determines whether *Escherichia coli* cells will elongate or divide. J. Bacteriol. 172, 6697-6703.

Bi, E and Lutkenhaus, J. (1991) FtsZ ring structure associated with division in *Escherichia coli.* Nature 354, 161-164.

Cullmann, W., Dalhoff, A. and Dick, W. (1984) Nonspecific induction of β-lactamase in *Enterobacter cloacae.* J. Gen. Microbiol. 130, 1781-1786.

Dai, K. and Lutkenhaus, J.(1991) *ftsZ* is an essential division gene in *Escherichia coli* J. Bacteriol. 173, 3500-3506.

Dalhoff, A. and Cullman, W. (1984) Specificity of β-lactamase induction in *Pseudomonas aeruginosa*. J. Antimicrob. Chemother. 14, 349-357

Findell, C.M. and Sherris, J.D. (1976) Susceptibility of *Enterobacter* to cefamandole: evidence for a high mutation rate for resistance. Antimicrob. Agents Chemother. 9, 970-974.

Gatus, B.J., Bell, S.M. and Jimenez A.J. (1986) Enhancement of β-lactamase production by glycine in *Enterobacter cloacae* ATCC 13047. Pathology 18, 145-147.

Gatus, B.J., Bell, S.M. and Jimenes A.J. (1988) Comparison of Glycine enhancement with cefoxitin induction of class 1 β-lactamase production in *Enterobacter cloacae*ATCC 13047. J. Antimicrob. Chemother., 21, 163-170.

Hammes, W., Schleifer, K.H. and Kandler, O. (1973) Mode of action of glycine on the biosynthesis of peptidoglycan. J. Bacteriol. 116, 1029-1053.

Korfmann, G. and Sanders, C.C. (1989) *ampG* is essential for high level expression of AmpC β-lactamase in *Enterobacter cloacae.* Antimicrob. Agents Chemother. 33, 1946-1951.

Lark, C. and Lark, K.G. (1959) The effects of D-amino acids on *Alcaligenes fecalis.* Can. J. Microbiol. 5, 369-379.

Letendre, E.D. and Turgeon, P.L. (1989) Production and induction of β-lactamase during growth of *Pseudomonas aeruginosa* in biological fluids. Antimicrob. Agents Chemother. 33, 776-777.

Lindberg, F., Westman, M. and Normark, S. (1985) Regulatory components in

Citrobacter freundii ampC β-lactamase induction. Proc. Natl. Acad.Sci. USA. 82, 620-4624.

Lindberg, F., Lindquist, S. and Normark, S. (1987) Inactivation of the *ampD* gene causes semiconstitutive overproduction of the inducible *Citrobacter freundii* β lactamase. J. Bacteriol. 169, 1923-1928

Martin, H.H., Schmidt, B., Bräutigam S., Noguchi, H. and Matsuhashi, M. (1988) Initiation of induction of chromosomal β-lactamase by binding of inducing β-lactam antibiotics to low molecular-weight penicillin binding proteins, in: "Antibiotic inhibition of bacterial cell surface assembly and function" (Actor, P., Daneo-Moore, L., Higgins, M.L., Salton, M.R.J. and Shockman, G.D., Eds.), pp. 494-451. American Society for Microbiology, Washington. D.C.

Nanninga, N. (1991) Cell division and peptidoglycan assembly in *Escherichia coli.* Mol. Microbiol. 5,791-795.

Oliva, B., Bennet, P.M. and Chopra, I. (1989) Penicillin binding protein 2 is required for induction of *Citrobacter freundii* class I chromosomal β-lactamase in *Escherichia coli* Antimicrob Agents Chemother. 33, 1116-1117.

Ottolenghi A, C. and Ayala, J.A., (1991) Induction of class I β lactamase from *Enterobacter freundii* in *Escherichia coli* requires active *ftsZ* but not *ftsA* or *ftsQ*, products. Antimicrob Agents Chemother. 35, 2359-2365

Ottolenghi, A.C., Caparrós, M. and de Pedro, M.A. (1991) Induction of class I β-lactamase by glycine and D-amino acids in *Enterobacter cloacae* ATCC 13047. Program Abstr. 31st Interscience Conf. Antimicrob. Agents. Chemother. abstr. 21, p 102.

Peter, K., Korfmann, G., and Wiedemann, B. (1989) Impact of the *ampD* gene and its product on β-lactamase production in *Enterobacter cloacae* Rev. Inf. Dis. 10, 800-805

Reeve, J.N. (1977) Mucopeptide biosynthesis in minicells *Escherichia coli.* J. Bacteriol. 131, 363-365.

Robin, A., Joseleau-Petit, D. and D'Ari, R. (1990) Transcription of the *ftsZ* gene and cell division in *Escherichia coli.* J. Bacteriol. 172, 1392-1399.

Torriani, A. (1960) Influence of inorganic phosphate in the formation of phosphatases by *Escherichia coli.* Biochim. Biophys. Acta 38, 460-469.

Wu, P.J. and Livermore, D.M. (1990) Response of chemostat cultures of *Pseudomonas aeruginosa* to carbamapens and other β-lactams. J. Antimicrob. Chemother. 25, 891-902.

FTSZ RINGS, POLAR MORPHOLOGY AND CELL LYSIS

Erfei Bi, Shambhavi Subbarao, Kang Dai and Joe Lutkenhaus

Department of Microbiology, Molecular Genetics and Immunology
University of Kansas Medical Center
Kansas City, KS 66103

The involvement of *ftsZ* in cell division was first suggested by study of the *ftsZ84* (Ts) mutation (Lutkenhaus et al. 1980). This mutation blocks cell division at the nonpermissive temperature leading to a "smooth" filamentous morphology; filaments without any sign of constriction. This phenotype contrasts with that induced by mutations in other *fts* genes which display an "indented" filamentous morphology (Begg and Donachie, 1985; Taschner et al., 1988). These results suggest that *ftsZ* acts earlier in the division pathway than other identified cell division genes. More recently, the construction of conditional *ftsZ* null mutations confirmed the notion that *ftsZ* is an essential cell division gene (Dai and Lutkenhaus, 1991; Wang et al. 1991; Pla et al. 1991). The involvement of *ftsZ* in cell division is also supported by studies that examined the effect of increasing *ftsZ* expression. Increases in the level of FtsZ leads to a hyperdivision activity in the form of a minicell phenotype (Ward and Lutknehaus, 1985). This result suggests that FtsZ is the rate limiting component for cell division.

Evidence suggesting a critical role for *ftsZ* in cell division also comes from studies of inhibitors of cell division. Underlying this conclusion is the assumption that specific inhibitors of cell division would target a critical component that acts early in the division pathway. Two inhibitors, *sulA*, a component of the SOS response (Huisman and D'Ari, 1981), and *minCD*, a bipartite inhibitor of the *min* system (de Boer et al., 1989) appear to target *ftsZ*. The evidence for an interaction between these inhibitors and *ftsZ* is largely genetic and consists of the determination that mutations within *ftsZ* can confer resistance to these inhibitors (Bi and Lutkenhaus, 1990 a, b).

Bacterial Growth and Lysis, Edited by M.A. de Pedro *et al.*, Plenum Press, New York, 1993

Recent studies using immunoelectromicroscopy demonstrate that during septation FtsZ localizes to a circumferential ring at the division site (Bi and Lutkenhaus, 1991). This ring is localized to the future division site at the midpoint of the cell before cell division is initiated and is therefore, an early step in cell division. During septation the FtsZ ring maintains its position at the leading edge of the septal invagination decreasing in diameter as septation ensues. Upon completion of septation the FtsZ-ring disassembles. Important to our understanding of cell division is the mechanisms of assembly and localization of the FtsZ ring during the cell cycle and the ultimate function of the FtsZ ring. The hypothesis is that formation of the FtsZ ring is a critical early step in cell division and that it functions as a cytoskeletal element that acts to mark the division site and to activate the septal biosynthetic machinery (Bi and Lutkenhaus, 1992). We would expect inhibitors of cell division to block ring formation, although they could act at a later step.

Cell division is spatially regulated such that division normally occurs at the midpoint of the long axis of the cell, however, under some conditions the localization of the division event is disturbed resulting in division occurring at the cell pole. Such division events are thought to involve the normal division machinery operating at previously used sites (Ward and Lutkenhaus, 1985). If so, minicell formation presumably involves FtsZ ring formation and implies that the *min* system is somehow involved in FtsZ localization. Overexpression of *ftsZ* as well as some mutations in *ftsZ* also induce minicell formation indicating that the *min* system can be bypassed. In addition, it has been reported that some *ftsZ* mutations affect cell morphology (Bi and Lutkenhaus, 1990a) indicating that the FtsZ ring may influence septal peptidoglycan biosynthesis. To further explore FtsZ function it is necessary to expand our collection of *ftsZ* mutations. At present the collection is limited to a conditional lethal, *ftsZ84*[Ts], a null mutation, and those mutations selected as resistant to *sulA*. By *in vitro* mutagenesis and selection for suppressors of existing *ftsZ* mutations we have obtained additional alleles of *ftsZ* that we have examined for effects on cell morphology. The results suggest that the FtsZ ring may dictate septal morphology and may have a role in the delicate balance between the synthetic and lytic activities operating during division.

RESULTS AND DISCUSSION

FtsZ Mutations

Several approaches were undertaken to increase our repetoire of *ftsZ* mutations. One approach involved an *in vitro* mutagenesis procedure developed in our laboratory (Dai and Lutkenhaus, unpublished). This method involves the semi random insertion of a kanamycin resistance determinant into the *ftsZ* gene contained on a plasmid. The insertion sites are a set of 24 different restriction sites scattered throughout the *ftsZ* gene. Since the *kan* element is flanked by symmetrical restriction sites it can be easily deleted such that two codons are retained at the site of the insertion. Two codon insertions were obtained at 21 of the 24 possible sites. At the other three sites either insertion of the *kan* element or subsequent deletion of the *kan* element resulted in a truncated *ftsZ* gene product that prevented cell survival. Previous results have shown that truncated derivatives of FtsZ can act as dominant negative inhibitors of cell division (Ward and Lutkenhaus, 1984). Approximately one half of the insertion alleles expressed some *ftsZ* function since they could complement the *ftsZ84* (Ts) mutation. Among these, one insertion allele, *ftsZ100*, expressed considerable resistance to SulA and expressed partial resistance to MinCD. The *ftsZ100* allele was of considerable interest since it caused an abnormal septal morphology indicating that *ftsZ* was acting at the septum and could influence septal biosynthesis (Table 1).

An additional *ftsZ* allele, *ftsZ26* (Ts), that affects polar morphology was isolated fortuitously in work involving the *ftsZ::kan* null allele (Bi and Lutkenhaus, submitted for publication). This allele has 6 codons inserted near the 5'end of the gene and confers temperature sensitive growth on the host. At the permissive temperature a strain containing this mutation has a normal cell size distribution but the polar morphology is affected in a majority of the cells and a few minicells are observed. Direct microscopic observation of this mutant indicates that this altered polar morphology is due to the division process; many dividing cells were observed that were misaligned in the division plane. At the nonpermissive temperature the cells filamented with "smooth" filamentous morphology supporting the view that the abnormal morphology at the permissive temperature was due to the division process. This allele was sensitive to the cell division inhibitors SulA and MinCD (Table 1).

By looking at revertants of *ftsZ26* (Ts) we isolated a cold sensitive allele of *ftsZ*, *ftsZ261* (Cs) [Table 1]. This allele contains two mutations; the 6 codon insertion and a nearby point mutation that suppresses the temperature sensitivity and either alone or in combination with the 6 codon insertion causes the cold sensitivity. Strains carrying this allele grow well at high temperature although they still show the altered polar morphology of the *ftsZ26* (Ts) mutation. Following a shift to low temperature cell growth stops and microscopic examination reveals that the cells lyse with a morphology suggesting that it is occurring at division sites. We believe that *ftsZ261* (Cs) causes a more dramatic division defect such that the altered polar morphology of

Table 1. Summary of the effects of *ftsZ* mutations

Mutation	Resistance		Effect on morphology
	SulA	MinCD	
ftsZ84	-	+	temperature sensitive filamentation
ftsZ2	+	+	minicells, altered polar morphology, temperature dedendent lysis
ftsZ100	+	+	altered polar morphology
ftsZ26	-	-	temperature sensitive filamentation, altered polar morphology
ftsZ261	NT[1]	NT	cold sensitive lysis

[1]Not tested

the *ftsZ26* (Ts) mutant is amplified leading to the cell lysis. Among previously isolated *ftsZ* mutations *ftsZ2* (Rsa) was of particular interest (Bi and Lutkenhaus, 1990ab). This mutation confered resistance to SulA, for which it was selected, and also resistance to MinCD. Introduction of the *ftsZ2* allele on a λ–transducing phage into wild type cells led to a significant minicell phenotype suggesting that this allele can ignor the *min* locus. This phenotype was even further exaggerated when the phage containing *ftsZ2* was introduced into an *ftsZ* null mutant. It appears that the presence of the wild type *ftsZ* gene can dampen the effect of the *ftsZ2* allele. Strains containing only the *ftsZ2* allele show growth defects including increased temperature sensitivity. Such strains exhibited some cell lysis, the extent of which, increased with increasing temperature. It was possible that the cell lysis is also connected to the division process in this mutant.

FtsZ Localization

The immunolabelling protocol we have established (Bi and Lutkenhaus, 1991) allows us to exaime the distribution of FtsZ in the mutants described above and to look at the effects of cell division inhibitors on FtsZ localization. These results are summarized in Table 2. One of the major conclusions from this work is that inhibitors or treatments that block cell divison and give a "smooth" filamentation phenotype prevent FtsZ ring formation. This includes induction of the inhibitors SulA and MinCD or incubation of temperature sensitive *fts*Z mutants at the nonpermissive temperature. This is in contrast to the results observed with *fts*A (Ts) mutants that cause an "indented" filamentous phenotype. With this mutant FtsZ rings are observed in the filaments although

Table 2. FtsZ ring formation

Mutation or Treatment	Effect on FtsZ ring formation
ftsZ84 (Ts)	FtsZ rings are not formed at the nonpermissive temperature
ftsZ2	FtsZ rings formed at internal or polar sites
ftsZ26 (Ts)	FtsZ rings formed at internal sites; often two rings that are adjacent No FtsZ rings formed at the nonpermissive temperature
ftsA12 (Ts)	FtsZ rings formed at the nonpermissive temperature
Δmin	FtsZ rings formed at internal or polar sites
sulA induction	FtsZ rings are not formed
minCD induction	FtsZ rings are not formed

[1]As determined by immunoelectromicroscopy (Bi and Lutkenhaus, 1991)

less than one per unit cell length. We interpret this to mean that in this mutant the FtsZ rings are assembled normally but are unstable due to the *fts*A block. Such transient ring formation may explain the indented morphology observed with various *fts* mutants other than *fts*Z mutants. We suspect cell division in initiated with FtsZ ring formation and then aborted accompanied with FtsZ disassembly. Another conclusion from these results is that divison, including mincell formation, is always associated with FJtsZ ring formation. This includes both minicells formed due to a *min* mutation and to the *ftsZ2* mutation (Table 2). This makes it likely that division always involves the FtsZ ring. Lastly, examination of the *ftsZ26* (Ts) mutant at the permissive temperature indicated that this mutation can affect the morphology of the FtsZ ring and ultimately polar morphology. In this case division in the middle of the cell if often associated with two closely positioned rings. Differential progression of these two rings would directly affect the polar morphology and explain why this mutation can cause minicell formation even though *ftsZ26* (Ts) is not resistant to *minCD*. Minicells are fromed, not because FtsZ rings are formed at the poles, but because two rings are formed near the midpoint of the cell.

DISCUSSION

From our studies on FtsZ we have proposed a model for cell division in which we suggest that FtsZ self assembles into a ring structure at the division site as an early step in the division process (Bi and Lutkenhaus, 1991; Fig. 1).

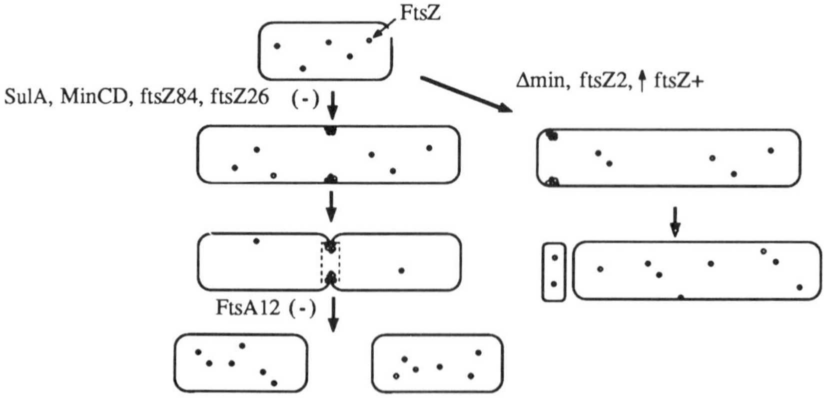

Figure 1. Model for FtsZ-ring formation and cell division. As cells double in size FtsZ accumulates in a circumferential ring at the midpoint of the cell. This localization is blocked by induction of SulA and excess MinCD. It is also blocked in ftsZ temperature sensitive mutants. Other conditions, including inaciivation of min, certain mutations in ftsZ and overexpression of ftsZ can lead to formai of the FtsZ-ring at midcell or the pole leading to minicell formation. Other fts mutants, such as ftsA1 block division after ring formation.

This ring functions as a cytoskeletal element to mark the division site and to activate the septal biosynthetic machinery, which includes the activities of the *ftsA,ftsQ* and *ftsl* (PBP3) gene products. The present results reveal that all operations that block cell division at an early step morphologically block FtsZ ring formation. This includes *ftsZ* mutations that block division and induction of the inhibitors SulA or MinCD. It is likely that in these cases FtsZ ring formation is blocked at the step on nucleation. In contrast, mutations that block division at a later step morphologically, do not block FtsZ ring formation but affect its stability. In this case the ring forms and cell division is initiated, but due to the missing division activitky, eventually disasembles. Our results also demonstrate that minicell production involves FtsZ ring formation at the poles. This altered localization of the ring can be achieved by loss of *min* function or alterations of *ftsZ* that override *min* fuction.

As part of our hypothesis we suggested that the FJtsZ ring would activate septal peptidoglycan biosynthesis and the study of novel *ftsZ* mutants supports this view. Thus, *ftsZ* mutants exist which modify the polar morphology and even lead to cell division-associated cell lysis. In vitro experiments are needed to futher test this hypothesis and the hypothesis that the FtsZ ring forms through self assembly.

ACKNOWLEDGEMENTS

This work was supported by grants from the National Institutes of Health and the Wesley Foundation.

REFERENCES

Begg, K. J., and Donachie, W. D., 1985, Cell shape and division in *Escherichia coli*: experiments with shape and division mutants.*J. Bacteriol.* 163:615-622.

Bi, E., and Lutkenhaus, J., 1990a, Analysis of *ftsZ* mutations that confer resistance to the cell division inhibitor SulA (SfiA), *J. Bacteriol.* 172:5602.

Bi, E., and Lutkenhaus, J., 1990b, Interaction between the *min* locus and *ftsZ*, J. Bacteriol. 172:5610.

Bi, E., and Lutkenhaus, J., 1991, FtsZ ring structure associated with division in *Escherichia coli*, Nature 354:161.

Dai, K. and Lutkenhaus, J., 1991, *ftsZ* is an essential cell division gene in *Escherichia coli*,*J. Bacteriol.*, 173:3500.

de Boer, P. A. J., Crossley, R., and Rothfield, L. I., 1989, A division inhibitor and a topological specificity factor coded for by the minicell locus determine proper placement of the division septum in *Escherichia coli*, *Cell* 56:641.

Huisman, O., and D'Ari, R., 1981, An inducible DNA replication-cell division coupling mechanism of *Escherichia coli*, *Nature* 290:797.

Lutkenhaus, J. F., Wolf-Watz, H., and Donachie, W. D., 1980, Organization of genes in the *ftsA-envA* region of the Escherichia coli gentic map and identification of a new *fts* locus (*ftsZ*), *J. Bacteriol.* 142:615.

Pla, J., Sanchez, M., Palacikos, P., Vicente, M., and Aldea, M., 1991, Preferential cytoplasmic location of FtsZ, a protein essential for *Escherichia coli* septation, *Mol. Microbiol.* 5:1681.

Taschner, P.E. M., Juis, P. G., Pas, E., and Woldringh, C. L., 1988, Division behavior and shape changes in isogenic *ftsZ*, *ftsQ*, *ftsA* and *ftsE* cell division mutants of *E. coli* during temperature shift experiments, *J. Bacteriol.* 170:1533-1540.

Ward Jr., J. E, and Lutkenhaus, J. F., 1984, A *lacZ-ftsZ* gene fusion is an analog of the cell division inhibitor SulA, *J. Bacteriol.* 157:815.

Ward Jr., J. E, and Lutkenhaus, J. F., 1985, Overproduction of FtsZ induces minicell formation in *E. coli*, *Cell* 42:941.

STABILITY OF COMPONENTS OF THE *Escherichia coli* SEPTATOR

Jesús Pla, Pilar Palacios, Manuel Sánchez, Teresa Garrido and Miguel Vicente

Centro de Investigaciones Biológicas
Consejo Superior de Investigaciones Científicas
C/ Velázquez 144. 28006 Madrid. Spain

INTRODUCTION

Most of the information concerning cell division in *Escherichia coli* derive from genetic observations. Mutations in *E. coli* genes causing inhibition of cell division and a filamentous phenotype have been isolated (Bachmann, 1990). In this way, several genes have been identified and among them, *ftsQ* (Begg *et al.*, 1980), *ftsA* (Donachie *et al.*, 1979) and *ftsZ* (Lutkenhaus *et al.*, 1980) seem specially relevant for the initial (*ftsQ* and *ftsZ*) and final (*ftsA*) steps of the process. FtsQ, FtsA and FtsZ have been located associated to different degrees to the cytoplasmic membrane, and have been postulated to form part of the septator, a structure active in cell division (Vicente *et al.*, 1991 *b*).

Proteins involved in septation may do so as building blocks, as modifiers of other molecules, or as catalysts. Which is the role of each component of the septator is likely to emerge in the future from the work currently underway. All the cellular elements that participate in division should anyway share one property, their periodic activation and/or inactivation, that allows the process to be driven in cycles. A mechanism, involving inhibitors such as MinC, operates on FtsZ (de Boer *et al.*, 1992), a protein that seems to be differentially located in the cytoplasm (Pla *et al.*, 1991), or associated to the septum (Bi and Lutkenhaus, 1991) at different stages of the division cycle.

FtsA has been shown to participate in several regulatory mechanisms such as the TER pathway of septation termination (Tormo *et al.*, 1985), and in the modification of the ampicillin-binding capacity of PBP3 (Tormo *et al.*, 1986). FtsA is perhaps one of the proteins of the septator for which more data are available, including now the existence of a predicted model for its structure (Bork *et al.*, in press) that makes it tempting to speculate on its biochemical functions. The *ftsA* gene, under normal conditions in which no excess of SdiA is present (Wang *et al.*, 1991), is expressed from a gearbox and a consensus promoter (Aldea *et al.*, 1990: Vicente *et al.*, 1991 *b*) located upstream from *ftsQ*, thus ensuring the presence of around 150 FtsA molecules per cell at any growth rate. It forms part of the structure of the septum (Tormo and Vicente, 1984), so that it may participate in division assuming several roles.

FtsQ is essential for cell division (Begg *et al.*, 1980) and has been identified as a cytoplasmic membrane protein with periplasmic domains (Storts *et al.*, 1989; Dopazo *et al.*, 1992).

In the present work, we have investigated if cell division proteins are susceptible to degradation *in vivo*. We have tested the effect of the ATP-dependent Lon protease

Bacterial Growth and Lysis, Edited by M.A. de Pedro *et al.*, Plenum Press, New York, 1993

(Gottesman, 1987) on the stability of cell division proteins FtsQ, FtsA and FtsZ and measured their half-life time. The amount of some of these proteins found per cell is low, and this study was done with the help of a temperature regulated expression system. Out of the three proteins, FtsA was the only one whose stability was shown to be dependent upon the Lon protease. In a *lon* mutant FtsA remained stable for over three doubling times.

CONSTRUCTION OF PLASMIDS THAT PRODUCE Fts PROTEINS IN CONTROLLED AMOUNTS

Plasmid pJPV8 has been described before (Pla *et al.*, 1990). Other plasmids used in this work are shown in Fig. 1. Plasmid pJPV7 was obtained by recircularization of plasmid pJPV4 (Pla *et al.*, 1990) following the disruption of the *ftsA* open reading frame by

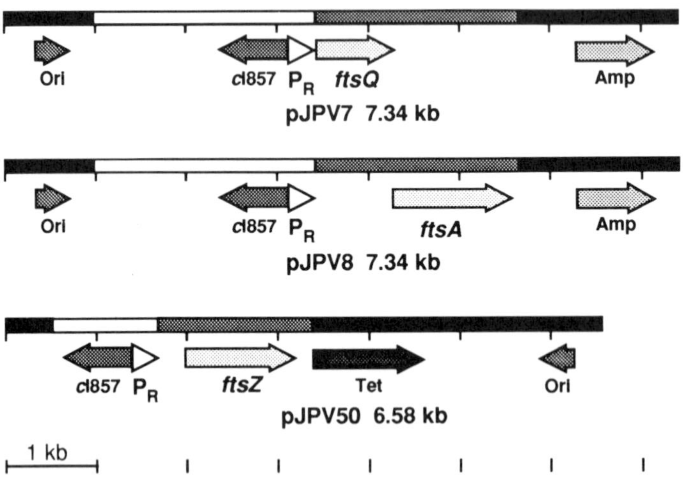

Figure 1 Plasmids used in this work. Abbreviations are: Ori, Origin of replication; *cI*857, lambda repressor (thermosensitive allele 857); P_R, lambda P_R promoter; Amp, ampicillin resistance gene; Tet, tetracyclin resistance gene. The location of the *ftsQ*, *ftsA* and *ftsZ* genes is indicated in each case. Other symbols: ■ vector derived fragments, □ lambda derived fragments, and ▩ chromosomal inserts.

digestion with *Bgl*II and subsequent Klenow filling of the 5' protruding ends. For the construction of pJPV50, a 1.8 kp *HindIII* fragment from plasmid pZAQ (Ward and Lutkenhaus, 1985) comprising the entire *ftsZ* gene was cloned in the *Hinc*II site of pUC19 (Yanisch-Perron *et al.*, 1985) to yield pLAR7. A 1.1 kb *Pst*I-*Eco*RI fragment from pJPV8 comprising the *c*I857 lambda repressor, the lambda P_R promoter and a fragment of the *cro* gene was cloned into the 5 kb *Pst*I-*Eco*RI fragment of plasmid pZAQ to yield pJPV35, in which a Cro-FtsZ fusion can be expressed. Finally, the 0.5 kb *EcoRI* fragment of pLAR7 was introduced in the *Eco*RI site of pJPV35 to yield pJPV50.

In this set of plasmids the *ftsQ*, *ftsA* and *ftsZ* genes were cloned under the control of the lambda P_R promoter (Fig 1). The presence of the *cI*857 thermosensitive repressor allele in the plasmids, allows then a controlled over-production of the Fts proteins depending on the extent of incubation at 42 ºC.

STRAINS AND GROWTH CONDITIONS

Strain SG1611: *Δlon510 Δgal165 Δ(lac-proAB) supE thi* (F' *traD36 proAB lacI*qZΔM15) (Trempy and Gottesman, 1989) and its parental strain JM101 (lon^+) (Messing *et al.*, 1981) were transformed with plasmids pJPV7, pJPV8, and pJPV50. To estimate the duration of the 42 ºC induction period best suited to our purposes, the amount of proteins expressed after different periods at 42 ºC was measured by using antiserum MVJ11, MVJ2 and MJV8 respectively. It was estimated that a 15 min period at 42 ºC was the minimum time required to produce enough quantity of protein to allow reliable detection without a drastic alteration of cell physiology (data not shown).

For the experiment, cells were then grown in LB broth at 30 ºC (under these conditions the doubling time was 42 minutes for all strains) until an $O.D._{540}=0.3$ was reached and transferred to 42 ºC for 15 minutes to allow expression of the *fts* genes from the P_R promoter. The cultures were then transferred back to 30 ºC and samples containing

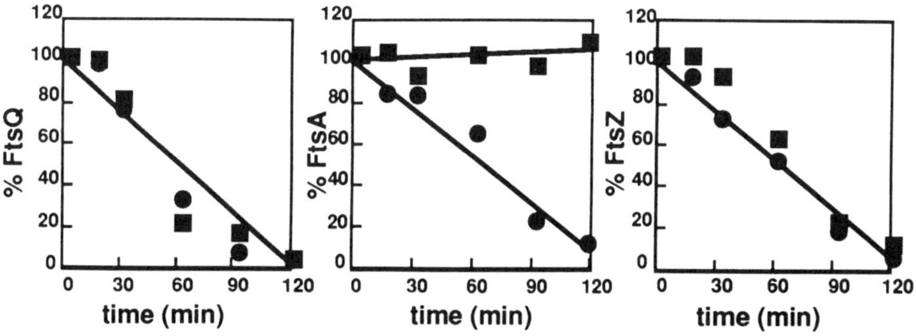

Figure 2. Dependence of the stability of Fts proteins on the presence of the Lon protease. Values obtained from densitometric analysis of western type blots are plotted versus time.. Cultures of lon^+ (●) and *lon* (■)strains containing pJPV7 (FtsQ), pJPV8 (FtsA), or pJPV50 (FtsQ) were shifted to 42ºC and after 15 min were transferred back to 30ºC (time 0) Values are represented as the percentage of protein present at each sampling time relative to the value corresponding to time 0.

2 x 10^8 cells were collected at intervals, prepared for SDS-PAGE by treatment with 30 μl of SDS-PAGE sample buffer at 100 ºC for 15 minutes, run on a 10% SDS-polyacrylamide gel (Laemli, 1970) and transferred to Immobilon (Millipore) membranes. FtsQ, A and Z were detected by indirect immunostaining using polyclonal antisera, protein A-horseradish peroxidase conjugate, and 4-chloro-1-naphtol as substrate (Towbin *et al.*, 1979). Quantification of bands was done by densitometry on positive films.

STABILITY OF Fts PROTEINS IN THE PRESENCE AND ABSENCE OF Lon

As observed in Fig. 2, FtsA was the only protein of those analyzed whose stability shows a *lon* dependence. While FtsA was degraded in 2 hours (with an estimated half-life time of 60 min.) to its normal levels in the lon^+ strain JM101, no degradation was observed in the *lon* strain SG1611. The absence of an active Lon protease had no effect on the degradation of FtsQ and FtsZ for which an estimated half-time life of 45 min. and 60 min. respectively, can be calculated. This result agrees with the time measured by Storts *et al* (16) for FtsQ from pulse and chase experiments in which the degradation of the protein labeled with ^{35}S-methionine was measured.

Our results show that, when present at higher than physiological levels, FtsA can be degraded by the Lon protease. This result agrees with previous observations that found a dependence of the ability to recover septation at sites blocked by the FtsA3 protein on the presence of an active Lon protease (Dopazo *et al.*, 1987). Yet, we cannot exclude the possibility that this dependence could be the result of an incorrect (i.e., non septal) location of the protein within the cell. This incorrect location could result in the overexpressed FtsA being more accessible to the cytoplasmically located Lon protease or, conversely, in a structure of FtsA different from that present at the septator and susceptible to degradation by Lon. This possibility would be specific for FtsA, as it is not observed for the other two Fts proteins. It has been already documented by genetic procedures that an altered form of FtsA, FtsA3, is likely to be degraded at faster rates in *lon*+ strains than in those in which this protease activity is impaired (Dopazo *et al.*, 1987); in this case, the permanence of the altered FtsA3 prevents septation to occur at those sites in which it is present. Susceptibility of FtsA to Lon agrees with the localizations of both proteins, as Lon is cytoplasmic and FtsA is associated to the cytoplasmic membrane facing towards the cytoplasm.

DISCUSSION

Degradation by Lon is not likely to be the sole mechanism regulating the levels of FtsA under physiological conditions. Expression of *ftsQ* and *ftsA* is under the control of a gearbox and a consensus promoter ensuring that they can be produced at levels stoichiometric with the septum (Vicente *et al.*, 1991 *a*). Physiological levels of FtsQ and FtsA are low, perhaps 25 FtsQ molecules per cell (Carson *et al.*, 1991) and 150 FtsA (Wang and Gayda, 1990; and our unpublished results) respectively, and are barely detectable by the antisera used in this work. The presence of a *lon* mutation is not followed by an increase in these levels (results not shown); an increase that should be expected, particularly for FtsA if Lon were the only mechanism to inactivate it. This observation, and the estimated half-life times of the FtsQ, FtsA and FtsZ proteins suggest that Lon does not play a crucial role in regulating the levels of septation proteins under physiological conditions. Lon has been involved in the degradation of abnormal proteins, from our results we can deduce that either it is also involved in the degradation of misplaced proteins, or that FtsA adopts a defective conformation when over-production leads to a wrong localisation. A mechanism to inactivate active septators should exist. The Min complex seems to exert its inhibitory role at the level of FtsZ (de Boer *et al.*, 1989). Fluctuation in the levels of FtsQ should have drastic effects on division given its low abundance, but we still have no clue regarding their regulation. The mechanism that inactivates FtsA as an active septator component may involve the Lon protease, but it is reasonable to assume that some of its biochemical features may be crucial for its biological activity. The predicted structure for FtsA (Bork *et al.*, in press) suggests the existence of ATP-binding and protein-interaction domains. We can now study the effects of specific mutations within these and other domains to test many of our hypothesis.

Acknowledgments

We thank Martí Aldea for helpful comments and discussion, and Susan Gottesman for providing strains. J.P., M.S. and T.G. acknowledge fellowships from Consejo Superior de Investigaciones Científicas, Ministerio de Educación y Ciencia, and Comunidad Autónoma de Madrid respectively. This work was funded by grant PB89-0030 from Dirección General de Investigación Científica y Técnica.

REFERENCES

Aldea, M., Garrido, T., Pla, J. and Vicente, M. (1990) Division genes in *Escherichia coli* are expressed coordinately to cell septum requirements by gearbox promoters. EMBO J. 9: 3787-3794.

Bachmann, B. Linkage Map of *Escherichia coli* K-12. Edition 8 (1990). Microbiol. Rev., 54: 130-197.

Begg, K.J., Hatfull, G.F., and Donachie, W.D. (1980). Identification of new genes in a cell envelope-cell division cluster of *Escherichia coli*: cell division gene product *ftsQ*. J. Bacteriol., 144: 435-437.

Bi, E., and Lutkenhaus, J. (1991) FtsZ ring structure associated with division in *Escherichia coli*, Nature (London) 354: 161-164.

Bork, P., Sander, C. and Valencia, A. (1992) An ATPase domain in prokaryotic cell cycle proteins related to actin, hsp70 and hexokinase. Proc, Nat. Acad. Sci. USA. in press.

Carson, M., Barondess, J.J., and Beckwith, J. (1991) The FtsQ protein of *Escherichia coli*: membrane topology, abundance, and cell division phenotypes due to overproduction and insertion mutations. J. Bacteriol. 173: 2187-2195.

de Boer, P.A.J., Crossley, R.E. and Rothfield, L.I. (1989). A division inhibitor and a topological specificity factor coded for by the minicell locus determine proper placement of the division septum in *E. coli*. Cell, 56, 641-649.

de Boer, P.A.J., Crossley, R.E., and Rothfield, L.I. (1992) Roles of MinC and MinD in the site-specific septation block mediated by the MinCDE system of *Escherichia coli*. J.Bacteriol. 174: 63-70.

Donachie, W.D. and Robinson, A.C. (1987). Cell division: parameter values and the process., in "*Escherichia coli* and *Salmonella typhimurium*: cellular and molecular biology" (Neidhardt, F.C., Ingraham, J.L., Low, K.B., Magasanik, B., Schaechter, M. and Umbarger, H.E. Ed.),.pp 1578-1593. American Society for Microbiology, Washington D.C.

Donachie, W.D., Begg K.J. and Sullivan, N.F. (1984). Morphogenes of *Escherichia coli*, in "Microbial Development" (Losick R. and Shapiro L., Ed.), pp.27-62.. Cold Spring Harbor Laboratory, Cold Spring Harbor, N.Y.

Donachie, W.D., Begg, K.J., Lutkenhaus, J.F., Salmond, G.P.C., Martínez-Salas, E., and Vicente, M. (1979) Role of the *ftsA* gene product in control of *Escherichia coli* cell division. J. Bacteriol. 140, 388-394.

Dopazo, A., Palacios, P., Sánchez, M., Pla, J. and Vicente, M. (1992) An amino-proximal domain requires for the localization of FtsQ in the cytoplasmic membrane, and for its biological function in *Escherichia coli*. Mol. Microbiol. 6: 715-722.

Dopazo, A., Tormo, A., Aldea, M. and Vicente, M. (1987) Structural inhibition and reactivation of *Escherichia coli* septation by elements of the SOS and TER pathways. J. Bacteriol. 169: 1772-1776.

Gottesman, S.(1987). Regulation by proteolysis, in "*Escherichia coli* and *Salmonella typhimurium*: cellular and molecular biology" (Neidhardt, F.C., Ingraham, J.L., Low, K.B., Magasanik, B., Schaechter, M. and Umbarger, H.E. Ed.),.pp 1313-1317. American Society for Microbiology, Washington D.C.

Laemli, U.K. (1970). Cleavage od structural proteins during the assembly of the head of bacteriophage T4. Nature (London), 227, 680-685.

Lutkenhaus, J.F. (1983). Coupling of DNA replication and cell division: *sulB* is an allele of *ftsZ*. J. Bacteriol. 154: 1339-1346.

Lutkenhaus, J.F., Moore, B.A., Masters, M. and Donachie, W.D. (1979). Individual proteins are synthesized continuously throughout the *Escherichia coli* cell cycle. J. Bacteriol. 138: 352-360.

Lutkenhaus, J.F., Wolf-Watz, H. and Donachie, W.D. (1980). Organization of genes in the *ftsA-envA* region of the *Escherichia coli* genetic map and identification of a new *fts* locus (*ftsZ*). J. Bacteriol. 142: 615-620.

Messing, J., Crea, R. and Seeburg, P.H. (1981) A system for shotgun DNA sequencing. Nucleic Acids Res. 9: 309-321.

Mizusawa, S. and Gottesman, S. (1983). Protein degradation in *Escherichia coli*: the *lon* gene controls the stability of SulA protein. Proc. Natl. Acad. Sci. USA. 80: 358-362.

Pla, J., Dopazo, A. and Vicente, M. (1990). The native form of FtsA, a septal protein of *Escherichia coli*, is located in the cytoplasmic membrane. J. Bacteriol. 172: 5097-5102

Pla, J.,. Sánchez. M., Palacios, P., Vicente, M. and Aldea, M. (1991) Preferential cytoplasmic location of FtsZ, a protein essential for *Escherichia coli* septation. Mol. Microbiol. 5: 1681-1686.

Robinson, A.C., Kenan, D.J., Hatfull, G.F., Sullivan, N.F., Spiegelberg, R. and Donachie, W.D. (1984). DNA sequence and transcriptional organization of essential cell division genes *ftsQ* and *ftsA* of *Escherichia coli*: evidence for overlapping transcriptional units. J. Bacteriol. 160: 546-555.

Robinson, A.C., Kenan, D.J., Sweeney, J. and Donachie, W.D. (1986). Further evidence for overlapping transcriptional units in an *Escherichia coli* cell envelope-cell division gene cluster: DNA sequence and transcriptional organization of the *ddl ftsQ* region. J. Bacteriol. 167: 809-817.

Storts, D.R., Aparicio, O., Shoemaker, J.M. and Markovitz, A. (1989). Overproduction and identification of the *ftsQ* gene product, an essential cell division protein in *Escherichia coli* K-12. J. Bacteriol. 171: 4290-4297.

Tormo, A. and Vicente, M. (1984). The *ftsA* gene participates in formation of the *Escherichia coli* septum structure. J. Bacteriol. 157: 779-784.

Tormo, A., Ayala, J.A., de Pedro, M.A., Aldea, M. and Vicente, M. (1986). Interaction of FtsA and PBP-3 proteins in the *Escherichia coli* septum. J. Bacteriol. 166: 985-992.

Tormo, A., Dopazo, A., de la Campa, A.G., Aldea, M. and Vicente, M. (1985). Coupling between DNA replication and cell division mediated by the FtsA protein in *Escherichia coli*: a pathway independent of the SOS response, the "TER" pathway. J. Bacteriol. 164: 950-953.

Towbin, H., T. Staehelin and J. Gordon. (1979). Electrophoretic transfer of proteins from polyacrylamide gels to nitrocellulose sheets: procedure and some applications. Proc. Natl. Acad. Sci. USA. 76: 4350-4354.

Trempy, J.E. and Gottesman, S. (1989). Alp, a Suppressor of Lon Protease Mutants in *Escherichia coli*. J. Bacteriol. 171: 3348-3353.

Vicente, M., Kushner, S.R., Garrido, T. and Aldea, M. (1991 *a*) The role of the 'gearbox' in the transcription of essential genes. Mol. Microbiol. 5: 2085-2091.

Vicente, M., Palacios, P., Dopazo, A., Garrido, T., Pla, J. and Aldea, M. (1991 *b*) On the chronology and topography of bacterial cell division. Res. Microbiol. 142: 253-257.

Wang, H. and Gayda, R.C. (1990) High-level expression of the FtsA protein inhibits cell septation in *Escherichia coli* K-12. J. Bacteriol. 172: 4736-4740.

Wang, X, de Boer, P.A.J. and Rothfield, L.I. (1991) A factor that positively regulates cell division by activating transcription of the major clusters of essential cell division genes of *Escherichia coli*. EMBO J. 11: 3363-3372.

Ward, J.E. and Lutkenhaus, J.F. (1985). Over-production of FtsZ induces minicell formation in *E. coli*. Cell 42: 941-949.

Yanisch-Perron, C., Vieira, I. and Messing, J, (1985) Improved M13 phage cloning vectors and host strains: nucleotide sequences of the M13mp18 and pUC19 vectors. Gene 33: 103-119.

CONSTRUCTION OF A TRIPLE DELETION OF PENICILLIN-BINDING PROTEINS 4, 5, AND 6 IN *ESCHERICHIA COLI*

David H. Edwards and William D. Donachie

Institute of Cell and Molecular Biology
Division of Biological Sciences
University of Edinburgh

INTRODUCTION

Historically seven Penicillin Binding Proteins (PBPs) have been recognised in *Escherichia coli*, though with recent work two more have been added. Two of the PBPs, PBP2 and PBP3 have distinct morphological roles (Spratt, 1983). PBP2 is involved in cell elongation and allows the rod shape to be maintained, while by contrast PBP3 is required for septum formation, and its inhibition leads to filamentation and lysis. Both these proteins have been demonstrated to have transpeptidase and associated transglycosylase activity (Ishino & Matsuhashi 1983. Ishino *et al* 1986).

In the case of PBP3 work on suppression of a temperature sensitive allele (TOE23) has shown two other PBPs to be implicated in its function (Begg *et al.* 1990). This suppression by increased levels of either PBP5 or PBP6; the two major DD-carboxypeptidases, has been used to support a model which ascribes tripeptides a role as preferred acceptors for PBP3. Recent work on another suppressor *sufI* (Kato *et al* 1988) of our PBP3 allele, has shown it too to increase DD-carboxypeptidase activity (unpublished). Further support for a role of this enzyme activity in septum formation has come from work predating our own, in *Streptococcus pneumoniae* (Schuster *et al.* 1990).

Of the three PBPs in *E.coli* to have been demonstrated to exhibit DD-carboxypeptidase activity, PBP5 and PBP6 are the most abundant. Yet a deletion of both these proteins can be tolerated. Under these circumstances it has been argued that residual activity by PBP4 could be sufficient for the cell to grow and divide. If the role of periplasmic DD-carboxypeptidases is essential then a triple deletion of PBP4,PBP5, and PBP6 would be lethal. Consequently we set out to construct a complete deletion of PBP4 and to use this to produce a strain lacking all three of the lower molecular weight PBPs, and DD-carboxypeptidase activity.

CONSTRUCTION OF THE TRIPLE DELETION

The construction of a PBP4 deletion involved the use of **pBK18-1** (supplied by W.Keck) which carries the 1.7kb coding region of *dacB*. The plasmid was restricted at a *ClaI* site approximately 500kb into its coding sequence. The 5' overhang was filled in using Klenow and an in vitro blunt ended ligation (Maniatas *et al.* 1989) was carried out with the

Bacterial Growth and Lysis, Edited by M.A. de Pedro
et al., Plenum Press, New York, 1993

omega fragment(Prentki *et al.* 1984). The omega fragment consists of an antibiotic resistance gene which carries transcriptional and translational stop signals in its flanking regions. The resulting ligation mixture was transformed into *E.coli* DH1 using suitable selection.

Once the deleted *dacB* had been constructed the plasmid was linearised and used to transform a *recD* deletion strain (Russell *et al* 1989). The spectinomycin resistant transformants were then selected and two used to produce P1 lysates. The putative disrupted chromosomal copy of *dacB* was then transduced into a **Sp1070** strain deleted for both PBP5 and PBP6 (Broome-Smith 1986).

For two separate transductants the deletion of PBP4 was then confirmed by southern analysis and the absence of all 3 PBPs by a Penicillin binding assay (Spratt 1977). The new strain was referred to as **D456**.

RESULTS

A triple deletion of PBP4,PBP5 and PBP6 was produced and proved viable. No obvious effects were apparent, apart from a slightly longer generation time than that of the double deletion, 36 minutes versus 32 minutes, and a slightly altered cell morphology. The phenotypic difference was slight but consistently observable during early and mid log phase of liquid cultures. Under the microscope the triple deletion appeared to contain a small percentage of aberrant cell shapes which had characteristically branched poles or centres. While this was as low as 2% it was enough to distinguish the triple deletion from the double, which produced aberrant cells less frequently.

Preliminary murein analysis of the two strains by the method of Glauner (1988), showed no major differences between the double and triple deleted strains. Similarly the assay of DD-carboxypeptidase activity carried out by Astrid Ursinus produced almost identical results for the two strains.

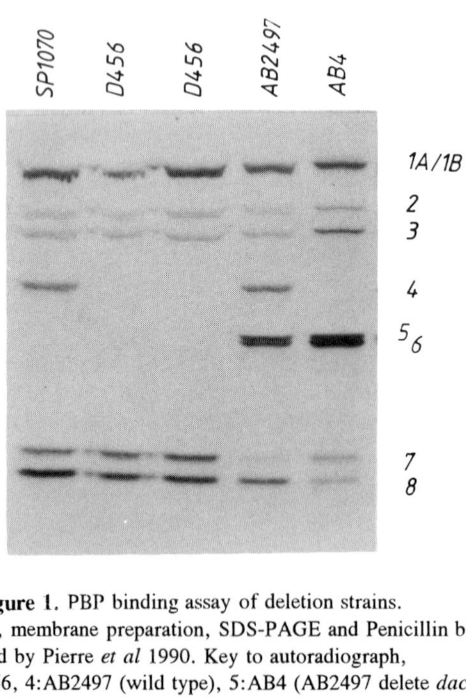

Figure 1. PBP binding assay of deletion strains.
Strains were grown in LB, membrane preparation, SDS-PAGE and Penicillin binding assays were carried out as previously described by Pierre *et al* 1990. Key to autoradiograph,
1:SP1070, 2:D456, 3:D456, 4:AB2497 (wild type), 5:AB4 (AB2497 delete *dacB*).

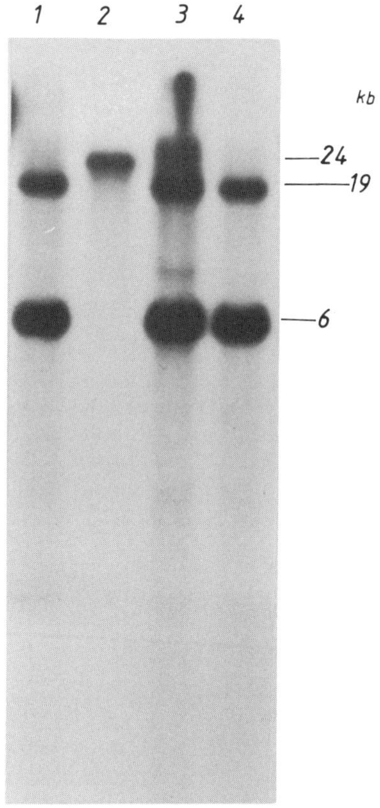

Figure 2. Southern analysis of strains.
Key, 1: D456, 2:SP1070, 3:D456, 5:AB4 .
Chromosomal DNA restricted with *HindIII*, separated by agarose gel electrophoresis, hybridised to Hybond-N and probed. The probe was constructed by random priming the 1.7kb fragment of **pBK18-1** encoding the *dacB* gene. Existence of an internal *HindIII* site in *dacB* is confirmed in the putative deletion strains by the presence of two bands compared with a single one in the wild type (**SP1070**).

Table 1

Muropeptides	Relative amounts (%)	
	SP1070	D456
Monomers	51.22	50.37
Dimers	44.16	44.62
Trimers	4.53	4.91
Tetramers	0.10	0.11
Total Pentapeptides	8.46	7.95
Anhydro-compounds	3.18	3.70

Abbreviations for muropeptides: monomer GlcNac-β-1,4-MurNac-peptide; dimer,bis(GlcNAc-β-MurNAc-peptide); trimer,tris(GlcNA-β-1,4-MurNAc-peptide); tetramers, tetrakis(GlcNAc-β-1,4-MurNAc-peptide); anhydro-compounds:1-6-anhydromuramic acid containing muropeptides form the end of glycan strands. Total Pentapeptides,cumulative percentage of all forms of complete side chains (L-ala-D-glu-*m*-A2pm-D-ala-D-ala).

Table 2

Strain	DD-Carboxypeptidase activity % Release of D-alanine From Muramyl-Pentapeptide
SP1070	2.58
D456	2.36
AB2497 (wild type)	35.73
C600 (wild type)	82.29

Note: The wild type background for SP1070 and D456 has proved unobtainable, so the C600 and AB2497 are included for reference.

DISCUSSION

The construction of a triple deletion of PBP4, PBP5, and PBP6 has shown all three of these lower molecular weight PBPs to be inessential. In our original model for a supply of preferred tripeptide acceptors to PBP3 we envisaged a significant role for periplasmic enzymes in supplying the precursor. However in the absence of all three enzymes which have been demonstrated to exhibit DD-carboxypeptidase activity, murein analysis has shown tripeptide containing bonds continue to be formed.

While it may still be that a putative Dala-Dala removing enzyme may be involved in supplying tripeptides, it would appear more appealing to consider the supply of tripeptides by a lipid carrier system to be sufficient for septum formation in the triple deletion strain. Such a supply of tripeptides could be closely linked to PBP3, and would therefore supply the estimated low numbers of this essential protein very efficiently.

To conclude, a deletion of PBPs 4, 5, and 6, is viable. Despite the absence of these three proposed DD-carboxypeptidases preliminary work shows tripeptides continue to be involved in cross linking reactions within the murein sacculus. Therefore the formation of tripeptides from pentapeptides under the control of periplasmic enzymes appears to be inessential during cell elongation and septation. It is also worth considering that with recent work suggesting PBP2 to be inessential, and the well established fact that either PBP1A or PBP1B (Spratt 1983) can be deleted, it may be possible to construct a strain with as few as two of the nine commonly recognised PBPs.

ACKNOWLEDGEMENTS

I would like to thank Stefan Korner and Astrid Ursinus for the murein and DD-carboxypeptidase activity assays, and in particular Joachim Holtje for presenting me with the opportunity to work in Tubingen.

REFERENCES

Begg.K.,Takasuga A.,Edwards D.,Dewar S.,Spratt B.G., Adachi H., Ohta T., Matsuzawa H., & Donachie W.D. (1990) The balance between different peptidoglycan precursors determines whether *E.coli* cells will elongate or divide. J. Bact 172, 6697-6703.

Broome-Smith J.K. (1985) Construction of a mutant of *E.coli* that has deletions of both the Penicillin-binding Protein 5 and 6 genes. J. of General Microbiol 131, 2115-2118.

Glauner B. (1988) Separation and quantitation of muropeptides with high performance liquid chromatography. Analyt.Biochem 172, 451-464.

Ishino F.,Park W.,Tomioka S.,Tamaki S.,Takase I.,Kunugita K., Matsuzawa H., Sadamitsu A., Takahisa O., Spratt B.G., & Matsuhashi M. (1986) Peptidoglycan Synthetic activities in membranes of *E.coli* caused by overproduction of Penicillin Binding Protein 2 and RodA protein. J.Biol.Chem. 261, 7024-7031.

Ishino F. & Matsuhashi M. (1981) Peptidoglycan synthetic enzyme activities of highly purified PBP3 in *E.coli* : A septum forming reaction sequence. Biochem & Biophys Research Communications pp. 905-911.

Kato J.,Nishimura Y.,Yamada M.,Suzuki H., and Hirota Y. (1988) Gene organisation in the region containing a new gene involved in chromosome partition in *E.coli* J.Bact. 170, 3967-3977.

Maniatas T.,Sambrook J., & Fritsch E. (1989) "Molecular Cloning a Laboratory Manual." Cold Spring Harbour Lab Press.

Pierre J., Boisivon A., and Gutmann L. (1990) Alteration of PBP3 entails resistance to Imipenem in *Listeria monocytogenes* . Antimicrobial Agents and Chemotherapy. 1990 Vol.34 No.9 pp.1695-1698.

Prentki P., & Krisch H.M. (1984) In vitro Insertional mutagenesis with a selectable DNA fragment. Gene 29, 303-313.

Russell C.D., Thaler D.s., & Dahlquist F.W. (1990) Chromosomal transformation of *E.coli recD* strains with linearised plasmids. J.Bact. 171, 2609-2613.

Schuster C.,Dobrinski B., & Hakenbach R. (1991) Unusual septum formation in *Streptococcus pnumoniae* mutants with an alteration in the D,D-Carboxypeptidase Penicillin Binding Protein 3. J.Bact. 172, 6499-6505.

Spratt B.G. (1983) Penicillin-binding Proteins and the future of beta-Lactam Antibiotics. J.G.M. 129, 1247-1260.

Spratt B.G. (1977) Properties of Penicillin Binding Proteins of *E.coli* K12. Eur.J.Biochem. 72, 341-352.

Spratt B.G. & Pardee, A.B. (1975) Penicillin Binding Proteins and cell shape in *E.coli*. Nature 254, 516-517.

DEFORMATIONS IN THE CYTOPLASMIC MEMBRANE OF *ESCHERICHIA COLI* DIRECT THE REPAIR OF PEPTIDOGLYCAN

Vic Norris and Sean Sweeney

Department of Genetics
University of Leicester
Leicester LE1 7RH, UK

INTRODUCTION

Turgor pressures of 3.5 Atmospheres in Gram negative bacteria (Stock *et al.*, 1977) and up to 20 Atmospheres in Gram positive bacteria (Mitchell and Moyle, 1956) mean that the weakening of any region of the force-bearing element of the bacterial wall, peptidoglycan, may result in lysis. A repair mechanism to strengthen such regions preferentially may therefore exist. In this communication, a model is proposed in which this mechanism involves local changes in the structure and composition of the cytoplasmic membrane, the requirements of the model are listed and supporting evidence and predictions are discussed.

THE HERNIA MODEL

A local weakening of the peptidoglycan sacculus by random cleavage, possibly involving autolysins, leads to a local decrease in the radius of curvature (in both longitudinal and circumferential directions) of the cytoplasmic membrane as, driven by the turgor pressure in exponential and stationary phase cells, this membrane begins to bulge through into the periplasm (Figure 1). Lateral diffusion of lipids leads to particular phospholipids congregating in this bulge or hernia: in its outer monolayer, phospholipids with large headgroups and saturated acyl chains are selected by the constraint that this monolayer be convex; conversely, in the inner monolayer small headgroups and bulky acyl chains are selected by the constraint that this monolayer be concave (Figure 2). This phospholipid composition prevents cleavage by autolysins either by excluding them from the region or by inhibiting their activity. At the same time, this composition favours the activity of the enzymes which thicken and elongate the peptidoglycan. These enzymes include those responsible for transpeptidation, transglycosylation and precursor synthesis. Hence weakening a region of the peptidoglycan leads to a hernia which in turn results in new peptidoglycan being laid down and lysis averted.

Bacterial Growth and Lysis, Edited by M.A. de Pedro
et al., Plenum Press, New York, 1993

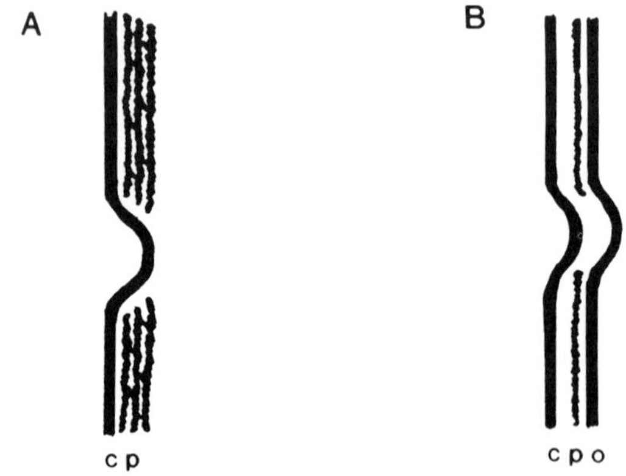

Figure 1. Weakened peptidoglycan results in a hernia. c, cytoplasmic membrane; p, peptidoglycan; o, outer membrane. A. Gram positive bacterium. B. Gram negative bacterium.

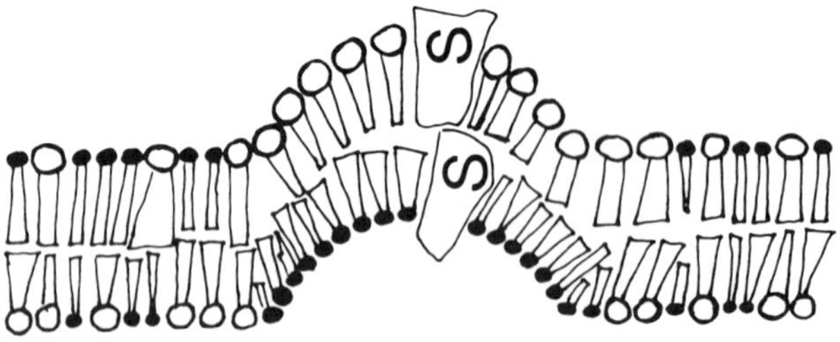

Figure 2. Altered monolayer composition in the hernia. S, enzymes synthesizing peptidoglycan.

REQUIREMENTS OF THE MODEL

1. Turgor pressure to generate the hernia at the site of the weakened wall.
2. A relationship between the curvature of membranes and phospholipid composition.
3. A relationship between phospholipid composition and enzyme activity.
4. A Gram negative sacculus more than one layer thick (at least temporarily or locally).
5. Growth of the peptidoglycan sacculus in patches.

1. TURGOR PRESSURE

In Gram negative bacteria turgor pressure has been calculated as around 3.5 Atmospheres (Stock *et al.*, 1977) whilst in the thicker-walled Gram positive bacteria turgor may be as high as 20 Atmospheres (Mitchell and Moyle, 1956). These latter values have been confirmed by observing the collapse of gas vacuoles in the Gram negative bacterium, *Ancylobacter aquaticus* (Pinette and Koch, 1987). Such pressures have to be resisted by a material with tensile strength, in the case of *E.coli* and almost all eubacteria, peptidoglycan.

In Gram positive bacteria in which the cytoplasmic membrane is restrained by the peptidoglycan wall, a weakening in this wall should lead to a deformation of the cytoplasmic membrane much as the inner tube of a bicycle tyre may protrude through a split in the casing. In Gram negative bacteria, the peptidoglycan sacculus is located in the periplasm between the inner, cytoplasmic, membrane and the outer membrane. The periplasm is believed to be isotonic with the cytoplasm and a Donnan equilibrium exists across the outer membrane (Stock *et al.*, 1977; Kennedy, 1982). Since the distance between cytoplasmic and outer membranes is only 12 nM or so (Burdett and Murray, 1974), a weakening in the peptidoglycan should result in a deformation in both cytoplasmic and outer membranes (Figure 1).

2. CURVATURE OF MEMBRANES AND PHOSPHOLIPID COMPOSITION

All biological membranes contain a variety of lipids which, when purified, adopt either a bilayer or a non-bilayer configuration in an aqueous environment. The molecular characteristics required for non-bilayer formation have been thoroughly studied (Israelachvili *et al.*, 1980; Cullis and de Kruijff, 1979; Gruner, 1985). The configurations adopted by phospholipids depend on the 'critical packing parameter' of the lipid in question (Israelachvili *et al.*, 1980). This parameter, v/a_0l_c describes the 'shape' characteristic of a particular phospholipid. In this expression, v is the volume of the fatty acid chain(s), l_c is the maximum length of the chains and a_0 is the optimal surface area per amphiphile as determined by the volume of the head-group, its hydration, charge, and hydrogen-bonding capabilities. Double-chained lipids with small head-group areas such as unsaturated phosphatidylethanolamine or cardiolipin (diphosphatidylglycerol) plus calcium resemble inverted truncated cones and form inverted micelles whilst single-chained lipids with large head-groups resemble cones and form micelles.

The membranes of *E. coli* and its close relative *Salmonella typhimurium* contain a wide range of lipids and proteins, including phosphatidylethanolamine, phosphatidylglycerol, cardiolipin, phosphatidlylserine, phosphatidic acid and diacylglycerol; the saturated fatty acids of the phospholipids consist of lauric, myristic, palmitic and stearic acids whilst the unsaturated fatty acids consist of palmitoleic and *cis*-vaccenic acids and neutral lipids; the phospholipids may also contain cyclopropane phospholipids (Ames, 1968; Raetz and Dowhan, 1990; Cronan and Vagelos, 1972; Goldfine, 1982). Generally, this heterogeneous population must adopt a bilayer structure. However, if the non-bilayer forming species undergoes lateral segregation, an alteration of normal bilayer structure will occur. A particular species of phospholipids has an intrinsic radius of curvature that represents the tendency of a monolayer of that species to curl in an aqueous environment (Gruner, 1985). Hence a monolayer of a species with a small headgroup and an unsaturated, bulky tail will have a relatively small radius of curvature. Given that it is energetically favourable for a particular phospholipid to form a structure with a small radius of curvature then, conversely, it should also be energetically favourable for a structure with a small radius of curvature to attract the appropriate species of phospholipid. Hence forming the hernia should impose its composition (Figure 2).

3. PHOSPHOLIPID COMPOSITION AND ENZYME ACTIVITY

In eukaryotic cells, the co-localization of particular proteins and phospholipids is a well-established phenomenon (for example, Haverstick and Glaser, 1989; Rodgers and Glaser, 1991). In *E. coli*, the localization of the penicillin binding proteins (PBPs) to particular membrane domains has also been suggested. These proteins direct the synthesis of peptidoglycan with one PBP, PBP3, essential for septation and others, such as PBP2, required for elongation (Spratt, 1975; Matsuhashi *et al.*, 1990). The localisation of different PBPs in different membrane domains is strongly suggested by their associations with different subsets of inner membrane vesicles following electrophoretic separation leading Leidenix *et al.* (1989) to propose that these associations reflect the existence of domains within the membrane specifically concerned with cell elongation and septation. Similar asymmetries in the lipid and protein content of *E. coli* vesicles have also been observed after sucrose gradient centrifugation. Other fractionation studies suggest a large amount of all PBPs in a labile structure which may correspond to the periseptal annuli (Barbas *et al.*, 1986), whilst immuno-electronmicroscopy suggests that PBP1b is localized in small regions which may correspond to adhesion sites (Bayer *et al.*, 1990).

Enzyme activity as well as localization is also sensitive to phospholipid composition (for references see Norris, 1992). This is exemplified by the sensitivity of enzymes synthesizing peptidoglycan precursors to phospholipid composition. The repeating unit of *E. coli* peptidoglycan consists of *N*-acetylglucosamine (NAcGlc) linked to *N*-acetylmuramic acid (NAcMur) pentapeptide (Park, 1987; Rogers *et al.*, 1980). The polysaccharide strands formed have β-(1-4) links between the sugars and are 30 dissaccharide units (30 nm) long on average. The strands are also cross-linked between the short peptides attached to NAcMur. The assembly of the repeating unit takes place on a lipid carrier, bactoprenyl phosphate (Umbreit and Strominger, 1972*a*). The first reaction of the bactoprenyl cycle is the transfer of P-NAcMur pentapeptide from UDP-NAcMur pentapeptide to bactoprenyl phosphate with the release of UMP. The second reaction is the transfer to NAcGlc from UDP-NAcGlc to bactoprenyl-P-P-NAcMur pentapeptide to form the disaccharide pentapeptide repeating unit. The first reaction is catalysed by phospho-NAcMur pentapeptide translocase and the second reaction by UDP-NAcGlc:NAcMur-(pentapeptide)-P-P-bactoprenyl-N-AcGlc transferase. Hence these enzymes play a key role in the initial stages of peptidoglycan synthesis. Both appear sensitive to their lipid microenvironment.

The solubilized translocase from *Micrococcus luteus* was stimulated by the addition of a neutral lipid fraction (Umbreit and Strominger, 1972*b*). The translocase from *Staphylococcus aureus* also shows a dependence on lipids and has been studied in detail by Neuhaus and co-workers. Firstly, they established the microenvironment of the fluorescent lipid intermediate, bactoprenyl-NAcMur-(N^{ϵ}-dansyl)-pentapeptide, with a variety of physical techniques (Weppner and Neuhaus, 1978); secondly, they perturbed the physical state of the membrane lipids by temperature changes and treatment with n-butanol (Weppner and Neuhaus, 1979; Lee *et al.*, 1980). For example, low concentrations (120-180 mM) of n-butanol increased the fluidity of the microenvironment of the enzyme and stimulated its transfer activity by 65%; concentrations outside this range inhibited the enzyme. The authors concluded that 'the physical state of the lipid matrix has a major effect on the catalytic activity of the translocase' (Weppner and Neuhaus, 1979). A similar relationship between n-butanol concentration and translocase activity was also observed by Strominger and co-workers (Matsuhashi *et al.*, 1967). The lipid dependence of the translocase from *E. coli* has also been examined (Geis and Plapp, 1978); in this case it was suggested that phospholipids sensitive to phospholipase A_2 and D are necessary for enzymic activity.

The second enzyme in the bactoprenyl cycle, the transferase, also has an activity stimulated by a crude lipid extract (Taku and Fan, 1976) and it has been speculated that this activity *in vivo* may require a lipid microenvironment similar to that of the translocase (Rogers *et al.*, 1980).

Finally, the synthesis of the lipid carrier, bactoprenyl phosphate, is sensitive to lipid composition (Rogers *et al.*, 1980). The phosphokinase responsible for the phosphorylation of bactoprenol in *S. aureus* revealed an activity stimulated by phosphatidylglycerol more than by phosphatidylethanolamine and stimulated more by cardiolipin with saturated fatty acid chains than by cardiolipin with unsaturated chains (Higashi and Strominger, 1970); subsequent studies showed that, in general, activity was stimulated by lipids which provided a hydrated, loosely packed, highly fluid environment (Gennis and Strominger, 1976).

A relationship between autolysin activity and phospholipid composition is also fundamental to the hernia model. It is therefore encouraging that the activity of a peptidoglycan hydrolase of *E. coli*, *N*-acetylmuramyl-L-alanine amidase, responds *in vitro* and *in vivo* to its phospholipid environment (Vanderwinkel and de Vlieghere, 1985): enzyme activity was increased by phosphatidylglycerol concentrations of up to 50 uM, decreased by phosphatidylglycerol concentrations over 300 uM and unaffected by cardiolipin; according to the authors, these results reflect an *in vivo* control of amidase activity by the surrounding phosphatidylglycerol (Vanderwinkel and de Vlieghere, 1985).

4. THE GRAM NEGATIVE SACCULUS MUST BE MORE THAN ONE LAYER THICK (at least temporarily or locally).

Where the weakened region is being reinforced, the sacculus is probably thicker than a single layer at a hernia. There is evidence to support multi-layered peptidoglycan (Braun *et al.*, 1973; Glauner *et al.*, 1988; Holtje and Glauner, 1990; Amako *et al.*, 1983; Dubochet *et al.*, 1983; Hobot *et al.*, 1984; Leduc *et al.*, 1985) including direct evidence from neutron small angle scattering on isolated sacculi of *E. coli*: this study suggests that 75 to 80% of the sacculus is single layered and 20 to 25% is triple layered (Labischinski *et al.*, 1991). Interestingly, no evidence was found for a region with a double layer or with more than three layers. The implication for the hernia model is that the formation of a hernia must stimulate the immediate synthesis of a triple layer of reinforcing peptidoglycan.

5. GROWTH OF THE PEPTIDOGLYCAN SACCULUS IN PATCHES

The subunits of peptidoglycan are disaccharides which polymerize to form the glycan chains. Each subunit has one short peptide which may cross-link to the peptide of another subunit to make the network of glycan strands cross-linked by dimeric peptide bridges that constitutes peptidoglycan. The fate of these bridges differs considerably depending on whether they are made up of tetramer-trimers or tetramer-tetramers (Holtje and Glauner, 1990). The tetra-tri dimers represent cross-linkage between old and new peptidoglycan and most, if not all, are turned over in a generation. The tetra-tetra dimers represent cross-linkage both between new and new and between new and old peptidoglycan and only half are turned over per generation. One interpretation is that tetra-tri dimers specifically cross-link one layer with another whilst the tetra-tetra dimers cross-link both within and between layers (Holtje and Glauner, 1990). These findings have been accomodated into the 'patches' model in which the synthesis and attachment of a non-stress bearing patch of peptidoglycan to the sacculus is followed by the cleavage of the old stress-bearing bonds (Holtje and Glauner, 1990).

Similar experiments by Burman and Park (1984) have led to the proposal that two glycan strands are incorporated side by side simultaneously along the circumference of a circle in the plane of the short axis of the cell. This process continues until a hoop completely encircles the cell; another pair of strands is then inserted next to the first pair. Cooper and co-workers, however, have proposed that in *S. typhimurium* only one glycan strand is incorporated at a time (Cooper *et al.*, 1988) whilst Nanninga and co-workers have proposed that in *E. coli* single strands are inserted during elongation and multiple strands during septation (De Jonge *et al.*, 1989). None of these mechanisms are incompatible with growth in patches.

DISCUSSION

Several predictions follow from the model. The initial membrane-associated steps in peptidoglycan are executed by the phospho-NAcMur-pentapeptide translocase and the UDP-N-AcGlc:N-AcMur-(penta-peptide)-P-P-bactoprenyl-N-Ac-Glc transferase; hydrolysis is mediated, in part, by *N*-acetyl-muramyl-L-alanine amidase. The activity or partitioning of these enzymes should be shown to depend on the radius of curvature of micelles. The feasability of this approach is demonstrated by studies of the activity of protein kinase C; activity was found to be dependent on micelle formation as determined by the length of the fatty acid chains of phosphatidylcholine (Walker *et al.*, 1990). It is tempting to speculate that the Bayer bridges (Bayer *et al.*, 1982; Kellenberger, 1990), junctions between the inner and outer membranes, correspond to hernias: direct evidence for the existence of hernias may come from improved techniques of electron microscopy. The study of peptidoglycan composition in mutants in which phospholipid composition has been altered may prove particularly rewarding: for example, the absence of the principal phospholipid of *E.coli*, phosphatidylethanolamine, should affect the length and cross-linking of the glycan chains. Altering the balance of phosphatidylglycerol to phosphatidylethanolamine or of saturated to unsaturated fatty acids should directly affect the formation of hernias; small angle neutron scattering of the corresponding sacculi should reveal an alteration in the proportion of the triple-layered structure. This structure should be absent following increases in the osmolarity of the growth medium sufficient to prevent the formation of hernias in mutants defective in osmoregulation.

Membrane phospholipids demonstrate a variety of physical properties including lateral and transverse asymmetry, curvature, and the formation of bilayer and non-bilayer structures. Such properties, which may regulate fundamental biological processes (Cullis and de Kruijff, 1979; Gruner, 1985; Lee *et al.*, 1980; Rand and Parsegian, 1986; Edidin, 1987; Tocanne *et al.*, 1989; Lindblom and Rifors, 1989; Zachowski and Devaux, 1990; Vaz *et al.*, 1989; Norris, 1989, 1990, 1992), include one rarely mentioned, the curvature of bilayers. In its present form the hernia model invokes curvature to explain only one aspect of wall synthesis: how the sacculus may be repaired. It is therefore compatible with the current paradigm for peptidoglycan synthesis, the surface stress theory (Koch, 1983, 1990) and its variants (Cooper, 1991). In later communications, we intend to explore the possibility that alterations in bilayer curvature and the activity of contractile proteins control cell shape itself.

ACKNOWLEDGMENTS

We thank H. Labischinski, J.-V. Holtje, A. Koch, A. Zachowski, S. Cooper, J. van Heijenoort, M. E. Bayer, P. H. Williams, T. J. Baldwin and M. Pocklington for encouragement and criticisms, and the Wellcome Trust and Cell Cycle Research for support.

REFERENCES

Amako, K., Murata, K. and Umeda, A. 1983. Structure of the envelope of *Escherichia coli* observed by rapid-freezing and substitution fixation method. Microbiol. Immunol. 27, 95-99.

Ames, G.F. 1968. Lipids of *Salmonella typhimurium* and *Escherichia coli*: structure and metabolism. J.Bacteriol. 95, 833-843.

Barbas, J. A., Diaz, J., Rodriguez-Tebar, A. and Vasquez, D. 1986. Specific location of penicillin-binding proteins within the cell envelope of *Escherichia coli*. J. Bacteriol. 165, 269-275.

Bayer, M. H., Costello, G. P. and Bayer, M. E. 1982. Isolation and partial characterization of membrane vesicles carrying markers of the membrane adhesion sites. J. Bacteriol. 149, 758-767.

Bayer, M. H., Keck, W. and Bayer, M. E. 1990. Localization of penicillin-binding protein 1b in *Escherichia coli*: immunoelectron microscopy and immunotransfer studies. J. Bacteriol. 172, 125-135.

Braun, V., Gnirke. H., Henning, U. and Rehn, K. 1973. Model for the structure of the shape-maintaining layer of the *Escherichia coli* envelope. J. Bacteriol. 114, 1264-1270.

Burdett, I.D.J. and Murray, R.G.E. 1974. Electron microscope study of septum formation in *Escherichia coli* strains B and B/r during synchronous growth. J. Bacteriol. 119, 1039-1056.

Burman, L.G. and Park, J.T. 1984. Molecular model for elongation of the murein sacculus of *Escherichia coli*. Proc. Natl. Acad. Sci. USA 81, 1844-1848.

Cooper, S. (1991) Synthesis of the cell surface during the division cycle of rod-shaped, gram-negative bacteria. Microbiol. Rev. 55, 649-674.

Cooper, S., Hsieh, M.-L. and Guenther, B. 1988. Mode of peptidoglycan synthesis in *Salmonella typhimurium*: single-strand insertion. J. Bacteriol. 170, 3509-3512.

Cronan, J.E. and Vagelos, P.R. 1972. Metabolism and function of the membrane phospholipids of *Escherichia coli*. Biochim. Biophys. Acta 265, 25-60.

Cullis, P.R. and de Kruijff, B. 1979. Lipid polymorphism and the functional roles of lipids in biological membranes. Biochim. Biophys. Acta 559, 399-420.

De Jonge, B.L.M., Wientjes, F.B., Jurida, I., Driehuis, F., Wouters, J.T.M. and Nanninga, N. 1989. Peptidoglycan synthesis during the cell cycle of *Escherichia coli*: composition and mode of insertion. J. Bacteriol. 171, 5783-5794.

Dubochet, J., McDowall, A.W., Menge, B., Schmid, E.N. and Lickfeld, K.G. 1983. Electron microscopy of frozen-hydrated bacteria. J. Bacteriol. 155, 381-390.

Edidin, M. 1987. Rotational and lateral diffusion of membrane proteins and lipids: phenomena and function. Curr. Top. Membr. Trans. 29, 91-127.

Geis, A. and Plapp, R. 1978. Phospho-*N*-Acetylmuramoyl-pentapeptide-tranferase of *Escherichia coli* K12: properties of the membrane-bound and the extracted and partially purified enzyme. Biochim. Biophys. Acta 527, 414-424.

Gennis, R.B. and Strominger, J.L. 1976. Activation of C_{55}-isoprenoid alcohol phosphokinase from *Staphylococcus aureus*. J. Biol. Chem. 251, 1264-1269.

Glauner, B., Holtje, J.-V. and Schwartz, U. 1988. The composition of the murein of *Escherichia coli*. J. Biol. Chem. 263, 10088-10095.

Goldfine, H. 1982. Lipids of prokaryotes - structure and distribution. In Curr. Top. Membr. Trans. Razin, S. and Rottem, S., editors. Academic Press, 1-43.

Gruner, S.M. 1985. Intrinsic curvature hypothesis for biomembrane lipid composition: a role for nonbilayer lipids. Proc. Nat. Acad. Sci. USA 82, 3665-3669.

Haverstick, D.M. and Glaser, M. 1989. Influence of proteins on the reorganization of phospholipid bilayers into large domains. Biophys. J. 55, 677-682.

Higashi, Y. and Strominger, J.L. 1970. Biosynthesis of the peptidoglycan of bacterial cell walls: identification of phosphatidylglycerol and cardiolipin as cofactors for isoprenoid alcohol phosphokinase. J. Biol. Chem. 245, 3691-3696.

Hobot, J.A., Carlemalm. E., Villiger, W. and Kellenberger, E. 1984. Periplasmic gel: new concept resulting from the reinvestigation of bacterial cell envelope ultrastructure by new methods. J. Bacteriol. 160, 143-152.

Holtje, J.-V. and Glauner, B. 1990. Structure and metabolism of the murein sacculus. Res. Microbiol. 141, 75-89.

Israelachvili, J.N., Marcelja, S. and Horn, R.G. 1980. Physical principles of membrane organization. Quart. Rev. Biophys. 13, 121-200.

Kellenberger, E. 1990. The "Bayer bridges" confronted with results from improved electron microscopy methods. Mol. Microbiol. 4, 697-705.

Kennedy, E.P. 1982. Osmotic regulation and the biosynthesis of membrane-derived oligosacchides in *Escherichia coli*. Proc. Natl. Acad. Sci. USA 79, 1092-1095.

Koch, A.L. 1983. The surface stress theory of microbial morphogenesis. Advances Microbiol. Physiol. 24, 301-366.

Koch, A. L. 1990. Additional arguments for the key role of "smart" autolysins in the enlargement of the wall of Gram-negative bacteria. Res. Microbiol. 141, 529-541.

Labischinski, H. Goodell, E.W., Goodell, A. and Hochberg, M.L. 1991. Direct proof of a "more-than-single-layered" peptidoglycan architecture of *Escherichia coli* W7: a neutron small-angle scattering study. J. Bacteriol. 173, 751-756.

Leduc, M., Frehel, C. and van Heijenoort, J. 1985. Correlation between degradation and ultrastructure of peptidoglycan during autolysis of *Escherichia coli*. J. Bacteriol. 161, 627-635.

Lee, P.P., Weppner, W.A. and Neuhaus, F.C. 1980. Initial membrane reaction in peptidoglycan synthesis: perturbation of lipid-phospho-*N*-acetylmuramyl-pentapeptide translocase interactions by *n*-butanol. Biochim. Biophys. Acta 597, 603-613.

Leidenix, M. J., Jacoby, G. H., Henderson, T. A. and Young, K. D. 1989. Separation of *Escherichia coli* penicillin-binding proteins into different membrane vesicles by agarose electrophoresis and sizing chromatography. J. Bacteriol. 171, 5680-5686.

Lindblom, G. and Rifors, L. 1989. Cubic phases and isotropic structures formed by membrane lipids - possible biological relevance. Biochim. Biophys. Acta 988, 221-256.

Matsuhashi, M. Dietrich, C.P. and Strominger, J.L. 1967. Biosynthesis of the peptidoglycan of bacterial cell walls: III the role of soluble ribonucleic acid and of lipid intermediates in glycine incorporation in *Staphylococcus aureus*. J. Biol. Chem. 242, 3191-3206.

Matsuhashi, M., Wachi, M. and Ishino, F. 1990. Machinery for cell growth and division: penicillin-binding proteins and other proteins. Res. Microbiol. 141, 89-103.

Mitchell, P. and Moyle, J. 1956. Osmotic structure and function in bacteria. Symp. Soc. Gen. Microbiol. 6, 150-180.

Norris, V. 1989. Phospholipid flip-out controls the cell cycle of *Escherichia coli.* J. Theor. Biol. 139, 117-128.

Norris, V. 1990. DNA replication in *Escherichia coli* is initiated by membrane detachment of *oriC.* J. Mol. Biol. 215, 67-71.

Norris, V. 1992. Phospholipid domains determine the spatial organization of the *E.coli* cell cycle: the membrane tectonics model. J. Theor. Biol., 154, 91-107.

Park, J.T. 1987. Murein synthesis In *Escherichia coli* and *Salmonella typhimurium.* F.C.Neidhardt editor. ASM, Washington, D.C. 663-671.

Pinette, M.F.S. and Koch, A.L. 1987. The stress in the Gram-negative wall is constant throughout the cell cycle of *Ancyclobacter aquaticus.* J. Bacteriol. 169, 4737-4742.

Raetz, C.R.H. and Dowhan, W. 1990. Biosynthesis and function of phospholipids in *Escherichia coli.* J. Biol. Chem. 265, 1235-1238.

Rand, R.P. and Parsegian, V.A. 1986. Mimicry and mechanism in phospholipid models of membrane fusion. Ann. Rev. Physiol. 48, 201-212.

Rodgers, W. and Glaser, M. 1991. Characterization of lipid domains in erythrocyte membranes. Proc. Natl. Acad. Sci. USA. 88, 1364-1368.

Rogers, H.J., Perkins, H.R. and Ward, J.B. 1980. Microbial cell walls and membranes. Chapman and Hall, London.

Spratt, B. G. 1975. Distinct penicillin binding proteins involved in the division, elongation, and shape of *Escherichia coli* K12. Proc. Natl. Acad. Sci. USA. 72, 2999-3003.

Stock, J.B., Rauch, B. and Roseman, S. 1977. Periplasmic space in *Salmonella typhimurium* and *Escherichia coli.* J. Biol. Chem. 252, 7850-7861.

Taku, A. and Fan, D.P. 1976. Identification of an isolated protein essential for peptidoglycan synthesis as the *N*-acetylglucosaminyltransferase. J. Biol. Chem. 251, 6154-6156.

Tocanne, J.-F, Dupou-Cezanne, L., Lopez, A. and Tournier, J-F. 1989. Lipid lateral diffusion and membrane organization. FEBS Lett. 257, 10-16.

Umbreit, J.N. and Strominger, J.L. 1972a. Isolation of polyisoprenyl alcohols from *Streptococcus faecalis.* J. Bacteriol. 112, 1302-1305.

Umbreit, J.N. and Strominger, J.L. 1972b. Complex lipid requirements for detergent-solubilized phosphoacetylmuramyl-pentapeptide translocase from *Micrococcus luteus.* Proc. Natl. Acad. Sci. USA. 69, 1972-1974.

Vanderwinkel, E. and de Vlieghere, M. 1985. Modulation of *Escherichia coli N*-acetylmuramyl-*L*-alanine amidase activity by phosphatidylglycerol. Biochim. Biophys. Acta 838, 54-59.

Vaz, W.L.C., Melo, E.C.C. and Thompson, T.E. 1989. Translational diffusion and fluid domain connectivity in a two-component, two-phase phospholipid bilayer. Biophys. J. 56, 869-876.

Walker, J. M., Homan, E. C. and Sando, J. J. 1990. Differential activation of protein kinase C isozymes by short chain phosphatidylserines and phosphatidylcholines J. Biol. Chem. 265, 8016-8021.

Weppner, W.A. and Neuhaus, F.C. 1978. Biosynthesis of peptidoglycan: definition of the microenvironment of undecaprenyl diphosphate-*N*-acetylmuramyl-(5-dimethyl-amino-naphthalene-1-sulfonyl) pentapeptide by fluorescence spectroscopy. J. Biol. Chem. 253, 472-478.

Weppner, W.A. and Neuhaus, F.C. 1979. Initial membrane reaction in peptidoglycan synthesis: interaction of lipid with phospho-*N*-acetylmuramyl-pentapeptide translocase. Biochim. Biophys. Acta 552, 418-427.

Zachowski, A. and Devaux, P.F. 1990. Transmembrane movements of lipids. Experientia 46, 645-656.

WALL TEICHOIC ACID, PEPTIDOGLYCAN SYNTHESIS AND MORPHOGENESIS IN *BACILLUS SUBTILIS*

Harold M. Pooley, François-Xavier Abellan and Dimitri Karamata

Institut de génétique et de biologie microbiennes
Rue César-Roux 19, 1005 Lausanne, Switzerland

INTRODUCTION

Ever since morphological mutants in several rod-shaped bacteria were isolated (Boylan and Mendelson, 1969 ; Rogers *et al.*, 1970 ; Satta *et al.*, 1969), the question of cell shape maintenance has been and remains a central one. We examine a possible role in cell morphogenesis of the interdependence of the syntheses of peptidoglycan and other cell wall components.

In *B. subtilis*, studies of the macromolecular biosynthesis and cell morphology in several such mutants (Rogers *et al.*, 1974 ; Boylan *et al.*, 1972 ; Cole *et al.*, 1970) did not clarify the picture of the shape maintenance mechanisms mainly, we believe, through lack of knowledge of the functions encoded by the mutated genes.

An interest of this laboratory in the wall teichoic acids led to evidence for their having a potentially essential role in normal cell growth (Karamata *et al.*, 1987 ; Briehl *et al.*, 1989 ; Mauël *et al.*, 1989). The previously observed (Boylan *et al.*, 1972 ; Rogers and Taylor, 1978) deficiencies in teichoic acid synthesis in many mutants was confirmed (Briehl *et al.*, 1989), and further work of this (Mauël *et al.*, 1991), as well as other laboratories (Honeyman and Stewart, 1989), led to the sequencing of six genes, organized in two divergently transcribed operons (Mauël *et al.*, 1991), which are involved in the synthesis of poly(glycerolphosphate) in *B. subtilis* 168.

Following the recent identification of the enzyme activities encoded by two of the genes, *tagD* (Pooley *et al.*, 1991) and *tagF* (Pooley *et al.*, 1992), a re-examination appeared worthwhile of the effect on macromolecular synthesis and on morphological parameters, of the collapse of an enzyme specifically involved in synthesis of the major cell wall teichoic acid. Rather than following the late, cumulative effects, we were interested in defining the immediate consequences of this block, both as a possible means of interpreting the complex morphological changes appearing after prolonged mutational expression, and more crucially, as an approach to understanding the biological functions of this polymer, particularly in overall cell surface expansion.

Bacterial Growth and Lysis, Edited by M.A. de Pedro
et al., Plenum Press, New York, 1993

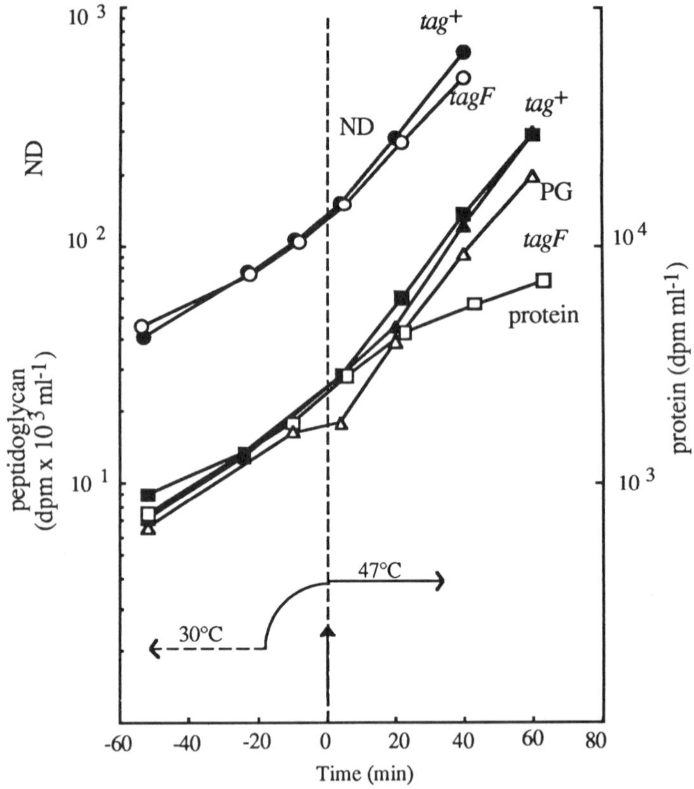

Figure 1. Cumulative peptidoglycan (PG) and protein synthesis in *B. subtilis* wild type L5087 *tag*$^+$ and an isogenic mutant L6477 *tagF1* strain during growth at 30 °C and following a shift to the non permissive temperature (47 °C). Cultures were inoculated at an ND of 5 (approximately 1 x 10^7 bacteria per ml) in SAT medium containing unlabelled N-acetyl glucosamine (GlcNAc, 300 μM) supplemented with either N-acetyl (^{14}C) glucosamine (1 μCi per ml) or ^{14}C-leucine (0.06 μCi/ml). After 4,5 generations at 30 °C, the growth temperature was raised to 47 °C over a ten to twelve minute period, and, upon reaching 47 °C, cultures were diluted to an ND of 10. Samples (1 ml) were taken at the indicated times and cells were removed by filtration, washed twice with medium containing carrier GlcNAc (10 mM) or leucine (1 mM) and twice with water. Hexosamine containing samples were treated to remove label in the galactosamine glucose phosphate teichoic acid, as previously described (Pooley *et al.*, 1987). Label remaining after this treatment was considered to be in peptidoglycan. Radioactivity (dpm) was determined as described (Pooley *et al.*, 1991). After correction for dilutions, peptidoglycan synthesis (Δ,▲) and protein synthesis (❑,■) are shown for mutant L6477 tagF1 (Δ,❑) and wild type (▲,■). Bacterial mass was measured as nephelometric density, ND (O,●).

RESULTS

Major macromolecular syntheses during the initial phase following a block in wall teichoic acid synthesis in *B. subtilis* 168

Immediately (T_0) following expression (at 47 °C) of the Ts mutations *tagF1, tagD11*, or *tagB1,* wall teichoic acid synthesis stops, as judged by [^{3}H-2C]glycerol incorporation into the cell wall fraction (not shown).

During the first three generations (60 min), following the block in synthesis of the major wall teichoic acid (Fig. 1, mutant L6477 *tagF1*), peptidoglycan synthesis continues at a rate that does not apparently deviate significantly from that of the wild type. However, a transient slowing down takes place immediately after shift to 47 °C.

Globally, mass increase continues, although there are indications of a slowing down towards the end of the 60 min period (at 47 °C). This is accounted for by an effect on the rate of protein synthesis which apparently does not increase after T_0 ; incorporation of label into protein proceeding approximately linearly. This is consistent with a total block of new ribosome formation and continuing synthesis from the existing ribosome pool.

The relative effects on protein and peptidoglycan synthesis revealed by this experiment are in agreement with an earlier report of the increase, at 47 °C, in the ratio of PG to protein synthesis (Rogers *et al.*, 1974). However, they do not give support to the conclusion of a derepression of PG synthesis. The changed ratio is accounted for rather by a failure of protein synthesis to increase in step with that of PG.

In view of the apparent pause in global peptidoglycan counts shown by the mutant L6477 *tagF1* at T_0 (Fig. 1), pulse labelling experiments were performed to measure the rate of peptidoglycan synthesis during the initial phase following arrest of teichoic acid synthesis.

Transfer of the wild type strain from 30 to 47 °C, over a 12 min period, was accompanied by an increased rate of PG synthesis measured at T_0 (Fig. 2). One generation time later, at T_{18}, the rate of PG synthesis was practically unchanged and at an intermediate time, T_9, synthesis showed a small decrease relative to T_0. In a second experiment (not shown), the decrease between T_0 and T_9 was 30 %. Subsequently, between T_{18} and T_{36}, the rate of PG synthesis increased in parallel with that of mass increase.

In two mutant strains, L6477 *tagF1* and L6476 *tagB1*, PG synthesis showed, relative to the wild type, a major 40 to 60 % inhibition at T_0 and T_9, in excellent agreement with the pause in total peptidoglycan labelling revealed for L6477 (Fig. 1), as well as for L6476 *tagB1*, and L6601 *tagD11* (not shown). Subsequently, there were marked increases in the rate of PG synthesis for both mutants. That for *tagB1* increased threefold between T_9 and T_{18}, and that for *tagF1* nearly doubled between T_{18} and T_{27}. The rates attained in the mutants became equal to that of the wild type at these latter times. These results, in agreement with the cumulative labelling experiment (Fig. 1), reveal that a block in teichoic acid synthesis is immediately followed by a phase lasting about one doubling time, during which there was no increase, even a slight fall, in PG synthesis.This period is followed by a very rapid increase in rate, which attains that of a similarly treated wild type strain.

To correlate the effects on macromolecular syntheses observed here with possible consequences for cell morphogenesis, we examined the changes in cell length and in cell number in the initial period following inhibition of wall teichoic acid synthesis, *i.e.* before the appearance of dramatic effects on cell morphology(Rogers *et al.*, 1970 ; Cole *et al.*, 1970).

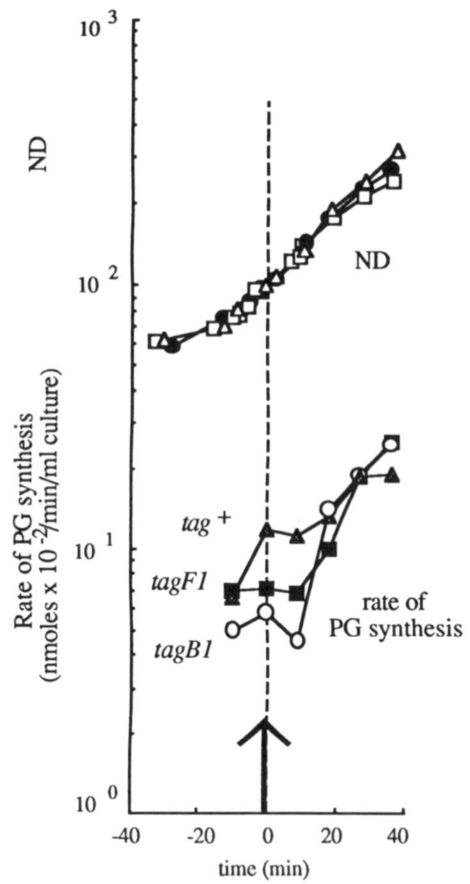

Figure 2. Rate of synthesis of PG following inhibition of teichoic acid synthesis in mutants shifted from 30 °C to 47 °C. 15 ml cultures of wild type L5087 *tag*$^+$ and mutants L6476 *tagB1* and L6477 *tagF1* were grown for 4 generations at 30 °C on SAT medium supplemented with N-acetyl glucosamine (GlcNAc, 60 μM) and $MgCl_2$ (2.5 mM). At ND of 80, the growth temperature was raised to 47 °C over a 10 min period, and growth continued. At the times shown, 200 μl samples were withdrawn and added to tubes containing 800 μl of preheated SAT medium supplemented with GlcNAc (20 μM) and [^{3}H]GlcNAc (3 μCi). After 180 seconds, incorporation was halted by immersion for 5 min in a boiling water bath. The radioactivity remaining on a membrane filter after washing twice with medium (2 ml) containing GlcNAc (1 mM) and with water, was considered to be in peptidoglycan. Approximately 5 % of the radioactivity enters protein (Pooley, 1976) and the fraction of the [^{3}H]GlcNAc label in the glucose galactosamine phosphate wall polymer is 2, and 20 %, at 47 °C, and 30 °C, respectively. Samples of cultures growing at 30 °C were incubated for 360 seconds in the presence of [^{3}H]GlcNAc. The mass doubling times for the *tag*$^+$ strain were approximately 38 and 16 min at 30 °C and 47 °C, respectively. To prevent GlcNAc exhaustion from the culture medium, 8.9 and 7.3 μl of GlcNAc (100 mM) were added (equivalent to 60 μM) at ND 90 and ND 190, respectively. The rate of PG synthesis is shown after correction for specific activity differences. The total GlcNAc concentration present during label uptake was estimated from standard curves for cell concentration dependence of GlcNAc incorporation ; 60 μM yielded a linear incorporation of label up to DN of 110. Label incorporated during pulses at the times indicated by L6476 *tagB1* (○), L6477 *tagF1* (■) and for L5087 *tag*$^+$ (▲) are shown. The bacterial mass, measured as ND, for the above strains, was indicated, respectively, by (●,❑ and Δ).

The consequences of an arrest of wall teichoic acid synthesis on cell number, mean cell length and overall cell length

Following transfer of the wild type to 47 °C, cell numbers per ml (Fig. 3) continued to increase in step with mass increase, while, for at least a generation after reaching 47 °C, mean cell length showed little variation from pre-shift values . Subsequently, by T_{38}, there may be a slight increase in cell length. As expected, the overall cell length per ml (the product of cell concentration and mean cell length) increased in parallel with cell numbers and mass.

From T_{-20} to T_9, when wall teichoic acid synthesis was blocked in mutant L6477 *tagF1* by transfer to 47 °C, cell numbers ml^{-1} increased more rapidly than for the wild type strain. After T_9, there appeared to have been a stagnation in cell numbers for about a generation before an increase was observed at T_{36}. Blocking teichoic acid synthesis had a marked effect on mean cell length, which fell from 2.8 µm at T_{-12} to 1.7 at T_{12}. Subsequently, up to T_{36}, there was little change. The rapid increase in cell numbers correlates well with the decrease in mean cell length. Both demonstrate that septum formation and cell division continue for at least half a generation following a block in teichoic acid synthesis. Thus, the cell has some potential for division that is not dependent upon continuing teichoic acid synthesis.

The rapid decrease in mean cell length, which accompanied continuing cell division in the mutant, suggested that cell elongation might not have been proceeding normally. When the total bacterial length is determined (as the product of cell numbers/ml and mean cell length), there appears to have been no increase, for more than one mass doubling time after T_0. This result, in clear contrast to the wild type, strongly suggests that cell elongation is blocked, either immediately or very shortly after the block in teichoic acid synthesis. The effects on cell numbers, mean cell length and overall bacterial length ml^{-1} are all accounted for by a block in cell elongation. It is noteworthy that, up to T_{18}, continuing mass increase and cell divisions with only limited increase in cell length produce little effect on cell morphology (not shown), whereas their prolonged action leads to dramatic deformations.

Table 1.

Sequence of events following collapse of poly(glycerolphosphate) polymerase (TagF) activity (transfer of L6477 *tagF1* to restrictive growth temperature, 47 °C)	References
(i) Arrest of teichoic acid polymerization	This study, data not presented.
(ii) Accumulation of CDP-glycerol	Pooley *et al.*, 1991.
(iii) Continuing septation and cell division	This study, Fig. 3.
(iv) Slow down or arrest of peptidoglycan synthesis involved in cylinder extension, blocking the latter process	This study, Figs. 1, 2, 3.
(v) After a delay of up to a generation, rapid increase in peptidoglycan synthesis	This study, Figs. 1, 2.
(vi) Can	
- the progressive slowing down in protein synthesis	This study.
and	
- the subsequent cell deformations	Rogers *et al.*, 1970 ; Cole *et al.*, 1970
be consequences of continuing mass increase but limited cell length increase ?	This study.

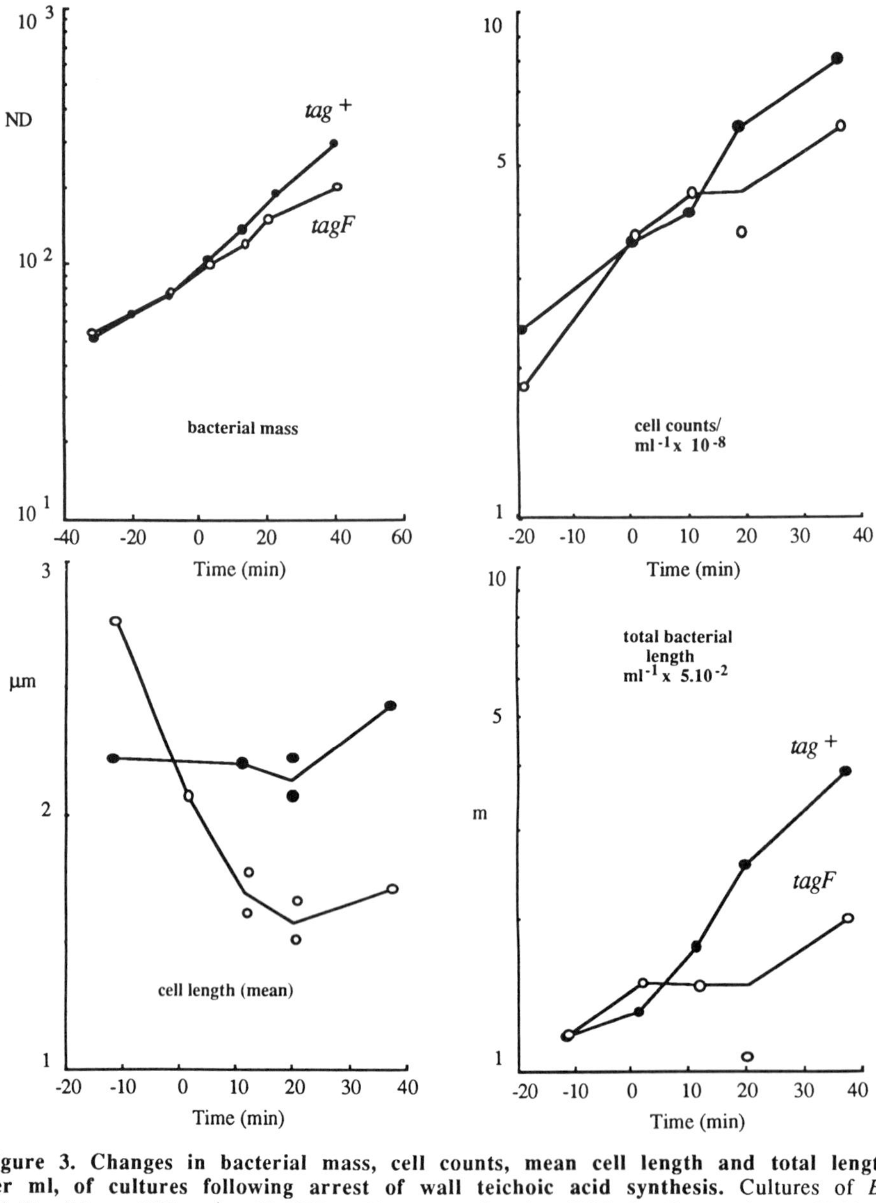

Figure 3. Changes in bacterial mass, cell counts, mean cell length and total length per ml, of cultures following arrest of wall teichoic acid synthesis. Cultures of *B. subtilis* wild type 5087 *tag*$^+$ and L6477 *tagF1* were transferred to 47 °C after 5 generations at 30 °C, as described in Fig. 1. At indicated times, samples were withdrawn to determine a) cell number, (900 µl were added to 100 µl of 20 % formaldehyde) and b) to measure cell length. Cells in 2 ml samples were filtered on a membrane filter (Sartorius), washed with water, then, having broken the vacuum, treated for 1 min with trichloro-acetic (TCA) acid (5 %) before being washed again with water. After resuspension in 500 µl of a trypsin (1 mg/ml) solution in phosphate buffer (0.05 M, pH 7.5), cells were incubated at 37 °C for 30 min, then spun down and resuspended in 30 µl of the same buffer. Cell counts were made immediately on formaldehyde fixed populations of at least 1500 cells with a Helber counter. Mean cell length for each sample was determined from light microscope photographs (Polaroid) of at least 300 cells after trypsin digestion, which allowed better visualisation of septa. Shown, for *tag*$^+$ (●) and mutant (○), respectively, are, ND, mean cell length, cell counts ml^{-1} and the total length of the population, determined as the product of the previous two measurements. Arrows show the time of reaching 47 °C.

DISCUSSION

At a general level, results presented here above provide a rather coherent image of the consequences for cell morphogenesis of arresting wall teichoic acid synthesis (Table 1). They strongly suggest that the key effect is that of inhibiting the activity of the cell wall elongation sites. It appears highly probable that the observed inhibition in overall PG synthesis represents a block in PG incorporation at these sites. This effect can be correlated with a linear and not an exponentially increasing rate of protein synthesis, by a block in the formation of new ribosomes, their synthesis being coupled to continuing volume increase through cell elongation. Septum formation and cell division, which continue after arrest of wall teichoic acid synthesis, possibly at an increased rate, can account for the important residual PG synthesis during this period.

The effect of stopping cell elongation could be accounted for in several ways. One, is that a stalled teichoic acid polymerase in *tagF* mutants could lead to sequestration of all the lipid carrier molecules, thus inhibiting PG incorporation. A second, could be that the cell wall biosynthetic machinary of the cell elongation sites, which links de novo teichoic acid chains to simultaneously made PG chains (Mauck and Glaser, 1972), is unable to function properly in the absence of teichoic acid synthesis. A third possibility is that the process of cell wall extension could require teichoic acid synthesis to achieve the spreading inherent to cell elongation (Pooley, 1976). The data available at present do not allow evaluation of these possibilities. Another way of accounting for our observations comes from the model of Satta and colleagues (Lleo *et al.*, 1990), involving competition between sites for cell wall elongation and that for septation. The effect of *tagF* or other *tag* mutations can be formally considered as an inhibition of sites for lateral wall formation or as a hyperactivity of septal formation sites. It has been shown that inhibition of the septal site by antibiotics or mutation, can favour lateral wall elongation. In the case of certain coccoid morphology mutants, such an inhibition can restore normal rod shaped morphology (Lleo *et al.*, 1990). Reeve *et al.* (1972) have performed an experiment of this type for *tagB1* by introducing a ts*dnaB* mutation ; the latter mutation blocks cell division, while allowing cylinder elongation. At the restrictive temperature, the double mutant formed elongated cells - but electron micrographs showed that both cell diameter and cell wall thickness were extremely irregular. Furthermore, the initial growth rate was at least 30 % slower than that conferred by the *tagB1* mutation alone, suggesting that cell volume increases more slowly when cell septation is inhibited. The results gave little support to the idea that inhibition of septation in these mutants facilitates lateral wall elongation.

CONCLUSIONS

During the first generation following collapse of a defective teichoic acid specific enzyme in *B. subtilis* 168, long before the spectacular effects on cell morphology are apparent, a block in cell elongation occurs, while cell division proceeds more or less normally. We envisage that, in the absence of the major wall teichoic acid synthesis, the process of surface enlargement in *B. subtilis* in the form of a cylinder of constant diameter, cannot take place.

ACKNOWLEDGEMENTS

We thank Natacha Ropiteaux for competent technical help.

REFERENCES

Boylan, R.J., and Mendelson, N.H. (1969) Initial characterization of a temperature-sensitive rod⁻ mutant of *Bacillus subtilis*. J. Bacteriol. 100, 1316-1321.

Boylan, R.J., Mendelson, N.H., Brooks, D., and Young, F.E. (1972) Regulation of the bacterial cell wall : analysis of a mutant of *Bacillus subtilis* defective in biosynthesis of teichoic acid. J. Bacteriol. 110, 281-290.

Briehl, M., Pooley, H.M., and Karamata, D. (1989) Mutants of *Bacillus subtilis* 168 thermosensitive for growth and wall teichoic acid synthesis. J. Gen. Microbiol. 135, 1325-1334.

Cole, R.M., Popkin, T.J., Boylan, R.J., and Medelson, N.H. (1970) Ultrastructure of a temperature-sensitive *rod⁻* mutant of *Bacillus subtilis*. J. Bact. 103, 793-810.

Honeyman, A.L., and Stewart, G.C. (1989) The nucleotide sequence of the *rodC* operon of *Bacillus subtilis*. Molec. Microbiol. 3, 1257-1268.

Karamata, D., Pooley, H.M., and Monod, M. (1987) Expression of heterologous genes for wall teichoic acid in *Bacillus subtilis* 168. Mol. Gen. Genet. 207, 73-81.

Lleo, M.M., Canepari, P., and Satta, G. (1990) Bacterial cell shape regulation : testing of additional predictions unique to the two-competing-sites model for peptidoglycan assembly and isolation of conditional rod-shaped mutants from some wild-type cocci. J. Bacteriol. 172, 3758-3771.

Mauck, J., and Glaser, L. (1972) On the mode of *in vivo* assembly of the cell wall of *Bacillus subtilis*. J. Biol. Chem. 247, 1180-1187.

Mauël, C., Young, M., Margot, P., and Karamata, D. (1989) The essential nature of teichoic acids in *Bacillus subtilis* as revealed by insertional mutagenesis. Mol. Gen. Genet. 215, 388-394.

Mauël, C., Young, M., and Karamata, D. (1991) Genes concerned with synthesis of poly(glycerol phosphate), the essential teichoic acid in *Bacillus subtilis* strain 168, are organized in two divergent transcription units. J. Gen. Microbiol. 137, 929-941.

Pooley, H.M. (1976) Turn-over and spreading of old wall during surface growth of *Bacillus subtilis*. J. Bacteriol. 125, 1127-1138.

Pooley, H. M., Paschoud, D., and Karamata, D. (1987). The *gtaB* marker in *Bacillus subtilis* 168 is associated with a deficiency in UDP-glucose pyrophosphorylase. *Journal of General Microbiology* **133**, 3481-3493.

Pooley, H.M., Abellan, F.X., and Karamata, D. (1991) A conditional-lethal mutant of *Bacillus subtilis* 168 with a thermosensitive glycerol-3-phosphate cytidylyl transferase, an enzyme specific for the synthesis of the major wall teichoic acid. J. Gen. Microbiol. 137, 921-928.

Pooley, H.M., Abellan, F.X., and Karamata, D. (1992) CDP-glycerol:poly(glycerolphosphate)phosphoglycerotransferase, involved in the synthesis of the major wall teichoic acid in *Bacillus subtilis* 168, is encoded by *tagF(rodC)*. J. Bacteriol. 174, 646-649.

Reeve, J.N., Mendelson, N.H., and Cole, R.M. (1972) Cell morphology of *Bacillus subilis* : the effect of genetic background on the expression of a Rod⁻ gene. Mol. Gen. Genet. 119, 11-26.

Rogers, H.J., McConnell, M., and Burdett, I.D.J. (1970) The isolation and characterization of mutants of *Bacillus subtilis* and *Bacillus licheniformis* with disturbed morphology and cell division. J. Gen. Microbiol. 61, 155-171.

Rogers, H.J., Thurman, P.F., Taylor, C., and Reeve, J.N. (1974) Mucopeptide synthesis by *rod* mutants of *Bacillus subtilis*. J. Gen. Microbiol. 85, 335-350.

Rogers, H.J., and Taylor, C. (1978) Autolysins and shape change in *rodA* mutants of *Bacillus subtilis*. J. Bacteriol. 135, 1032-1042.

Satta, G., Schito, G.C., and Meloni, G.A. (1969) Transizione bastoncinosfera in un ceppo di *Klebsiella pneumoniae*. Ultrastruttura delle forme coccoidie tipiche ed anormi. Atti del XV° Congresso Nazionale di Microbiologia Torina-Saint Vincent 2, 247-253.

PENICILLIN-INDUCED BACTERIOLYSIS OF STAPHYLOCOCCI AS A POST-MORTEM CONSEQUENCE OF MUROSOME-MEDIATED KILLING VIA WALL PERFORATION AND ATTEMPTS TO IMITATE THIS PERFORATION PROCESS WITHOUT APPLYING ANTIBIOTICS

Peter Giesbrecht, Thomas Kersten, Kazimierz Madela, Harald Grob
Peter Blümel, and Jörg Wecke

Robert Koch-Institute
of the Federal Health Office
Nordufer 20
D-1000 Berlin 65, Germany

In the early years of microbiological research, scientists coined the term "bacteriolysis" to describe the loss of turbidity of a bacterial culture. Viewing such a population under a microscope revealed that most or even all of the bacteria had disappeared. Today, the definition of bacteriolysis is basically still the same. It is understood as the disintegration of the bacterial shape which is accompanied by a reduced optical density of the culture and by a decline of the number of cells. In this contribution, we are using the term "bacteriolysis" strictly in this sense.

Bacteriolysis ("lytic death") is often considered to be the most characteristic aspect of penicillin action. Since a considerable number of rather different ideas has been published about the possible mechanism of this event (see Shockman et al., 1981 for review), we have recently analyzed it in more detail.

INHIBITION OF BACTERIOLYSIS, DUE TO WALL STABILIZATION

During treatment of staphylococci with 0.1 µg of penicillin G per ml, an optimal bacteriolysis took place already 70 - 90 min after application of the drug (Wecke et al., 1987). To analyze this effect, experiments were performed with substances that proved to inhibit the penicillin-induced bacteriolysis without interfering with the growth of untreated staphylococci. Especially some anionic polyelectrolytes (Liquoid, Evans blue, Heparin, Suramin) inhibited the penicillin-induced bacteriolysis of staphylococci instantly and very effectively and at any stage during this process (Wecke et al., 1986, 1990). After treatment with penicillin alone, the disintegration of the staphylococcal cells could easily be observed in the electron microscope (Wecke et al., 1987). However, the simultaneous addition of an anionic polyelectrolyte did not only inhibit but even prevent bacteriolysis so that hardly any

Bacterial Growth and Lysis, Edited by M.A. de Pedro
et al., Plenum Press, New York, 1993

disintegrated staphylococci could be detected (Giesbrecht et al., in press). It could be shown that this inhibition of bacteriolysis was the result of wall stabilization owing to a reduced degradability of the cell wall against autolytic wall enzymes (Wecke et al. 1987). A similar inhibition of penicillin-induced bacteriolysis could also be achieved by the addition of monovalent or divalent cations like sodium, calcium or magnesium (Kersten and Wecke, in press).

In order to analyze the influence of these lysis protectants on the killing rate of penicillin-treated staphylococci, the viability of such cells was determined by CFU tests. Surprisingly, the killing rate was only rather slightly reduced and the number of living cells dropped following a similar kinetics than with penicillin alone (Giesbrecht et al., in press). This finding indicated that the penicillin-induced killing effect is apparently not connected with bacteriolysis.

Of great interest was the observation, that all substances that had the capacity of inhibiting the penicillin-induced bacteriolysis resulted in another characteristic effect: the inhibition of cell separation (Wecke et al., 1987). Since also cell separation is known to be dependent on the activity of autolytic wall enzymes, the inhibitory effects of anionic polyelectrolytes on penicillin-induced bacteriolysis and on cell separation were attributed to a direct inhibition of autolytic wall enzymes or to an indirect inhibition of the target structures for these autolysins in the cell wall (Wecke et al., 1987).

ENHANCEMENT OF BACTERIOLYSIS, DUE TO MORE INTENSE WALL DISINTEGRATION

If autolytic wall enzymes really do play a central role in penicillin-induced bacteriolysis, it should be possible not only to inhibit but also to enhance penicillin-induced bacteriolysis. In fact, the addition of some cationic proteins like cytochrome, ribonuclease or lysozyme, known to be activators of the autolytic wall enzymes, proved to be very effective in this concern, which resulted in a more rapid bacteriolysis than with penicillin alone (Wecke et al., 1982; Wecke et al., in press). It could be shown that enhancement of bacteriolysis was the result of a more intense wall disintegration (c. f. Fig. 1). However, the corresponding determination of the viability by CFU tests revealed that even lysis enhancement did not essentially influence the killing rate, and the number of living cells dropped following again a similar kinetics than with penicillin alone. This finding indicated again that the penicillin-induced killing effect is apparently not connected with bacteriolysis.

ONSET OF KILLING AND ONSET OF BACTERIOLYSIS

The next rational step was to find out, by determining the very moments of the onset of the killing event and of the onset of bacteriolysis, if killing and bacteriolysis take place simultaneously. For this purpose, the kinetics of heat production, colony-forming units (CFU), cell numbers, and optical density of staphylococci in the presence of penicillin were compared with each other. Most interesting was the finding that there was no coincidence between killing and bacteriolysis: The killing process, detected by thermography and by CFU, started already 50 minutes after application of the drug while bacteriolysis, in terms of decline of cell number, started only about 40 minutes later (Labischinski et al., 1988; Maidhof et al., 1989). Hence, killing had started already more than one generation time before bacteriolysis occurred, indicating that killing and bacteriolysis must be considered to be two completely different, consecutive events. Consequently, the staphylococci had apparently not died from bacteriolysis but due to an earlier killing process. This killing process was considered to be identical with the murosome-induced death indicating that only those staphylococci underwent bacteriolysis which were already dead. Therefore, it was concluded that not bacteriolysis but the murosome-induced "nonlytic death" is the universal lethal effect of penicillin (Giesbrecht et al., 1992).

INHIBITION OR ENHANCEMENT OF AUTOLYTIC WALL PROCESSES BY PENICILLIN ?

Very interesting was the finding that the simultaneous application of penicillin and lysozyme (or ribonuclease or cytochrome), which have proven to enhance bacteriolysis, resulted in a very intense wall disintegration comprising multiple breakages and the final disintegration of the entire cell wall (Fig. 1).

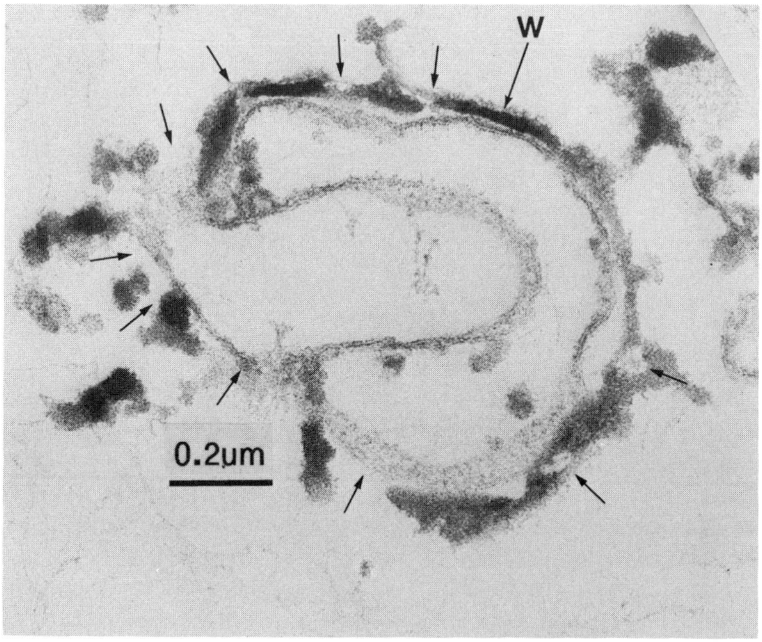

Figure 1. Thin section of S. aureus after simultaneous application of 0.1 μg penicillin per ml and 100 μg lysozyme per ml (2 h); intense wall disintegration comprising multiple breakages (arrows) of the cell wall (W) already during rather early stage of bacteriolysis.

Possibly, most of the wall autolysins present in the cells were activated under these conditions. These findings are in contrast to the bacteriolytic effect induced by penicillin alone which always resulted only in the disintegration of the protoplast, while the cell walls of such staphylococci showed only one or a few breakages but often their shape remained rather intact (Fig. 2). This observation led to the assumption that penicillin-induced bacteriolysis might not be the consequence of an overall activation of autolytic wall enzymes, but could rather be the effect of a local wall perforation (Giesbrecht et al., 1985; 1992).

The distinct preservation of large parts of the perforated bacterial cell walls after penicillin-induced bacteriolysis, even at late stages of this process where the protoplasts were already completely disintegrated (Fig. 2), is rather astonishing. It is in contrast to former findings on Escherichia coli and Klebsiella pneumoniae (Fontana et al., 1977) and streptococci (Tomasz, 1979) which pointed to the capacity of penicillin to activate directly or indirectly (via LTA-release: "triggering") the autolytic wall enzymes. However, it was shown earlier that in staphylococci lytic concentrations of penicillin (e.g., 0.1 μg/ml) failed to activate autolytic wall processes (Reinicke et al., 1985), while indications have been

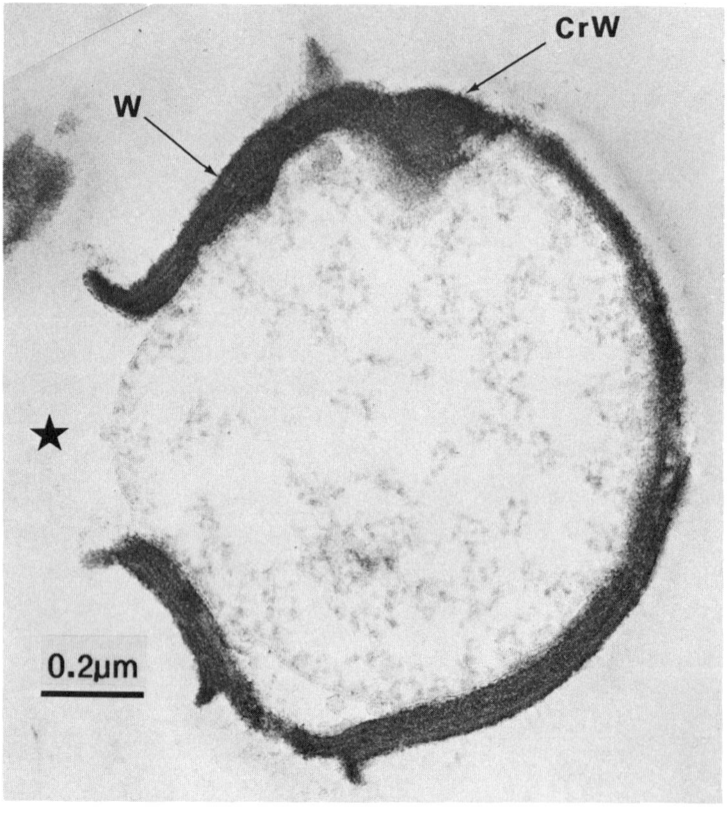

Figure 2. Thin section of S. aureus after 4 h application of 0.1 μg penicillin per ml alone; the shape of the cell wall (W) seems to be more or less intact (CrW = cross wall) even at this late stage of bacteriolysis. Asterisk = site of the single, murosome-induced wall perforation at the second division plane.

presented that higher concentrations of penicillin even inhibited them (Giesbrecht et al., 1983). Therefore, these contradictory data about the effect of penicillin had to be investigated in more detail.

First of all, the intensity of the autolytic wall processes was analyzed by monitoring the wall turnover of staphylococci growing in the presence of different concentrations of this drug. In this way, the inhibitory influence of penicillin on autolytic wall processes could be shown directly (Fig. 3): The higher the concentration of the drug, the lower the turnover rates of labelled wall material. Furthermore, also the amount of labelled wall material was determined which was released from staphylococci during autolysis in buffer after precultivation in the presence of different concentrations of penicillin. Again it could be shown that high concentrations of this drug inhibited the release of wall material more intensely than did normal, low concentrations (data not shown). All these findings indicate that the missing disintegration of the staphylococcal cell walls after penicillin-induced bacteriolysis would best be interpreted by assuming that penicillin itself had inhibited some autolytic wall processes. This conclusion might also provide an explanation of the fact that rising concentrations of penicillin inhibit more and more those autolytic wall processes which are responsible for bacteriolysis till, finally, in the presence of extremely high concentrations of penicillin (1.0 or 10 µg/ml) hardly any bacteriolysis could be observed ("nonlytic death": c.f. Shockman et al., 1981). Characteristic for such extremely high penicillin concentrations is the missing decline of the optical density.

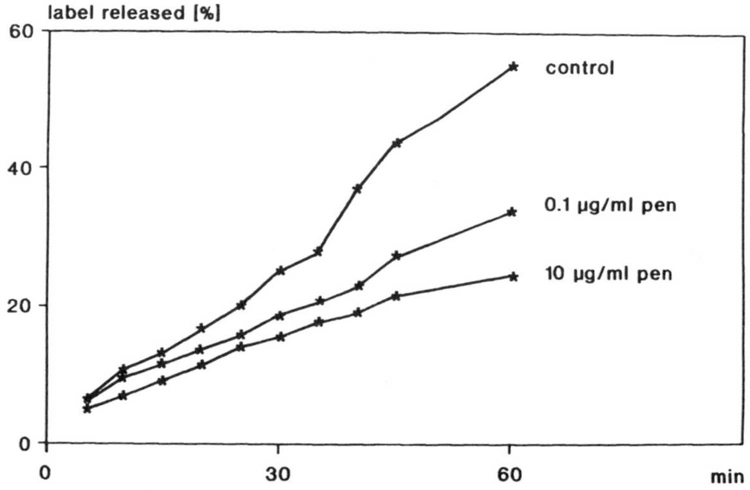

Figure 3. Release of wall label (acetylglucosamine) from S. aureus growing in the presence of different concentrations of penicillin G (pen).

However, if such an inhibitory effect of penicillin was really involved also in nonlytic death, it would also be conceivable that this resulted from a penicillin-mediated inhibition of the formation of wall autolysins (Rogers, 1967) rather than from a direct effect of the drug on these autolytic wall processes. In the latter case, however, this inhibitory effect might be only a transitory reaction which could be compensated by the subsequent or the simultaneous addition of activators of autolytic wall enzymes during penicillin treatment. Consequently, attempting to inhibit the autolytic wall processes, we treated staphylococci with different concentrations of penicillin and, simultaneously, with activators of autolytic wall enzymes like ribonuclease or cytochrome. In the presence of 0.1 µg of penicillin per ml, the addition of cationic proteins resulted in a very rapid bacteriolysis and a high degree of wall degradability. However, in the presence of 1.0 µg of penicillin per ml only a slight wall degradability could be induced under these conditions of "nonlytic death", while in the presence of 10 µg of penicillin per ml, the capacity of cationic proteins to activate autolytic wall processes, which would result in wall degradation, was further reduced. Probably, in the presence of high penicillin concentrations, the cationic proteins were no longer able to compensate the penicillin-induced inhibition of autolytic wall processes. These findings indicated again that the higher the penicillin concentration applied, the more intense the penicillin-mediated inhibitory capacity of this drug for autolytic wall processes.

All these data show that, in contrast to former considerations, penicillin may act as an effective inhibitor of autolytic wall processes. Therefore, it was no surprise that high concentrations of penicillin also exerted another characteristic effect being induced by some inhibitors of autolytic wall processes (like anionic polyelectrolytes), which is the inhibition of cell separation (Fig. 4). The same observation was made with cells of some MRSA-strains, the cell separation of which was intensely disturbed only in the presence of high concentrations of penicillin (see Labischinski et al., this Volume).

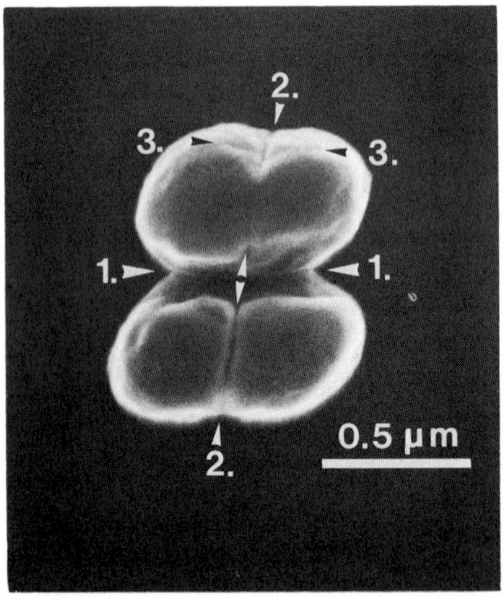

Figure 4. Scanning electron micrograph of S. aureus , treated with 10 µg penicillin per ml. Inhibition of cell separation : even three division planes can be recognized (1., 2.,3.; arrows)

Since we could prove that in penicillin-treated staphylococci neither lysozyme nor ribonuclease were capable of breaking the bonds between penicillin and the PBPs (data not shown), one can exclude a reverse reaction of the synthetic capacity of the PBPs that could otherwise be made responsible for the reactivation of the capacity for wall disintegration by cationic proteins. Consequently, the conclusion might be drawn that penicillin has not only one target, the PBPs, as is presently believed, but that there is a "second target", the autolytic wall processes. We have, however, indications that the mechanisms of penicillin-induced inhibition of autolytic wall processes are different, depending on whether (i) high penicillin concentrations or (ii) low penicillin concentrations combined with anionic polyelectrolytes are applied. It is of interest to mention that the effect of high doses of penicillin resembled to some extent the action of another potent inhibitor of autolytic wall processes, namely chloramphenicol.

The recognition of a "second target" for penicillin could be of some importance in view of attempts to explain the hitherto hardly understood fact that highly resistant staphylococci (MRSA-strains; Berger-Bächi et al., 1986) do not exert a typical bacteriolysis (strain BB 270) or that bacteriolysis takes place only in the presence of rather high concentrations of this drug (strain Col). These facts are especially strange since we could show that also the MRSA-strain BB 270 did contain autolytic wall enzymes, the effectivity of which was even higher than those of the corresponding wild-type strain BB 255 (data not shown). The new "two-target-model" of penicillin action could possibly explain these findings: In wild-type strains, rather low concentrations of penicillin are sufficient to inhibit the first target of penicillin, the PBPs, effectively while the "second target", the autolytic wall processes, is only slightly inhibited so that wall perforations, some breakages of the cell wall and subsequent bacteriolysis remain possible. However, in the case of MRSA-strains, in which considerable higher concentrations of ß-lactams are necessary to block the acquired PBP2a, these concentrations would also be sufficient to inhibit intensely the "second target", the autolytic wall processes, so that hardly any bacteriolysis can take place. But this would be true only in those cases, in which the components, which are involved in the autolytic wall processes, were not changed along with the acquirement of PBP2a. However, if there were any changes in the biochemical architecture of the cell wall (see Labischinski et al., this Volume), it is conceivable that also the penicillin sensitivity of the "second target", the autolytic wall processes, had changed simultaneously. In this case, probably a considerable higher penicillin concentration as that necessary to block PBP2a and to induce bacteriolysis would finally be essential to inhibit bacteriolysis. In fact, this proved to be true at least for the MRSA-strain Col, since 30 µg per ml penicillin G induced bacteriolysis whereas 100 µg per ml were ncecessary to inhibit bacteriolysis.

THE PARADOXICAL EFFECT OF LYSOZYME: INHIBITION OF BACTERIOLYSIS IN SPITE OF WALL DISINTEGRATION

In order to analyze the reduced wall degradability of staphylococcal cell walls by ribonuclease or cytochrome in the presence of high concentrations of penicillin, we applied lysozyme as a more potent activator. This cationic protein is not only considered to activate autolytic wall enzymes but has been also widely understood to be an effective muralytic agent in many gram-positive bacteria although, in staphylococci, the latter was shown to hold true only to a very limited extent (Westmacott et al., 1979; Blümel et al., 1983). As expected, in comparison to ribonuclease, low concentrations of lysozyme (100 µg/ml) further accelerated penicillin-induced bacteriolysis (c. f. Fig. 1) and wall degradation but, surprisingly, high concentrations of lysozyme (500 or 1000 µg/ml) resulted seemingly in an intense inhibition of bacteriolysis, as deduced from monitoring the optical density (Fig. 5). Even after application of an extremely high concentration of penicillin (10 µg/ml) the optical density went on rising after addition of high concentrations of lysozyme. Only by electron microscopy, these puzzling results could be cleared up: High concentrations of lysozyme were not only capable of disintegrating the cell walls during penicillin action but also of "stabilizing" the protoplasts, thus preventing bacteriolysis (Fig. 6). The same effect could be observed after a simultaneous application of the extremely high concentration of 10 µg of penicillin per ml and high concentrations of lysozyme. The mechanism by which lysozyme exerted its protoplast-stabilizing capacity even after an entire disintegration of the bacterial cell wall, is still hardly understood. However, there is some indication that under these experimental conditions, the viscosity of the cytoplasm rose (sol-gel transformation ?) which might have stabilized the protoplast.

These findings have led to several conclusions: (i) inhibition of bacteriolysis can be achieved not only via wall stabilization but also via wall disintegration and simultaneous stabilization of the bacterial protoplast, (ii) a missing decline of the OD is not a reliable indicator of the integrity of the bacterial cell, (iii) an increase of the OD is not a reliable indicator of growth.

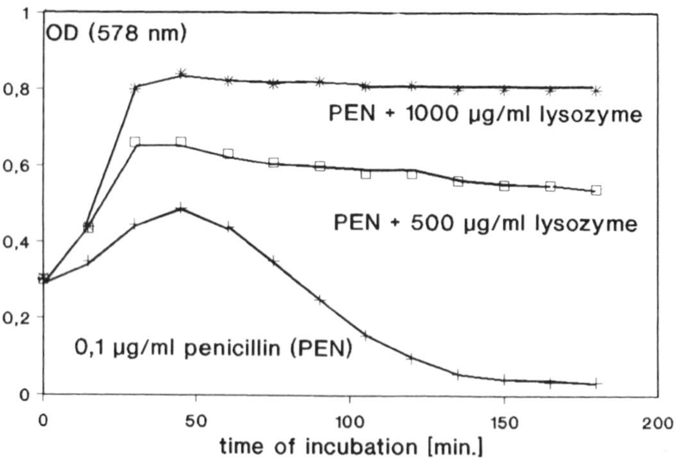

Figure 5. Paradoxical effect of high concentrations of lysozyme on staphylococci growing in the presence of penicillin: increase of OD in spite of the addition of an activator of autolytic wall enzymes.

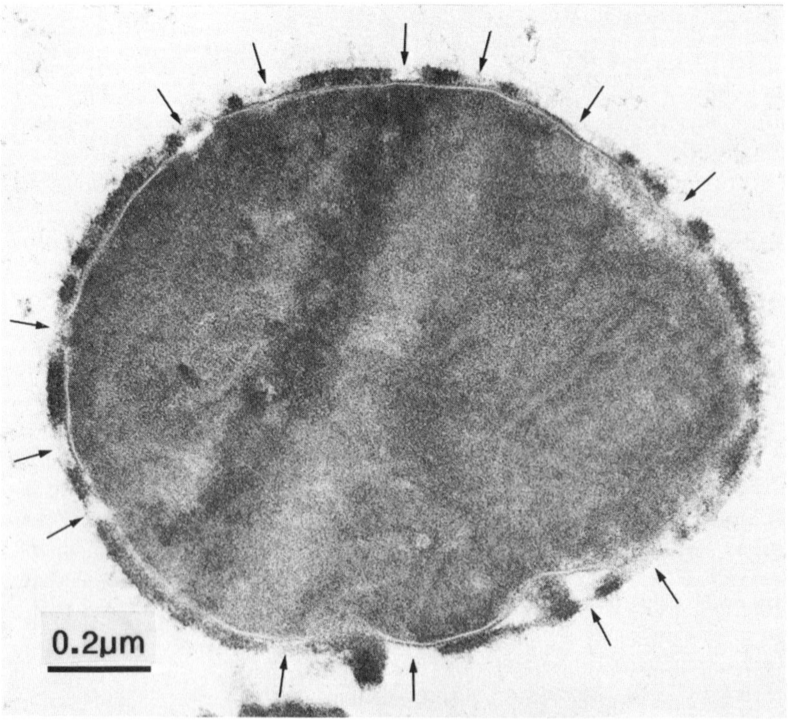

Figure 6. Thin section of S. aureus after 30 min of simultaneous application of penicillin (0.1 µg/ml) and high concentration of lysozyme (1000 µg/ml). In spite of multiple breakages of the cell wall (arrows) the protoplast remains "stabilized".

ENHANCEMENT OF KILLING AND BACTERIOLYSIS BY INCREASING THE NUMBER OF KILLING MUROSOMES ?

Since neither inhibition nor enhancement of bacteriolysis did essentially influence the killing effect of penicillin on staphylococci, the influence of other methods on penicillin-induced alterations in the viability of these bacteria was investigated.

For a better understanding of these procedures it should be mentioned that, in the presence of penicillin, normally only one "killing murosome" perforated the pressure-stabilized peripheral cell wall in the second division plane, where cell separation is no longer preceded by the assembly of sufficient cross-wall material (Giesbrecht et al., 1985). However, since it could be proven, that approximately 20 murosomes are located in the second division plane, it was attempted to involve more murosomes in this killing event. Therefore, penicillin-treated staphylococci were cultivated in the presence of 3 % sodium chloride which established an external osmotic pressure almost equalling the internal pressure of a staphylococcal cell, namely 20 to 25 atmospheres. It was expected that, in this way, most of the murosomes would be activated which, however, could not yet be ejected because of the missing internal overpressure. At the very moment, when untreated staphylococci would initiate cell separation in the second division plane, the high-pressure growth medium was replaced by a normal medium and sufficient fixation fluid was added simultaneously to preserve the expected effect. In fact, by applying scanning electron microscopy, it was possible to show that a row of murosomes was ejected from the staphylococci like a fan-shaped fountain (Giesbrecht et al., 1992).

However, in spite of such a drastical increase of the number of killing murosomes, neither the killing rate nor the degree of bacteriolysis could be enhanced under these experimental conditions.

ATTEMPTS TO IMITATE THE PENICILLIN-INDUCED WALL PERFORATION WITHOUT APPLYING ANTIBIOTICS

In order to promote the killing process by other means, several experiments were performed to imitate the ability of murosomes to induce wall perforation without, however, applying antibiotics. Since the wall-perforation process is the result of the lytic activity of the murosomes, and since these murosomes were shown to be vesicular derivatives of the cytoplasmic membranes, the creation of additional vesicular derivatives of the cytoplasmic membrane of staphylococci was attempted.

The rationale of these considerations was based on the observation that the staphylococcal cell wall, the elasticity of which was estimated to be approximately 200 % (Labischinski et al., 1983), could easily be osmotically compressed while the cytoplasmic membrane exhibited only a rather low compressibility. Consequently, after osmotic compression of staphylococci (via sucrose), many vesicles, folds and mesosomes, being derivatives of the cytoplasmic membrane, could be observed (Giesbrecht, 1991). Freeze-etching experiments with unfixed staphylococci proved that these "mesosomes" were no fixation artifacts as has been hitherto considered (Dubochet et al., 1983). Interestingly, it could be shown that in those cases, in which such mesosomes came into contact with the peripheral cell wall, perforation events were observed at these sites of the cell wall (Fig. 7). Consequently, the staphylococci were lysed at such sites (Fig. 8; c. f. Hirai, 1979). These processes were reminiscent to murosome-mediated wall perforation and subsequent bacteriolysis under penicillin. In many cases the mesosome formation started after a slit-like affection of the interlayer, located between the cell wall proper and the cytoplasmic membrane, in which vesicular structures appeared. Especially interesting was the fact, that

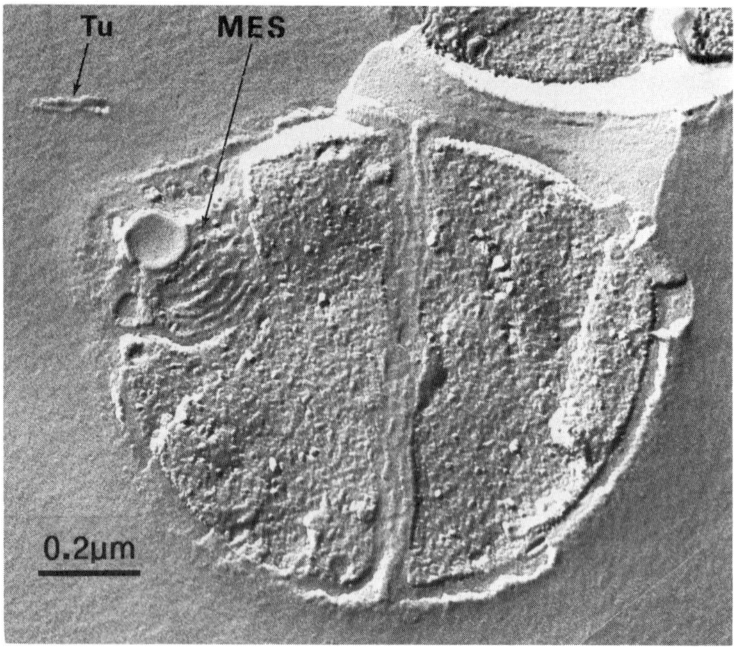

Figure 7. Freeze fracture of an unfixed cell of S. aureus after osmotic compression by 1.8 M sucrose; a mesosome (MES) is in contact with the peripheral cell wall and perforates it . The first tubuli (Tu) are released from the site of wall perforation.

by inside-to-inside contact of the folds of the cytoplasmic membrane, such "non-morphogenetic" vesicles and tubuli were created (Giesbrecht, 1991).

In order to find out if such slit-like affections were really the consequence of the sucrose-induced osmotic pressure, surface drying was applied which is known to produce extremely high pressures. For this purpose, the cells were thoroughly dried on a supporting film and embedded for thin sectioning (Fig. 9). In fact, under these conditions, exactly the same slit-like affections in the interlayer could be observed. However, very interesting was the finding that, by this drying procedure, also the formation of mesosomes could be observed (Fig. 10). But these mesosomes did not only consist of membraneous structures but contained also considerable amounts of substances resembling lipid-containing constituents of apparently the same type as in the slits of the interlayer. These findings indicated that during compression of the staphylococci by an intense drying, the compression-resistant cytoplasmic membrane had reduced its surface not only by folding processes but also, to a considerable extent, by releasing lipid-containing components. This conclusion would also mean that mesosomal vesicles, derived from folding processes of the cytoplasmic membrane, must not be chemically identical with those vesicles derived from released lipid-containing components. Furthermore, since these drying experiments resulted basically in the same type of morphological variations as have been observed in the presence of sucrose we have to conclude that, in fact, in both cases volume reduction by compression is the very reason for the creation of "non-morphogenetic" vesicles.

At least some of these vesicular structures in the interlayer were apparently involved in wall perforation and, consequently, in a compression-induced killing process. Whether these non-murosomal "killing vesicles" were derived from folding processes of the cytoplasmic membrane or from released lipid-containing components of this membrane is presently not known. However, it is extremely important to note, that at least sucrose-induced killing processes started much earlier than those induced by penicillin.

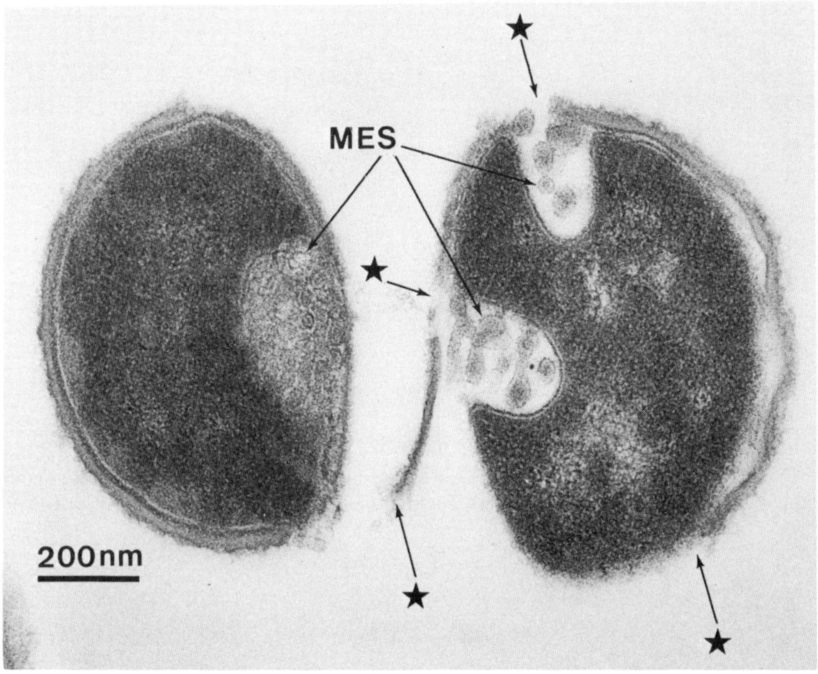

Figure 8. Thin section of S. aureus after osmotic compression by 1.8 M sucrose: mesosomes (MES), being in contact with the cell wall, have perforated it. Asterisks = sites of wall breakages.

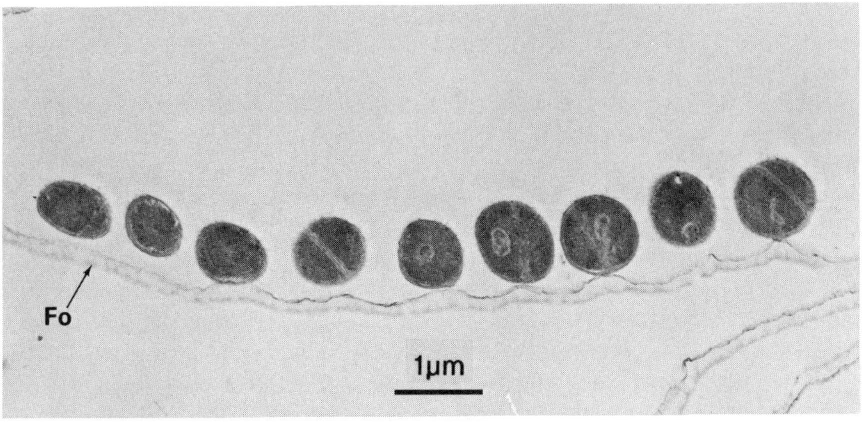

Figure 9. Thin section of staphylococci after intense drying on a foil (Fo).

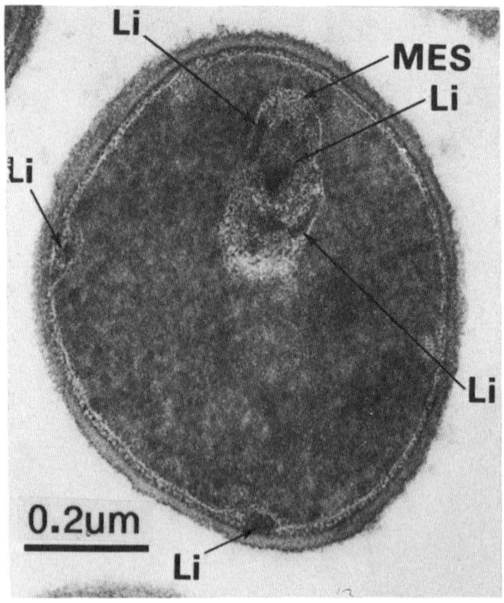

Figure 10. Thin section of S. aureus after intense drying on a foil; formation of a mesosome (MES) and lipid-containing components (Li) in the slits of the wall-membrane-interlayer and in the mesosome.

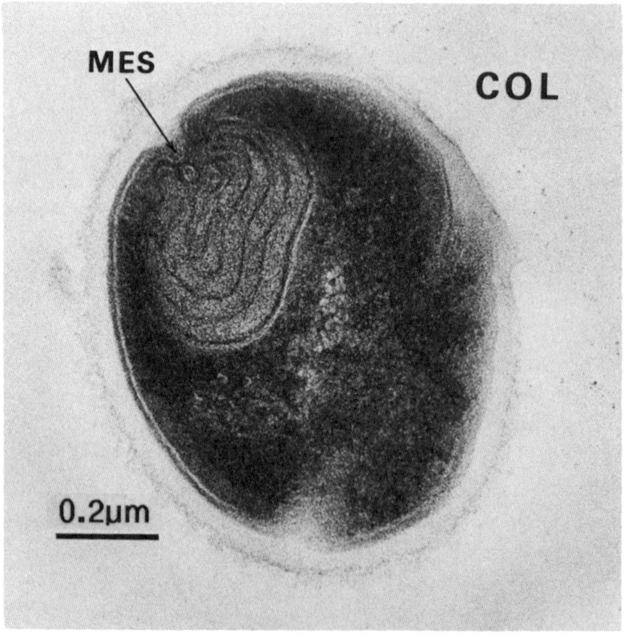

Figure 11. Thin section of methicillin resistant S. aureus strain Col after osmotic compression by sucrose; formation of a mesosome (MES).

Consequently, the killing must have taken place already during the first rather than during the second cell separation. Therefore, it was no surprise that the sucrose-induced wall perforation processes were localized at any site of the cell wall (compare Figs. 7 and 8) and were not restricted to the second division plane (see Fig. 2) as it was the case in penicillin-induced wall perforation processes (Giesbrecht, 1991). Hence, these sucrose-induced effects cannot be considered to be identical with the murosome-mediated wall perforations and killing procedures induced by penicillin. Unfortunately, it is presently only possible to kill varying numbers of the sucrose-treated staphylococci within the first 30 minutes and we have no convincing explanation for this fact.

Most important, however, was the observation, that not only penicillin-sensitive staphylococci could be killed by this sucrose-induced compression procedure but also cells of the highly resistant MRSA-strains. Also, by electron microscopy, the same process of mesosome-formation, wall perforation and subsequent bacteriolysis could be observed in some cells of such resistant staphylococcal strains like BB270 or Col (Fig. 11), as it was the case in cells of sensitive strains.

SUMMARY

- Penicillin-induced killing and bacteriolysis of staphylococci have been shown to be two completely different, consecutive events.

- The universal lethal effect of penicillin is not bacteriolysis ("lytic death") but a murosome-mediated wall perforation ("non-lytic death").

- Bacteriolysis can be inhibited (via wall stabilization) or enhanced (via wall disintegration) without essentially influencing the penicillin-induced killing rate.

- Bacteriolysis has proven to be nothing but a postmortem effect which, depending on the experimental conditions, may or may not occur in the dead bacteria after penicillin-induced killing via murosome-mediated wall perforation.

- Several concentrations of penicillin have been shown to be effective inhibitors of some autolytic wall processes (wall turnover, wall autolysis, bacteriolysis, wall degradability, and cell separation) indicating that penicillin has a "second target" besides the PBPs.

- A novel paradoxical effect of lysozyme on penicillin-induced bacteriolysis is described, comprising inhibition of bacteriolysis in spite of an intense wall disintegration (protoplast stabilization).

- Formation and activity of murosomal "killing vesicles" could be imitated to a limited extent by sucrose-induced osmotic compression resulting in the formation of non-murosomal vesicles. Some of them might be involved in wall perforation (non-murosomal "killing vesicles") and subsequent bacteriolysis not only in sensitive staphylococci but also in cells of MRSA-strains.

ACKNOWLEDGMENTS

We thank Dominique Krüger, and Bernhard Reinicke for their cooperation and Gerta Tochtermann, Bernd Nürnberg, Chang Han, Erich Mertens, Georg Kniffke, Matthias Flötenmeyer, and Ellen Albrecht for their excellent technical assistance.

REFERENCES

Blümel, P., Reinicke, B., Lahav, M. and Giesbrecht, P. (1983) Cell wall degradation of Staphylococcus aureus by autolysins and lysozyme, in "The Target of Penicillin: The

Murein Sacculus of Bacterial Cell Walls, Architecture and Growth" (Hakenbeck, R., Höltje, J.-V. and Labischinski, H., Eds.), pp. 323-328. Walter de Gruyter, Berlin.
Berger-Bächi, B., Strässle, A. and Kayser, F.H. (1986) Characterization of an isogenic set of methicillin-resistant and susceptible mutants of *Staphylococcus aureus*. Eur. J. Clin. Microbiol. 5, 697-701.
Dubochet, J., McDowall, A.W., Menge, B., Schmid, E.N. and Lickfeld, K.G. (1983) Electron microscopy of frozen-hydrated bacteria. J. Bacteriol. 155, 381-390.
Fontana, R., Satta, G. and Romanzi, C.A. (1977) Penicillins activate autolysins extracted from both Escherichia coli and Klebsiella pneumoniae envelopes. Antimicrob. Agents Chemother. 12, 745-747.
Giesbrecht, P. (1991) Perturbation of the membrane-wall-interlayer by volume reductions of staphylococci results in wall disintegration and bacteriolysis after formation of mesosomal vesicles, being no fixation artifacts but rather multiplied <<exosomes>>, in "The staphylococci" (Jeljaszewicz, J. and Ciborowski, J., Eds.), pp. 173-177. Gustav Fischer Verlag, Stuttgart.
Giesbrecht, P., Franz, M., Krüger, D., Labischinski, H. and Wecke, J. Bacteriolysis of staphylococci is only a side-effect of penicillin-induced death, in "50 years of penicillin application - history and trends" (Kleinkauf, H and von Döhren, H., Eds.), in press. Walter de Gruyter, Berlin.
Giesbrecht, P., Kersten, T. and Wecke, J. (1992) Fan-shaped ejections of regularly arranged murosomes involved in penicillin-induced death of staphylococci. J. Bacteriol. 174, 000-000.
Giesbrecht, P., Labischinski, H. and Wecke, J. (1985) A special morphogenetic wall defect and the subsequent activity of "murosomes" as the very reason for penicillin-induced bacteriolysis in staphylococci. Arch. Microbiol. 141, 315-324.
Giesbrecht, P., Morioka, H., Krüger, D., Kersten, T. and Wecke, J. (1983) Restoration of penicillin-damaged cell walls by de novo murein synthesis and successive murein degradation in staphylococci, revealing a hitherto unknown mechanism of penicillin action: blockage of autolytic wall processes by penicillin, in "The Target of Penicillin: The Murein Sacculus of Bacterial Cell Walls, Architecture and Growth" (Hakenbeck, R., Höltje, J.-V. and Labischinski, H., Eds.), pp. 243-248. Walter de Gruyter, Berlin.
Hirai, Y. (1979) The development of mesosomes of Staphylococcus aureus during autolysis. Acta Med. Okayama 33, 219-238.
Kersten, T. and Wecke, J. Delayed penicillin-induced bacteriolysis of staphylococci by some cations, in "50 years of penicillin application - history and trends" (Kleinkauf, H. and von Döhren, H., Ed.), in press. Walter de Gruyter, Berlin.
Labischinski, H., Barnickel, G. and Naumann, D. (1983) The state of order of bacterial peptidoglycan, in "The Target of Penicillin: The Murein Sacculus of Bacterial Cell Walls, Architecture and Growth" (Hakenbeck, R., Höltje, J.-V. and Labischinski, H., Eds.), pp. 49-54. Walter de Gruyter, Berlin.
Labischinski, H., Hochberg, M., Sidow, T., Maidhof, H., Henze, U., Berger-Bächi, B. and Wecke, J. Biophysical and biochemical studies on the fine structure of the sacculi from Escherichia coli and Staphylococcus aureus. In this Volume.
Labischinski, H., Maidhof, H., Franz, M., Krüger, D., Sidow, T. and Giesbrecht, P. (1988) Biochemical and biophysical investigations into the cause of penicillin-induced lytic death of staphylococci: checking predictions of the murosome model, in "Antibiotic inhibition of bacterial cell surface assembly and function" (Actor, P., Daneo-Moore, L., Higgins, M.L., Salton, M.R.J. and Shockman, G.D., Eds.), pp. 242-257. American Society for Microbiology, Washington D.C..
Maidhof, H., Johannsen, L., Labischinski, H. and Giesbrecht, P. (1989) Onset of penicillin-induced bacteriolysis in staphylococci is cell cycle dependent. J. Bacteriol. 171, 2252-2257.

Reinicke, B., Blümel, P., Labischinski, H. and Giesbrecht, P. (1985) Neither an enhancement of autolytic wall degradation nor an inhibition of the incorporation of cell wall material are pre-requisites for penicillin-induced bacteriolysis in staphylococci. Arch. Microbiol. 141, 309-314.

Rogers, H.J. (1967) Killing of staphylococci by penicillins. Nature 213, 31-33.

Shockman, C.D., Daneo-Moore, L., McDowell, T.D. and Wong, W. (1981) Function and structure of the cell wall - its importance in the life and death of bacteria, in "ß-Lactam antibiotics" (Salton, M. and Shockman, G.D., Eds.), pp. 31-65. Academic Press, New York.

Tomasz, A. (1979) The mechanism of the irreversible antimicrobial effects of penicillin: How beta-lactam antibiotics kill and lyse bacteria. Annu. Rev. Microbiol. 33, 113-137.

Wecke, J., Franz, M. and Giesbrecht, P. (1990) Inhibition of the bacteriolytic effect of ß-lactam antibiotics on Staphylococcus aureus by the polyanionic drugs suramin and Evans Blue*. APMIS 98, 71-81.

Wecke, J., Kersten, T. and Giesbrecht, P. The modulation of the bacteriolytic effect of ß-lactam antibiotics by non-antibiotics. APMIS, in press.

Wecke, J., Kwa, E., Lahav, M., Ginsburg, I. and Giesbrecht, P. (1987) Suppression of penicillin-induced bacteriolysis of staphylococci by some anticoagulants. J. Antimicrob. Chemother. 20, 47-55.

Wecke, J., Lahav, M., Ginsburg, I. and Giesbrecht, P. (1982) Cell wall degradation of Staphylococcus aureus by lysozyme. Arch. Microbiol. 131, 116-123.

Wecke, J., Lahav, M., Ginsburg, I., Kwa, E. and Giesbrecht, P. (1986) Inhibition of wall autolysis of staphylococci by sodium polyanethole sulfonate "liquoid". Arch. Microbiol. 144, 110-115.

Westmacott, D. and Perkins, H.R. (1979) Effects of lysozyme on Bacillus cereus 569: rupture of chains of bacteria and enhancement of sensitivity to autolysins. J. Gen. Microbiol. 115, 1-11.

REGULATION OF THE MORPHOGENETIC CYCLE OF *ESCHERICHIA COLI*: 1992

William D. Donachie

Institute of Cell and Molecular Biology
University of Edinburgh
Mayfield Road, Edinburgh EH9 3JR
Scotland

INTRODUCTION

In exponentially increasing populations, the individual rod-shaped cells of *E. coli* grow by continuous elongation (without significant change in diameter) and divide by the formation of a centrally placed septum after each successive doubling in length and volume. Many of the genes and proteins which are specifically required for this morphogenetic cycle are now known (see Table) but the way in which their activities are regulated remains largely unknown. What follows is a brief synopsis of the present picture. (Only a few more recent and key references are given; earlier references can be obtained from recent reviews, e.g. Matsuhashi et al. 1990, Donachie & Begg 1990.)

CONTROL OF CELL SHAPE DURING ELONGATION

The activity of two membrane proteins, RodA and PBP2, is dedicated to the maintenance of the cylindrical shape of the cell and ensures that doubling in cell volume during growth is accompanied by doubling in cell length. PBP2 is a mecillinam-sensitive peptidoglycan(PG)-transpeptidase (TPase). PBP2 and RodA together have both TPase and PG-transglycosylase (TGase) activity. In the absence of either one of these proteins, cells grow and divide as near-spherical cocci. The two proteins are each present in extremely low number (ca. 20/cell). (In order to survive, cells lacking either of these proteins require to be able to produce high levels of ppGpp, for as yet unknown reasons: Vinella et al. 1991.)

FORMATION OF THE SEPTUM

The first known stage in septum formation is the assembly

Bacterial Growth and Lysis, Edited by M.A. de Pedro *et al.*, Plenum Press, New York, 1993

of a ring of FtsZ protein on the inside of the cell membrane at the site at which the septum will form (Bi & Lutkenhaus 1991). Before this stage, FtsZ protein is distributed throughout the cytoplasm. FtsZ protein is present in up to 50,000 molecules/cell and, during septation, most or all of it appears to be located in the ring.

The appearance of the Z-ring is associated with the formation of a constriction of the cell at that site, but further stages in septum formation also require the septum-specific proteins FtsA, FtsQ, PBP3, FtsW, MraR, and perhaps others (e.g. FtsE, FtsX, FtsY, SecA, see below). PBP3 is a septum-specific TPase/TGase, which is preferentially inhibited by certain β-lactams (e.g. furazlocillin) which block septation: it is present in the cell membrane at about 50 molecules/cell. FtsW shows sequence similarity to RodA and may therefore act together with PBP3; as RodA does with PBP2 (Matsuhashi et al. 1990). FtsA is a membrane-associated protein, present at about 150 molecules/cell, which is absolutely required for septation. (It shows sequence similarities to the MreB regulator protein, see below.) FtsQ is a transmembrane protein with a large periplasmic domain and a small cytoplasmic domain (Carson et al. 1991). It is present at only about 25 molecules/cell but is absolutely essential for septation. The MraR protein has not yet been identified but it seems to be required both for septation and for maintenance of the integrity of the PG-sacculus. The septum forms as a continually closing annulus with a contracting ring of FtsZ protein associated with the inner edge. (The locations of other septal proteins are unknown, but their low abundance suggests that they are likely also to be associated with the Z-ring.) The growing septum consists of two PG-layers, covalently bonded together, which appear to be split apart soon after formation, so as to peel back to form the curved surfaces of the new cell poles. The EnvA protein seems to be essential for this splitting-apart of the septal layers.

The genes coding for the FtsZ, A, Q, W, MraR and EnvA proteins are all part of a single multi-promoter operon (together with other genes coding for enzymes of PG-synthesis; fig.1). Other proteins may be required less directly for septation: some of these (such as the chaperonin GroESL and the secretion protein SecA) may be required for correct folding and insertion into the cell membrane of septation proteins, such as FtsQ, PBP3 and FtsW, while affecting many other proteins at the same time, but the FtsX, E and Y proteins may be "dedicated" to septation. The role of these proteins is unknown but their deduced primary structures suggest the possibility that they are signal-recognition and docking proteins for some sort of membrane proteins, perhaps solely those taking part in septation? The genes for these three proteins form a separate operon (fig.1).

Cell division also requires that the inner and outer cell membranes be divided. The way in which this comes about is not understood: possibly the inner membrane is attached to a contracting Z-ring (so that FtsZ protein may be the only protein specifically required for cell division in L-forms), while the outer membrane is attached to the PG-sacculus through the lipoprotein and may therefore by passively "pasted" to the PG of the nascent cell poles by this means. (However, lipoproteinless mutant cells appear to be able to cover their poles with outer-membrane, so that other factors may also be involved.)

CONTROL OF THE MORPHOGENETIC CYCLE

What causes the Z-ring to form when and where it does? A currently popular hypothesis is that septa can form wherever and whenever there is a part of the sacculus which is more than a certain distance from the point of attachment of a chromosome: thus, septa form exactly between sister nucleoids in normal cells but form at apparently random positions in the DNA-free regions of elongated cells which have failed to partition their DNA properly. Because septa do not normally form near cell poles

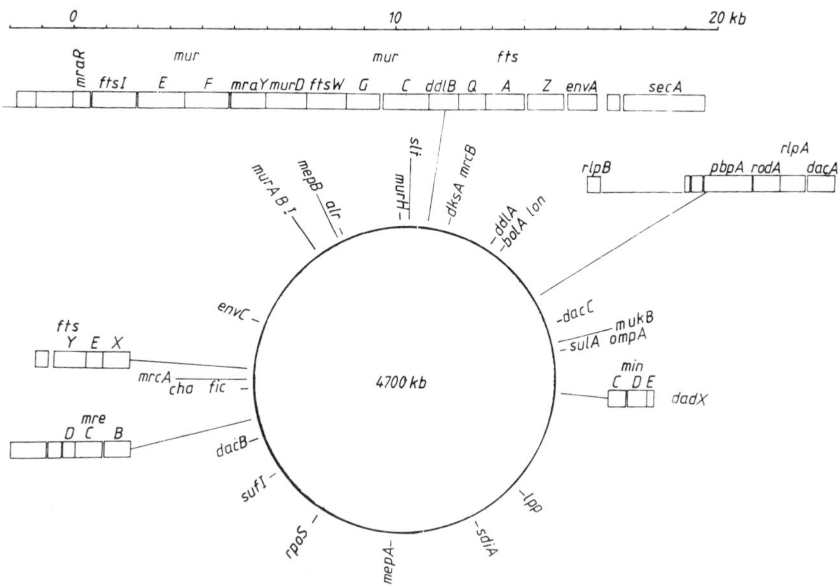

Figure 1. **GENETIC MAP OF MORPHOGENES OF *E. COLI*.** Gene symbols are explained in the Table.

(which will be beyond the proposed minimum inhibitory distance of a chromosomal attachment point in cells which have not yet completed chromosome replication and partition) it is necessary to assume that formation of septa is blocked at or near cell poles. The block to septation at cell poles requires the MinC, D & E proteins (coded by the genes of the *minB* operon: fig.1). MinC is an inhibitor of septation when activated by MinD (which may be a kinase) and is somehow directed to the cell poles by the action of MinE (de Boer et al. 1989). Thus, in $minB^+$ cells, the earliest time that a septum can form is immediately after partition of sister chromosomes and the only place is the point which is as far away as possible from both of these, i.e. the cell centre. (In *minB*⁻ mutants, of course, cells can also form septa near the poles, so as to produce minicells.)

Partition of newly-replicated sister chromosomes is therefore the normal signal for septation, according to this model. Almost all that is known about partition at present is that it takes place within a few minutes of the completion of chromosome replication, that it requires de novo protein synthesis during this short post-replication period, that it normally takes place when cells have reached a certain minimum length, and that the sister chromosomes then appear to move rapidly apart by a fixed distance, which normally places them in the centres of the future sister cells (Donachie & Begg 1989, 1991). Mutations in many genes can inhibit partition: some of these genes encode subunits of DNA topoisomerases (e.g. *gyrA*, *gyrB*, *parC*, *parE*) and may be necessary for partition because they are required for successful decatenation of sister DNA circles. The *dif* locus, at the replication terminus, and the XerC protein seem similarly to be required to resolve chromosomal dimers which can form by recombination between homologous regions of partly-replicated chromosomes (Clerget 1991, Blakely et al. 1991, Kuempel et al. 1991). Mutations in genes (e.g. *dnaG*, *dnaE*) coding for other proteins required for DNA synthesis can also affect the distribution of DNA and the locations of septa in the cell. It is not known for certain whether there are additional proteins which are part of a partition apparatus, although this might seem likely. The best candidate is the 177kD MukB protein, which has a proposed secondary structure resembling eukaryotic kinesins. A mutant lacking MukB protein shows many abnormal cells (without DNA, or elongated with abnormal numbers, sizes or locations of nucleoids) but also many normal cells, and is viable, at least at 30C (Niki et al. 1991). A specific role for this protein in partition is therefore not yet proven, especially as similar phenotypes can be produced by interfering with the normal coordination between DNA synthesis and cell growth (Begg & Donachie 1991).

Are other factors, apart from chromosome partition, required to turn on septation (or to turn it off)? It has been shown that increases or decreases in certain PG-modifying enzymes can either potentiate or inhibit septation or cell elongation. Thus, increase in either of the two main DD-carboxypeptidases (DD-CPases) restores division to mutant cells with very low levels of PBP3 (Begg et al. 1990) but cause cells with normal PBP3 levels to lose their cylindrical shape and form cocci (Markiewicz et al. 1982). Increased activity of these enzymes should increase the proportion of PG with tetrapeptide and tripeptide side chains (at the expense of pentapeptide). In accord with this interpretation, partial inhibition of pentapeptide formation (by D-cycloserine) also causes wild-type cells to become spherical but restores cell division to the PBP3-deficient mutant. Similarly, increased expression of the D-alanine:D-alanine adding enzyme (*murF* product: fig.1) inhibits cell division but allows cells to elongate into long cylinders (Mengin-Lecreulx et al. 1989). It is now known (Edwards & Donachie, this volume) that deletion of all 3 known genes coding for proteins with DD-CPase activity is non-lethal and has only subtle effects on cell morphology. If periodic changes in the activities of any of these enzymes normally assist the switch from cell elongation to septation (& back) then they must play only a minor role in morphogenesis. There remains the possibility that the activity of the D-alanine:D-alanine adding enzyme plays a more critical role in this switch, and this is currently under investigation.

Transcription of the consecutive *ftsQ, A, Z* and *envA* genes takes place from many promoters (located within the *ddlB, ftsQ* and *ftsA* genes, and between *ftsZ* and *envA*: Robinson et al. 1984, 1986, Yi et al. 1985, Aldea et al. 1990). Transcription from at least four of these promoters is strongly growth rate and culture-phase dependent, such that the number of transcripts produced per cell cycle is approximately constant under all conditions (Donachie et al. 1984, Dewar et al. 1989, Aldea et al. 1990, Wang et al. 1991, R. Smith & WDD unpublished). Such "gearbox" (Aldea et al. 1989) regulation may have evolved so as to produce a sufficient number of division proteins to produce a septum after each round of chromosome replication, even when cells are making other proteins very slowly (Donachie et al. 1984, Dewar et al. 1989). A new sigma factor (product of the *rpoS/katF* gene: Lange & Hengge-Aronis 1991) is produced in a similar fashion (i.e. its concentration is higher as cells grow slower and is highest in cells entering stationary phase) and has been shown to be required for prolonged cell division during entry into stationary phase (i.e. for the conversion of all cells to a minimum size, each with a single copy of the chromosome). The RpoS protein is also required for growth-phase dependent transcription of *bolA*, another morphogene (Aldea et al. 1988) but is not required for growth phase dependent transcription of at least two of the promoters upstream of *ftsZ* (R. Smith & WDD, unpublished). Another transcription factor (product of the *sdiA* gene: Wang et al. 1991) increases transcription specifically from one of the two promoters within the *ddlB* gene, but does not affect the growth phase dependence of transcription from this promoter. The *rpoS*, *bolA* and *sdiA* genes are all dispensible and only the RpoS-deficient cells show any deficiency in growth or survival (namely, larger cells and reduced long-term viability in stationary phase). A mutant deficient in FtsA also shows increased transcription from at least one of these same promoters, and this transcription retains the same growth rate dependence as in wild-type cells (Dewar et al. 1989). It has been suggested that the level of free FtsA protein itself may govern the rate of transcription from these promoters, because FtsA protein is known to be somehow "used up" or sequestered during septation: the onset of septation would therefore cause derepression of the cell division gene promoters at the appropriate time in the cycle (Dewar et al. 1989). However this hypothesis has not been tested further and the controlling factors for "gearbox" transcription are still uncertain.

The products of the *mreBCD* operon (Wachi et al. 1989: fig.1) appear to affect expression of genes coding for some PBPs. The MreB protein represses transcription of the *ftsI* (PBP3-coding) gene (Wachi & Matsuhashi 1989) while deletion of the whole operon also causes an increase in the level of PBPs1B. *MreB* mutants have a coccal morphology, like *rodA* or *pbpA* mutants, or cells overproducing DD-CPases. (It would be interesting to know whether *mreB* mutants can restore division in PBP3-deficient cells.) The genes in this operon may therefore play an important role in the elongation-septation switch. If they do however, it is not at all clear how such a control would work: in order to turn on *ftsI* transcription, preexisting MreB protein would require to be inactivated (and resynthesised or reactivated after sufficient PBP3 had been produced); alternatively, MreB protein would have to be labile and transcription of *mreB* itself controlled. In either case, other factors would be required to control MreB-activity. Despite these problems, the MreB protein seems to be an exciting candidate for a true cell cycle-regulating protein. (It is interesting to note here that MreB has sequence similarity to FtsA and Fic proteins, both of which have also been implicated in control of septation: Dewar et al. 1989, Kawamukai et al. 1989. Even more speculatively, sequence similarity exists between a part of FtsA and the key cell cycle control protein p32$^{cdc2/28/ns}$ of eukaryotes: Robinson et al. 1987.)

In summary, the main outlines of morphogenetic control are becoming clear (fig.2) and most of the key proteins and genes have been identified. A small number of "dedicated" cell cycle proteins are probably enough to run the basic morphogenetic cycle, while numbers of others increase the efficiency of the process under special conditions (Table). (A complication which has not been mentioned here is the presence in many strains of genes of extrinsic origin (e.g. *dicABF*, *sfiC*) which now appear to play no part in the cell cycle but can interfere with essential steps in it under special circumstances. Their original roles may have been in viral or plasmid replication.)

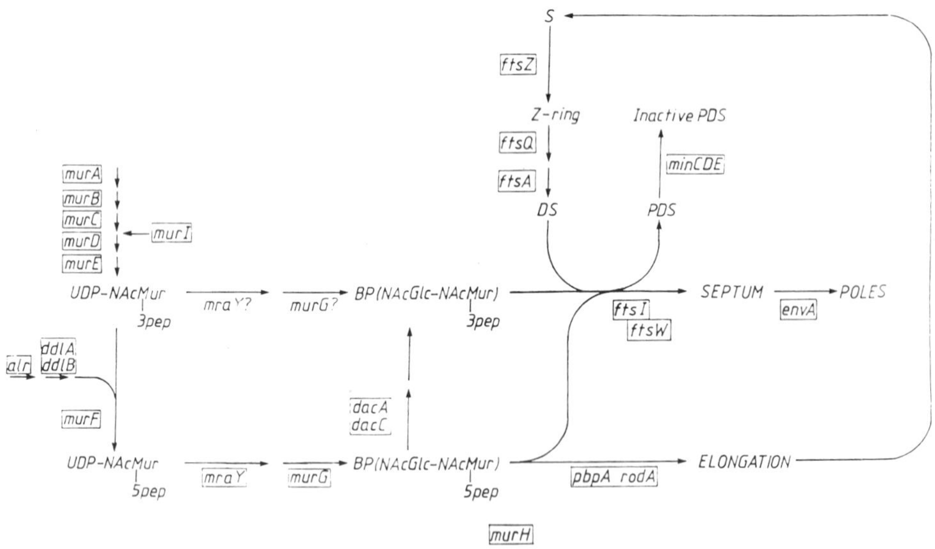

Figure 2. **MORPHOGENETIC SEQUENCE IN THE *E. COLI* CELL CYCLE.**

The following is a possible sequence of events giving rise to the normal growth and division of *E. coli* cells.

Lipid-II (BP[NAcGlc-NAcMur-5pep]) is formed in a series of reactions catalysed by the enzymes coded by the genes shown in boxes. Net PG synthesis requires the MurH protein and PBPs1A & 1B but the action of the *pbpA-rodA* proteins modifies the formation of the PG sacculus to give it a cylindrical shape and allow cells to grow by elongation at constant diameter. When cells reach a critical length ("S"), chromosome replication is complete and sister chromosome separation (partition) can take place. The FtsZ protein then forms a membrane-associated ring in the cell centre (at the position vacated by the replicated sister chromosomes). A PG-septum then forms at this site, following behind the contracting Z-ring. Formation of the septum requires the FtsQ, FtsA, PBP3(FtsI) and FtsW proteins; PBP3 being a septum-specific TPase-TGase which may use tripeptide (-3pep) side chains as preferential acceptors in transpeptidation. The -3pep side chains are produced in two ways: by successive removal of D-alanines from existing -5pep side chains by DD- and DL-CPases, and by direct incorporation of -3pep precursor into PG. FtsQ and FtsA are proposed to form a "division site" (DS) at which PBP3 and FtsW can act to form a septum. The newly-formed septum consists of two covalently-bonded PG-layers which are split apart soon after formation by a process which requires the EnvA protein. Internal pressure in the cell causes the new cell ends to bulge into their final hemispherical shape. After the septum has formed, this location in the cell remains available for further septum formation (i.e. is a "potential division site", PDS) until inactivated by the action of the Min proteins. No further septation is then possible until the cell has once again reached the critical length S and produced a central chromosome-free space.

TABLE 1. "MORPHOGENES" (key to genes in figure 1)

GENE	**LOCUS**	**KBP**	**FUNCTION**
mraR	2	91	Septation/PG synthesis?
ftsI	2	92	Septation: PBP3/TPase-TGase
murE	2	94	PG synthesis: mDAP-adding enzyme
murF	2	96	PG synthesis/septation switch? D-ala:D-ala adding enzyme
mraY	2	97	PG synthesis: lipid-I synthesis
murD	2.1	98	PG synthesis: D-ala adding enzyme
ftsW	2.1	99	Septation: "RodA-like" protein
murG	2.1	100	PG synthesis: lipid-II synthesis
murC	2.2	102	PG synthesis: L-ala adding enzyme
ddlB	2.2	103	PG synthesis: D-ala:D-ala ligase
ftsQ	2.2	104	Septation: transmembrane protein
ftsA	2.2	105	Septation:
ftsZ	2.3	106	Septation/division: Z-ring
envA	2.3	107	Division/pole formation
(secA	2.3	110	Secretion of proteins/secondary effects on septation)
dksA	3.4	160	Suppressor of *dnaK*-septation block
mrcB	3.5	166	PG synthesis: PBP1B/TPase-TGase
ddlA	8.7	407	PG synthesis: D-ala:D-ala ligase
bolA	9.9	461	Enhanced cell division in slow growth
(lon	10	467	Degrades SulA: ATP-dependent protease)
dacA	14.4	673	PG modification/septation switch?: PBP5/DD-CPase
(rlpA	14.4	674	"rare lipoprotein")
rodA	14.4	675	Cell shape: membrane protein
pbpA	14.5	677	Cell shape: PBP2/TPase-TGase
(rlpB	14.6	682	"rare lipoprotein")
dacC	18	890	PG modification/septation switch?: PBP6/DD-CPase
mukB	21.1	986	DNA partition?: "kinesin-like"
(ompA	22	1027	Cell shape (with lpp)?: outer-membrane)
sulA	22	1028	Inhibitor of FtsZ: SOS-regulon
(sfiC	25		Inhibitor of FtsZ: phage-origin)

TABLE 1 (continued)

minE	26.3	1230	Localisation of MinC
minD	26.3	1231	Activation of MinC
minC	26.3	1232	Inhibitor of septation (FtsZ?)
dadX	~26		Alanine racemase
(dif	34.4	1608	XerC-dependent resolution site)
(dicA	35.4	1656	Repressor of dicB(F?): phage origin
dicB	35.5	1658	Inhibitor of FtsZ through MinC
dicF	35.5		Antisense RNA - blocks *ftsZ*)
(lpp	37.8	1767	PG-OM link [cell shape with OmpA?])
sdiA	43	2010	Transcription activator of *ftsQAZ*
mepA	52.7	2463	Penicillin-insensitive DD-endopeptidase
rpoS	61.7	2885	Enhanced division in slow growth:σ^S
sufI	65.3		Suppressor of *ftsI*
dacB	71.7	3349	PG modification: PBP4/endopeptidase
mreD	73.1	3418	Regulation of PBP1B?
mreC	73.1	3419	Regulation of PBP1B?
mreB	73.2	3420	Repressor of *ftsI*
(cha	~75		Division)
(fic	75.1	3512	Interaction between cAMP & septation?)
mrcA	75.9	3546	PG synthesis: PBP1A/TPase-TGase
ftsX	77.5	3624	Septation
ftsE	77.5	3625	Septation (processing Fts proteins?)
ftsY	77.6	3626	Septation?
(envC	~81		Septum localisation/separation?)
(xerC	86.1	4025	dif-specific recombinase)
murI	89.8	4195	PG synthesis: D-glutamic synthesis
murB	89.9	4202	PG synthesis: UDP-NAcGln enolpyruvate reductase
murA	~90		PG synthesis: UDP-NAcGln transferase
mepB	~92		Penicillin-insensitive DD-endopeptidase
alr	91.9	4296	PG synthesis: alanine racemase
murH	99.2		PG synthesis: very late step
slt	99.8	4664	PG modification: soluble lytic TGase

(Loci are given positions to the nearest 0.1 map units and 1 kbp, using a recent compilation of restriction and sequence data: Rudd, K.E.(1992) Alignment of *E. coli* DNA sequences to a revised, integrated genomic restriction map. In "A Short Course in Bacterial Genetics" by J.H. Miller. Cold Spring Harbor Laboratory Press, NY.)

REFERENCES

Aldea, M., Hernandez-Chico, C., de la Campa, A.G., Kushner, S.R. and Vicente, M. (1988) Identification, cloning and expression of *bolA*, an *ftsZ*-dependent morphogene of *Escherichia coli*. J. Bacteriol. 170, 5169-5176.

Aldea, M., Garrido, T., Hernandez-Chico, C., Vicente, M. and Kushner, S.R. (1989) Induction of growth-phase-dependent promoter triggers transcription of *bolA*, an *E. coli* morphogene. EMBO J 8, 3923-3931.

Aldea, M., Garrido, T., Pla, J. and Vicente, M. (1990) Division genes in *Escherichia coli* are expressed coordinately to cell septum requirements by gearbox promoters. EMBO J 9, 3787-3794.

Begg, K.J., Takasuga, A., Edwards, D.H., Dewar, S.J., Spratt,

B.G., Adachi, H., Ohta, T., Matsuzawa, H. and Donachie, W.D. (1990) The balance between different peptidoglycan precursors determines whether *E. coli* cells will elongate or divide. J. Bacteriol. 172, 6697-6703.

Begg, K.J. and Donachie, W.D. (1991) Experiments on chromosome separation and positioning in *Escherichia coli*. New Biol. 3, 1-11.

Bi, E. and Lutkenhaus, J.F. (1991) FtsZ ring structure associated with cell division in *Escherichia coli*. Nature 354, 161-164.

Blakely, G., Colloms, S., May, G., Burke, M. and Sherratt, D. (1991) *Escherichia coli* XerC recombinase is required for chromosomal segregation at cell division. New Biol. 3, 789-798.

Carson, M.J., Barondess, J. and Beckwith, J. (1991) The FtsQ protein of *Escherichia coli*: membrane topology, abundance and cell division phenotypes due to overproduction and insertion mutations. J. Bacteriol. 173, 2187-2195.

Clerget, M. (1991) Site-specific recombination promoted by a short DNA segment of plasmid R1 and by a homologous segment in the terminus region of the *Escherichia coli* chromosome. New Biol. 3, 780-788.

de Boer, P.A.J., Crossley, R.E. and Rothfield, L.I. (1989) A division inhibitor and a topological specificity factor coded for by the minicell locus determine proper placement of the division septum in *E. coli*. Cell 56, 641-649.

Dewar, S.J., Kagan-Zur, V., Begg, K.J. and Donachie, W.D. (1989) Transcriptional regulation of cell division genes in *Escherichia coli*. Mol. Microbiol. 3, 1371-1377.

Donachie, W.D., Begg, K.J. (1989) Chromosome partition in *Escherichia coli* requires post-replication protein synthesis. J. Bacteriol. 171, 5405-5409.

Donachie, W.D., Begg, K.J. (1990) Genes and the replication cycle of *Escherichia coli*. Res. Microbiol. 141, 64-75.

Donachie, W.D., Begg, K.J. and Sullivan, N.F. (1984) The morphogenes of *E. coli*. In "Microbial Development", ed. Losick, R. and Shapiro, L. Cold Spring Harbor Laboratory, NY, USA.

Kawamukai, M., Matsuda, H., Fujii, W., Utsumi, R. and Komano, T. (1989) Nucleotide sequence of *fic* and *fic-1* genes involved in cell filamentation induced by cAMP in *Escherichia coli*. J. Bacteriol. 171, 4525-4529.

Kuempel, P.L., Henson, J.M., Dircks, L., Tecklenburg, M. and Lim, D.F. (1991) *dif*, a *recA*-independent recombination site in the terminus region of the chromosome of *Escherichia coli*. New Biol. 3, 799-811.

Lange, R. and Hengge-Aronis, R. (1991) Growth phase-regulated expression of *bolA* and morphology of stationary-phase *Escherichia coli* cells are controlled by the novel sigma factor σ^S. J. Bacteriol. 173, 4474-4481.

Markiewicz, Z., Broome-Smith, J., Schwarz, U. and Spratt, B.G. (1982) Spherical *E. coli* due to elevated levels of D-alanine carboxypepetidase. Nature 297, 702-704.

Mengin-Lecreulx, D., Parquet, C., Desviat, L.R., Pla, J., Flouret, B., Ayala, J.A. and van Heijenoort, J. (1989) Organization of the *murE-murG* region of *Escherichia coli*: identification of the *murD* gene coding for the D-glutamic

acid-adding enzyme. J. Bacteriol. 171, 6126-6134.
Matsuhashi, M., Wachi, M. and Ishino, F. (1990) Machinery for cell growth and division: penicillin-binding proteins and other proteins. Res. Microbiol. 141, 89-102.
Niki, H., Jaffe, A., Imamura, R., Ogura, T. and Hiraga, S. (1991) The new gene *mukB* codes for a 177 kDa protein with coiled- coil domains involved in chromosome partitioning in *E. coli*. EMBO J 10, 183-193.
Robinson, A.C., Kenan, D.J., Hatfull, G.F., Sullivan, N.F., Spiegelberg, R. and Donachie, W.D. (1984) DNA sequence and transcriptional organization of essential cell division genes *ftsQ* and *ftsA* of *E. coli*: evidence for overlapping functional units. J. Bacteriol. 160, 546-555.
Robinson, A.C., Kenan, D.J., Sweeney, J. and Donachie, W.D. (1986) Further evidence for overlapping transcriptional units in an *Escherichia coli* cell envelope-cell division gene cluster: DNA sequence and transcriptional organization of the *ddl ftsQ* region. J. Bacteriol. 167, 809-817.
Robinson, A.C., Collins, J.F. and Donachie, W.D. (1987) Prokaryotic and eukaryotic cell cycle proteins. Nature 328, 766.
Vinella, D., D'Ari, R. and Bouloc, P. (1991) Penicillin-binding protein 2 is dispensible in *Escherichia coli* when ppGpp is induced. EMBO J 11, 1493-1501.
Wachi, M., Doi, M., Okada, Y. and Matsuhashi, M. (1989) New *mre* genes *mreC* and *mreD*, responsible for formation of the rod shape of *Escherichia coli* cells. J. Bacteriol. 171, 6511-6516.
Wachi, M. and Matsuhashi, M. (1989) Negative control of cell division by *mreB*, a gene that functions in determining the rod shape of *Escherichia coli* cells. J. Bacteriol. 171, 3123-3127.
Wang, X, de Boer, P.A.J. and Rothfield, L.I. (1991) A factor that positively regulates cell division by activating transcription of the major cluster of essential cell division genes of *Escherichia coli*. EMBO J 10, 3363-3372.
Yi, Q-M., Rockenbach, S., Ward, J.E.,Jr. and Lutkenhaus, J.F. (1985) Structure and expression of the cell division genes *ftsQ, ftsA* and *ftsZ*. J. Mol. Biol. 184, 399-412.

"THREE FOR ONE" - A SIMPLE GROWTH MECHANISM THAT GUARANTEES A PRECISE COPY OF THE THIN, ROD-SHAPED MUREIN SACCULUS OF *ESCHERICHIA COLI*

Joachim-Volker Höltje

Max-Planck-Institut für Entwicklungsbiologie
Abteilung Biochemie
Spemannstraße 35
7400 Tübingen, FRG

INTRODUCTION

During growth and division of *Escherichia coli* the stress-bearing rod-shaped murein sacculus necessarily has to be elongated and divided into two intact daughter sacculi. For Gram-positive rods such as *Bacillus subtilis* it has been shown that this is accomplished by an inside-to-outside growth mechanism (Koch and Doyle, 1985). It is based on the presence of a thick multi-layered murein shell which is characteristic for this group of bacteria. The thin murein layer of Gram-negative bacteria demands for a different, more sophisticated mechanism.

CURRENT GROWTH MODELS

Clearly, an insertion mechanism where bond hydrolysis precedes covalent intercalation of new glycan strands would be very risky. It is more likely that growth of the Gram-negative sacculus follows the save strategy of "make before break" (Koch, 1990). Although still a matter of debate whether there are one or two complete layers of murein it clearly is a thin, highly vulnerable structure. Recent experimental results obtained by neutron small-angle scattering studies indicate a value of about 1.5 layers for the average thickness of the murein sacculus (Labischinski *et al.*, 1991). Therefore we proposed a patching model for the growth of the murein sacculus of *E. coli* (Höltje and Glauner, 1990). Accordingly, patches of nascent, relaxed murein are hooked underneath the stress-bearing layer via crosslinks to existing dimeric peptide bridges yielding trimeric crossbridges. The patches will then be pulled into the layer under stress by specific cleavage of the correct bond in the trimeric structures.

However, as pointed out by Arthur Koch the patching model suffers from being incompatible with a mechanism that preserves the given rodshape (Koch, 1990). Diffuse insertion of patches of murein would cause bulging and finally rounding off to a sphere. Arthur Koch, therefore, presented a model that was derived from a growth mechanism postulated by Ted Park (Burman and Park, 1984) and earlier from Helmut Pelzer (1983), in which new glycan strands become inserted between the existing strands which are arranged in a ring-like structure perpendicular to the long axis of the cylinder. This attractive hypothesis could explain cell elongation while maintaining the given shape, that is the diameter of the cylindrical part of the rod. However, the important finding of murein

Bacterial Growth and Lysis, Edited by M.A. de Pedro
et al., Plenum Press, New York, 1993

turnover cannot be explained by such a mechanism. To explain turnover of murein, which has been reported to be in the range of about 40 - 50% per generation (Goodell, 1985), Arthur Koch (1990) suggested that two new strands become inserted at the expense of one old strand. This mechanism resembles the patching model with the patch more precisely defined as a double strand. It was postulated that "smart enzymes" recognize the presence of unstressed new murein underneath the stress-bearing murein layer. Molecular model considerations, however, seem to favour a model by which three new glycan strands are inserted all in a body at the expense of one old strand. The "three for one" model does not depend on smart enzymes and could explain not only cell elongation but cell division as well. Finally, it is compatible with a mechanism that precisely replicates the parent murein sacculus into two identical daughter sacculi.

THE BASICS OF THE "THREE FOR ONE MODEL"

It is proposed that enlargement of the murein sacculus of *E. coli* takes place in three steps (Fig. 1). Firstly, a prefabricated triplet of crosslinked glycan strands is hooked underneath one of the strands in the stress-bearing layer of the murein sacculus. The chosen strand to which the new glycan chains are attached will be called the template strand, because it determines the lengths of the newly synthesized glycan chains. Due to the flexibility of the peptide bridges, which in an extended conformation can span more than twice the distance than in a relaxed state (Braun *et al.*, 1973; Labischinski *et al.*, 1983) three glycan strands crosslinked with each other can easily be covalently attached to the stressed crossbridges on both sides of a template glycan strand thereby forming trimeric crossbridges. Specific removal of the template strand would immediately pull in the triple pack of glycan chains into the stress-bearing layer. This can be done by enzymes specifically recognizing the trimeric crossbridges which would function as specific signals. Therefore, the smart enzymes that have been postulated by Arthur Koch (1990) need not be too smart. Simple enzymes with a specificity for trimers would do. For a complete degradation of the template strand endopeptidase and muramidase activity must cooperate. The question is, which enzyme makes the first cut. The two possible pathways are shown in Fig. 1. Since an arrangement of the glycan strands perpendicular to the long axis of the rod-shaped murein sacculus has been demonstrated (Verwer *et al.*, 1978), this will result in elongation of the sacculus. The final outcome of this mechanism would be the insertion of three new glycan strands at the expense of one old strand, namely the template strand. This is not only a safe but also a quite elegant process.

In order to make the insertion mechanism a real safe device to enlarge the murein we have to postulate that the template strand is substituted with acceptor peptides only (Fig. 2) An acceptor peptide is characterized by the fact that the ε-amino group of its diaminopimelic acid (A_2pm) residue is part of the crosslinking peptide bond. By contrast the donor peptide carries a free ε-amino group and therefore can function as an acceptor for the formation of trimeric and higher oligomeric crossbridges. As illustrated in Fig. 2, the incoming triplet of glycan strands must be linked to the dimeric crossbridges on both sides of the template strand in such a way that the newly formed crosslinks enclose the pre-existing stress-bearing D-Ala-A_2pm crosslinks. This is only possible if the template strand is an acceptor strand interlinked to two donor strands. If we further assume that donor and acceptor strands alternate in the murein it follows that the incoming triple pack consists of a donor, an acceptor and a second donor strand. Since the template strand is removed and only one new acceptor strand is inserted one may conclude at first sight that the number of growing points remains constant throughout. However, by the conversion of the dimeric crossbridges into trimeric crossbridges the two old donor strands on both sites of the incoming triple pack are transformed into mixed acceptor/donor strands. During further growth they finally will become plain acceptor strands. Therefore the number of possible growing points, that is the number of acceptor strands, increases with growth and may double during one cell cycle.

CELL DIVISON BY THE "THREE FOR ONE" MECHANISM

By this very same "three for one" mechanism also constriction of the cell can easily be envisaged. During cell division the addition of tripletts of glycan strands would be

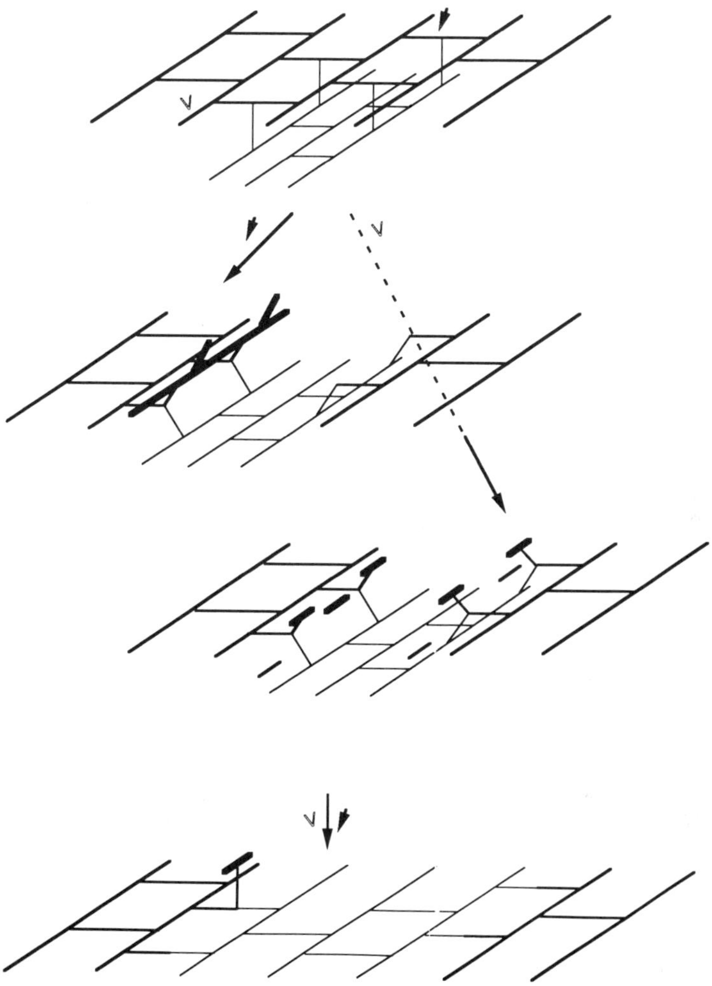

Figure 1. Schematic representation of the hypothetical "three for one" growth mechanism. Two possible pathways are shown. The arrow pointing to the left illustrates the mechanism that is started by the action of a specific endopeptidase (→); the arrow pointing to the right shows the insertion reaction that begins with the specific cleavage of the acceptor strand by the action of an exo-muramidase (>). Both pathways finally lead to the result shown at the botton of the drawing.

directed excusively to a group of template strands which in its entirety form a ring at the future site of cell division. The assembly of the *ftsZ* gene product into a ring structure, which has recently been reported, may define this site (Bi and Lutkenhaus, 1991). Due to an increase in the rate of murein synthesis during septum formation (Wientjes and Nanninga, 1989) cleavage of the trimeric crosslinks is delayed compared with the rate of addition of further triple packs of glycan chains. As a result of this a three-layered ingrowing septum is formed (Fig. 3A). Unlike in Gram-positive bacteria, in *E. coli* synthesis of the septum overlaps with its hydrolytic cleavage. By specific recognition of the

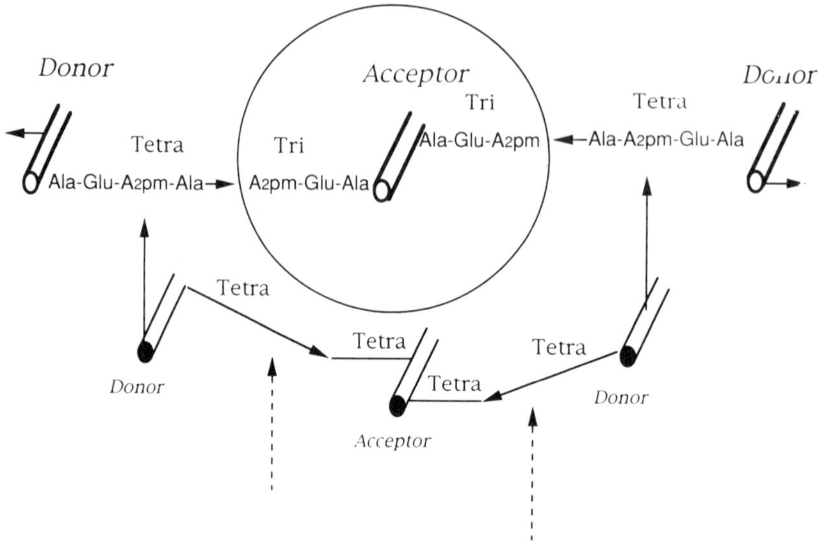

Figure 2. Attachment of the incoming triplet of crosslinked glycan chains to the template strand in the stress-bearing murein layer by the intermediate formation of trimeric crossbridges followed by cleavage of the pre-existing peptide link of the accepting tetra-tri dimer.

trimeric crosslinks would the the middle layer composed of the template strands be removed progessively (Fig. 3A-C). Such a zipper-like cutting process by murein hydrolases (Kohlrausch and Höltje, 1991) is reminiscent of the splitting system described for *Staphylococci* (Giesbrecht et al., 1976). The tripletts added to the leading edge of the inwards growing septum will be made of shorter and shorter glycan strands because of the ever-decreasing circumference this leads to the formation of the typical constriction (Fig. 3B). Cell division proceeds until the aperture is almost closed. In this situation further addition of triplets of glycan strands to the leading edge is hindered sterically, whereas the glycan strands on both sides of the merging aperture are close enough to each other to allow lateral attachment of triple packs of glycan chains to two dimeric crossbridges residing at the opposite sides of the orifice. Cleavage of the trimeric crossbridges would finally fuse the diaphragm giving rise to the formation of two new polar caps with concomittant release of the separated daughter cells (Fig. 3D). At no time point during this operation is the stress-bearing sacculus disrupted. Thus a safe division of the cell is guaranteed.

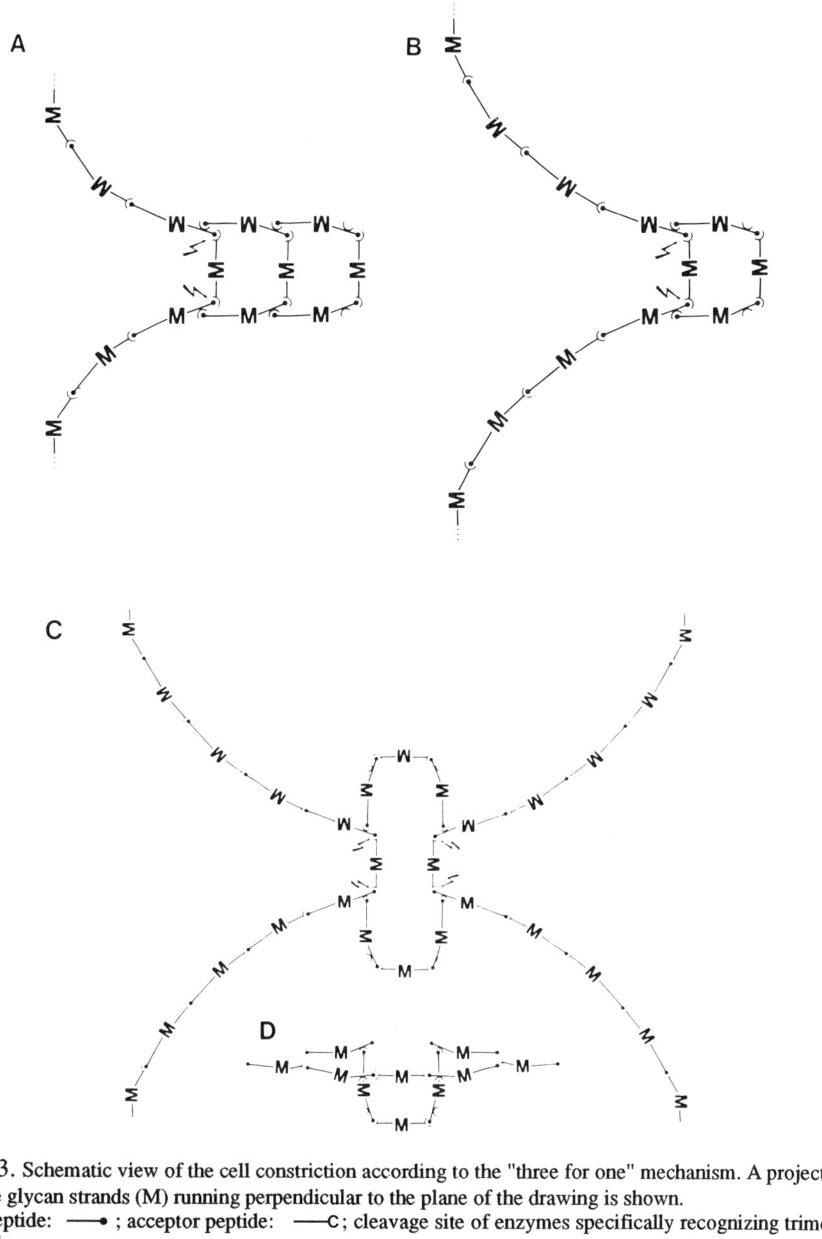

Figure 3. Schematic view of the cell constriction according to the "three for one" mechanism. A projection along the glycan strands (M) running perpendicular to the plane of the drawing is shown.
Donor peptide: —• ; acceptor peptide: —c; cleavage site of enzymes specifically recognizing trimeric crossbridges: ↯

DATA SUPPORTING THE MODEL

Although the model describes a safe mechanism for cell elongation and division, it remains to be shown that this model is supported by experimental facts. The "three for one" hypothesis as a matter of fact is compatible with a number of fundamental observations. Firstly, it explains murein turnover as an inherent feature of the growth strategy. It should be stressed that turnover must not be considered a waste of energy; this energy is well invested since it allows growth of the sacculus according to a rather safe mechanism. Secondly, since trimers would function exclusively as intermediate structures they will have a rather short half-life. Hence, the low amounts of trimers which have been determined (2.33% for the tetra-tetra-tetra and 0.28% for the tetra-tetra-tri; Glauner *et al.*, 1988) are consistent with the model. Furthermore, all trimers will be cleaved during growth. Pulse and pulse-chase experiments clearly demonstrated that in accordance with the model all trimers are cleaved with time (Glauner and Höltje, 1990).

Interestingly, pulse pulse-chase experiments revealed that almost all tetra-tri dimers but only a fraction of the tetra-tetra dimers are cleaved during growth (Glauner and Höltje, 1990). This is a peculiar finding because *in vitro* the endopeptidase present in *E. coli* makes no difference between these two dimers. According to the "three for one" growth mechanism these two types of dimers follow different metabolic routes. Tetra-tri dimers known to be present in old murein only are the preferred acceptors for the attachment of new glycan strands during cell elongation (see Fig. 2). After the intermediate formation of a trimer structure is the tetra-tri linkage being cleaved in order to insert the new strands into the plane under stress. By contrast tetra-tetra dimers are likely to be trimmed to tetra-tri by carboxypeptidase action before getting also involved in accepting further incoming glycan strands. However, during cell division also tetra-tetra dimers will function as acceptors for the triple packs of glycan chains because of the increase in the rate of murein synthesis and will be cleaved during cell separation.

Finally, according to the model the formation of the constriction is a result of the locally restricted insertion of glycan strands with decreasing lengths due to sterical hindrance in the closing septum. Indeed analysis of the length distribution of the glycan chains in the murein of mini cells, which may be considered to consist exclusively of two polar caps, revealed a shift to the shorter glycan strands as compared to cylindrical murein (W. Obermann and J.-V. Höltje, unpublished results).

IDENTICAL REPLICATION OF THE MUREIN SACCULUS

Growth and division of a bacterial cell takes place with its specific shape maintained. Therefore the shape determining murein sacculus must be replicated identically during the cell cycle. The "three for one" growth mechanism proposes that the parent murein sacculus acts as a template for the formation of two identical daughter sacculi. This is accomplished by precisely copying the different lengths of the glycan strands from which the parent sacculus is made (Harz et al., 1990). To insure a co-ordinated synthesis of a triplet of glycan strands one can visualize a complex of three synthesizing enzymes. By moving along the strand to be replicated the enzyme complex synthesizes three copies until reaching the end of the template where the complex would dissociate off. Thus, the length of the newly synthesized strands would be determined by the pre-existing ones.

Finally one could think of co-ordinating the synthesis of new glycan chains with the degradation of the template strand by the formation of a complex that consists not only of the three synthesizing enzymes but also of the hydrolytic activity. In the model shown in Fig. 4, the template strand is chopped up by an exo-muramidase. As discussed in this volume (Romeis and Höltje; Templin and Höltje) the *E. coli* muramidase, called soluble lytic transglycosylase, binds strongly to murein. It is therefore speculated in this model that the muramidase attaches the complex to the template strand and by processively degrading the template pushes the machinery forwards. A proper arrangement and possibly an allosteric interaction of the subunits in such a hypothetical multienzyme complex could warrant that hydrolysis of covalent bonds at no circumstance takes place before new subunits have been covalently linked to the growing murein sacculus. In analogy to the complex process of DNA replication carried out by an intricate interplay of many proteins it

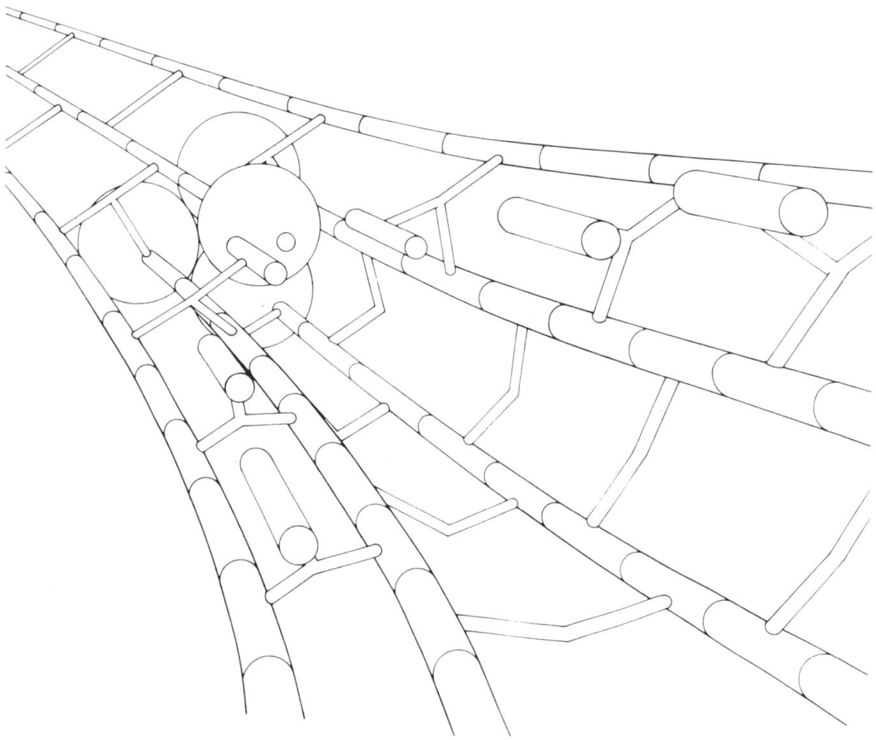

Figure 4. Artists view of a hypothetical multi-enzyme complex engaged in murein synthesis by the insertion of three new glycan strands at the expense of one pre-existing one.

is quite likely that the replication process of the murein sacculus will turn out to be by far more complicated as proposed in this simple, hypothetical "three for one" growth model.

REFERENCES

Bi, E. and Lutkenhaus, J. (1991) FtsZ ring structure associated with division in *Escherichia coli*. Nature 354, 161-164.

Braun, V., Gnirke, H., Henning, U. and Rehn, K. (1973) Model for the structure of the shape-maintaining layer of *Escherichia coli* cell envelope. J. Bacteriol. 114, 1264-1270.

Burman, L.G. and Park, J.T. (1984) Molecular model for elongation of the murein sacculus of *Escherichia coli*. Proc. Natl. Acad. Sci. USA 81, 1844-1848.

Giesbrecht, P., Wecke, J. and Reinicke, B. (1976) On the morphogenesis of the cell wall of *Staphylococci*. Int. Rev. Cytol. 44, 225-318.

Glauner, B. and Höltje, J.-V. (1990) Growth pattern of the murein sacculus of *Escherichia coli*. J. Biol. Chem. 265, 18988-18996.

Glauner, B., Höltje, J.-V. and Schwarz, U. (1988) The composition of the murein of *Escherichia coli*. J. Biol. Chem. 263, 10088-10095.

Goodell, E.W. (1985) Recycling of murein by *Escherichia coli*. J. Bacteriol. 163, 305-310.

Harz, H., Burgdorf, K. and Höltje, J.-V. (1990) Isolation and separation of the glycan strands from murein of *Escherichia coli* by reversed-phase high-performance liquid chromatography. Anal. Biochem. 190, 120-128.

Höltje, J.-V. and Glauner, B. (1990) Structure and metabolism of the murein sacculus. Res. Microbiol. 141, 75-89.

Koch, A.L. (1990) Additional arguments for the key role of "smart" autolysins in the enlargement of the wall of Gram-negative bacteria. Res. Microbiol. 141, 529- 541.

Koch, A.L. and Doyle, R.J. (1985) Inside-to-outside growth and turnover of the wall of Gram-positive rods. J. Theor. Biol. 117, 137-157.

Kohlrausch, U. and Höltje, J.-V. (1991) Analysis of murein and murein precursors during antibiotic-induced lysis of *Escherichia coli.* J. Bacteriol. 173, 3425-3431.

Labischinski, H., Barnickel, G. and Naumann, D. (1983) The state of order of bacterial peptidoglycan, in "The target of Penicillin" (Hackenbeck, R., Höltje, J.-V. and Labischinski, H., Eds.) pp. 49-54. de Gruyter, Berlin.

Labischinski, H., Goodell, E.W., Goodell, A. and Hochberg, M.L. (1991) Direct proof of a "more-than single-layered" peptidoglycan architecture of *Escherichia coli* W7: a neutron small-angle scattering study. J. Bacteriol. 173, 751-756.

Verwer, R.W.H., Nanninga, N., Keck, W. and Schwarz, U. (1978) Arrangement of glycan chains in the sacculus of *Escherichia coli.* J. Bacteriol. 136, 723-729.

Pelzer, H. (1983) Murein ring structure and biosynthesis in gram-negative bacteria: some tentative contemplations, in "The target of Penicillin" (Hackenbeck, R., Höltje, J.-V. and Labischinski, H., Eds.) pp. 105-111. de Gruyter, Berlin.

Wientjes, F.B. and Nanninga, N. (1989) Rate and topography of peptidoglycan synthesis during cell division in *Escherichia coli* : concept of a leading edge. J. Bacteriol. 171, 3412-3419.

STRESSES ON THE SURFACE STRESS THEORY

Arthur L. Koch

Biology Department
Indiana University
Bloomington, Indiana, 47405

INTRODUCTION

An understanding of wall growth in prokaryotes will only be achieved by the combined assault of microbial physiology, biochemistry, genetics, and biophysics. Essentially the problem is one in engineering, material science, and architecture; these subjects provide the paradigms that we must used. On this basis, the surface stress theory (SST) was formulated twelve years ago (Koch *et al.,* 1981). Over time it has been variously refined, modified, reviewed, quoted, incorporated into textbooks, restated, attacked, criticized, misunderstood, rejected, and overlooked. I hope here to present it in a fully defendable mature form.

The prokaryotic strategy proffered by the SST (Koch *et al.,* 1981 *a*; Koch, 1983, 1988 *a*) represents an alternative to the eukaryotic process of cellular growth and is one that could be imagined to have been built up, step by step, by organisms evolving before the split into the three kingdoms of eubacteria, archaebacteria, and eukaryotes (Koch, 1992). I have argued that the completion of a mechanism allowing strong external walls was the feature that led to the splitting off of the eubacteria from the remaining groups. The main biophysical problem is to comprehend how rod-shaped bacteria can elongate instead of rounding up and growing bigger. Although control of shape and ability to divide is achieved with contractile elements in the cells of eukaryotes (see Koch, 1991), this is not the way of prokaryotes.

The major misunderstanding concerning the surface stress theory has to do with its use of the analogy of wall growth to the blowing up a soap bubble (Koch *et al.*, 1982; Koch, 1982, 1984, 1988 *b*). Given the appropriate conditions, the soap bubble analogy can explain elongation of cylindrical cells without the ncecessity of postulating high regularity to the peptidoglycan wall structure. On the alternative view of the wall as an elastic solid, there is

Bacterial Growth and Lysis, Edited by M.A. de Pedro
et al., Plenum Press, New York, 1993

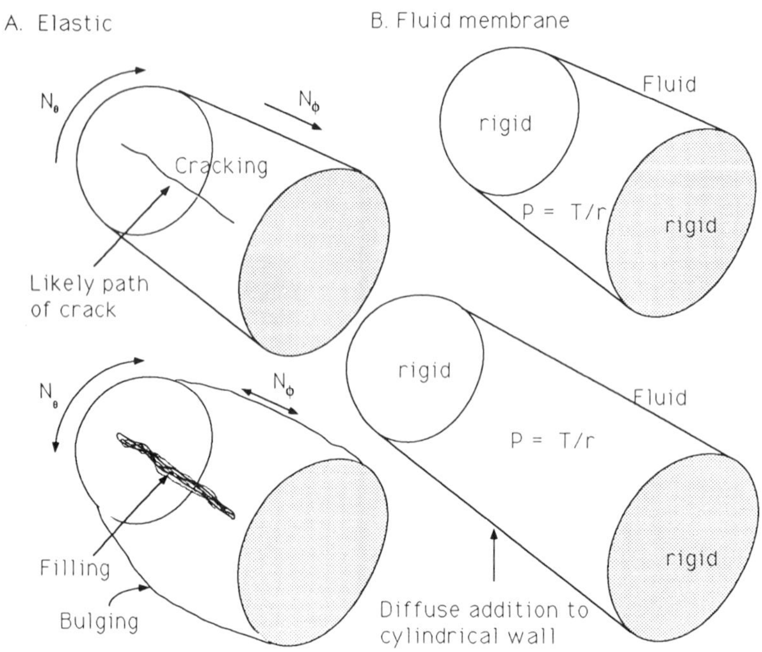

Fig. 1 An elastic solid versus a fluid membrane. Panel A. Cracking and filling of an elastic solid leading to bulging. At the top, a closed tube or pipe of an elastic material is subject to an internal pressure resulting in hoop and axial stresses. For this geometry the stresses are given by: $N_\theta = Pr$ and $N_\varnothing = Pr/2$, where r is the radius of the cylinder and P is the turgor pressure. Because of Hooke's law, a unit length on the cylinder circumference is enlarged two times more than a unit length in the axial dimension. With sufficiently high pressure the solid will exceed its elastic limit and split and cause preferential splitting to occur axially. If the splitting occurs slowly enough so that complete destruction does not occur, growth by repair (shown at the bottom) can take place and bulging will occur which will lead to an increase in diameter, eventually to a spherical form. Panel B. Diffuse addition to a fluid membrane. A fluid soap bubble film between two rings is shown with pressure leading its enlargement. For the soap film, water molecules from the interior are forced to enter the surface in the energy-requiring step; this is the basis of the surface tension. Surface tension, T, is the work required to increase the surface area by a unit amount. The relationship between pressure, tension, and the radius of the cylindrical film is given by the Young-Laplace equation: $P = T(1/r_1 + 1/r_2)$, where the r's are the principle radii of curvature. For a cylinder this reduces to $P = T/r$. If cylindrical wall of a bacterium grows by diffuse enlargement then the physical situation is analogous to that of the soap film. Consequently, at constant pressure under conditions where the poles are rigid then stable cylindrical elongation can occur and any bulging or constriction will self-correct with further growth.

no way that the biophysical processes alone could lead to cylindrical growth of either Gram-positive or negative cells (see Fig. 1). The stresses on a rigid cylinder would lead to cracks developing in the longitudinal direction, if these were filled with new material as they formed the consequence would be an enlargement and gradually the cell would become more spherical. On the other hand, when viewed as a process in which pressure-volume work forces wall growth, many of the features of bacterial growth can be understood. What is required that the wall formation process satisfies the equations that lead to theYoung-LaPlace equation: $P = T(1/r_1 + 1/r_2)$ (Koch, 1984) that also applies to soap bubbles and that several additional conditions and constraints are also met.

The next section review the evidence against contractile elements as determinants of cell shape, the following sections deals with the ways in which a wall growth process can function to satisfy the requirements of fluid membrane for the Gram-negative and the Gram-positive organisms. Then the possibility suggested by Höltje (Höltje and Glauner, 1990; Glauner and Höltje, 1990 for the elongation of the wall of *Escherichia coli* in a manner similar to the inside-to-outside process (Koch *et al.*, 1981*b*, 1982) which does function for growth in rod-shaped cells with a multilayered wall is reviewed.

IS THE SHAPE CONTROLLED BY AN INTERNAL CYTOSKELETON?

At present the major laboratory trying to find mechano-proteins that could render the problem of bacterial shapes equivalent to those pursued by eukaryotes is that of Ian Holland and Vic Norris at Leicester (Holland *et al.*, 1990; Casaregola *et al.*, 1990) They have found evidence for a myosin-like protein and a protein kinase C-like protein (Norris *et al.*, 1991) in *Escherichia coli* and cite examples in the literature of prokaryotic proteins that appear to be similar to those in higher organisms. On the other hand, many of the similarities are not so extensive as to suggest that the eukaryotic kinds of apparatus (involving families of actins, dynines, kinesins, and tubulins) are present and functioning in prokaryotes (Koch, 1991). For example, a similarity was noted between actin and a protein of E. coli; this protein turned out to be EF-T_u. The similarity of the amino acid sequence of a portion of the E. coli protein is real, but there are several reasons to believe that it does not function as a contractile element in the bacteria. First, it has a known role in protein synthesis that does not involve the transduction of chemical into mechanical work. Second, EF-T_u is closer in function to G-proteins of eukaryotes and are quite unrelated to any mechano-protein. The third reason is that while EF-T_u does polymerize in the test tube it has not been demonstrated to contract or expand in response to ATP or other chemical signals. In my literature survey of this question (Koch, 1991), carried out in the Spring of 1990, many citations to mechano-proteins in prokaryotes were found; most were studies concerned eukaryotic genes that had been cloned and did lead to the formation of functional proteins without disrupting their bacterial host. I found some suggestive evidence that in some of the 'higher' prokaryotes that there were proteins that might be closer to contractile proteins. I have extended the literature review to the Spring of 1992 without finding any report of a bona fide mechano-protein or mechano-enzyme. As the number of unassigned open reading frames in the *E. coli* continues to decrease, the chances that contractile proteins would have been overlooked is decreasing. If such proteins truly served a role in bacterial wall growth and division, they should have been found in screens such as those that produced temperature sensitive mutations in cell division; i.e., *fts* mutations.

FLUID MEMBRANE GROWTH OF THE GRAM-NEGATIVE WALL

Growth models for stress-bearing cylindrical walls can logically be divided into two categories. In one, the inside-to-outside model (Koch *et al.*, 1981*a*, 1982), a layer of unstressed wall is laid down in apposition to the stress-bearing wall. It is cross-linked to itself and the older wall and through further growth and turnover comes to stretch and bear stress. This process requires a multilayered wall structure and is described in the next section.

For organisms like *E. coli*, the evidence is now clear that most of the wall is single layered; indeed, the stress-bearing layer may be only one molecule thick throughout. This was established by neutron studies (Labischinski *et al.*, 1991) and are consistent with only 20-25% of the surface being multilayered, but even from that we must subtract away the portion in which wall has been added that does yet not bear stress, wall that no longer bears stress, wall regions at the crevasses of the constrictions formed as part of the cell division process, and wall wrinkled due to the shrinkage of the wall of the bacteria, once killed.

The basic strategy for this second case is considered in this section and is that of insertion and linkage of new material into a stress-bearing wall that is then is by followed by cleavages destroying old stress-bearing bonds permitting the surface area to increase. The SST Variable-T model for enlargement of prokaryotic organism (Koch and Burdett, 1984) with a wall one stress-bearing layer thick is that the insertion of new disaccharide pentapeptide elements need not, and probably does not occur, in a regular systematic way, but are inserted at random (Koch, 1988 *a*). This gives the process the physical properties of a fluid film and the stresses and strains in a non-growing elastic solid of the same shape are totally irrelevant. The disaccharide pentapeptides after translocation through the wall become attached to existing stress-bearing peptidoglycan chains by both glycosyl and peptidyl bonds. (Actually these units have six ways to be linked to existing wall. While two are sufficient to form part of a chain, more are required to be part of a structural two-dimensional fabric.) For this random insertion strategy to work successfully, there are three requirements: i) The cross linkages be formed first; ii) the cross-links be capable of elastic stretching; and iii) stress-bearing glycan be selectively cleaved so that the newly inserted murein is extended and comes to bear the surface stress. It was argued that three assumptions (monolayer, random insertion, elastic stretching) are the minimum set of assumption which are *a priori* sufficient to allow the physics of the situation to result in stable cylindrical growth. With the addition

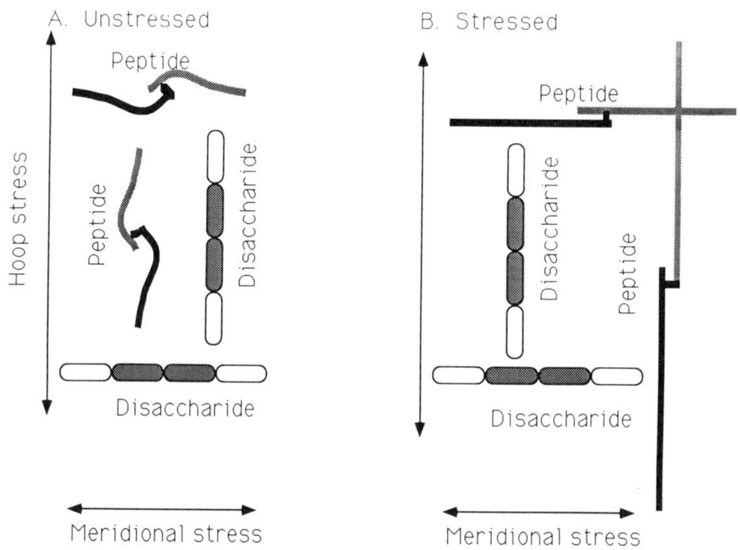

Fig. 2 Extensibility of the glycan and peptide chains. Two principles are illustrated in the figure. The first is that the glycan chain, whose sugar components are shown by the ovals, cannot be elongated appreciably, without rupture, by stress due to turgor pressure, whereas the peptide chain can be. The second principle is that the stress in the meridional direction is less that in the hoop direction. It has been argued, see text, that for the case of a monolayer (but not for the thicker inside-to-outside mode of growth for the Gram-positive rod-shaped walls) enlarging by diffuse insertion that the new wall elements will gradually become orientated so that the peptide chains are roughly parallel to the long axis. This will occur even though it appears contrary to the magnitude of stresses. This is because orientation of the inserted chains with the saccharide align in the meridional direction will be gradually selected against. Then as new wall is added, only if not oriented in the hoop direction can the units be inserted readily between earlier wall with partial orientation.

of the assumption that the surface tension can be varied by the cell from location to location, they can account for cell division.

How is the new wall added?

The evidence for systematic order in addition peptidoglycan is not strong, but neither is the evidence for random addition. The evidence in favor of order is from the studies of the Dutch group (Verwer *et al*., 1978, 1980). They used sonication and endopeptidase treatment and published electron microscope photographs that can be interpreted as showing insertion of glycan hoops around the cylindrical part of the cell. The strictest possible interpretation of glycan chains entirely encircling the cell would have little backing because the glycan chains are much too short to reach. One is on stronger grounds if the data are interpreted as showing that there is a preferential alignment of the peptide cross-bridges parallel to the cell axis. Even without a systematic insertion it is possible that the orientation will become ordered to some degree during the insertion process (Koch, 1988 *b*). Such partial ordering could take place because the glycan chains are almost inextensible while the crosslinked peptide chains have a range of conformations with quite variable end-to-end distances (Braun *et al.*, 1973; Burge *et al.* 1977 *a,b*); Labischinski *et al.*, 1979, 1983, 1985). Consequently if the units are installed preferentially in the orientation that allows them to be able to expand axially when they come to bear stress it follows that the subsequent units inserted even at random positions in the wall will be attached to the existing wall in a way to reinforce and maintain this order (see Fig. 2). There seems to be little evidence from the photographs that the orientation is highly regular and none that the wall is inserted in a helical path. (Below, the latter point is important when the Gram-positive rod is considered.)

From the SST point of view it is the transpeptidase action that is most vital to the elongation process. This follows (again) from the fact that only the relaxed peptide chains are capable of considerable extension and from the necessity of linking the new wall to the old in such a way that as an old bond is cleaved the new peptide crossbridges extend, by conformational changes, to enlarge the surface. This means a new oligopeptidoglycan must be initially linked to connect one old glycan chain to a second. Whether turnover does not occur or is obligatory will determine if the new glycan is inserted between existing chains (allosteric model) (Koch, 1990 *a*) or whether several nascent chains must be pre-linked and then attached across one or more glycan chains and then the old chains excised from the stress-bearing structure and subsequently detached from the wall, leading to turnover (Fig. 3) (multienzyme model) (Koch, 1990 *a*) .

How extensible is new wall?

There are questions about the stretchability and about the actual degree of stretch of the peptidoglycan fabric as it exists in the walls of a living bacterium. To measure the strechability, Koch and Woeste (1992) made preparations of purified peptidoglycan sacculi

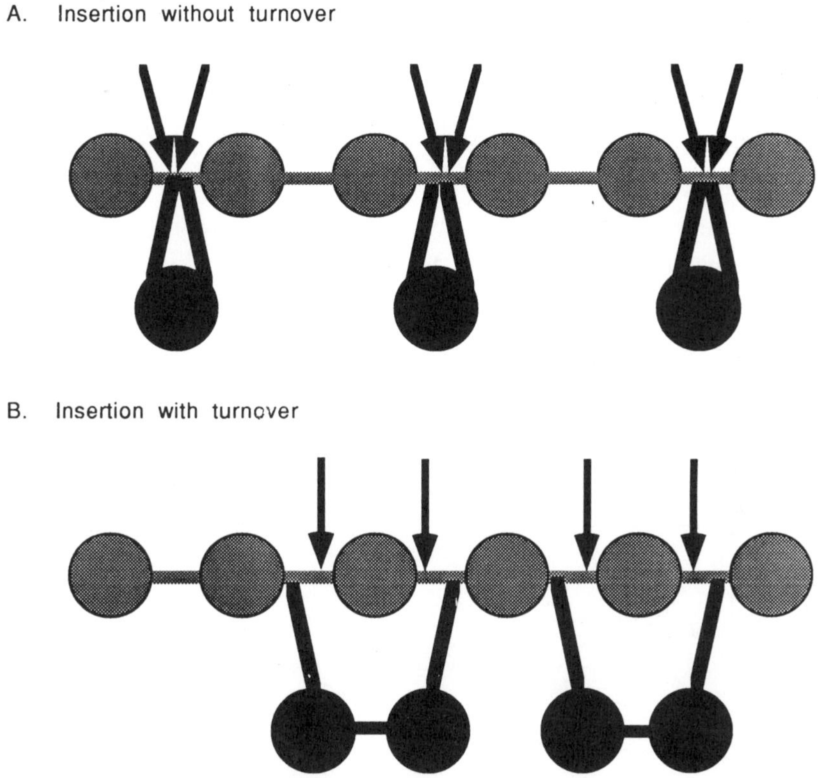

Fig. 3 Two modes for Gram-negative monolayer growth. With only a single stress-bearing layer it is necessary for insertion of new wall, in a covalently linked form, precede cleavage. Two strategies can be considered depending on whether turnover of the old wall does or does not take place. The figure shows two cross sections of the growing wall. At the left, the insertion without turnover model, is shown. In this case (the allosteric model), the new oligopeptidoglycan strand must be covalently linked, via transpeptidation, to groups on adjacent existing stress-bearing strands. Once linked, autolysins can selectively cleave the bonds holding these two existing strands. At the right, the insertion with turnover model is shown. In it two strands of newly secreted oligopeptidolglycan are first linked together then linked around a stress-bearing chain and then the autolysins cleave and release only the previously stress-bearing oligopeptidoglycan strand. This case (the'multienzyme model), is limiting case for the 'patches' model of Höltje and Glauner, see text.

of *Escherichia coli* and studied them by low-angle laser-light scattering to estimate the mean sacculi surface area (MSSA). Control experiments and theoretical calculations based on the Rayleigh-Gans theory showed the MSSA could be accurately inferred from measurements with our apparatus using computer routines developed previously (Koch, 1989). The saccular shape was modeled on the assumption that sacculi preparations consisted of a population drawn from balanced growth and each sacculus had the shape of an ellipsoidal shell of revolution. Large changes in MSSA resulted from varying the net charge on the sacculi. The net charge was affected by altering the pH of the suspending medium causing carboxyl and amino groups in the peptidoglycan to gain or lose protons. It was also varied by acetylation and succinylation of the amino groups. Acetylation prevents protonation of the amino groups and succinylation replaces potentially positively charged amino groups with potentially negatively charged carboxylic acid groups. Either a preponderance of plus or a minus charges caused an expansion of the MSSA. The largest increase in area probably represents the elastic limit of the peptidoglycan and was 300% above that of isoionic sacculi

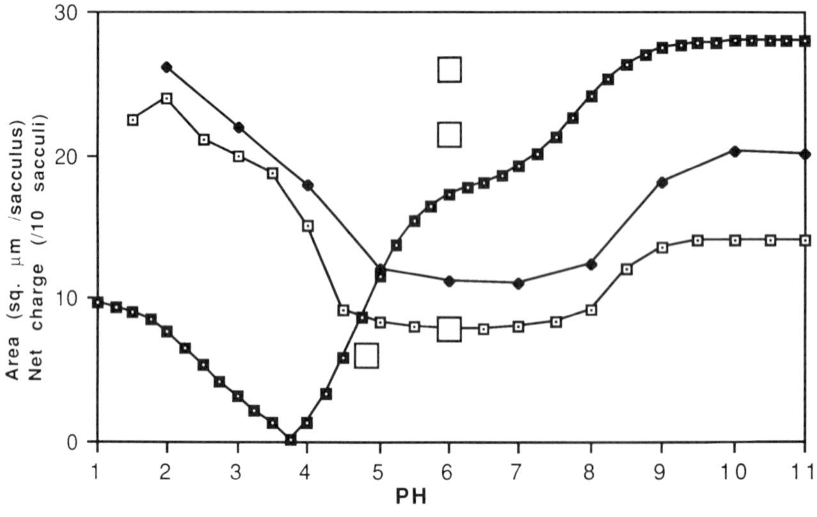

Fig. 4. The mean sacculus surface area (MSSA) as a function of pH. As the experimental data were recorded, acidic titrations started at pH 6.0 and ended at pH 1.5 and then the samples were back titrated. Similarly, basic titrations started at pH 6.0 and ended at pH 12.5 and then were back titrated. The light scattering measurements under different conditions were converted into areas of ellipsoidal shells of revolution and are reported as μm^2. The mean sacculus surface area as a function of pH in the absence and presence of 10.1 M urea are shown by the small squares and the filled diamonds. The large empty square ranging from top to bottom are: succinylated sacculi; acetylated sacculi; untreated sacculi at the same pH; and untreated sacculi in 1 M KCl. The filled small squares are for the theoretical net charge are the average charge per sacculus calculated from the composition of the sacculi and the pK values of the ionizable groups.

(Fig. 4). This degree of expansion is consistent with possible conformations of the intact peptidoglycan structure achievable without necessitating any rupture of the wall fabric. Further enlargement would probably require a very large stress to break covalent bonds. Thus these findings mean that newly-formed fabric can stretch sufficiently to allow the monolayer, as formed, to expand a good deal.

The second question concerns the degree of expansion of the fabric as it exists in the growing cell and is supporting the normal turgor pressure. This required a different kind of experiment. Koch *et al.* (1987) grew a mutant TOE13 of *E. coli* at its non-permissive temperature. Under these conditions this *fts* mutant grows as a filament. Growth was permitted until the filaments were long enough to be accurately measured in the phase microscope. The cell were attached to the glass of rectangular capillaries with polylysine. We could follow the elongation of the filaments on the heated stage. The critical experiment was to flow detergent solutions through the capillary to cause the breakdown of the phospholipid bilayers and the dissipation of the turgor pressure. The filaments were observed to shrink in length by 17%. This corresponds to the MSSA of the growing cell being 45% greater that of the relaxed conformation in turgor-free cells.

The results with the first type of experiment with the isolated sacculi *in vitro* showed that the sacculi can reversibly increase in area 300% above the relaxed conformation much more than the 45% *in vivo*. The disparity of the *in vivo* results with growing cells with the *in vitro* studies of intact sacculi could have several explanations. The least probable is that the wall is in a partially extended configuration due to interactions with the inner and outer membrane even after disruption with detergent and it is more likely that the sacculus under growing conditions never is stretched enough to approach its elastic limit. These findings obviously will be very important in further studies of the mechanics of bacterial growth.

INTERPLAY OF STRESSES FOR THE CYLINDRICAL ELONGATION OF THE GRAM-POSITIVE ROD

Because the gram-positive rod wall is multilayered, only the wall near the cell membrane can serve as a 'primer' for the addition of the new peptidoglycan. Unless some quite remarkable process intervenes, the layer to which the new material is being crosslinked can have itself very little order. Even if the primer wall had some order at one time, it would gradually be lost as successive layers were added because there can be only very little stress and strain on the wall immediately outside the cell membrane of the nascent chains to direct the orientation and it would be difficulty to imagine that the lipid bilayer is sufficiently rigid to maintain the directionality of the crosslinking machinery in relation to the cells axis.

Because the chemical structure of the peptidoglycan of *Bacillus subtilis* is the same as that of *E. coli* (Rogers *et al.*, 1980), it can be presumed that the former murein like the latter is capable of a good deal of expansion beyond the state of the newly polymerized layer. In the previous section, the evidence that the wall of the *E. coli* can expand by 300% in vitro was discussed. The degree of stretch of the *B. subtilis* wall has been estimated in two ways. First, based on the finding that a delay of one generation after pulse-labeling occurs before there is any release wall label. One doubling in biomass corresponding to an increase in surface area of 100%. The increase in area of the intact layer from the time of pulse to the time of the first cleavage of a layer must have been smaller and have occurred earlier (Koch and Doyle, 1985). Thus the 100% must be an overestimate of the strechability since several bonds must probably be cleaved in order for a fragment to be released,

The ability of *B. subtilis* wall to expand can be estimated in a second way from the degree that the planar septum stretches as it is split and bulged out to the shape of the mature pole. This estimate is that there is an increase in surface area of 50% in the transition from septum to pole which occurs without new wall introduction and without covalent cleavages of murein (Koch and Burdett, 1986 *a,b*). Putting these two estimates together it appears that a layer of wall of the Gram-positive organism expands 50% or somewhat more as an elastic fabric. It is possible that the multilayer character of the wall of *B. subtilis* accounts for the fact that these factors are smaller than the 300% observed for the isolated sacculi of *E. coli.*

It is characteristic of a fabric that the ability to expand can be partitioned differently depending upon the distribution of stresses. A fabric may have elastic properties, but if fully stretched in one direction, without ripping, it cannot be stretched in other directions. This is in accord with the early ideas of Previc (1970) who used the analogy of the Chinese finger puzzle.

It must necessarily be the case that the wall, as polymerized, is in a relaxed conformation. As it moves outward after a newer layer is attached inside of it, the side wall becomes more and more stretched, mainly because it is being elongated. Finally, the effective elastic limit is approached. Thus, the elasticity is 'spent' mostly in elongating the cylinder because that is the direction it is being stretched by the expansion of layers above it. None is left over to be stretched later to cause potentially lethal dilation, and this is the basis for the wall not expanding at a later time to form bulges. Thus the outermost *intact* layer in the wall is like a cylindrical section cut from a stocking which has then been stretched to the limit, mainly longitudinally and only a little bit laterally. Stresses due to the turgor pressure are felt by this layer, but not in the simple way indicated in Fig. 1A for a wall of constant properties throughout its thickness. This is because the Gram-positive sidewall is a composite of layers laid down at different times and differentially cracked and stressed and wall more peripheral to it bear some of the stresses.

Wall more peripheral than the outermost intact layer must have earlier been ruptured by autolysins thereby reducing the strain and permitting elongation. Autolysis in this region is favored for several reasons: i) increased stress that cannot be dissipated by further stretching; ii) the pH is no longer as acidified due to the chemiosmotic processes nearer the membrane; and iii) there is surface accumulation of released and rebound autolysins which were formed in earlier generations (Koch, 1988 *c*) If the outermost intact layer of the cylinder bore all the stress, axial cracks would form. But of course, some of the stress is still supported by the more peripheral wall (Koch, 1988 *b*). Because the cylinder is elongating, the stable structure for the wall material outside the outermost intact layer would be a series of ribbons of peptidoglycan encircling the cylinder in a helical path. It is well known in engineering that a supporting structure of this kind can give much greater strength to a pipe or hose. The outermost layer of the ribbon would have greatest strain. There autolysins would more rapidly remove the murein, allowing the truncated ribbon to extend further along the helical path. As this process occurs, the thickness of the ribbons would be maintained constant because still newer wall would be recruited to the ribbons by cleavages of intact wall along the helical crevasse in the outermost intact layer.

The net effect is to support the cylindrical wall and to give the whole dynamic growing system the properties of a fluid membrane.

DOES THE INSIDE-TO-OUTSIDE PROCESS FUNCTION FOR THE GRAM-NEGATIVE WALL?

Just as a tire patch glued onto a punctured inner tube distorts its shape when the tube is inflated, a patch of new murein attached to the old would stretch in an asymmetric wall when the overlying stress-bearing bonds become cleaved to allow wall expansion (Koch, 1990 *a*). For this reason I have criticized the 'Patches' model of Höltje and Glauner (1990). Such a process would lead to bulging; repetition of the process would lead to large spherical, non-dividing cell. Of course, the inner-tube tire patch does not stretch asymmetrically as much when placed in the less elastic tire and pressurized. This combination is equivalent to part of the process of Gram-positive rod-shaped bacteria described above. But that is not likely to be the case in the thin walled Gram-negative cell, given the facts about the strechability of the sacculus discussed above.

If the inside-to-outside model functions for the Gram-negative rod-shaped bacteria, one expectation is that the two ends would have to rotate, like their Gram-positive counterpart, is as they grew (Mendelson, 1976; Koch, 1989 *a*, 1990 *c*). This does not appear to be the case. In our earlier studies of the stretch of the peptidoglycan in vivo of the Gram-negative rod, in order to achieve filaments of the temperature sensitive mutants, we had them grow attached to the glass of the square capillaries. Had they rotated as do the gram-positive chains of cells we would have observed it during the initial phases and in control aspect of those studies, however we did not.

THE WEAVING OF THE OLIGOPEPTIDOGLYCAN INTO A FABRIC

To put the assumptions of the surface stress theory in terms of what it does not require or assume, it is presumed that: It does not require the presence of mechano-proteins; it could function in very primitive organisms and does not require sophisticated regulatory mechanism; and it does not require crystalline order in the peptidoglycan as organized in the sacculus. Thus it basically simpler than DNA replication, the control of transcription, or mechanism of protein synthesis. With these exclusions few, in fact only one, *a priori* mechanisms are conceivable. The presumption of the SST is that in given regions of the cell, peptidoglycan precursor molecules on their lipid precursor are available on the outside of the wall and without being subject to further control and regulation; they are polymerized into oligopeptidoglycan chains and crosslinked to other chains. Transpeptidases and transglycosylases can only function at the cell membrane while the autolysin may or may not have such restriction. However during dissolution of wall, autolysin action depends on the local stresses. To the degree that autolysis creates new wall susceptible to addition of new peptidoglycan, this, in turn, can control the rate of linkage of nascent wall.

The main topic of this paper is that it is absolutely necessary to assume that it is the biophysical characteristic of the growth processes that make the wall behave as a plastic and not as an elastic material. Thwaites and Mendelson (1991) have carried out an excellent analysis that elastic processes by themselves could not account for bacterial growth and form. For the Gram-negative organism the fluid properties occur if the wall is only one stress-bearing layer thick. So that the addition of new wall will have the equivalent energetics of a water molecule that is forces into the surface of a soap film. For the case of the Gram-positive rod-shaped organism, where most of the wall at any time is not growing and expresses stress and strain properties of normal substances, it is the fact that the wall is added diffusely inside and the stresses develop asymmetrically and variably through the thickness of the wall that give the growth of the cylindrical surface the properties of fluid film and not simply those of a solid. The difference between the two processes are listed in Table 1.

SUMMARY

Growth of bacteria cannot be explained by assuming that the mechano-proteins control shape, that the cell has the equivalent of guy wires, that the cell has control systems to direct wall enlargement in particular regions, or that the wall has crystalline properties so that it can serve as a precise template for insertion of new wall. Moreover, viewing the wall as a solid membrane that cracks under stress and then new wall is inserted in the cracks or viewing the membrane as the full equivalent of a fluid soap bubble film is not statisfactory because it would predict that rod-shaped organisms would round up. At present the only viable explanation is that wall growth is coupled to increase in cytoplasmic synthesis in such a way that the pressure-volume work forces the surface expansion due to the stress-strain relations of autolysin action. But because the new wall is added before the old is cut the distinction of hoop and axial stress is irrelevant and the process follow that equivalent to the physics of a fluid membrane. The shapes, stability, and regions of wall growth predicted by the surface stress theory agree with observations of a large variety of types.

Table 1. Comparison of insertional *versus* layered models for rod-shaped cells.

'Variable-T' for single stress-bearing layer	'Inside-to-outside' for multi-layered wall
MODEL	**MODEL**
Multiple linkage of a few new chains to old stress-bearing chains,	Linkage of new to other new oligopeptido-glycan to form a continuous layer,
	Linkage to the older continuous layer above it,
Cleavage of only 'protected' stress-bearing bonds,	As layer is stretched axially it loses ability to stretch circumferentially,
Expansion pulling new chains into the plane of fabric.	Stresses in different layers result from support from adjacent layers,
	Growth leads to cracks due to autolysin action,
	Crack form helically and propagate because of differential stresses,
	The innermost non-intact layer gives helical support for the pressurized cylinder,
Division by constriction resulting from local decrease in T.	Division by formation and splitting of a septum.
PROPERTIES	**PROPERTIES**
No crystalline wall,	No crystalline wall,
Monolayer of stress-bearing wall, with small amount of non-stress bearing nascent and no longer functioning wall,	Many continuous and non-continuous layers,
	Turnover common,
Partial orientation during insertion,	Little orientation during covalent linkage,
No rotation during elongation.	Rotation during growth.

REFERENCES

Braun, V., Gnirke, H., Henning, U. and Rehn, K. (1973) Model for the structure of the shape-maintaining layer of the *Escherichia coli* cell envelope. J. Bacteriol. 114, 1264-1270.

Burge, R.E., Fowler, A.G. and Reaveley, D.A. (1977*a*) Structure of the peptidoglycan of bacterial cell walls. I. J. Mol. Biol. 117, 927-953.

Burge, R.E., Adams, R., Balyuzi, H.H.M. and Reaveley, D.A. (1977*b*) Structure of the peptidoglycan of bacterial cell walls. II. J. Mol. Biol. 117, 955-974.

Casaregola, S., Norris, V., Goldberg, M. and Holland, I.B. (1990) Identification of a 180 kD protein in *Escherichia coli* related to a yeast heavy-chain myosin. Mol. Microbiol. 4, 505-511.

Glauner, B. and Höltje, J.-V. (1990) Growth pattern of the murein sacculus of *Escherichia coli*. J. Biol. Chem. 265, 18988-18996.

Holland, I.B., Casaregola, S. and Norris, V. (1990) Cytoskeletal elements and calcium: do they play a role in *Escherichia coli*? Res. Microbiol. 41, 131-136.

Höltje, J.-V. and Glauner, B. (1990) Structure and metabolism of the murein sacculus. Res. Microbiol. 141, 75-89.

Koch, A.L. (1982) On the growth and form of *Escherichia coli*. J. Gen. Microbiol. 128, 2527-2540.

Koch, A.L. (1983) The surface stress theory of microbial morphogenesis. Adv. Microbial Physiol. 24, 301-366.

Koch, A.L. (1984) How bacteria get their shapes: the surface stress theory. Com. Mol. Cell Biophys. 2, 179-196.

Koch, A.L. (1988 *a*) Biophysics of bacterial wall viewed as a stress-bearing fabric. Microbiol. Rev. 52, 337-353.

Koch, A.L. (1988 *b*) The sacculus, a nonwoven, carded, stress-bearing fabric. p. 43-59. (Actor, P., Daneo-Moore, L., Higgins, M.L., Salton, M.R.J. and Shockman, G.D., Eds.), Antibiotic inhibition of bacterial cell surface assembly and function. Amer. Soc. Microbiol., Washington.

Koch, A.L. (1988 *c*) Partition of autolysins between the medium, the internal part of the wall, and the surface of the wall of Gram-positive rods. J. Theor. Biol. 134, 463-472.

Koch, A.L. (1989 *a*) The origin of the rotation of one end of a cell relative to the other end during the growth of Gram-positive rods. J. Theor. Biol. 141, 391-402.

Koch, A.L. (1989 *b*) Calculation of surface area of sacculi from low-angle light scattering measurements. J. Microbiol. Methods 9, 139-150.

Koch, A.L. (1990 *a*) Additional argument for the key role of "smart" autolysins in the enlargement of the wall of gram-negative bacteria. Res. Microbiol. 141, 529-541.

Koch, A.L. (1990 *b*) Recent extensions of the surface stress theory, p. 39-63. "Microbial Growth Dynamics" (Poole, R.K. and Bazin, M.J. Eds.). Oxford University Press, Oxford.

Koch, A.L. (1990 *c*) The relative rotation of the ends of *Bacillus subtilis* Arch. Microbiol. 153, 569-573.

Koch, A.L. (1991) The wall of bacteria serves the role that mechano-proteins do in eukaryotes. FEMS *Microbiol. Rev.* 88, 15-26.

Koch, A.L. (1992 *a*) Differences in the formation of the poles of *Enterococcus* and *Baccilus*. J. Theor. Biol. 154, 205-217.

Koch, A.L. (1992 *b*) From before the first cell to after the last universal ancestor. J. Mol. Evol. (in press).

Koch, A.L. and Burdett, I.D.J. (1984) The Variable-T model for Gram-negative morphology. J. Gen. Microbiol. 130, 2325-2338.

Koch, A.L. and Burdett, I.D.J. (1986 *a*) Normal pole formation during total inhibition of wall synthesis. J. Gen. Microbiol. 132, 3441-3449.

Koch, A.L. and Burdett, I.D.J. (1986 *b*) Biophysics of pole formation of Gram-positive rods. J. Gen. Microbiol. 132, 3451-3457.

Koch, A.L. and Doyle, R.J. (1985) Inside-to-outside growth and the turnover of the Gram-positive rod. J. Theor. Biol. 117, 137-157.

Koch, A.L., Higgins, M.L. and Doyle, R.J. (1981 *a*) Surface tension-like forces determine bacterial shapes: *Streptococcus faecium*. J. Gen. Microbiol. 123, 151-161.

Koch, A.L., Mobley, H.L.T., Doyle, R.J. and Streips, U.N. (1981 b) The coupling of wall growth and chromosome replication in Gram-positive rods. FEMS *Microbiology Letters* 12, 201-208.

Koch, A.L., Higgins, M.L. and Doyle, R.J. (1982) The role of surface stress in the morphology of microbes. J. Gen. Microbiol. 128, 927-945.

Koch, A.L., Lane, S.L., Miller, J. and Nickens, D. (1987) Contraction of filaments of *Escherichia coli*after disruption of the cell membrane by detergent. J. Bacteriol. 166, 1979-1984.

Koch, A.L. and Woeste, S.W. (1992) The elasticity of the sacculus of *Escherichia coli*. J. Bacteriol. (in press).

Labischinski, H., Barnickel, G., and Naumann, D. (1983) The state of order of bacterial peptidoglycan, p. 49-54. "Target of penicillin" (Hakenbeck, R. Höltje J.-V. and Labischinski, H. Eds.), Walter de Gruyter, Berlin-New York.

Labischinski, H., Barnickel, G., Naumann, D. and Keller, P. (1985) Conformational and topological aspects of the three-dimensional architecture of bacterial peptidoglycan. Ann. Inst. Pasteur/Microbiol. 136A, 45-50.

Labischinski, H., Barnickel, G., Bradaczek, H. and Giesbrecht, P. (1979) On the secondary and tertiary structure of murein. Eur. J. Biochem. 95, 147-155.

Labischinski, H., Goodell, E.W., Goodell, A. and Hochberg, M.L. (1991) Direct proof of a 'more-than-single-layered' peptidoglycan architecture of *Escherichia coli* W7: a neutron small angle scattering study. J. Bacteriol. 173, 751-756.

Leps, B., Barnickel, G., Bradaczek, H. and Labischinski, H. (1983) Comparison between the sugar chain conformation in murein and pseudomurein, p. 67-72. "Target of penicillin" (Hakenbeck, R. Höltje J.-V. and Labischinski, H. Eds.), Walter de Gruyter, Berlin-New York.

Mendelson, M.H. (1976) Helical growth of *Bacillus subtilis*: A new model for cell growth. Proc. Natl. Acad. Sci. U.S.A. 73, 1740-1744.

Norris, V., Baldwin, T.J., Sweeney, S.T., Williams, P.H. and Leach, K.L. (1991) Evidence for protein Kinase C in *Escherichia coli.* Mol. Microbiol. 5, 2977-2982.

Previc, E. (1970) Biochemical determination of the bacterial morphology and the geometry of cell division. J. Theor. Biol. 27, 471-497.

Rogers, H.J., Perkins, H.R. and Ward, J.B. (1980) Microbial cell walls and membranes. Chapman and Hall, New York.

Thwaites, J.J. and Mendelson, N.H. (1991). Mechanical behaviour of bacterial cell walls. Adv. Microbial Physiol. 32, 174-222.

Verwer, R.W.H., Nanninga, N., Keck, W. and Schwarz, U. (1978) Arrangement of glycan chains in the sacculus of *Escherichia coli.* J. Bacteriol. 136, 723-729.

Verwer, R.W.H., Beachey, E.H., Keck, W., Stoub, A.M. and Poldermans, J.E. (1980) Orientated fragmentation of *Escherichia coli* sacculi by sonication. J. Bacteriol. 141, 327-333.

GROWTH AND CONTROL OF THE CELL WALL: A MECHANICAL MODEL FOR *BACILLUS SUBTILIS*

John J. Thwaites

Engineering Department
Cambridge University
Cambridge, UK, CB2 1PZ

INTRODUCTION

Turgor pressure in bacterial cells has been known for a long time (Mitchell and Moyle, 1956). It is therefore surprising that the first attempt to relate it to cell shape was not made until 12 years ago (Koch *et al.*, 1981). The basic idea, as later expressed by Koch (1983), is that growth and division are not primarily the result of molecular arrangement but are driven by the cell wall stresses due to turgor. Subsequent attempts to develop the idea have not been made with the thoroughness the subject requires. The so-called 'surface-stress theory', despite the large number of papers written about it, remains little more than a good idea. The main reasons for this are that the problems considered have not been posed properly mechanically and that no stress analysis has been done.

In this paper a stress analysis is given for an apparently simple geometry, that is the cylinder wall of a rod-shaped cell. The model developed applies to any Gram-positive rod, but more particularly to *Bacillus subtilis*, for which there is most data. This, because of the growth feature, is a novel problem in applied mechanics; it is far from easy, but it is soluble. The main aim is to explain why cell diameter and wall thickness remain constant even though there is substantial elongation. Is this an autonomous process driven by stress or does the bacterium exercise some control? If the latter, how effective are the biologically more plausible control strategies? The solution involves quantitative argument, a by-product of which is the calculation of stress distribution through the wall. A subsidiary aim is to explain, again quantitatively, a mechanism for the twisting of rod-shaped cells as they elongate.

THE BASIS FOR THE MODEL

The walls of rod-shaped bacteria are, to all intents and purposes circular cylinders with polar caps. The model pertains only to the cylinder wall and the influence, if any, of

Bacterial Growth and Lysis, Edited by M.A. de Pedro *et al.*, Plenum Press, New York, 1993

the poles is neglected. The assumptions made about the mechanical properties of wall material and the forces involved are mostly based on direct experimental evidence. Where simplifications are made, in order to make the model more tractable, they are not intrinsic.

The cylinder wall is subjected to a net internal pressure (turgor) and to other forces due to electrostatic repulsion. These arise because the high negative charge on the wall is not completely neutralised *in vivo*. It appears that only a small fraction of the total charge is required in order to produce stresses of the same order as those due to turgor (Thwaites and Mendelson, 1991). Support for the mechanical significance of charge is given by Cole *et al.* (1970) who report the frequent irregular shape of mutants deficient in teichoic acid, which is the seat of most wall charge. Because the poles close the cylinder ends, turgor produces both longitudinal and hoop (i.e. circumferential) stress components. Electrostatic repulsion produces essentially longitudinal stress which, if the charge distribution were uniform, would vary slightly over the cylinder. There is some evidence for an asymmetric charge distribution (Sonnenfeld *et al.*, 1985) but in the model the electrostatic stress is given an average value as if it were due to pressure on the poles only.

The wall material must be highly anisotropic. Despite the fact that longitudinal and hoop stresses are of the same order, the cell steadily increases in length while the diameter remains remarkably constant (Trueba and Woldringh, 1980). A likely situation is that, because of the average orientation of the peptidoglycan backbone, the material is much stiffer in the hoop direction than longitudinally. This is assumed in the model. Furthermore since *Bacillus subtilis* cells appear to twist as they elongate (Mendelson, 1976; Fein, 1980), it is plausible to assume that the material axis of greatest stiffness is inclined slightly to the hoop direction. Wall material is also, like nearly all polymers, visco-elastic, that is it exhibits a time-varying response to stress (Thwaites and Mendelson, 1985; Thwaites and Surana, 1991). When held at constant tension, the material 'creeps', i.e. it extends over a period of time. Likewise when held at a constant extension the stress 'relaxes', i.e. it decays asymptotically to a final value. The asymptotic behaviour is in fact not a simple exponential decay, but in order to facilitate analysis it is assumed to be so in the model. The corresponding creep time constant is smaller than the cell doubling time, but it is of the same order. Finally, the material is taken to be homogeneous and incompressible, i.e. that its volume does not change with stress. Neither of these is essential, but many polymers are approximately incompressible and for cell wall there is no accepted view on homogeneity. There may be changes in cross-linking through the wall, which would affect mechanical properties. The average orientation might, because of increasing deformation, change. Both features could be included without altering the basic idea, but they would lead to more complex analysis.

Growth of the cell wall is represented by elongation, and (possible) diametral swelling and wall thickening. However these are not the only causes of material deformation. Growth is made possible by the addition at the plasma membrane of new wall material. The material wells up through the wall and is then turned over at or near the outer surface (Archibald, 1976; Archibald and Coapes, 1976; Pooley, 1976*a*). Available evidence indicates uniform distribution of new wall synthesis and also uniform upwelling, so that at a given radius in the cylinder, the material is of the same age (Merad *et al.* 1989; Pooley, 1976*b*). This is assumed in the model, with new synthesis and turnover rates proportional to surface area.

MODEL DEVELOPMENT

Analysis of Stress

The upwelling of the wall material engenders increasing deformation. Stresses are accordingly not uniform through the wall. Nonetheless the mathematical complexity can be avoided by dealing with averages over the wall cross-section. The average longitudinal and hoop components of stress σ_l and σ_h must balance the turgor and electrostatic pressures p_1 and p_2 (Figure 1). They are simple functions of the pressures, but because the latter act over surface area whereas wall stresses act over cross-sectional area, σ_h and σ_l are larger by a factor cylinder radius divided by wall thickness a/w, which is of the order of 10.

The corresponding average strain components (strictly direct strain, i.e. extension/length) each depend not only on the stress component in that direction but also on a fraction ν (Poisson's ratio) of the perpendicular one. They are given by

$$\left.\begin{aligned} \bar{\varepsilon}_l &= (\sigma_l - \nu\sigma_h)/E = \{(1-2\nu)p_1 + p_2\}a/2wE \\ \bar{\varepsilon}_h &= (2\nu\sigma_h - \nu\sigma_l)/E = (3p_1 - p_2)\nu a/2wE \end{aligned}\right\} \tag{1}$$

where E is the relaxed tensile modulus in the least-stiff direction, i.e. approximately longitudinal. The 'relaxed' modulus is the ratio of stress to asymptotic final strain. For an incompressible material the modulus in the stiff direction is E/2ν. Average strain here

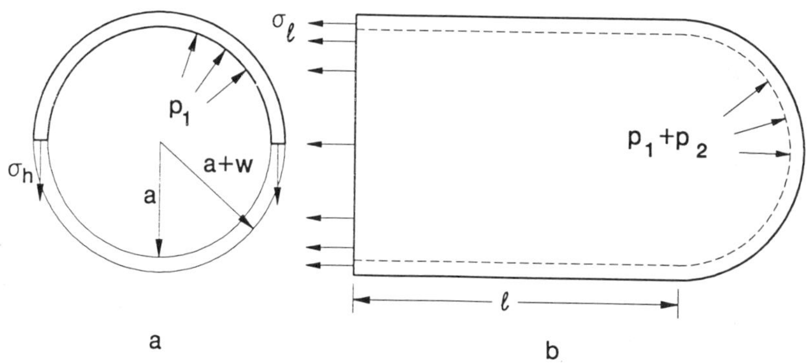

Figure 1. Average hoop and longitudinal stresses σ_h and σ_l in a cylinder subject to internal and end pressures p_1 and p_2. (a) Cross-section; (b) longitudinal view.

must, because of the visco-elastic property, be understood as the average over the wall cross section of a weighted average of the strain-rate over the time since the material was newly inserted. Since this time depends on position in the wall cross section, there is scope for some mathematical difficulty!

Fortunately, for steady growth, which in model terms means constant diameter and wall thickness, the averages can be stated succinctly. The longitudinal strain-rate η is the same at all points. This leads to

$$\bar{\varepsilon}_l = 1 - c + \eta T \tag{2a}$$

where c is the proportional wall turnover rate times the time taken for the material to well up from the membrane to the outersurface and T is the material creep time constant. 1-c is in fact the (simple) average strain; it represents the elastic component, whereas ηT is simply what would be obtained if the wall material were a viscous fluid.

The hoop strain in steady growth presents more difficulty. At radius r in the wall, the strain rate $\dot{r}/r$ depends on r and must be found by considering mass balance for the material within radius r at a given time and its changes in a short interval thereafter (Figure 2). The averaging results in

$$\bar{\varepsilon}_h = (1 + 2\lambda T - \eta T) w / 2a \tag{2b}$$

where λw is the rate of new wall synthesis per unit area. The strain due solely to being pushed from the membrane to the outer surface is w/a, so the first term represents the average elastic strain. The viscous term is just T times the average strain rate, which in this case involves a new-wall-insertion-rate component.

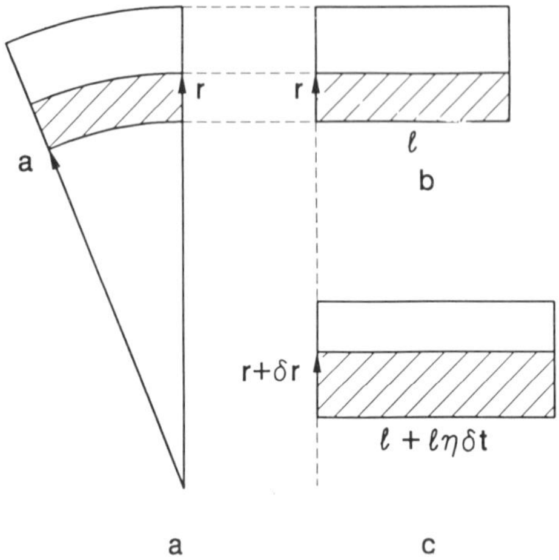

Figure 2. Wall upwelling, (a) transverse section, (b) longitudinal section, (c) longitudinal section after time interval δt. The rate of new wall insertion is λw per unit area and the elongational strain rate is η.

Combining (1) and (2) there are two equations relating cylinder radius a, wall thickness w and the elongational strain rate η with 'input' parameters p_1, p_2 and λ. A third is obtained by considering mass balance for the whole wall. It is

$$\eta = \lambda - \mu \tag{3}$$

where μw is the turnover rate per unit area.

Steady Wall Growth

In principle, the three equations give the cell geometry in terms of the input parameters, but in practice the latter are less firmly established experimentally than the former. For a given ratio of turnover rate to growth rate, the equations reduce to two, involving five parameters: the wall thickness/cylinder radius ratio w/a; the product of elongation strain-rate and creep time constant ηT; the ratio of turgor pressure to tensile modulus p_1/E; the pressure ratio p_2/p_1 and Poisson's ratio ν. Reasonable estimates of the first three exist for *Bacillus subtilis* at room temperature: w/a is about 0.1; ηT is about

0.4 (i.e. 0.02 $min^{-1} \times$ 20 min) and p_1/E is about 0.1. This implies a turgor pressure of about 1 MPa which is less than commonly quoted, but if the longitudinal stress is not to exceed the measured tensile strength, that is not unreasonable (Thwaites and Surana, 1991). The electrostatic pressure is unknown, but likely to be of the same order as turgor pressure (it may even be a contributor to measured values of turgor!). A mere 3.5% of wall charge unneutralised would achieve equality in the pressures (Marquis, privately communicated, based on Doyle *et al.*, 1980). Poisson's ratio is also unknown, for the transverse modulus of the wall has yet to be measured. Other highly-oriented polymers have a modulus ratio of the order of 10. Using the stated values for the first three parameters and taking the turnover rate to be 10% of the growth rate, the equations yield approximately 0.1 for ν and 1.5 for p_2/p_1. Both of these are reasonable. Changing the given parameters clearly alters the result; for example taking a smaller wall thickness, so that w/a is 0.08, the equations yield 0.045 for ν and 0.9 for p_2/p_1. While these values are acceptable, there are limits beyond which the model does not provide reasonable answers.

Further quantitative results can be obtained. The stress components increase through the wall, but by how much? The appropriate equations are like the set (1)-(2) but at different radii in the wall not for averages. For the set of parameter values given above, Figure 3 shows the variation. The material on the inside of the wall is not, as sometimes stated, stress-free. It is in a state of 'hydrostatic' stress, i.e. all three stress components are compressive and equal to turgor pressure. The greater stress component, always the longitudinal one, rises to a maximum of 29 MPa, whereas the measured tensile strength lies between 10 and 20 MPa. However the tensile strength is an average over the wall cross section, which leaves open the possibility that the mechanical properties vary, the material becoming stronger toward the outside. On the whole, this seems unlikely and it is worth noting that a turgor pressure of 0.5 MPa would put the maximum stress within the tensile strength range. A final point is that, since strength and modulus are usually related, the wall material is far more likely to crack, if it cracks at all on the plane perpendicular to the (maximum) longitudinal stress i.e. approximately circumferentially.

Twisting-with-Elongation

If a cylinder with helical anisotropy simply extends due to internal pressures an axial torque (a couple) is required in order to prevent it twisting. However, if there is relative rotation of the ends, a compensating shear stress is introduced which, provided the rotation is of the correct magnitude, leaves the cylinder torque-free. The average of the corresponding shear strain γ, as understood above for direct strain, is given by

$$(\bar{\varepsilon}_h - \bar{\varepsilon}_l)2\alpha - \bar{\gamma} = (\sigma_h - \sigma_l)\alpha / G \qquad (4)$$

where G is the relaxed shear modulus and α is the (small) inclination of the material axis of greatest stiffness to the hoop direction. In the steady state $\bar{\gamma} / \bar{\varepsilon}_l$ is the angle ψ of twisting, i.e. the angle of the helix along which a point on one end of the cylinder moves with respect to a point on the other. This leads a formula for ψ/α in terms of only three ratios: E/G, ν and p_2/p_1 (Thwaites and Mendelson, 1991, equation (4)).

Recent experiments by the author, on the torsional properties of bacterial thread, show the shear modulus of *Bacillus subtilis* cell wall to be about 0.125 of the extension modulus. Taking ν to be 0.1 the model predicts a change in twist angle of 0.4α for a change in p_2/p_1 of 10% from 1.5. Twist angles inferred from macrofibre measurements on FJ7 are in the range +6° to -5° (+ indicating right hand) (Mendelson *et al.*, 1984). It is clear that only a very slight degree of helical anisotropy is required for such a range to result from small changes in unneutralised charge due to the presence of ions in the medium. However, ions also influence the mechanical properties of the wall material, so the situation is unclear

(Thwaites *et al.*, 1991). What is clear is that the model shows how changes in twisting, including changes in helix hand, might be achieved without changing wall structure during growth.

CONTROL OF WALL GROWTH

Stability of Steady Growth

In order to investigate stability, it must be supposed that the wall thickness and cylinder diameter are not at their steady-growth values. Differential equations can then be derived showing whether or not the rates of change of these dimensions indicate a return to

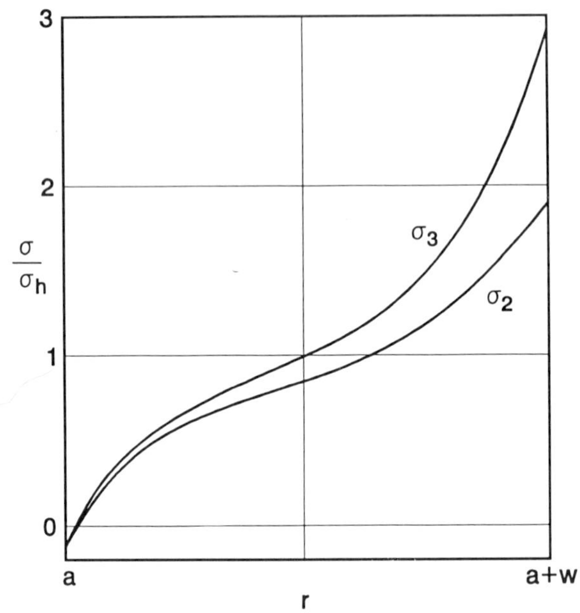

Figure 3. Variation through the wall of local hoop stress σ_2 and longitudinal stress σ_3 as proportions of average hoop stress σ_h. Turgor pressure: 0.1 σ_h; electrostatic pressure: 0.15 σ_h. Other parameters are given in the text.

the steady state. For a visco-elastic wall material these equations are extremely complicated owing to the time averaging required. However, if the elastic terms are insignificant, that is the material is effectively a viscous gel, the equations are much simpler, and the conclusions are qualitatively the same as for the material in its true state. The expressions for average strain are the viscous terms in equations (2) with the addition to the hoop strain of T times the diametral strain rate $\dot{a}/a$. The mass-balance equation also contains both $\dot{a}/a$ and $\dot{w}/w$, but it is clear without writing it in full that if the other parameters do not change an increasing wall thickness must be accompanied by a decreasing cylinder radius. The hoop stress-strain equation gives

$$\dot{a}/a = (3p_1 - p_2)va/2wET - \left(\lambda - \frac{1}{2}\eta\right)w/a \qquad (5)$$

in which $3p_1>p_2$ and $\lambda > \frac{1}{2}\eta$. Thus if a increases, so that a/w increases more so, $\dot{a}$ increases and the cylinder diameter increases, in theory, without limit. A full analysis, which involves considering small changes in the elongational strain rate η as well as those in w and a confirms this rough demonstration for all reasonable values of the various parameters involved. The autonomous system is unstable and, although wall stresses can be said to drive growth, they cannot do so without some form of control.

Possible Control Mechanisms

What can the bacterium do about this? On the basis of the model (i.e. mathematically) the rates of new material synthesis and of turnover, the turgor and electrostatic pressures, and even the material properties, could be varied, using information about dimensional changes and employing the negative feed-back principle. Biologically, it is difficult to see how material properties, wall charge and lytic enzyme concentration could be changed rapidly enough for effective control. Such changes could only be made on the time scale of inside to outside growth, which is longer than, but of the same order as, generation time. The rate of synthesis of new wall can presumably be changed more rapidly, but changes in turgor appear to be the most likely means of control.

Turgor is known to be controlled by changing the concentration in the cytoplasm of certain ions which influence osmotic flow. But what is 'measured' in order to do so? It cannot be turgor pressure itself because this is the difference between internal and external pressures and while the bacterium may be ableto sense internal changes, at the plasma membrane, it cannot sense external pressure. It is suggested that turgor is controlled in response to changes in cytoplasmic dimensions which, since the wall encloses the cytoplasm, are also wall dimensions. That is that control of turgor and wall growth are intimately linked. There is support for this idea in the results of experiments in which protoplasts swell in response to slow changes in external pressure (Corner and Marquis, 1969). Specifically, it is suggested that either the elongational strain-rate η or the diametral swelling strain-rate $\dot{a}/a$ could be used as control 'signals', with turgor pressure changed proportionally (via ion concentration) in order to effect control.

The diametral swelling strain-rate is the obvious candidate, since it must be made zero on average irrespective of disturbance to the system. There are two difficulties, the first essentially a biological one (though put in model terms). The sensor is most likely at or near the plasma membrane. The material hoop-strain rate there is not $\dot{a}/a$ but $\dot{a}/a + \lambda w/a$ the second part of which has a substantial steady component during growth. Indeed there is no part of the wall where the material has the strain rate $\dot{a}/a$ alone. The second difficulty is that simple negative feed-back using $\dot{a}/a$ does not ensure stability. One needs the actual changes in diameter to do so. This requires integration, essentially a memory term. While this is no doubt quite possible, many organisms use so-called (by engineers) 'integral control', it is an added complication which turns out to be quite unnecessary if η is used instead. Mathematically, changes in the elongational strain rate provide the most effective feed-back 'signal'. This is not obvious physically, but if changes in p_1 are made proportionately equal to minus K times those in η, it follows from the longitudinal stress equation (1)-(2a) that, for large enough values of K, changes in p_1 become proportionally equal to changes in the ratio w/a. This removes the main de-stabilising term in equation (5). A full analysis considering, as before, small changes from steady-growth values, confirms this and shows that K need not be very large. Since K is an 'amplification' it represents energy expended in effecting control, so that smaller values are more economic.

Showing, on the basis of a plausible model, that a given control strategy is possible, even that it is theoretically the most effective, does not of course demonstrate that it is actually employed. There are several biological examples of apparently uneconomic regulation. There are also cases in which several control mechanisms are applied simultaneously giving a robustness against otherwise damaging environmental changes. It is possible to use the model to investigate the simultaneous control of two parameters using two feed-back 'signals', but the author has so far not been able to improve significantly on the mechanism outlined.

This model of regulation based upon a mechanical model needs to be tested experimentally. Unfortunately the theoretically obvious way, that of 'disconnecting' the feed-back link, does not appear to be possible (but the author would welcome suggestions). A more profitable procedure must be to compare with experiment the dynamic changes predicted by the model, due to disturbances in various parameters; this is currently being done for dynamic changes in twist which are readily observed in filaments and can be measured in macrofibres using techniques employed earlier (Mendelson *et al.*, 1984; Favre *et al.*, 1985). A major problem in such comparisons is that the prime cause of change in the experiment is often not related directly to one model parameter e.g. non-lethal doses of antibiotics affect diameter, wall thickness and elongation rate. One of the main agents of change, temperature, affects almost every variable with a weighting that it is almost impossible to determine.

REFERENCES

Archibald, A.R. (1976) Cell wall assembly in *Bacillus subtilis*: development of bacteriophage - binding properties as a result of the pulsed incorporation of teichoic acid, J. Bacteriol. 127, 956-960.

Archibald, A.R. and Coapes, H.E. (1976) Bacteriophage SP50 as a marker for cell wall growth in *Bacillus subtilis*. J. Bacteriol. 125, 1195-1206.

Cole, R.M., Popkin, T.J., Boylan, R.J. and Mendelson, N.H. (1970) Ultrastructure of a temperature-sensitive *rod*⁻mutant of *Bacillus subtilis*. J. Bacteriol. 103, 793-810.

Corner, T.R. and Marquis, R.E. (1969) Why do bacterial protoplasts burst in hypotonic solutions? Biochim. Biophys. Acta 183, 544-558.

Doyle, R.J., Matthews, T.H. and Streips, U.N. (1980) Chemical basis for selectivity of metal ions by the *Bacillus subtilis* cell wall. J. Bacteriol. 143, 471-480.

Favre, D., Thwaites, J.J. and Mendelson, N.H. (1985) Kinetic studies of temperature-induced helix hand inversion in *Bacillus subtilis* macrofibers. J. Bacteriol. 164, 1136-1140.

Fein, J.E. (1980) Helical growth and macrofiber formation of *Bacillus subtilis 168* autolytic enzyme deficient mutants. Can J. Microbiol. 26, 330-337.

Koch, A.L. (1983) The surface stress theory of microbial morphogenesis. Adv. Microbial Physiol. 24, 301-366.

Koch, A.L., Higgins, M.L. and Doyle, R.J. (1981) Surface tension-like forces determine bacterial shapes: *Streptococcus faecium*. J. Gen. Microbiol. 123, 151-161.

Mendelson, N.H. (1976) Helical growth of *Bacillus subtilis*: a new model for cell growth. Proc. Natl. Acad. Sci. USA. 73, 1740-1744.

Mendelson, N.H., Favre, D. and Thwaites, J.J. (1984) Twisted states of *Bacillus subtilis* macrofibers reflect structural states of the cell wall. Proc. Natl. Acad. Sci. USA. 81, 3562-3566.

Merad, T., Archibald, A.R., Hancock, J.C., Harwood, C.R. and Hobot, J. (1989) Cell wall assembly in *Bacillus subtilis*: vizualisation of old and new wall material by electron-microscopic examination of samples stained selectively for teichoic acid and teichuronic acid. J. Gen. Microbiol. 135, 645-655.

Mitchell, P. and Moyle, J. (1956) Osmotic function and structure in bacteria. Symp. Soc. Gen. Microbiol. 6, 150-180.

Poole, H.M. (1976a) Turnover and spreading of old wall during surface growth of *Bacillus subtilis*. J. Bacteriol. 125, 1127-1138.

Pooley, H.M. (1976b) Layered distribution according to age within the cell wall of *Bacillus subtilis*. J. Bacteriol. 125, 1139-1147.

Sonnenfeld, E.M., Beveridge, T.J., Koch, A.L. and Doyle, R.J. (1985) Asymmetric distribution of charge on the cell wall of *Bacillus subtilis*. J. Bacteriol. 163, 1167-1171.

Thwaites, J.J. and Mendelson, N.H. (1985) Biomechanics of bacterial walls: studies of bacterial thread made from *Bacillus subtilis*. Proc. Natl. Acad. Sci. USA. 82, 2163-2167.

Thwaites, J.J. and Mendelson, N.H. (1991) Mechanical behaviour of bacterial cell walls. Adv. Microbial Physiol. 32, 173-222.

Thwaites, J.J. and Surana, U.C. (1991) Mechanical properties of *Bacillus subtilis* cell walls: effects of removing residual culture medium. J. Bacteriol. 173, 197-203.

Thwaites, J.J. Surana, U.C. and Jones, A.M. (1991) Mechanical properties of *Bacillus subtilis* cell walls: effects of ions and lysozyme. J. Bacteriol. 173, 204-210.

Trueba, F.J. and Woldringh, C.L. (1980) Changes in cell diameter during the division cycle of *Escherichia coli*. J. Bacteriol. 142, 869-879.

CELLULAR GROWTH WITHOUT A MUREIN SACCULUS - THE NUCLEOID-ASSOCIATED COMPARTMENTATION CONCEPT

Johannes Gumpert

Institut für Molekulare Biotechnologie, Beutenbergstrasse 11
D-O-6900 Jena, Germany

INTRODUCTION

Cell wall-less bacterial strains, isolated by experimental treatment from normal bacteria, are designated as L-forms. Several of them are genetically stable organisms showing extreme pleotropic changes in their properties. During numerous years of cultivation, some L-form strains could be adapted to grow in media without supplements of serum or additional osmotic stabilizers (Gilpin and Patterson, 1976; Gumpert and Taubeneck, 1983; Leon and Panos, 1976; Madoff, 1986).

For example, protoplast type L-form strains of *Proteus mirabilis* and *E. coli* were able to grow as submerged cultures in complex media like Meat Extract broth or Brain Heart Infusion broth under physiological isotonic conditions. Growth curves showed an exponential phase during 6 - 12 hrs, followed by a stationary phase. The cell number doubled 4 - 6 times and reached maximal values of 10^9 - 10^{10} cells per ml. The doubling times ranged between 60 and 180 min. Some strains were able to grow under semitechnical fermentation parameters in 12 - 150 L fermentors with high hydro-dynamic stress forces (blade turbine impellers, agitation rate 300 - 600 rpm, air flow rate 0,8 : 1). Under these rather hard environmental conditions, recombinant L-form strains of *P. mirabilis* synthesized heterologous foreign proteins like hu interferon $\alpha 1$ and bovine prochymosin and secreted them into the medium in considerable amounts (Klessen et al., 1989; Laplace et al., 1989).

The question arises, how can L-form cells manage this when they have no murein sacculus, no other cell wall structures, and no periplasmic compartment?

There exists an interior turgor pressure of 2 - 5 atmospheres in bacterial cells. The murein sacculus represents the bacterial exoskeleton, acting as a barrier against this pressure and preventing extension and disruption of the cytoplasmic membrane. Destruction of the rigid murein armour normally leads to an immediate lysis of the cell. The adapted stable protoplast type L-forms of *P. mirabilis, E. coli*, and other bacteria, however, do not lyse. Not only are they able to survive, but they carry out intense metabolic activities and transport processes. In addition, they divide and propagate without help from a murein sacculus or other cell wall components for DNA replication, nucleoid segregation, cell division, and cell separation.

During 30 years of experience with L-forms, a concept was elaborated to understand the mechanisms of cellular growth and division in these organisms. This concept might contribute to a better understanding of cell cycle events in normal bacteria.

Bacterial Growth and Lysis, Edited by M.A. de Pedro *et al.*, Plenum Press, New York, 1993

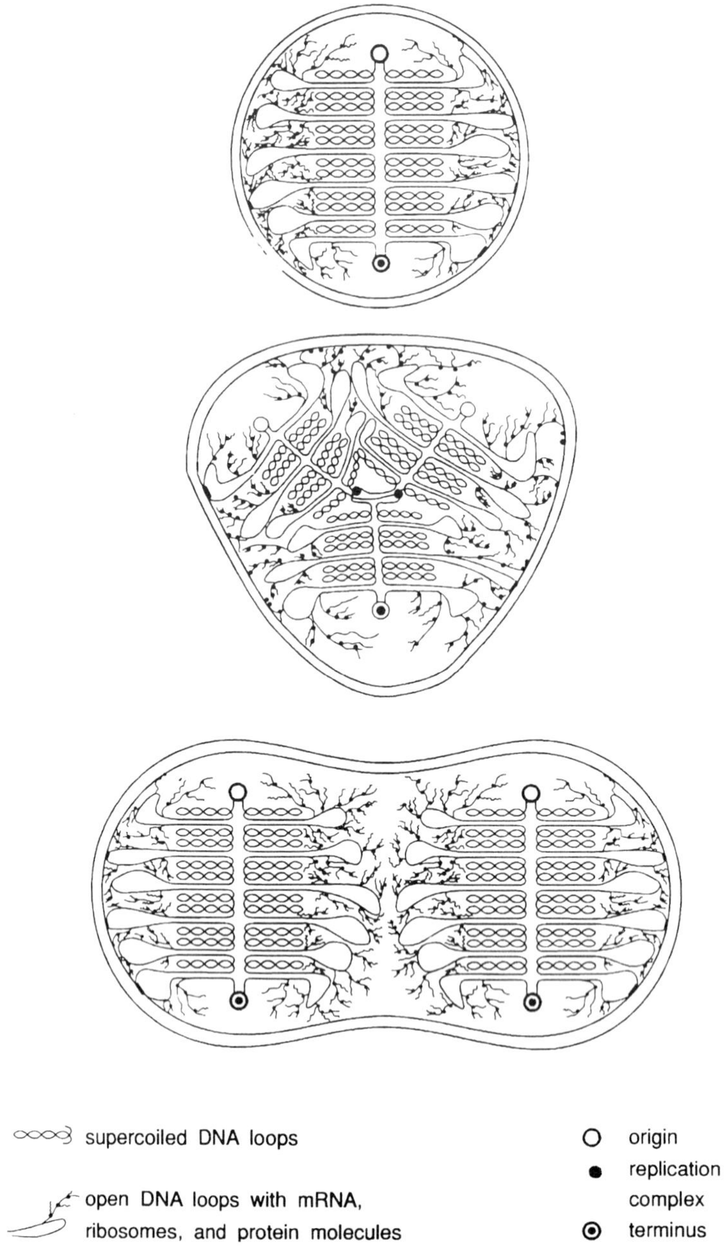

Figure 1. Schematic illustration of the nucleoid-associated compartmentation concept for an explanation of mechanisms of DNA replication, nucleoid segregation, and cell division in cell wall-less L-forms.

THE NUCLEOID-ASSOCIATED COMPARTMENTATION (NAC) CONCEPT

The concept (Figure 1) states that macromolecular components of the cytoplasm also represent structural elements. They are connected directly or indirectly with the chromosomal DNA of the nucleoid and may have contacts with the cytoplasmic membrane. A kind of framework is formed which determines a compartmentation of the cell contents during nucleoid segregation and cell division.

The concept is based on well known and accepted facts concerning structural organization of the chromosomal DNA, mechanisms of replication, transcription, and translation, properties of the cytoplasmic membrane, and morphological and physiological behaviour of dividing cells (Cooper, 1991; Cooper and Helmstetter, 1968; Drlica and Riley, 1990; Gualerzi and Leon, 1986; Nanninga, 1985). It has also been derived from light and electron microscopic studies of nucleoid segregation and cell division in *E. coli*, from biochemical and electron microscopical investigations of isolated nucleoids, from many years experience with wall-less L-forms of *P. mirabilis, E. coli, Bac. subtilis*, and *Streptomyces hygroscopicus*, and from experiments on protoplasting, carried out in our laboratory (Gumpert, 1983; Gumpert and Taubeneck, 1966; Gumpert et al. 1980, 1986).

The main statements of the NAC-concept are:

- the nucleoid is both the informational and the structural center of the cell
- DNA, RNA, and proteins form a discrete nucleoid body which appears condensed or more dispersed, depending on the physiological state of the cell
- both halves of the circular DNA molecule between the origin and the terminus of replication are packaged separately, forming a cylindrical nucleoid body with the origin at one pole and the terminus region at the opposite end
- DNA packing is determined by domains of supercoiled DNA loops
- the two replication complexes migrate to the terminus in or on the nucleoid. This allows replication and segregation without direct help, or connections with components of the cell envelope
- the newly replicated DNA is packed immediately into two daughter chromosomes; this leads to an increase in volume in that nucleoid region
- the supercoiled DNA domains are able to relax and form open loops, which can reach far into the peripheric regions of the cytoplasm
- open DNA loops, RNA polymerase, mRNA, ribosomes, translated proteins, and various DNA-binding proteins form a nucleoid-associated framework
- this nucleoid-associated framework plays an active role in chromosome segregation, nucleoid separation, and compartmentation of the cell content
- the nucleoid-mediated compartmentation in dividing cells determines the positioning of the nucleoids and might control time and site of septation and cell division

The NAC-concept has similarities to the transcription-translation-mediated segregation model and to the nucleoid oclusion model of Woldringh et al. (1985, 1990). In the following sections, data will be summarized which support this concept.

ANALYSIS OF THE NUCLEOID SHAPE DURING CELL CYCLE IN *E. COLI*

When cells of *E. coli* grow on media supplemented by 20 % gelatine, nucleoids become visible by positive phase contrast microscopy as light areas within the cell. By using techniques of microculture, phase contrast microscopy, and time-lapse photography, the changes of these nucleoid areas were followed during cell growth and cell division.

In cells of *E. coli*, growing with generation times less than 100 min, nucleoids were condensed and appeared as light rod-like or bar-like areas placed in transverse direction to the length axis of the cell.

An analysis of several hundreds of cells showed that nucleoids change their shape in a

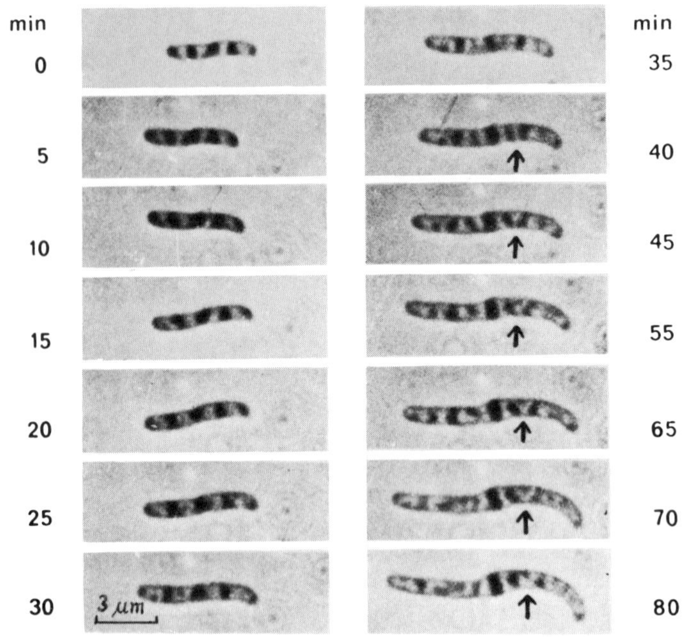

Figure 2. Changes of the nucleoid shape in a cell of *E. coli* 15 T⁻ growing on Nutrient broth supplemented by 20 % gelatine. The septation is inhibited by 40 μg/ml Cryptocillin. The arrows mark characteristic changes in the shape of a nucleoid which has a position parallel to the optical plane. Phase contrast micrographs.

characteristic manner (Figure 2). The rod-shaped area becomes enlarged at one end, followed by a division process starting at that side and resulting in Y, V, and U forms. The division is finished by separation of the two arms of the U form. This is a quick process like a jumping to their centres of the developing daughter cells.

Most nucleoids did not show these characteristic changes. One reason is that the sequence of changes can be observed only if nucleoid bodies are placed in a position parallel to the optical plane (Figure 3). Because this occurs by a change of about 20 %, the majority were seen as spherical or ellipsoidal shapes only. Furthermore, the picture depends upon the level of focus and a certain flexibility of the nucleoid parts.

Nucleoids in slow-growing cells (generation times greater than 100 min) were more dispersed. They appeared as larger oval areas and only allowed distinction of V and U forms during their division. Similar nucleoid configurations were recognized in papers of other authors (Fuhs, 1969; Mason and Powelson, 1956; Murray, 1960).

Rod-like and U-shaped nucleoid areas were obtained also by phase contrast microscopy in L-form cells of *P. mirabilis* and *E. coli* (Strunk and Schuhmann, 1973). In cells larger than 2 μm in diameter, several separate nucleoids or ring-like arrangements could be distinguished. Unfortunately, L-forms grow very slowly and they very rarely propagate under these conditions. This does not allow a similar microphotographic analysis of nucleoid shapes.

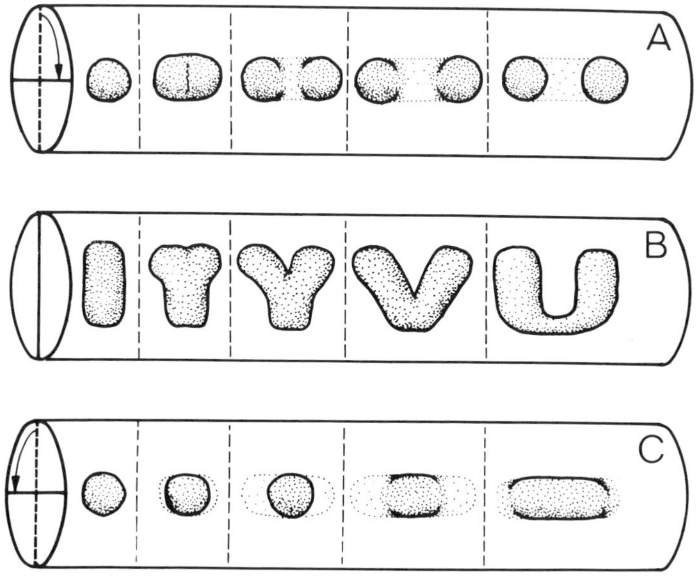

Figure 3. A diagram showing the dependence of optical recognition of the nucleoid shape from the position toward the optical axis in phase contrast micrographs. The characteristic sequence of changes can be followed only if the nucleoid body remains permanently placed parallel to the optical plane.

ELECTRON MICROSCOPIC ANALYSIS OF SHAPE AND FINE STRUCTURE OF NUCLEOIDS IN *E. COLI*

After fixation with osmium tetroxide and staining by uranyl acetate, condensed nucleoids could be observed in ultrathin sections. An analysis of reconstructed serial sections was carried out to get more detailed information about nucleoid shapes.

During a 100 min period, 23 samples were taken from an *E. coli* 15 T^- culture synchronized by thymine starvation. Samples were fixed by the Ryter-Kellenberger procedure and serial sections were made from 10 - 15 cells of every sample. Most nucleoids in the 200 cells, reconstructed by foil and plasteline, showed the typical I, Y, V, and U forms. However, in many cases they were more irregular and complex.

The analysis of these 4000 serial sections brought no clear information concerning fine structure and principles of DNA packing. Furthermore, there were no hints of the existence of a central protein core, of specific attachment sites of the nucleoid to the cytoplasmic membrane, or of replication complexes at the membrane or at the nucleoid body.

ELECTRON MICROSCOPICAL AND BIOCHEMICAL DATA OF ISOLATED NUCLEOIDS

Nucleoids have been isolated from *E. coli* and several other bacteria. We investigated the folded chromosome of streptomycetes. Using the DNA spreading technique according to Kleinschmidt, membrane-free nucleoids showed a characteristic arrangement (Gumpert et al., 1986). About 90 - 110 loops of the DNA strand emerged from a more or less dense core or fiber structure. The average length of the loops was 2 - 4 μm.

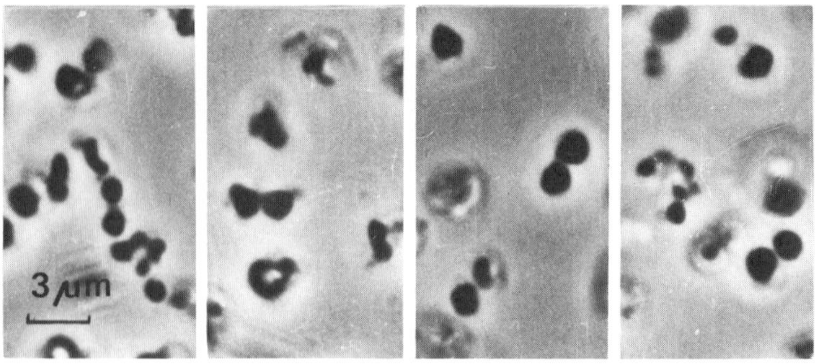

Figure 4. Typical shapes of propagating L-form cells of a protoplast type L-form of *Proteus mirabilis* taken from a culture, which was incubated at 37°C for 6 hrs under shaking conditions. Note the characteristic 3-sided spheroidal cell forms which represent early stages in the division process. Phase contrast micrographs taken with flash-light equipment.

The loop number was in the same order of magnitude as the number of domains of supercoiling present in the bacterial chromosome. Data from relaxation experiments (Drlica and Riley, 1990; Pettijohn, 1982; Sinden and Pettijohn, 1981) led to the conclusion that the whole chromosome is subdivided into 100 ± 30 separate topological units, representing domains of supercoiling. The ends of these domains are fixed sites on the DNA double helix that limit its rotation. The DNA strand of such domains can fold and unfold, independent from the folded state of adjacent domains.

Biochemical studies demonstrated that the isolated membrane-free nucleoids of *S. hygroscopicus* consist of 55 % DNA, 20 % RNA, and 25 % protein. Ethidium bromide intercalation experiments documented the superhelical state of DNA. Furthermore, specific histone-like DNA-binding proteins (e.g. S-protein of 15 kDa) have been detected in isolated nucleoids (Gumpert et al., 1986; Sarfert et al., 1983).

The results obtained with isolated nucleoids of streptomycetes are in agreement with those of other bacteria. Thus allow the conclusion, that the structural organization of the chromosomal DNA seems to follow the same principle in different bacteria.

If one supposes that both halves of the circular DNA thread between the origin and the terminus are packed separately as domains of supercoiling within one nucleoid body, and that during replication both replication forks move along the nucleoid from the origin side to the terminus side, then the changes in nucleoid shape become understandable (Figure 1). This model implies an immediate packing of the newly replicated DNA into two separate daughter chromosomes. Then a segregation of the daughter nucleoids follows the replication process. The model allows a multifork replication according to the same principle, resulting in more complex nucleoid bodies.

CELLULAR GROWTH AND DIVISION OF CELL WALL-LESS L-FORM CELLS

Cell populations in L-form cultures appear rather inhomogeneous when they have reached the stationary growth phase, or if they are growing slowly. All cells have a spherical shape and can vary in diameter between 0,3 and 5 µm.

In exponentially growing cultures however, the population looks more homogeneous. Cell diameters vary between 0,3 and 2 µm only. Cells show a spherical shape, but many of them are spheroidal and rod-shaped (Figures 4,5). These nonspherical cells represent division stages.

Growth and division of individual L-form cells of *P. mirabilis* were followed by using a special microculture technique and flash-light microphotographic equipment (Figure 5).

The first indication of a division start is the formation of characteristic spheroidal cell forms (Figures 4,5). A clear length axis is formed and the cell looks like a rod, about 0,8 µm in diameter and 2 - 3 µm in length. In these rod-shaped cells, a differentiation into two portions or masses takes place. Both parts of the cell become more loose and flexible and a division zone is formed. Finally, two individual spherical cells are born from the rod-shaped cell. Many of these daughter cells remain joined by a thin thread-like connection, which seems to consist of twisted membraneous material. Chains of 3 to 6 cells are produced in this way.

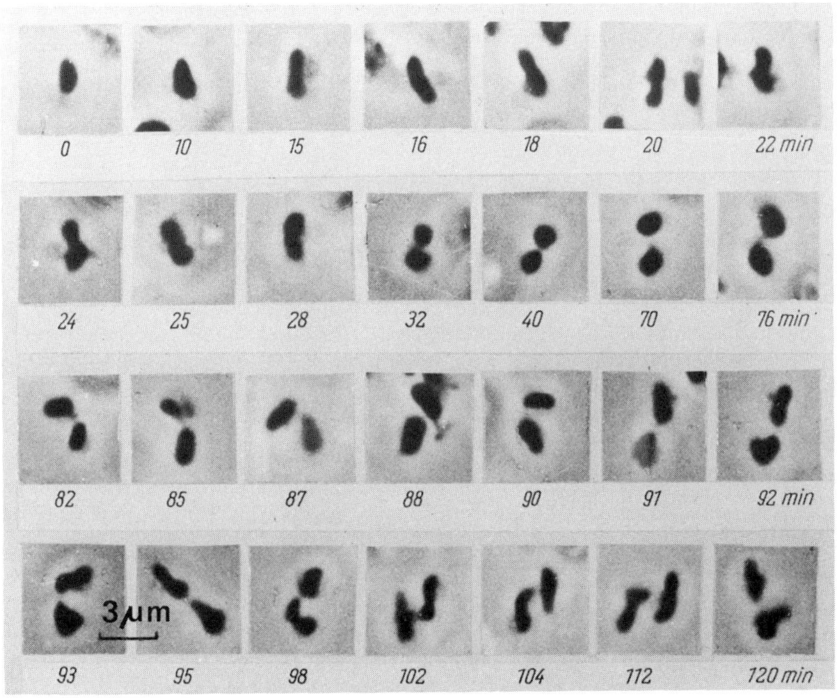

Figure 5. Growth and division of a wall-less L-form cell of *Proteus mirabilis* L VI in Nutrient broth. The changes in cell shape where followed using a special microculture technique, phase contrast microscopy, and flash-light microphotographic equipment. The spheroidal cell at 0 min finished its division after 32 min. Both daughter cells remain spherical during the following 40 min. After that, they start to grow and divide again, showing characteristic nonspherical and rod-like division stages.

This characteristic behaviour of propagating L-form cells can be observed only with well-adapted strains growing under favourable circumstances. If strains were poorly adapted or if the growth conditions were insufficient, then larger cells arosed which propagated by irregular division processes. However, also in these cases the formation of a length axis and a differentiation into cell masses seem to be prerequisites to cell division and cell separation.

These processes of cellular growth and division become understandable if one assumes an internal compartmentation into two or more individual cell units, equivalent to the unit

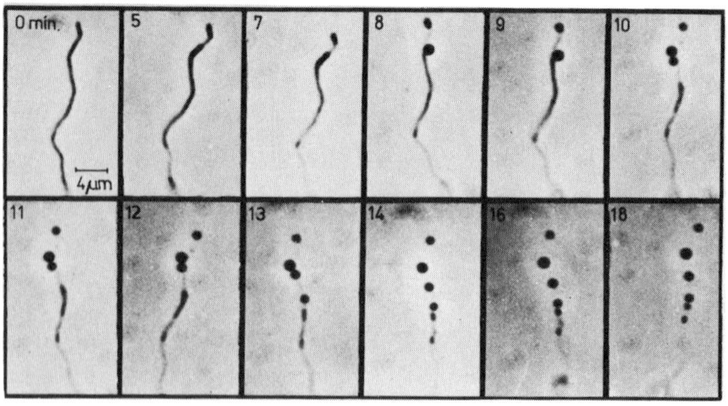

Figure 6. Fragmentation of the contents of a hyphal unit from *Streptomyces hygroscopicus* into individual protoplasts during lysozyme treatment (1000 µg/ml) on AL 58 medium containing 10 % sucrose. Phase contrast micrograph.

cell in the meaning of Donachie and Begg (1970). Each of them has its own structural coherence or integrity.

According to the model of the structural organization of the nucleoid, the process of compartmentation starts with the beginning of a replication round. The newly replicated parts of the chromosome are packed into separate nucleoid bodies. Both nucleoid bodies form their own structural framework consisting of open DNA loops, mRNA, ribosomes, and proteins. As a consequence, the cellular content increases in that region, which leads to characteristic nonspherical cell shape, a kind of three-sided spheroid (Figure 4).

The more the replication goes forward, the more the two compartments increase and the more a length axis is formed. When the replication has terminated, two nucleoids and the associated cytoplasmic compartments are placed side by side. The cell has a rod-like shape.

Because the nonreplicating cell has the tendency to form a sphere, the two unit cell compartments become more and more differentiated and demarcated. The contact zone between them loses its stability and becomes the division zone. The concept of a nucleoid-associated compartmentation would explain why nucleoid segregation in L-form cells does not need direct help by the murein sacculus, by other cell wall components, by the cytoplasmic membrane, or by contractile proteins.

FORMATION OF UNIT CELL PROTOPLASTS IN HYPHAE OF *STREPTOMYCES*

Streptomycetes are characterized by their mycelial growth. The mycelium is subdivided into hyphal units. In *S. hygroscopicus* hyphal units are 5 to 20 µm long sections, separated from adjacent units by a cross wall. Within these units 3 - 15 nucleoids can be seen by HCl-Giemsa staining or by phase contrast microscopy. The contents of a hyphal unit is surrounded by a common cytoplasmic membrane.

If the cell wall is removed by lysozyme treatment, one would expect the whole content of a hyphal unit to form one protoplast, containing all nucleoids and the cytoplasm associated with them. Under appropriate conditions however, the hyphal content becomes fragmentated into several individual spherical protoplasts which correspond in size to one or two unit cells (Figure 6). Obviously, these completely separated protoplasts contain only one or two nucleoids.

These results allow the conclusion that also in hyphal cells, an internal compartmentation that is genetically, physiologically, and structurally more or less independent unit cells exist. A destruction of the cell wall and the periplasmic milieu in an osmotically protected environment allows these cellular compartments to form individual cells. It is a rather surprising aspect of this process that the cytoplasmic membrane invaginates circumferentially and separates both cells by a final fusion step. No structural components of the cell wall participate in this fragmentation process.

CONCLUSIONS

Which forces might be responsible for final division and separation in L-form cells?

For normal cells of *E. coli*, the circumferential inward growth of the rigid murein layer is believed to be the driving force in formation of the division septum. The cytoplasmic membrane is passively pushed ahead of the ingrowing murein, while the outer membrane is pulled behind (Nanninga et al., 1990).

Without doubt, this mechanism does not work in L-form cells. There are three other mechanisms which might play a role in division of wall-less cells: an action of physical forces from outside, a passive invagination of the cytoplasmic membrane, or an active invagination of the cytoplasmic membrane.

L-form cells growing in liquid medium show an intensive Brownian movement. The formation of the two unit cell compartments within the L-form cell is simultaneously a differentiation into two masses which are open for attack by these multivectoral forces. By following the division process under the microscope, one has the impression that the progressive loosening of the region between the two cell masses may be the result of these attacking forces, and that they indeed play a role in the final division and separation process.

A passive invagination seems to be imaginable if one considers the numerous connections between the nucleoid and the cytoplasmic membrane. About 20 - 80 DNA-membrane attachment sites per nucleoid were described (Schaechter, 1990) and many more connections should be formed via mRNA, ribosomes, and proteins. The tendency of a unit cell containing one resting chromosome to form a sphere might pull the membrane inward. The formation of individual mono- or binucleated protoplasts in polyenergidic hyphal cells of streptomycetes by lysozyme treatment would support a possible role of such passive membrane invagination in division processes.

An active constriction of the cytoplasmic membrane implies the presence of a contractile principle, such as myosin-like or actin-like proteins. Yet, there is no clear evidence for a function of contractile proteins in bacteria. More data about identification and function of myosin-like proteins in *E. coli*, as it has been described recently (Holland et al., 1990; Casaregola et al., 1990) would indeed bring important progress for the explanation of L-form propagation and the understanding of septation and division in normal bacteria.

Another principle which might play an active role in division could be structural changes in the cytoplasmic membrane. In freeze-fracture electron micrographs of L-form membranes from *E. coli* and *P. mirabilis*, protein-free lipid domains were observed. In some of these domains, small granular and strand-like structures of about 12 nm in diameter could be recognized. They had a similar appearance as the lipidic particles which are characteristic for lipid phase transitions (Verkleij, 1984). Whether such local transitions in the lipid matrix, expressed as cylindrically or spherically oriented cubic phases, are involved in membrane invaginations remains speculative at the moment. Local changes in the cytoplasmic membrane, e.g. a segregation of phospholipids, were also supported by other authors (Norris, 1990).

Can we learn something from the behaviour of propagating wall-less L-form cells which may contribute to a better understanding of the division process in normal *E. coli* cells? The fundamental events, namely DNA replication, nucleoid segregation, and separation of daughter cells are the same in both cell types. From this, one can conclude that the cell wall and especially the murein sacculus is not absolutely necessary for either cellular growth or cell division. However, one has to keep in mind that the L-forms have learned to divide in a more or less regular mode during a long adaptation period. Furthermore, they do so only under sufficient growth conditions, and their toporegulation is not so perfect in comparison to normal walled bacteria. From that one may argue that L-forms have developed alternative mechanisms, different from those which manage and regulate nucleoid segregation, septation, and separation in normal bacteria. The formation of a stronger internal integrity by nucleoid-associated compartments and new stabilizing properties of the cytoplasmic membrane might play a role in such alternative mechanisms.

Concerning future work, one should pay more attention to the internal cellular architecture and the existence and function of nucleoid-associated compartments. More data about the manyfold structural interactions between the components of the nucleoid, the cytoplasm, and the membrane at the molecular and subcellular level are necessary. That data will surely contribute to a better understanding of many unsolved questions in cellular growth and cell division.

REFERENCES

Casaregola, S., Norris, V., Goldberg, M. and Holland, I.B. (1990) Identification of a 180 kD protein in *Escherichia coli* related to a yeast heavy-chain myosin. Mol. Microbiol. 4, 505-512.

Cooper, S. (1991) "Bacterial Growth and Division, Biochemistry and Regulation of Prokaryotic and Eukaryotic Division Cycles". Academic Press, New York.

Cooper, S. and Helmstetter, C.E. (1968) Chromosome replication and the division cycle of *Escherichia coli* B/r. J. Mol. Biol. 31, 519-540.

Donachie, W.D. and Begg, K.J. (1970) Growth of the bacterial cell. Nature 227, 1220-1224.

Drlica, K. and Riley, M. (1990) "The Bacterial Chromosome". Americ. Soc. for Microbiology, Washington.

Fuhs, G.W. (1969) "The Nuclear Structure of Prokaryontic Organisms/Bacteria and *Cyanophyceae*". Springer Verlag, Wien.

Gilpin, R.W. and Patterson, S.K. (1976) Adaptation of a stable L-form of *Bacillus subtilis* to minimal salts medium without osmotic stabilizers. J. Bacteriol. 125, 845-849.

Gualerzi, C.O. and Pon, C.L. (1986) "Bacterial Chromatin". Springer Verlag, Berlin.

Gumpert, J. (1983) A model of the organization of the bacterial nucleoid and consequences for replication and segregation, in "Progress in Cell Cycle Controls", (Chaloupka, J., Kotyk, A. and Streiblova, E., Edts.) pp 153-154. Institute of Microbiology. Czechoslovak. Academy of Sciences, Prag.

Gumpert, J. und Taubeneck, U. (1966) Beobachtungen über die Vermehrung der stabilen L-Form von *Proteus mirabilis* in flüssigen Nährmedien und die Funktion der Zellwand bei der Zellteilung. Z. Allg. Mikrobiol. 6, 211-218.

Gumpert, J. and Taubeneck, U. (1983) Characteristic properties and biological significance of stable protoplast type L-forms. Experientia Suppl. Vol. 46, 227-241.

Gumpert, J., Sarfert, E. and Zimmer, C. (1986) Chromosome structure and DNA-binding proteins in streptomycetes, in "Biological, Biochemical and Biomedical Aspects of Actinomycets Part A", (Szabo, G., Biro, S. and Goodfellow, M., Edts.) pp 453-463. Akademiai Kiado, Budapest.

Gumpert, J., Todorov, T. and Toshkov, A. (1980) Light microscopic investigations on lysozyme- and penicillin-induced morphological changes in *Erysipelothrix rhusiopathiae* and on propagation of its protoplast type L-form. Z. Allg. Mikrobiol. 20, 431-440.

Holland, I.B., Casaregola, S. and Norris, V. (1990) Cytoskeletal elements and calcium: do they play a role in the *Escherichia coli* cell cycle? Research in Microbiol. 141, 131-139.

Klessen, C., Schmidt, K.H., Gumpert, J., Grosse, H.H. and Malke, H. (1989) Complete secretion of activable bovine prochymosin by genetically engineered L-forms of *Proteus mirabilis*. Appl. Environ. Microbiol. 55, 1009-1015.

Laplace, F., Egerer, R., Gumpert, J., Kraft, R., Kostka, S. and Malke, H. (1989) Heterologous signal peptide processing in fusion interferon synthesis by engineered L-forms of *Proteus mirabilis*. FEMS Microbiol. Letters 59, 59-64.

Leon, O. and Panos, C. (1976) Adaptation of an osmotically fragile L-form of *Streptococcus pyogenes* to physiological osmotic condition and its ability to destroy human heart cells in tissue culture. Infect. Immun. 13, 252-262.

Madoff, S. (1986) "The Bacterial L-Forms". Marcel Dekker Inc., New York.

Mason, D.J. and Powelson, D.M. (1956) Nuclear division as observed in live bacteria by a new technique. J. Bacteriol. 71, 474-479.

Murray, R.G.E. (1960) The internal structure of the cell, in "The Bacteria Vol. I, Structure", (Gunsalus, I.C. and Stanier, R.Y., Edts.). Academic Press, New York.

Nanninga, N. (1985) "Molecular Cytology of *Escherichia coli*". Academic Press, London.

Nanninga, N., Wientjes, F.B., de Jonge, B.L.M. and Woldringh, C.L. (1990) Polar cap formation during cell division in *Escherichia coli*. Research in Microbiol. 141, 103-118.

Norris, V. (1990) DNA replication in *Escherichia coli* is initiated by membrane detachment of oriC. A model. J. Mol. Biol. 215, 67-72.

Pettijohn, D.E. (1982) Structure and properties of the bacterial nucleoid. Cell 30, 667-669.

Sarfert, E., Zimmer, C., Gumpert, J. and Lang, H. (1983) Folded chromosome structure and DNA-binding protein of *Streptomyces hygroscopicus*. Biochim. Biophys. Acta 740, 118-124.

Schaechter, M. (1990) The bacterial equivalent of mitosis, in "The Bacterial Chromosome", (Drlica, K. and Riley, M., Edts.), pp 313-322, American Society for Microbiology, Washington.

Sinden, R.R. and Pettijohn, D.E. (1981) Chromosomes in living *Escherichia coli* cells are segregated into domains of supercoiling. Proc. Natl. Acad. Sci. USA 78, 224-228.

Strunk, C. und Schuhmann, E. (1973) Cytomorphologische Untersuchungen an einer stabilen L-Form von *Escherichia coli*. Z. Allg. Mikrobiol. 13, 597-612.

Verkleij, A.J. (1984) Lipidic intramembranous particles. Biochim. Biophys. Acta 779, 43-63.

Woldringh, C.L. and Nanninga, N. (1985) Structure of nucleoid and cytoplasm in the intact cell, in "Molecular Cytology of *Escherichia coli*", (Nanninga, N., Ed.), pp 616-697. Academic Press, London.

Voldringh, C.L., Mulder, E., Valkenburg, J.A.C., Wientjes, F.B., Zaritzky, A. and Nanninga, N. (1990) Role of the nucleoid in the toporegulation of division. Research in Microbiol. 141, 39-49.

CONTRIBUTORS

PHOTO OF EDITORS

INDEX